Solitons in optical fibres, theoretically proposed in 1973, have now developed to the point where they are being seriously examined for their future application to all-optical high-bit-rate communications. This book describes both the theoretical and experimental aspects of optical soliton generation, soliton properties and soliton applications.

The intention of the book is to provide an overview of our current understanding of optical soliton properties, introducing the subject for the student and reviewing the most recent research. Only temporal optical solitons in fibres are considered. Each chapter has been written by experts, indeed chapters 1 and 2 have been contributed by the pioneers of theoretical and experimental optical soliton research – Dr A. Hasegawa and Dr L. F. Mollenauer respectively. A glance at the contents list will further reveal the wide range of topics covered.

The book will be of importance to graduate students and researchers in optics, optical engineering and communications science, providing a useful introduction for those who are entering the field. It will provide an up-to-date summary of recent research for the expert, who will also find the references to each chapter extremely valuable.

CAMBRIDGE STUDIES IN MODERN OPTICS: 10

Series Editors

Optical solitons

TITLES IN THIS SERIES

1 Interferometry (second edition)
W. H. Steel

2 Optical Holography – Principles, Techniques and Applications
P. Hariharan

3 Fabry–Perot Interferometers
G. Hernandez

4 Optical Instabilities
edited by R. W. Boyd, M. G. Raymer and L. M. Narducci

5 Coherence, Cooperation and Fluctuations
edited by F. Haake, L. M. Narducci and D. Walls

6 Holographic and Speckle Interferometry (second edition)
R. Jones and C. Wykes

7 Laser Chemical Processing for Microelectronics
edited by K. G. Ibbs and R. M. Osgood

8 Multiphoton Processes
edited by S. J. Smith and P. L. Knight

9 The Elements of Nonlinear Optics
P. N. Butcher and D. Cotter

10 Optical Solitons – Theory and Experiment
edited by J. R. Taylor

11 Particle Field Holography
C. S. Vikram

Optical solitons–
theory and experiment

Edited by

J. R. TAYLOR

Professorial Research Fellow
Femtosecond Optics Group, Department of Physics
Imperial College, University of London

The right of the
University of Cambridge
to print and sell
all manner of books
was granted by
Henry VIII in 1534.
The University has printed
and published continuously
since 1584.

CAMBRIDGE UNIVERSITY PRESS

CAMBRIDGE

NEW YORK PORT CHESTER

MELBOURNE SYDNEY

CAMBRIDGE UNIVERSITY PRESS
Cambridge, New York, Melbourne, Madrid, Cape Town, Singapore, São Paulo

Cambridge University Press
The Edinburgh Building, Cambridge CB2 2RU, UK

Published in the United States of America by Cambridge University Press, New York

www.cambridge.org
Information on this title: www.cambridge.org/9780521405485

First published 1992
This digitally printed first paperback version 2005

A catalogue record for this publication is available from the British Library

Library of Congress Cataloguing in Publication data

Optical solitons / edited by J. R. Taylor.
p. cm. – (Cambridge studies in modern optics ; 10)
Includes bibliographical references.
ISBN 0 521 40548 3 (hardback)
1. Solitons. 2. Optical fibers. I. Taylor, J. R. (James Roy),
1949– . II. Series.
QC174.26.W28068 1992
530.1′4–dc20 91-20776 CIP

ISBN-13 978-0-521-40548-5 hardback
ISBN-10 0-521-40548-3 hardback

ISBN-13 978-0-521-01779-4 paperback
ISBN-10 0-521-01779-3 paperback

Contents

Contributors

Chapter 1
Optical solitons in fibers: theoretical review

AKIRA HASEGAWA
AT & T Bell Laboratories, Murray Hill, NJ 07974 USA

Chapter 2
Solitons in optical fibers: an experimental account

L. F. MOLLENAUER
AT & T Bell Laboratories, Holmdel, NJ 07733 USA

Chapter 3
All-optical long-distance soliton-based transmission systems

L. F. MOLLENAUER
AT & T Bell Laboratories, Holmdel, NJ 07733 USA

K. SMITH
British Telecom Research Laboratories, Martlesham Heath, Ipswich, England

Chapter 4
Non-linear propagation effects in optical fibres: numerical studies

K. J. BLOW AND N. J. DORAN
British Telecom Research Laboratories, Martlesham Heath, Ipswich, England

Chapter 5
Soliton–soliton interactions

C. DESEM AND P. L. CHU
The University of New South Wales, School of Electrical Engineering and Computer Science, PO Box 1, Kensington, New South Wales, Australia 2033

Chapter 6
Soliton amplification in erbium-doped fiber amplifiers and its application to soliton communication

MASATAKA NAKAZAWA
NTT Transmission Systems Laboratories, Lightwave Communication Laboratory, Tokai, Ibaraki-Ken 319-11, Japan

Chapter 7
Non-linear transformation of laser radiation and generation of Raman solitons in optical fibers

E. M. DIANOV, A. B. GRUDININ, A. M. PROKHOROV AND V. N. SERKIN
General Physics Institute USSR Academy of Sciences, 38 Vavilov Street, Moscow, 117942, USSR

Chapter 8
Generation and compression of femtosecond solitons in optical fibers

P. V. MAMYSHEV
General Physics Institute, Academy of Sciences of the USSR, 38 Vavilov Street, Moscow 117942, USSR

Chapter 9
Optical fiber solitons in the presence of higher order dispersion and birefringence

CURTIS R. MENYUK
Department of Electrical Engineering, University of Maryland, Baltimore, MD 21228, USA

PING-KONG A. WAI
Applied Physics Operation, Science Applications International Corporation, 1710 Goodridge Drive, McLean, VA 22102, USA

Chapter 10
Dark optical solitons

A. M. WEINER
Bellcore, RedBank, NJ 07701-7040, USA

Chapter 11
Soliton-Raman effects

J. R. TAYLOR
Femtosecond Optics Group, Physics Department, Imperial College, London SW7 2BZ

Preface

Although the concept of solitons has been around since Scott Russell's various reports on solitary waves between 1838 and 1844, it was not until 1964 that the word 'soliton' was first coined by Zabusky and Kruskal to describe the particle-like behaviour of the solitary wave solutions of the numerically treated Korteweg–deVries equation. At present, more than one hundred different non-linear partial differential equations exhibit soliton-like solutions.

However, the subject of this book is the optical soliton, which belongs to the class of envelope solitons and can be described by the non-linear Schrödinger (NLS) equation. In particular, only temporal optical solitons in fibres are considered, omitting the closely related work on spatial optical solitons.

Hasegawa and Tappert in 1973 were the first to show theoretically that, in an optical fibre, solitary waves were readily generated and that the NLS equation description of the combined effects of dispersion and the non-linearity self-phase modulation, gave rise to envelope solitons. It was seven years later, in 1980 before Mollenauer and co-workers first described the experimental realisation of the optical soliton, the delay primarily being due to the time required for technology to permit the development of low loss single-mode fibres.

Over the past ten years, there has been rapid developments in theoretical and experimental research on optical soliton properties, which hopefully is reflected by the contents of this book. Primarily, research was driven by the promise of massively increased bit rates, through the application of ultrashort soliton pulses in long-distance optical communication networks, and although additional non-linearities which had not been originally considered, but which are described here, have put limits on potential systems, the application to communication remains the principal objective of many laboratories. In addition, the ideal phase property of solitons makes them ideal bits for application to ultrafast switching and processing, based on interferometric techniques and, in the past few years, significant advance has been made in this area. Some of the major developments are described herein. Aside from communications, soliton techniques have also been used to generate femtosecond pulses, frequently using very simple experimental procedures, and it is often neglected to note that although modified fibre-grating compressors have generated the shortest pulses to date, based on 'soliton-like' pulse shaping, multi-soliton break-up effects in fibres alone have allowed the generation of pulses of the same number of optical cycles in the near infrared (3–4) and tunable sub 100 femtosecond pulses can be readily generated throughout the near infrared based on soliton Raman effects.

The intention of this book is to review the past, and the current position of experimental and theoretical understanding of the multifarious soliton properties. It is a pleasure to have the pioneers of theoretical and experimental optical soliton research – Dr Akira Hasegawa and Dr Linn Mollenauer respectively, review their work. These two introductory chapters provide much of the background for the subject of the book.

In Chapter 3, Smith and Mollenauer update Mollenauer's introductory chapter and present current results on long-distance soliton-based transmission systems, carrying out the definitive experiment in this field and point the way forward in future all-optical communication. In Chapter 4, Blow and Doran discuss the application of numerical techniques to the problems of non-linear pulse propagation in optical fibres and describe their original work on non-linear (soliton-based) fibre devices and switches. The particle behaviour and mutual interaction of solitons, which is of major consideration in a soliton-based communication network is described in a self-contained chapter by Desem and Chu. Two of the major technological influences on solitons in future telecommunications have been, the introduction of efficient erbium-doped fibre amplifiers operating in the telecommunications window around 1.55 μm, and the generation of solitons from semiconductor lasers. Dr Masataka Nakazawa describes the dramatic advances and pioneering work of his group in these areas, highlighting the current state of the art of high-repetition-rate soliton transmission and giving insight into future developments of telecommunications systems.

In Chapters 7 and 8 soliton research is presented from the perspective of the group headed by Professor E. M. Dianov from the General Physics Institute, Moscow. They describe theoretical and experimental aspects of soliton self-action and the generation of controlled femtosecond pulses using soliton effects while describing related non-linear processes. The effects of higher order dispersion in the region of negligible second-order dispersion and of birefringence on soliton propagation is theoretically treated by Menyuk and Wai, relating their results to experimental work and proposed transmission systems, as well as application of polarisation effects to soliton-based ultrafast switching mechanisms.

It is often forgotten that at the same time as proposing the generation of bright optical solitons, Hasegawa and Tappert indicated the possibility of developing dark solitons. Dr Andrew Weiner in Chapter 10 discusses aspects of dark soliton solutions of the non-linear Schrödinger equation and based on novel pulse shaping techniques, developed in his laboratory, describes the experimental realisation of true dark solitons and presents verification of their properties.

Finally in Chapter 11, J. R. Taylor describes his group's work on the generation of tunable sources of ultrashort pulses based on soliton-shaping and Raman generation in fibres.

It is hoped that the reader will find in this book, a review of optical soliton principles and research as it currently stands. Since development in the field is rapid, no doubt, research milestones presented here will be surpassed even before publication. This will undoubtedly be so in the topics of long distance propagation, application of erbium amplification schemes, the generation of solitons from semiconductor and erbium fibre lasers and the application of solitons to switching processes. However, it is hoped that this volume will provide a useful source book for researchers and graduate students entering the field, providing the basics of soliton generation and their properties, while hopefully being of interest to the expert as well; allowing reference to the major part of optical soliton research as it currently stands to be readily accessed, and perhaps stimulating further research.

Finally, I would like to express my sincere gratitude to all those who have contributed to this book, for finding time from their busy research schedules to collate their results and for bearing with me in this project.

J. R. Taylor
London

1

Optical solitons in fibers: theoretical review

AKIRA HASEGAWA

Theoretical properties of light wave envelope propagation in optical fibers are presented. Generation of bright and dark optical solitons, excitation of modulational instabilities and their applications to optical transmission systems are discussed together with other non-linear effects such as the stimulated Raman process.

1.1 Introduction

The envelope of a light wave guided in an optical fiber is deformed by the dispersive (variation of the group velocity as a function of the wavelength) and non-linear (variation of the phase velocity as a function of the wave intensity) properties of the fiber. The dispersive property of the light wave envelope is decided by the group velocity dispersion (GVD) which may be described by the second derivative of the axial wavenumber k $(= 2\pi/\lambda)$ with respect to the angular frequency ω of the light wave, $\partial^2 k/\partial\omega^2$ $(= k'')$. k'' is related to the coefficient the group velocity delay D in ps per deviation of wavelength in nm and per distance of propagation in km, through $k'' = D\lambda^2/(2\pi c)$ where λ is the wavelength of the light and c is the speed of light. For a standard fiber, D has a value of approximately -10 ps/nm · km for the wavelength of approximately 1.5 μm. D becomes zero near $\lambda = 1.3$ μm for a standard fiber and near $\lambda = 1.5$ μm for a dispersion-shifted fiber.

The non-linear properties of the light wave envelope are determined by a combination of the Kerr effect (an effect of the increase in refractive index n in proportion to the light intensity) and stimulated Brillouin and Raman scatterings. The Kerr effect is described by the Kerr coefficient n_2 $(\simeq 1.2 \times 10^{-22}$ m^2/V$^2)$ which gives the change in the refractive index, n, in proportion to the electric field intensity $|E|^2$ of the light wave, $n = n_0 + n_2|E|^2$.

The model equation which describes the light wave envelope in the anomalous dispersion regime, $k'' < 0$, is given by the non-linear Schrödinger (NLS) equation or the cubic Schrödinger equation. Using the complex amplitude $q(x, t)$, the equation is expressed by

$$\mathrm{i}\frac{\partial q}{\partial Z} + \frac{1}{2}\frac{\partial^2 q}{\partial T^2} + |q|^2 q = 0$$

In this equation, Z represents the distance along the direction of propagation, and T represents the time (in the group velocity frame). The second term originates from the GVD, and the third term originates from the Kerr effect.

Hasegawa and Tappert (1973a) were the first to show theoretically that an optical pulse in a dielectric fiber forms a solitary wave based on the fact that the wave envelope satisfies the NLS equation.

However, at that time, neither a dielectric fiber which had small loss, nor a laser which emitted a light wave at the appropriate wavelength ($\simeq 1.5\ \mu$m), was available. Furthermore, the dispersion property of the fiber was not known. Consequently, it was necessary to consider a case where the group dispersion, k'', is positive, that is when the coefficient of the second term in the NLS equation is negative, in which case the solitary wave appears as the absence of a light wave (Hasegawa and Tappert, 1973b). Two years prior to the publication of the paper by Hasegawa and Tappert, Zakharov and Shabat (1971) showed that the NLS equation can also be solved using the inverse scattering method in a way analogous to the Korteweg–de Vries equation (Gardner *et al.*, 1967). According to this theory, the properties of the envelope soliton of the NLS equation can be described by the complex eigenvalues of Dirac-type equations, the potential being given by the initial envelope wave form. Because of this, the solitary wave solution derived by Hasegawa and Tappert could, in fact, be called a soliton.

Although the electric field in a fiber has a relatively large magnitude ($\simeq 10^6$ V/m) (for an optical power of a few hundred mW in a fiber with a cross section of 100 μm^2), the total change in the refractive index $n_2|E|^2$ is still 10^{-10}, and this seems to be negligibly small. The reason why such a small change in the refractive index becomes important is that the modulation frequency $\Delta\omega$, which is determined by the inverse of the pulsewidth, is much smaller than the frequency of the wave ω and secondly the GVD, which is produced by $\Delta\omega$, is also small. For example, the angular frequency ω for light wave with wavelengths λ of 1.5 μm is 1.2×10^{15} s^{-1}. If a pulse modulation with a pulsewidth of 10 ps is applied to this light wave, the ratio of the modulation frequency $\Delta\omega$ to the carrier frequency ω becomes approximately 10^{-4}. As will be shown later, the amount of wave distortion due to the GVD is proportional to $(\Delta\omega/\omega)^2$ times the coefficient of the group dispersion $k''(= \partial^2 k/\partial\omega^2)$. Consequently, if the group dispersion coefficient is of the order of 10^{-2}, the relative change in the wavenumber due to the group dispersion becomes comparable with the non-linear change.

Seven years after the prediction by Hasegawa and Tappert, Mollenauer *et al.* (1980) succeeded in the generation and transmission of optical solitons in a fiber for the first time.

The fact that the envelope of a light wave in a fiber can be described by the NLS equation can also be illustrated by the generation of modulational instability when a light wave with constant amplitude propagates through a fiber (Akhmanov *et al.*, 1968; Hasegawa and Brinkman, 1980; Anderson and Lisak, 1984; Hermansson and Yevick, 1984). Modulational instability is a result of the increase in the modulation amplitude when a wave with constant amplitude propagates through a non-linear dispersive medium with anomalous dispersion, $k'' < 0$. The origin of modulational instability can be identified by looking at the structure of the NLS equation. If we consider the third term in the NLS equation as a potential which traps the quasi-particle described by the NLS equation, the fact that the potential is proportional to the absolute square of the wavefunction indicates that the potential depth becomes deeper in proportion to the density of the quasi-particle. Consequently, when the local density of the quasi-particle increases, the trapping potential increases further thus enhancing the self-induced increase of the quasi-particle density.

A further interesting fact regarding the nature of an optical soliton in a fiber is the observation of the effect of higher order terms which cannot be described by the NLS equation. The small parameter $(\Delta\omega/\omega)$ used for the derivation of the NLS equation is of the order of 10^{-4}. Therefore, the description of a process which is of higher order in this small parameter would seem to require an extremely accurate experiment.

An example of this higher order effect is the Raman process which exists within the spectrum of a soliton. When the central spectrum of a soliton acts as a Raman pump, amplifying the lower sideband spectra within the soliton spectra, the frequency spectrum gradually shifts to the lower frequency side. This effect was first observed in an experiment by Mitschke and Mollenauer (1986), and was theoretically explained by Gordon (1986) in terms of the induced Raman process. Kodama and Hasegawa (1987) have identified that Mollenauer's discovery is due to the higher order term which represents the Raman effect. It is thus shown that the Raman process in the spectrum of a soliton produces a continuous shift of the soliton frequency spectrum to the lower frequency side, without changing the pulse shape.

The self-induced Raman process can be utilised to split two or more solitons which are superimposed in a fiber. The reason why it is possible to detect a process which is dependent on such a small parameter is that the light frequency, which is of order 10^{14}, is very large and, therefore, even if the perturbation is of the order 10^{-10}, the modification in the light frequency becomes 10^4 Hz and is consequently readily observable.

One important application of optical solitons is for a high bit-rate optical transmission system. Since solitons are not distorted by fiber dispersion, they can be transmitted for an extended distance (beyond several thousands of kilometres) only by providing amplification to compensate the fiber loss. Since fibers can be converted to amplifiers by themselves, this property of a soliton can be used to construct an all-optical transmission system. Such a system is much more economical and reliable than a conventional system which requires repeaters involving both photonics and electronics in order to reshape the optical pulse distorted by the fiber dispersion.

In this chapter, a theoretical review of these interesting phenomena is presented. In Section 1.2 the derivation of the model non-linear wave equation for the light wave envelope in a cylindrical dielectric guide is presented starting from the first principle. In Section 1.3 examples of non-linear light wave behavior in a fiber are presented including the optical soliton solution (1.3.1), the modulational instability (1.3.2) and the effect of higher-order terms on the soliton propagation (1.3.3). In Section 1.4, technical issues related to the application of optical solitons to all-optical transmission systems are discussed.

1.2 Derivation of the envelope equation for a light wave in a fiber

In this section, by introducing an appropriate scale of coordinates based on the physical setting of a cylindrical dielectric guide we reduce the Maxwell equation (three-dimensional vector equations) into the NLS equation with the higher-order terms describing the linear and the non-linear dispersion, as well as dissipation effects. The method used here is based on the asymptotic perturbation technique developed by Taniuti (1974), and gives a consistent scheme for the derivation of the NLS equation and the higher-order terms. The derivation follows closely Kodama and Hasegawa (1987).

The electric field $\boldsymbol{E}$, with the dielectric constant $\varepsilon_0\chi$, satisfies the Maxwell equation,

$$\boldsymbol{\nabla} \times \boldsymbol{\nabla} \times \mathbf{E} = -\frac{1}{c^2}\frac{\partial^2}{\partial t^2}\boldsymbol{D} \tag{1.1}$$

Here, the displacement vector $\boldsymbol{D} = \chi_* \boldsymbol{E}$ is the Fourier transform of $\hat{\boldsymbol{D}}(\omega)$ defined by

$$\chi_*\boldsymbol{E}(t) = \int_{-\infty}^{t} \mathrm{d}t_1\, \chi^{(0)}(t - t_1)\boldsymbol{E}(t_1)$$

$$+ \int_{-\infty}^{t}\mathrm{d}t_1\int_{-\infty}^{t}\mathrm{d}t_2 \int_{-\infty}^{t} \mathrm{d}t_3 \cdot \chi^{(2)}(t - t_1, t - t_2, t - t_3)\cdot\{\boldsymbol{E}(t_1)\cdot\boldsymbol{E}(t_2)\}\boldsymbol{E}(t_3)$$

$$+ \text{(higher non-linear terms)} \tag{1.2}$$

Here, $*$ indicates convolution integral, the second term describing the non-linear polarisation includes the Kerr and Raman effects with proper retardation. The dielectric tensors $\chi^{(0)}$, $\chi^{(2)}$ are dependent on the spatial coordinates in the transverse direction of the fiber axis. We write equation (1.1) in the following form:

$$\nabla^2\boldsymbol{E} - \frac{1}{c^2}\frac{\partial^2}{\partial t^2}\boldsymbol{D} = \boldsymbol{\nabla}(\boldsymbol{\nabla}\cdot\boldsymbol{E}) \tag{1.3}$$

It should be noted that $\boldsymbol{\nabla}\cdot\boldsymbol{E}$ in equation (1.3) is not zero, since $\boldsymbol{\nabla}\cdot\boldsymbol{D} = 0$ (the constraint for $\boldsymbol{D}$ in Maxwell's equation) implies that $\chi_*(\boldsymbol{\nabla}\cdot\boldsymbol{E}) = -(\boldsymbol{\nabla}\chi_*)\cdot\boldsymbol{E} \neq 0$, namely, the electric field cannot be described by either the TE or TM modes.

Since our purpose is to reduce (1.3) in the sense of an asymptotic perturbation it is convenient to write it in the following matrix form,

$$L\mathbf{E} = 0 \tag{1.4}$$

where $\mathbf{E}$ represents a column vector, i.e. $(E_r, E_\theta, E_z)^t$. In cylindrical coordinates, where the z-axis is the axial direction of the fiber, the matrix L consisting of the three parts $L = L_a + L_b - L_c$ is defined by

$$L_a = \begin{pmatrix} \nabla_\perp^2 - \dfrac{1}{r^2} & -\dfrac{2}{r^2}\dfrac{\partial}{\partial\theta} & 0 \\ \dfrac{2}{r^2}\dfrac{\partial}{\partial\theta} & \nabla_\perp^2 - \dfrac{1}{r^2} & 0 \\ 0 & 0 & \nabla_\perp^2 \end{pmatrix} \tag{1.5}$$

$$L_b = \left(\frac{\partial^2}{\partial z^2} - \frac{1}{c^2}\frac{\partial^2}{\partial t^2}\cdot *\right)\begin{pmatrix} 1 & 0 & 0 \\ 0 & 1 & 0 \\ 0 & 0 & 1 \end{pmatrix} \tag{1.6}$$

$$L_c = \begin{vmatrix} \dfrac{\partial}{\partial r}\dfrac{1}{r}\dfrac{\partial}{\partial r}r & \dfrac{\partial}{\partial r}\dfrac{1}{r}\dfrac{\partial}{\partial \theta} & \dfrac{\partial^2}{\partial r\partial z} \\ \dfrac{1}{r^2}\dfrac{\partial^2}{\partial r\partial\theta}r & \dfrac{1}{r^2}\dfrac{\partial^2}{\partial\theta^2} & \dfrac{1}{r}\dfrac{\partial^2}{\partial\theta\partial z} \\ \dfrac{1}{r}\dfrac{\partial^2}{\partial r\partial z}r & \dfrac{1}{r}\dfrac{\partial^2}{\partial\theta\partial z} & \dfrac{\partial^2}{\partial z^2} \end{vmatrix} \tag{1.7}$$

It should be noted that these matrices imply that

$$L_a\mathbf{E} = \nabla_\perp^2\mathbf{E} \equiv \left(\frac{1}{r}\frac{\partial}{\partial r}r\frac{\partial}{\partial r} + \frac{1}{r^2}\frac{\partial^2}{\partial\theta^2}\right)\boldsymbol{E}$$

$$L_b\mathbf{E} = \left(\frac{\partial^2}{\partial z^2} - \frac{1}{c^2}\frac{\partial^2}{\partial t^2}\right)\boldsymbol{E} \qquad \text{and} \qquad L_c\mathbf{E} = \nabla(\nabla\cdot\boldsymbol{E})$$

We consider the electric field as an almost monochromatic wave propagating along z-axis with the wavenumber k_0 and angular frequency ω_0, i.e. the field $\mathbf{E}$ is assumed to be in the expansion form,

$$\mathbf{E}(r, \theta, z, t) = \sum_{l=-\infty}^{\infty} \mathbf{E}_l(r, \theta, \xi, \tau; \varepsilon)\exp\{i(k_l z - \omega_l t)\} \tag{1.8}$$

with $\mathbf{E}_{-l} = \mathbf{E}_l^*$ (complex conjugate). $k_l = lk_0$, $\omega_l = l\omega_0$ and the summation is taken over all the harmonics generated by the non-linear response of the polarisation, $\boldsymbol{E}_l(r, \theta, \xi, \tau; \varepsilon)$ being the envelope of the lth harmonic which changes slowly in z and t. Here, the slow variables ξ and τ are defined by

$$\xi = \varepsilon^2 z \qquad \tau = \varepsilon\left(t - \frac{z}{v_g}\right) \tag{1.9}$$

where v_g is the group velocity of the wave given later. Since the radius of the fiber has the same order as the wavelength $(2\pi/k_0)$, the scale for the transverse coordinates (r, θ) is of order 1. In this scale of the coordinates of equation (1.9), the behavior of the field can be deduced from the balance between the non-linearity and dispersion which results in the formation of optical solitons confined in the transverse direction.

If we proceed from equations (1.2) and (1.9) we find that the displacement, $\boldsymbol{D} = \chi * \boldsymbol{E} = \sum \boldsymbol{D}_l \exp\{\mathrm{i}k_l z - \omega_l t)\}$, is given by

$$\mathbf{D}_l = \chi_l^{(0)}\mathbf{E}_l + \varepsilon \mathrm{i} \frac{\partial \chi_l^{(0)}}{\partial \omega_l} \frac{\partial \mathbf{E}_l}{\partial \tau} - \varepsilon^2 \frac{1}{2} \frac{\partial^2 \chi_l^{(0)}}{\partial \omega_l^2} \frac{\partial^2 \mathbf{E}_l}{\partial \tau^2}$$
$$- \varepsilon^3 \frac{\mathrm{i}}{6} \frac{\partial^3 \chi_l^{(0)}}{\partial \omega_l^3} \frac{\partial^3 \mathbf{E}_l}{\partial \tau^3} + \sum_{l_1+l_2+l_3=l} \{\chi_{l_1 l_2 l_3}^{(2)} (\mathbf{E}_{l_1} \cdot \mathbf{E}_{l_2})\mathbf{E}_{l_3}$$
$$+ \mathrm{i} \sum_{i=1}^{3} \left(\frac{\partial}{\partial \omega_{l_i}} \chi_{l_1 l_2 l_3}^{(2)} \right) \frac{\partial}{\partial \tau_i} (\mathbf{E}_{l_1} \cdot \mathbf{E}_{l_2})\mathbf{E}_{l_3}\} + \ldots \tag{1.10}$$

where $\chi_l^{(0)}$ is the Fourier coefficient $\hat{\chi}^{(0)}(\Omega)$ of $\chi^{(0)}(t)$ at $\Omega = \omega_l$, i.e. $\chi_l^{(0)} = \hat{\chi}^{(0)}(\omega_l)$, and $\chi_{l_1 l_2 l_3}^{(2)}$ is the Fourier coefficient $\hat{\chi}^{(2)}(\Omega_1, \Omega_2, \Omega_3)$ of $\chi^{(2)}(t_1, t_2, t_3)$ at $\Omega_1 = \omega_{l_1}$, $\Omega_2 = \omega_{l_2}$, $\Omega_3 = \omega_{l_3}$, and

$$\partial(\mathbf{E}_{l_1} \cdot \mathbf{E}_{l_2})\mathbf{E}_{l_3}/\partial \tau_1 = (\partial \mathbf{E}_{l_1}/\partial \tau \cdot \mathbf{E}_{l_2})\mathbf{E}_{l_3}$$
$$\partial(\mathbf{E}_{l_1} \cdot \mathbf{E}_{l_2})\mathbf{E}_{l_3}/\partial \tau_2 = (\mathbf{E}_{l_1} \cdot \partial \mathbf{E}_{l_2}/\partial \tau)\mathbf{E}_{l_3}$$

and so on. The last term in equation (1.10) represents the retarded response of the non-linear polarisation which gives both the higher-order non-linear dispersion and dissipation. We note that owing to the dispersion properties of the dielectric constant, the real space wave equation contains terms with higher-order time derivatives. From equations (1.9) and (1.10), equation (1.4) can be written in the following expansion form:

$$L_l \mathbf{E}_l + \mathrm{i}\varepsilon \frac{\partial L_l}{\partial \omega_l} \frac{\partial \mathbf{E}_l}{\partial \tau} - \frac{\varepsilon^2}{2} \frac{\partial^2 L_l}{\partial \omega_l^2} \frac{\partial^2 \mathbf{E}_l}{\partial \tau^2} - \frac{\mathrm{i}}{6} \varepsilon^3 \frac{\partial^3 L_l}{\partial \omega_l^3} \frac{\partial^3 \mathbf{E}_l}{\partial \tau^3}$$

$$c^2 \left\{ \frac{\partial}{\partial k_l} \left(\mathbf{L}_l - \frac{\omega_l^2 \chi_l^{(0)}}{c^2} \right) \right\} \cdot \left(\mathrm{i} \frac{\partial \mathbf{E}_l}{\partial \xi} - \frac{1}{2} \frac{\partial^2 k_l}{\partial \omega_l^2} \frac{\partial^2 \mathbf{E}_l}{\partial \tau^2} - \varepsilon \frac{\mathrm{i}}{6} \frac{\partial^3 k_l}{\partial \omega_l^3} \frac{\partial^3 \mathbf{E}_l}{\partial \tau^3} \right)$$

$$- \varepsilon^3 \left\{ \frac{\partial^2}{\partial \omega_l \partial k_l} \left(L_l - \frac{\omega_l^2 \chi_l^{(0)}}{c^2} \right) \right\} \mathrm{i} \frac{\partial}{\partial \tau} \left(\mathrm{i} \frac{\partial \mathbf{E}_l}{\partial \xi} - \frac{1}{2} \frac{\partial^2 k_l}{\partial \omega_l^2} \frac{\partial^2 \mathbf{E}_l}{\partial \tau^2} \right)$$

continued overleaf

$$+\frac{1}{c^2}\sum_{l_1+l_2+l_3=l}\left[\omega_l^2\chi_{l_1 l_2 l_3}^{(2)}(\mathbf{E}_{l_1}\cdot\mathbf{E}_{l_2})\mathbf{E}_{l_3}\right.$$

$$\left.+\,\mathrm{i}\varepsilon\sum_{i=1}^{3}\left\{\frac{\partial}{\partial\omega_{l_i}}(\omega_l^2\chi_{l_1 l_2 l_3}^{(2)})\right\}\frac{\partial}{\partial\tau_i}(\mathbf{E}_{l_1}\cdot\mathbf{E}_{l_2})\mathbf{E}_{l_3}\right]$$

$$+\,\mathrm{i}\varepsilon\left(\frac{1}{v_g}-\frac{\partial k_l}{\partial\omega_l}\right)\left\{\frac{\partial}{\partial k_l}\left(L_l-\frac{\omega_l^2\chi_l^{(0)}}{c^2}\right)\right\}\frac{\partial\mathbf{E}_l}{\partial\tau}$$

$$+\,\varepsilon^2\left(\frac{1}{v_g}-\frac{\partial k_l}{\partial\omega_l}\right)\begin{bmatrix}1&0&0\\0&1&0\\0&0&0\end{bmatrix}\frac{\partial}{\partial\tau}\left\{\left(\frac{1}{v_g}+\frac{\partial k_l}{\partial\omega_l}\right)\frac{\partial\mathbf{E}_l}{\partial\tau}-2\varepsilon\frac{\partial\mathbf{E}_l}{\partial\xi}\right\}$$

$$+\ldots=0 \tag{1.11}$$

where L_l is L in equations (1.5) to (1.7) with the replacement $\partial/\partial z=\mathrm{i}k_l$, $\partial/\partial t=-\mathrm{i}\omega_l$ and $\chi_*=\chi_l^{(0)}$. It should be noted that the operator L_l is self-adjoint in the sense of the following inner product:

$$(\mathbf{U},\,L\mathbf{V})\equiv\int\mathbf{U}^*\cdot L\mathbf{V}\mathrm{d}S$$

$$=\int L^*\mathbf{U}^*\cdot\mathbf{V}\mathrm{d}S=(L\mathbf{U},\,\mathbf{V}) \tag{1.12}$$

where $\mathrm{d}S=r\mathrm{d}r\mathrm{d}\theta$ and $\mathbf{U},\,\mathbf{V}\to 0$ as $\|x_\perp\|\to\infty$.

We now assume that $\mathbf{E}_l(r,\,\theta,\,\xi,\,\tau,;\,\varepsilon)$ can be expanded in terms of ε,

$$\mathbf{E}_l(r,\,\theta,\,\xi,\,\tau;\,\varepsilon)=\sum_{n=1}^{\infty}\varepsilon^n\mathbf{E}_l^{(n)}(r,\,\theta,\,\xi,\,\tau) \tag{1.13}$$

Then, from equations (1.4), (1.8), (1.9) and (1.10) we have, at order ε,

$$L_l\mathbf{E}_l^{(1)}=0 \tag{1.14}$$

In equation (1.14) we consider a mono-mode fiber in which there is only one bound state with eigenvalue k_0^2 (i.e. $l=\pm 1$) and the eigenfunction $\mathbf{U}=\mathbf{U}(r,\,\theta)$ (which is the mode-function describing the confinement of the pulse in the transverse direction and, in general, consists of two parts corresponding to the right and left polarisations). We further assume that the fiber maintains the polarisation (a polarisation preserving fiber). Without this assumption the resultant equation for the wave envelope becomes a coupled equation between the left and the right polarised waves. The solution to equation (1.14) may then be written as

$$\mathbf{E}_l^{(1)}(r, \theta, \xi, \tau) = \begin{cases} q_1^{(1)}(\xi, \tau)\mathbf{U}(r, \theta) & \text{for } l = 1 \\ 0, & \text{for } l \neq \pm 1 \end{cases} \tag{1.15}$$

Here, the coefficient $q_1^{(1)}(\xi, \tau)$ with $q_{-1}^{(1)} = q_1^{(1)*}$ is a complex, scalar function satisfying certain equations determined by the higher-order equation of equation (1.11). From the expression $L_1\mathbf{U} = 0$, the inner product $(\mathbf{U}, L_1\mathbf{U}) = 0$ gives the linear dispersion relation $k_0 = k_0(\omega_0)$,

$$\frac{1}{4} k_0^2 S_0 = \frac{\omega_0^2}{c^2}(\mathbf{U}, n_0^2\mathbf{U}) + (\mathbf{U}, L_0\mathbf{U}) + \mathrm{i}k_0 \int (U_z \boldsymbol{\nabla}_\perp \cdot \mathbf{U}^* - U_z^* \boldsymbol{\nabla}_\perp \cdot \mathbf{U})\mathrm{d}S \tag{1.16}$$

where $L_0 = L_l(l = 0)$, $n_0 = (\chi_0^{(0)})^{1/2}$ is the refractive index, and we have assumed the normalisation for $\mathbf{U}$ to be $\int (|U_r|^2 + |U_\theta|^2)\mathrm{d}S = S_0/4$.

At order ε^2, we have

$$L_l\mathrm{E}_l^{(2)} = \mathrm{i}\left\{-\frac{\partial L_l}{\partial \omega_l} - \left(\frac{1}{v_\mathrm{g}} - \frac{\partial k_0}{\partial \omega_0}\right)\frac{\partial}{\partial k_l} \cdot \left(L_l - \frac{\omega_l^2}{\mathrm{c}^2}\chi_l^{(0)}\right)\right\}\frac{\partial \mathrm{E}_l^{(1)}}{\partial \tau} \tag{1.17}$$

from which we obtain $\mathbf{E}_l^{(2)} = 0$ if $l \neq \pm 1$. In the case where $l = 1$, it is necessary that the inhomogeneous equation (1.17) satisfies the compatibility condition (the condition required for equation (1.17) to be solvable, which is known as the Fredholm alternative)

$$(\mathbf{U}, L_1\mathbf{E}_1^{(2)}) = 0 \tag{1.18}$$

This gives the group velocity v_g in terms of the linear dispersion relation equation (1.16)

$$\frac{1}{v_\mathrm{g}} = \frac{\partial k_0}{\partial \omega_0} \tag{1.19}$$

and for $l = 1$ equation (1.17) becomes

$$L_1\mathbf{E}_1^{(2)} = -\mathrm{i}\frac{\partial L_1}{\partial \omega_0}\frac{\partial \mathbf{E}_1^{(1)}}{\partial \tau} = -\mathrm{i}\frac{\partial L_1}{\partial \omega_0}\frac{\partial q_1^{(1)}}{\partial \tau}\mathbf{U} \tag{1.20}$$

From equation (1.13) for $l = 1$, the solution of equation (1.20) may be found in the form

$$\mathbf{E}_1^{(2)} = \mathrm{i}\frac{\partial q_1^{(1)}}{\partial \tau}\frac{\partial \mathbf{U}}{\partial \omega_0} + q_1^{(2)}\mathbf{U} \tag{1.21}$$

where $q_1^{(2)} = q_1^{(2)}(\xi, \tau)$ with $q_{-1}^{(2)} = q_1^{(2)*}$ is a scalar function to be determined in the higher-order equation. As we will see later, the first term in equation (1.21) represents the effect of wave guide dispersion in the coefficient of the non-linear dispersion terms in the NLS equation.

At order ε^3, we have

$$L_l \mathbf{E}_l^{(3)} \begin{cases} = 0 \qquad \text{if } l \neq \pm 1, \pm 3 \\ = -\dfrac{27\omega_0^2}{c^2} \chi_{111}^{(2)} q_1^{(1)3} (\mathbf{U}\cdot\mathbf{U})\mathbf{U} \qquad \text{if } l = 3 \\ = \mathrm{i}\dfrac{\partial L_1}{\partial \omega_0}\dfrac{\partial \mathbf{E}_1^{(2)}}{\partial \tau} + \dfrac{1}{2}\dfrac{\partial^2 L_1}{\partial \omega_0^2}\dfrac{\partial^2 \mathbf{E}_1^{(1)}}{\partial \tau^2} \\ \quad + \left(\mathrm{i}\dfrac{\partial q_1^{(1)}}{\partial \xi} - \dfrac{1}{2}\dfrac{\partial^2 k_0}{\partial \omega_0^2}\dfrac{\partial^2 q_1^{(1)}}{\partial \tau^2}\right)\left\{\dfrac{\partial}{\partial k_0}\left(L_1 - \dfrac{\omega_0^2 n_0^2}{c^2}\right)\right\}\mathbf{U} \\ \quad - |q_1^{(1)}|^2 q_1^{(1)} \dfrac{\omega_0^2}{c^2} \mathbf{F}(\mathbf{U}, \mathbf{U}^*; \chi^{(2)}) \qquad \text{if } l = 1 \end{cases} \tag{1.22}$$

where the column vector $\mathbf{F}$ is given by

$$\mathbf{F}(\mathbf{U}, \mathbf{U}^*; \chi^{(2)}) = \chi_{-111}^{(2)}(\mathbf{U}^*\cdot\mathbf{U})\mathbf{U} + \chi_{1-11}^{(2)}(\mathbf{U}\cdot\mathbf{U}^*)\mathbf{U} + \chi_{11-1}^{(2)}(\mathbf{U}\cdot\mathbf{U})\mathbf{U}^*$$

Note if $\mathbf{U}$ is real and $\chi^{(2)}$ is symmetric, $\mathbf{F} = \chi^{(2)}(\mathbf{U}\cdot\mathbf{U})\mathbf{U}$.) From equation (1.22) one can obtain the solutions $\mathbf{E}_l^{(3)} = 0$ for $l \neq \pm 1$ or ± 3, and since L_3 does not have an eigenmode,

$$\mathbf{E}_3^{(3)} = -\frac{27\omega_0^2}{c^2} q_1^{(1)3} L_3^{-1}\{\chi_{111}^{(2)}(\mathbf{U}\cdot\mathbf{U})\mathbf{U}\} \tag{1.23}$$

which is the harmonic generated by the non-linearity. For $l = 1$, we again require the compatibility condition

$$(\mathbf{U}, L_1\mathbf{E}_1^{(3)}) = 0 \tag{1.24}$$

from which we obtain the NLS equation for $q_1^{(1)}(\xi, \tau)$,

$$\mathrm{i}\frac{\partial q_1^{(1)}}{\partial \xi} - \frac{1}{2}\frac{\partial^2 k_0}{\partial \omega_0^2}\frac{\partial^2 q_1^{(1)}}{\partial \tau^2} + \nu |q_1^{(1)}|^2 q_1^{(1)} = 0 \tag{1.25}$$

Here, the Kerr coefficient ν is a positive real number given by

$$\nu = \frac{2\omega_0^2}{\tilde{k}_0 c^2 S_0}(\mathbf{U}, \mathbf{F}(\mathbf{U}, \mathbf{U}^*; \chi^{(2)})) \tag{1.26}$$

where $\tilde{k}_1 = k_1 - (2\mathrm{i}/S_0)\int (U_z \boldsymbol{\nabla}_\perp \cdot \mathbf{U}^* - U_z^* \boldsymbol{\nabla}_\perp \cdot \mathbf{U})\mathrm{d}S$. It is worth noting that the explicit form given in equation (1.21) for $\mathbf{E}_1^{(2)}$ is unnecessary in the calculation of the compatibility condition equation (1.24), and that equation (1.24) can be obtained directly from the equations for $\mathbf{E}_1^{(1)}$ and $\mathbf{E}_1^{(2)}$, i.e. equations (1.14) and (1.20).

In order to see the effect of the higher-order terms, one needs to find the equation for $q_1^{(2)}$ in (1.21). For this purpose, we have, at order ε^4, the expression $L_l\mathbf{E}_l^{(4)}$ for $l = 1$,

$$L_1\mathbf{E}_1^{(4)} = -\mathrm{i}\frac{\partial L_1}{\partial \omega_0}\frac{\partial \mathbf{E}_1^{(3)}}{\partial \tau} + \frac{1}{2}\frac{\partial^2 L_1}{\partial \omega_0^2}\frac{\partial^2 \mathbf{E}_1^{(2)}}{\partial \tau^2} + \frac{\mathrm{i}}{6}\frac{\partial^3 L_1}{\partial \omega_0^3}\frac{\partial^3 \mathbf{E}_1^{(1)}}{\partial \tau^3}$$

$$+ \left\{\frac{\partial}{\partial k_0}\left(L_1 - \frac{\omega_0^2 n_0^2}{c^2}\right)\right\}\left(\mathrm{i}\frac{\partial \mathbf{E}_1^{(2)}}{\partial \xi} - \frac{1}{2}\frac{\partial^2 k_0}{\partial \omega_0^2}\frac{\partial^2 \mathbf{E}_1^{(2)}}{\partial \tau^2} - \frac{\mathrm{i}}{6}\frac{\partial^3 k_0}{\partial \omega_0^3}\frac{\partial^3 \mathbf{E}_1^{(1)}}{\partial \tau^3}\right)$$

$$+ \mathrm{i}\left\{\frac{\partial^2}{\partial \omega_0 \partial k_0}\left(L_1 - \frac{\omega_0^2 n_0^2}{c^2}\right)\right\}\cdot\frac{\partial}{\partial \tau}\left(\mathrm{i}\frac{\partial \mathbf{E}_1^{(1)}}{\partial \xi} - \frac{1}{2}\frac{\partial^2 k_0}{\partial \omega_0^2}\frac{\partial^2 \mathbf{E}_1^{(1)}}{\partial \tau^2}\right)$$

$$- \frac{1}{c^2}\sum_{l_1+l_2+l_3=1}\left[\omega_0^2 \sum_{i+j+k=4} \chi^{(2)}_{l_1 l_2 l_3}(\mathbf{E}_{l_1}^{(i)}\cdot\mathbf{E}_{l_2}^{(j)})\mathbf{E}_{l_3}^{(k)}\right.$$

$$\left. + \mathrm{i}\sum_{i=1}^{3}\left\{\left(2\omega_0 + \omega_0^2\frac{\partial}{\partial \omega_{l_i}}\right)\chi^{(2)}_{l_1 l_2 l_3}\right\}\frac{\partial}{\partial \tau_{\mathrm{i}}}(\mathbf{E}_{l_1}^{(1)}\cdot\mathbf{E}_{l_2}^{(1)})\mathbf{E}_{l_3}^{(1)}\right] \qquad (1.27)$$

Using equations (1.14), (1.20), (1.22) and the remark following equation (1.26), one can calculate the compatibility condition for equation (1.27), i.e. $(\mathbf{U}, L_1\mathbf{E}_1^{(4)}) = 0$. The resulting equation for $q_1^{(2)}$ is

$$\mathrm{i}\frac{\partial q_1^{(2)}}{\partial \xi} - \frac{1}{2}\frac{\partial^2 k_0}{\partial \omega_0^2}\frac{\partial^2 q_1^{(2)}}{\partial \tau^2} + 2\nu|q_1^{(1)}|^2 q_1^{(2)} + \nu q_1^{(1)2} q_1^{(2)*}$$

$$- \frac{\mathrm{i}}{6}\frac{\partial^3 k_0}{\partial \omega_0^3}\frac{\partial^3 q_1^{(1)}}{\partial \tau^3} + \mathrm{i}a_1\frac{\partial}{\partial \tau}(|q_1^{(1)}|^2 q_1^{(1)}) + \mathrm{i}a_2 q_1^{(1)}\frac{\partial}{\partial \tau}|q_1^{(1)}|^2 = 0 \qquad (1.28)$$

Here the coefficients of the non-linear dispersion terms a_1 and a_2 are

$$a_1 = \frac{\partial \nu}{\partial \omega_0} \qquad (1.29)$$

$$a_2 = \frac{2\omega_0^2}{\tilde{k}_1 c^2 S_0}(\mathbf{U}, \mathbf{G}(\mathbf{U}, \mathbf{U}^*; \chi^{(2)})) \qquad (1.30)$$

where

$$\mathbf{G} = \partial\mathbf{F}/\partial\omega_0 + \chi^{(2)}_{-\dot{1}11}(\dot{\mathbf{U}}^*\cdot\mathbf{U})\mathbf{U} - \chi^{(2)}_{-\dot{1}11}(\mathbf{U}^*\cdot\mathbf{U})\mathbf{U}$$
$$+ \chi^{(2)}_{1-11}(\mathbf{U}\cdot\dot{\mathbf{U}}^*)\mathbf{U} - \chi^{(2)}_{1-\dot{1}1}(\mathbf{U}\cdot\mathbf{U}^*)\mathbf{U}$$
$$+ \chi^{(2)}_{11-1}(\mathbf{U}\cdot\mathbf{U})\dot{\mathbf{U}}^* - \chi^{(2)}_{11-\dot{1}}(\mathbf{U}\cdot\mathbf{U})\mathbf{U}^*$$

with $\dot{\mathbf{U}} = \partial\mathbf{U}/\partial\omega_0$. A 'dot' over the subscript indicates the partial derivative with respect to that component of frequency, i.e. $\chi_{\dot{l}_1 l_2 l_3} = \partial\chi_{l_1 l_2 l_3}/\partial\omega_{l_1}$, and so on. The equation for $\mathbf{E}_1$ cannot be separated in each order of ε, and one should use the following equation for the combined variable $q_1 = \varepsilon q_1^{(1)} + \varepsilon^2 q_1^{(2)}$ (which is given by the projection of $\mathbf{E}_1$ onto $\mathbf{U}$, i.e. $q_1 = 4\int \mathbf{E}_1\cdot\mathbf{U}\mathrm{d}S/S_0$),

$$\mathrm{i}\frac{\partial q_1}{\partial \xi} - \frac{1}{2}\frac{\partial^2 k_1}{\partial \omega_0^2}\frac{\partial^2 q_1}{\partial \tau^2} + \tilde{\nu}|q_1|^2 q_1 - \varepsilon\frac{\mathrm{i}}{6}\frac{\partial^3 k_1}{\partial \omega_0^3}\frac{\partial^3 q_1}{\partial \tau^3} \tag{1.31}$$

$$+ \varepsilon \mathrm{i}\tilde{a}_1 \frac{\partial}{\partial \tau}(|q_1|^2 q_1) + \varepsilon \mathrm{i}\tilde{a}_2 q_1 \frac{\partial}{\partial \tau}(|q_1|^2) = \mathrm{O}(\varepsilon^3)$$

with $\tilde{\nu} = \nu/\varepsilon^2$, $\tilde{a}_1 = a_1/\varepsilon^2$, $\tilde{a}_2 = a_2/\varepsilon^2$. Equation (1.31) is the desired equation for the envelope function q_1.

The coefficients of the non-linear envelope equation (1.31) may be simplified for a weakly guided fiber. From here on, we remove from ω and k the subscripts 0 which were used to designate the carrier frequency and wavenumber. In order to derive the simplified expression, we first calculate the approximated mode function $\mathbf{U}$ satisfying equation (1.14), $L_1\mathbf{U} = 0$, assuming weak guiding, i.e.

$$|\lambda \nabla_\perp \ln n_0| \simeq \mathrm{O}(\delta) \ll 1 \tag{1.32}$$

where λ is the wavelength of the carrier, and then we evaluate the coefficients given by equations (1.16), (1.26), (1.29) and (1.30).

We write the electric field, $\hat{\boldsymbol{E}} = 2q_1\boldsymbol{U}$, as

$$\hat{\boldsymbol{E}} = 2q_1\boldsymbol{U} = \nabla\phi \times \hat{z} + \mathrm{V} \qquad \text{with } (\nabla\phi \times \hat{z})\cdot \boldsymbol{V} = 0 \tag{1.33}$$

(Note that if $\boldsymbol{V} = 0$, an exact TE mode is implied.) Then, equation (1.14) becomes

$$\hat{z} \times \nabla_\perp\left\{(\nabla_\perp^2 - k^2 + \frac{\omega^2}{c^2} n_0^2)\phi\right\} - \frac{\omega^2}{c^2}\phi\hat{z} \times \nabla_\perp n_0^2 - L_1\mathbf{V} = 0 \tag{1.34}$$

where the column vector $L_1\mathbf{V}$ is considered to be the usual vector. Using the small parameter δ in equation (1.32), we solve equation (1.34) by the perturbation expansion, i.e. $k = k^{(0)} + \delta k^{(1)} + \ldots$ and $V = \delta V^{(1)} + \ldots$. Then, in leading order, we have

$$\nabla_\perp^2\phi - k^{(0)2}\phi + \frac{\omega^2}{c^2} n_0^2\phi = 0 \tag{1.35}$$

where we have assumed that $|\phi| \to 0$ as $||\boldsymbol{x}_\perp|| \to \infty$, i.e. the bound state. Here, the leading order eigenvalue $k^{(0)}$ is obtained by

$$k^{(0)2} = \frac{(\omega^2/c^2)\int |\phi|^2 n_0^2 \mathrm{d}S - \int |\nabla_\perp \phi|^2 \mathrm{d}S}{\int |\phi|^2 \mathrm{d}S} \tag{1.36}$$

At the order of δ, we have

$$L_1\mathbf{V} = -2k^{(0)}k^{(1)}\hat{z} \times \nabla_\perp\phi - \frac{\omega^2}{c^2}\phi(\hat{z} \times \nabla_\perp n_0^2)\delta^{-1} \tag{1.37}$$

From the compatibility condition (see the statement preceding

(1.18)), i.e. $(\mathbf{U}, L_1\mathbf{V}) = 0$, we obtain the equation for $k^{(1)}$,

$$2\delta k^{(0)} k^{(1)} = -\frac{(\omega^2/c^2)\int \phi^*(\nabla_\perp\phi \cdot \nabla_\perp n_0^2)\mathrm{d}S}{\int |\nabla_\perp\phi|^2\mathrm{d}S} + \mathrm{O}(\delta^2) \qquad (1.38)$$

where we have used the condition that equation (1.35) has only one bound state, i.e. the single-mode fiber. From equations (1.35), (1.36) and (1.38), we obtain the approximated eigenvalue k, i.e. the linear dispersion relation,

$$\begin{aligned} k^2 &= k^{(0)2} + 2\delta k^{(0)} k^{(1)} + \mathrm{O}(\delta^2) \\ &= \frac{(\omega^2/c^2)\int |\nabla_\perp\phi|^2 n_0^2\mathrm{d}S - \int |\nabla_\perp^2\phi|^2\mathrm{d}S}{\int |\nabla_\perp\phi|^2\mathrm{d}S} + \mathrm{O}(\delta^2) \qquad (1.39) \end{aligned}$$

We note that equation (1.39) can be obtained from $(\mathbf{U}, L_1\mathbf{U}) = 0$ with $2q_1U = \nabla_\perp\phi \times \hat{z}$ where ϕ satisfies equation (1.35). Thus, we can assume the TE mode up to order δ, and the effect of an axial component of the electric field appears at order δ^2. However, equation (1.39) indicates that the eigenvalue for the transverse electric field gives a better approximation than that for the potential field as given by equation (1.36).

Let us now evaluate the first term in the coefficient of the non-linear dispersion, v, equation (1.26). When equation (1.33) is used in equation (1.26), the coefficient v becomes

$$\begin{aligned} v &= \frac{\omega^2 S_0}{8kc^2}\frac{3\int \chi_2|\nabla_\perp\phi|^4\mathrm{d}S}{(\int |\nabla_\perp\phi|^2\mathrm{d}S)^2} + \mathrm{O}(\delta^2) \\ &= \frac{\omega^2}{kc^2 S_0 E_0^4}\int n_0 n_2|\nabla_\perp\phi|^4\mathrm{d}S + \mathrm{O}(\delta^2) \qquad (1.40) \end{aligned}$$

where we have assumed the medium to be isotropic and symmetric in the instantaneous non-linear response, i.e. $\chi^{(2)}_{-111} = \chi^{(2)}_{1-11} = \chi^{(2)}_{11-1} = \chi_2$. This non-linear dielectric constant χ_2 may be given by the following form of the real non-linear response function $\chi^{(2)}$ in equation (1.2):

$$\chi_2(\omega) = \int_{-\infty}^{0}\mathrm{d}t_1\int_{-\infty}^{0}\mathrm{d}t_2\int_{-\infty}^{0}\mathrm{d}t_3\chi^{(2)}(t_1, t_2, t_3)\exp(-\mathrm{i}\omega(t_1 - t_2 - t_3)) \qquad (1.41)$$

where the imaginary part of χ_2 may be negligible for most practical fibers. Equation (1.40) can be evaluated given the eigenfunction ϕ. In particular, for a dielectric wave guide with step dielectric profile, ϕ is given by a Bessel function and v is reduced to the value obtained by Hasegawa and Tappert (1973a).

Given these assumptions, the coefficients of the higher-order non-linear dispersion and dissipation terms a_1 and a_2 are obtained in the same way. Using equations (1.29) and (1.40)

$$a_1 = \frac{\partial v}{\partial \omega}$$

$$= \frac{\partial}{\partial \omega}\left[\frac{\omega^2}{kc^2 S_0 E_0^4}\int n_0 n_2 |\nabla_\perp \phi|^4 \mathrm{d}S\right] + \mathrm{O}(\delta^2) \qquad (1.42)$$

Similarly, the coefficient a_2 is obtained from equation (1.30)

$$a_2 = \frac{\omega^2}{kc^2 S_0 E_0^4}\int\left[n_0 n_2 \frac{\partial |\nabla\phi|^4}{\partial \omega} + \frac{3}{4}(\chi_1^{(2)} - \chi_{-1}^{(2)})|\nabla\phi|^4\right]\mathrm{d}S + \mathrm{O}(\delta^2) \qquad (1.43)$$

where

$$\chi_1^{(2)} \equiv \chi_{-1\mathrm{i}1}^{(2)} = \chi_{-1\mathrm{1i}}^{(2)} = \chi_{\mathrm{i}-11}^{(2)} = \chi_{1-1\mathrm{i}}^{(2)} = \chi_{\mathrm{i}1-1}^{(2)} = \chi_{1\mathrm{i}-1}^{(2)}$$

and

$$\chi_{-1}^{(2)} \equiv \chi_{-\mathrm{i}11}^{(2)} = \chi_{1-\mathrm{i}1}^{(2)} = \chi_{11-\mathrm{i}}^{(2)}$$

are derivatives of $\chi^{(2)}$ with respect to ω and $-\omega$. From the definitions of $\chi_{-1}^{(2)}$ and $\chi_1^{(2)}$, we have the formula for $\chi_1^{(2)} - \chi_{-1}^{(2)}$ in terms of the real non-linear response function $\chi^{(2)}$ in equation (1.2)

$$\chi_1^{(2)}(\omega) - \chi_{-1}^{(2)}(\omega) = \frac{\mathrm{i}}{2}\int_{-\infty}^{0}\mathrm{d}t_1\int_{-\infty}^{0}\mathrm{d}t_2\int_{-\infty}^{0}\mathrm{d}t_3 \cdot \chi^{(2)}(t_1, t_2, t_3)(2t_1-t_2-t_3) \cdot \exp(-\mathrm{i}\omega(t_1-t_2-t_3)) \qquad (1.44)$$

which is imaginary. We note that the term a_2 can be obtained by the retarded non-linear response which has the form $q_1(t) \int_{-\infty}^{t} \mathrm{d}s f(t-s)|q_1(s)|^2$ assuming short delays (this assumption is consistent with the quasi-monochromatic approximation). As has been shown by Gordon (1986), this term corresponds to the retarded Raman effect in which the higher-frequency components of the soliton spectrum pump energy to the lower-frequency components. Consequently, this term produces a non-linear dissipation and, as will be discussed in Section 1.3.3, leads to a down-shift in the carrier frequency of the soliton.

1.3 Properties of non-linear Schrödinger equation

The non-linear wave equation (1.31) can be put into a normalised form,

$$\mathrm{i}\left(\frac{\partial q}{\partial Z} + \Gamma q\right) + \frac{1}{2}\frac{\partial^2 q}{\partial T^2} + |q|^2 q$$

$$+ \varepsilon\mathrm{i}\left\{\beta_1 \frac{\partial^3 q}{\partial T^3} + \beta_2 \frac{\partial}{\partial T}(|q|^2 q) + \beta_3 q \frac{\partial}{\partial T}|q|^2\right\} = 0 \tag{1.45}$$

Here,

$$q = \frac{(g\lambda)^{1/2}}{\varepsilon} E \tag{1.46}$$

$$T = \frac{\tau}{T_0} = \frac{\varepsilon(t - x/v_\mathrm{g})}{(-\lambda k'')^{1/2}} \tag{1.47}$$

$$Z = \frac{\xi}{\lambda} \tag{1.48}$$

$$T_0 = (-\lambda k'')^{1/2} \tag{1.49}$$

$$\Gamma = \frac{\gamma\lambda}{\varepsilon^2} \tag{1.50}$$

γ is the fiber loss rate per unit distance, $k''(= \partial^2 k/\partial\omega^2)$ is given by the second derivative of the wavenumber k in the fiber given by equation (1.39), and

$$g = \frac{\omega^2}{kc^2 S_0 E_0^4} \int n_0 n_2 |\nabla\phi|^4 \mathrm{d}S \simeq \frac{\pi n_2}{\lambda} \tag{1.51}$$

The coefficients of the higher-order terms are given by

$$\beta_1 = \frac{1}{6}\frac{k'''\lambda}{T_0^3} \qquad \lambda = \frac{2\pi}{k} \tag{1.52}$$

$$\beta_2 = \frac{1}{gT_0}\frac{\partial}{\partial\omega}\left[\frac{\omega^2}{kc^2 S_0 E_0^4}\int n_0 n_2 |\nabla\phi|^4 \mathrm{d}S\right] \tag{1.53}$$

$$\beta_3 = \frac{1}{gT_0}\frac{\omega^2}{kc^2 S_0 E_0^4}$$

$$\cdot \int \left[n_0 n_2 \frac{\partial|\nabla\phi|^4}{\partial\omega_0} + \frac{3}{4}(\chi_1^{(2)} - \chi_{-1}^{(2)}|\nabla\phi|^4\right]\mathrm{d}S. \tag{1.54}$$

Here β_1 represents the higher-order linear dispersion, β_2 the non-linear dispersion of the Kerr coefficient, and β_3 (which is imaginary) the non-linear dissipation due to the Raman process in the fiber. In this section, we discuss the mathematical property of equation (1.45) and its physical consequences.

1.3.1 *Soliton solutions*

1.3.1.1 Bright soliton Equation (1.45) without the higher-order term (terms with the coefficient ε) and the loss,

$$\mathrm{i}\frac{\partial q}{\partial z} + \frac{1}{2}\frac{\partial^2 q}{\partial T^2} + |q|^2 q = 0 \tag{1.55}$$

was found to be integrable for localised initial wave envelope $q_0(T)$ by Zakharov and Shabat (1972).

Following the method of Gardner *et al.* (1967) and of Lax (1968), they discovered that the eigenvalue λ of Dirac-type eigenvalue equations,

$$L\psi = \lambda\psi \tag{1.56}$$

$$\psi = \begin{pmatrix} \psi_1 \\ \psi_2 \end{pmatrix} \tag{1.57}$$

with

$$L = \mathrm{i}\begin{pmatrix} 1+\beta & 0 \\ 0 & 1-\beta \end{pmatrix}\frac{\partial}{\partial x} + \begin{bmatrix} 0 & u^* \\ u & 0 \end{bmatrix}$$

become time-invariant if u evolves in accordance with the NLS equation of the form,

$$\mathrm{i}\frac{\partial u}{\partial t} + \frac{\partial^2 u}{\partial x^2} + Q|u|^2 u = 0 \tag{1.58}$$

where $\beta^2 = 1 - 2/Q = \text{constant}$, and the time evolution of the eigenfunction ψ is given by

$$\mathrm{i}\frac{\partial \psi}{\partial t} = A\psi \tag{1.59}$$

where

$$A = -\beta\begin{bmatrix} 1 & 0 \\ 0 & 1 \end{bmatrix}\frac{\partial^2}{\partial x^2} + \begin{bmatrix} |u|^2/(1+\beta) & \mathrm{i}\partial u^*/\partial x \\ -\mathrm{i}\partial u/\partial x & -|u|^2/(1-\beta) \end{bmatrix} \tag{1.60}$$

Once the structure of the eigenvalue equation which satisfies the Lax criteria has been discovered, one can apply the inverse scattering technique to obtain the time evolution of the potential q, and the NLS equation can then be solved for a localised initial condition. As in the case of the KdV equation (Gardner *et al.*, 1967), the time invariance of the eigenvalues provides those properties of solitons which are created from the initial condition in terms of the eigenvalues of the initial potential shape. For the NLS equation (1.55) the appropriate structure of the eigenvalue equation becomes (Satsuma and Yajima, 1974),

$$\mathrm{i}\,\frac{\partial \psi_1}{\partial T} + q_0(T)\psi_2 = \zeta\psi_1 \tag{1.61}$$

$$-\mathrm{i}\,\frac{\partial \psi_2}{\partial T} - q_0^*(T)\psi_1 = \zeta\psi_2 \tag{1.62}$$

If we write the eigenvalue of this equation as

$$\zeta_n = \frac{\kappa_n + \mathrm{i}\eta_n}{2} \tag{1.63}$$

the n-soliton solutions which arise from this initial wave form are

$$q(T, Z) = \sum_{j=1}^{N} \eta_j \operatorname{sech} \eta_j(T + \kappa_j Z - \theta_{0j}) \cdot \exp\left\{-\mathrm{i}\kappa_j T + \frac{\mathrm{i}}{2}(\eta_j^2 - \kappa_j^2)Z - \mathrm{i}\sigma_{0j}\right\}. \tag{1.64}$$

We note that the (η_j) amplitude and speed (κ_j) of the soliton are characterised by the imaginary and real parts of the eigenvalue (1.63).

For example, if we approximate the input pulse shape of a mode-locked laser as

$$q_0(T) = A \operatorname{sech} T \tag{1.65}$$

the eigenvalues of equations (1.61) and (1.62) are obtained analytically and the number of eigenvalues N is given by (Satsuma and Yajima, 1974),

$$A - \tfrac{1}{2} < N \leqslant A + \tfrac{1}{2} \tag{1.66}$$

where the corresponding eigenvalues are imaginary and given by

$$\zeta_n = \mathrm{i}\,\frac{\eta_n}{2} = \mathrm{i}(A - n + \tfrac{1}{2}) \qquad n = 1, 2, \ldots, N \tag{1.67}$$

If A is exactly equal to N, the solution can be obtained in terms of N solitons and their amplitudes are then given by

$$\eta_n = 2(N - n) + 1 = 1, 3, 5, \ldots, (2N - 1) \tag{1.68}$$

We should note here that, in this particular case, all the eigenvalues are purely imaginary. Consequently, $\kappa_n = 0$, and all the soliton velocities in the frame of reference of the group velocity are 0.

While the speed of the solitons in the KdV equation are proportional to the amplitude, those in the Schrödinger equation have no such dependence. In general, if the input pulse shape has a symmetric shape, as in this example of sech T, the eigenvalues of equations (1.61) and (1.62) can be shown to be purely imaginary, and the output solitons propagate at exactly the same speed (Kodama and Hasegawa, 1987). When a number of solitons propagate at the same speed, the superimposed pulse shape oscillates due to the phase interference among the solitons.

If we remember that the eigenvalues are given by equation (1.67), we can obtain the amplitude of the output soliton when the initial amplitude is slightly different from that which corresponds to the one-soliton solution. For example, if we write the input pulse shape as

$$q_0(T) = (1 + \Delta)\,\mathrm{sech}\,T \tag{1.69}$$

equation (1.67) gives $\eta = 1 + 2\Delta$ for the value of $A = 1 + \Delta$. From this, the output soliton shape is given by

$$q(T) = (1 + 2\Delta)\,\mathrm{sech}\,(1 + 2\Delta)T \tag{1.70}$$

Since the soliton energy is given by

$$\varepsilon = \int_{-\infty}^{\infty} |q|^2 \mathrm{d}T \tag{1.71}$$

the difference in energy between the input and output becomes

$$\begin{aligned}\Delta\varepsilon &= \int_{-\infty}^{\infty} (1 + \Delta)^2\,\mathrm{sech}^2\,T\mathrm{d}T \\ &\quad - \int_{-\infty}^{\infty} (1 + 2\Delta)^2\,\mathrm{sech}^2\,(1 + 2\Delta)T\mathrm{d}T \\ &= \Delta^2\end{aligned} \tag{1.72}$$

i.e. the soliton energy is smaller by a factor Δ^2 compared with the input pulse energy. This means that if the input amplitude is not exactly an integer N, part of the energy in the input pulse is transferred to other waves, namely to a linear dispersive wave which does not form a soliton.

N solitons which propagate at the same speed propagate together with phase interactions. In general, the periodicity of this oscillation is given by the lowest common beat frequency,

$$Z_0 = \frac{2\pi}{|\omega_i - \omega_j|} \tag{1.73}$$

where $\omega_j = \eta_j^2/2$, η_j being the imaginary part of the eigenvalues of the initial pulse shape. In particular, when the pulse shape is given by equation (1.65), the period of oscillation of n solitons reduces to the simple form,

$$Z_0 = \frac{\pi}{2} \tag{1.74}$$

The quantity Z_0 is often referred to as a soliton period.

The NLS equation is integrable owing to the fact that it has an infinite number of conserved quantities. We write here the three lowest conserved quantities;

$$C_1 = \int_{-\infty}^{\infty} |q(Z, T)|^2 \mathrm{d}T \tag{1.75}$$

$$C_2 = \int_{-\infty}^{\infty} \left(q^* \frac{\partial q}{\partial T} - q \frac{\partial q^*}{\partial T} \right) \mathrm{d}T \tag{1.76}$$

and

$$C_3 = \int_{-\infty}^{\infty} \left(\left| \frac{\partial q}{\partial T} \right|^2 - |q|^4 \right) \mathrm{d}T \tag{1.77}$$

We note that the one-soliton solution (equation (1.65)) is obtained by minimising C_3, with constraints of C_1 and C_2 being constant.

1.3.1.2 Dark soliton In the wavelength range shorter than the zero group dispersion point where k'' is zero, the soliton solution does not exist. However, even in this range of wavelengths, the portion without light, which is produced by chopping a continuous light wave, is shown to form a soliton (Hasegawa and Tappert, 1973b). Such a soliton solution is often called a dark soliton. We present here the theoretical and experimental properties of a dark soliton.

When $k'' > 0$, (1.55) can be written by using a new time variable T,

$$T = \frac{\tau}{(\lambda k'')^{1/2}} \tag{1.77'}$$

as

$$\mathrm{i}\frac{\partial q}{\partial Z} - \frac{1}{2}\frac{\partial^2 q}{\partial T^2} + |q|^2 q = 0 \tag{1.78}$$

In this expression, the definition of q and Z are the same as those of (1.46) and (1.48), respectively. The dark soliton solution is obtained by introducing variables ρ and σ,

$$q = \rho^{1/2} \exp(\mathrm{i}\sigma) \tag{1.79}$$

and by using the condition that ρ becomes a function of T only (i.e. a stationary solution).

$$\rho = \rho_0[1 - a^2 \operatorname{sech}^2(\rho_0^{1/2} a T)] \qquad a^2 = \frac{\rho_0 - \rho_s}{\rho_0} \leqslant 1$$

$$\sigma = \int \frac{c_1}{\rho}\,\mathrm{d}T + \Omega Z$$

$$= \int \frac{\rho_0^{3/2}(1 - a^2)^{1/2}}{\rho}\,\mathrm{d}T - \frac{\rho_0(3 - a^2)}{2}\,Z$$

$$= [\rho_0(1 - a^2)]^{1/2} T + \tan^{-1}\left[\frac{a}{(1 - a^2)^{1/2}} \tanh(\rho_0^{1/2} a T)\right]$$

$$- \frac{\rho_0(3 - a^2)}{2}\,Z \qquad (1.80)$$

$$c_1^2 = \rho_0^3(1 - a^2)$$

$$\Omega = -\tfrac{1}{2}\rho_0(3 - a^2)$$

Unlike a bright soliton, a dark soliton has an additional new parameter, a, which designates the depth of modulation. We should also note the fact that at $T \to \pm\infty$, the phase of q changes. Such a soliton is called a *topological soliton* while the bright soliton, which has no phase change at $T \to \pm\infty$, is called a *non-topological soliton*. When $a = 1$, the depth approaches 0 and the solution becomes

$$q = \rho_0^{1/2} \tanh(\rho_0^{1/2} T) \qquad (1.81)$$

As in the case of a bright soliton solution, a general dark soliton solution can be obtained by a Galilei transformation of (1.80) and is given by

$$\rho' = \rho_0\{1 - a^2 \operatorname{sech}^2 [\rho_0^{1/2} a(T - UZ)]\} \qquad (1.82)$$

$$\sigma' = \sigma + uT - \tfrac{1}{2}u^2 Z$$

1.3.2 *Modulational instability*

As has been shown, the light wave in a fiber can be described by the NLS equation. When the input wave has a pulse shape, the output can be described in terms of a set of solitons and a dispersive wave as shown from the inverse scattering calculation. The question we should like to discuss here is: what happens if the input light wave has a continuous amplitude? Let us start with the description of the wave envelope q given by equation (1.55)

$$\mathrm{i}\frac{\partial q}{\partial Z} + \frac{1}{2}\frac{\partial^2 q}{\partial T^2} + |q|^2 q = 0 \qquad (1.55)$$

Here, we show that the input light wave becomes unstable for a small perturbation around the initial amplitude q_0 (Akhmanov *et al.*, 1968).

This instability is called the modulational instability. To show the instability, we introduce new real variables ρ and σ through.

$$q = \rho^{1/2} \exp(\mathrm{i}\sigma) \tag{1.83}$$

and substitute equation (1.83) into the NLS equation (1.55),

$$\frac{\partial \rho}{\partial Z} + \frac{\partial \rho}{\partial T}\frac{\partial \sigma}{\partial T} + \rho \frac{\partial^2 \sigma}{\partial T^2} = 0 \tag{1.84}$$

and

$$\rho - \frac{\partial \sigma}{\partial Z} + \frac{1}{4\rho}\frac{\partial^2 \rho}{\partial T^2} - \frac{1}{2}\left(\frac{\partial \sigma}{\partial T}\right)^2 - \frac{1}{8\rho^2}\left(\frac{\partial \rho}{\partial T}\right)^2 = 0 \tag{1.85}$$

We consider a small modulation of ρ and σ with the sideband frequency given by Ω, such that

$$\rho(T, Z) = \rho_0 + \mathrm{Re}\, \rho_1 \exp(\mathrm{i}(KZ - \Omega T)) \tag{1.86}$$

and

$$\sigma(T, Z) = \sigma_0 + \mathrm{Re}\, \sigma_1 \exp(\mathrm{i}(KZ - \Omega T)) \tag{1.87}$$

If we substitute equations (1.86) and (1.87) into (1.84) and (1.85) and linearise the results, we obtain the following dispersion relation for the wavenumber K and frequency Ω:

$$K^2 = \tfrac{1}{4}(\Omega^2 - 2\rho_0)^2 - \rho_0^2 \tag{1.88}$$

This expression gives the spatial growth rate $\mathrm{Im}\, K$ which achieves its maximum value at

$$\Omega \equiv \Omega_{\mathrm{m}} = (2\rho_0)^{1/2} = 2^{1/2}|q_0| \tag{1.89}$$

and the corresponding growth rate becomes

$$\mathrm{Im}\, K = \rho_0 = |q_0|^2 \tag{1.90}$$

If we write the variables in terms of the original parameters, the frequency that shows the maximum growth rate is given by

$$\omega_{\mathrm{m}} = \sqrt{2} E_0 \pi n_2 / (-\lambda k'')^{1/2} \tag{1.91}$$

and the corresponding spatial growth rate γ_{g} is given by

$$\gamma_{\mathrm{g}} = \pi n_2 |E_0|^2 / \lambda \tag{1.92}$$

If we recognise that ω_{m}^{-1} corresponds approximately to the pulsewidth of a soliton with amplitude E_0, we can see a close relationship between the formation of a soliton and modulational instability.

1.3.3 *Effects of higher-order terms*

In this section we discuss the influence of the higher-order terms in the NLS equation on the soliton propagation in a fiber. In particular we consider the effect of the self-induced Raman effect on the soliton transmission.

First we put the full envelope equation (1.45) into the following form,

$$\mathrm{i}\frac{\partial q}{\partial Z} + \frac{1}{2}\frac{\partial^2 q}{\partial T^2} + |q|^2 q + h = 0$$

where h represents the higher-order effects,

$$h = \mathrm{i}\varepsilon\left\{\beta_1 \frac{\partial^3 q}{\partial T^3} + \beta_2 \frac{\partial}{\partial T}(|q|^2 q) + \mathrm{i}\sigma_3 q \frac{\partial}{\partial T}|q|^2\right\} \tag{1.93}$$

Here, the last term represents the self-induced Raman effect which produces the down-shift of the soliton spectrum by the Raman-induced spectral decay. The coefficient of this term is given by

$$\sigma_3 = -\mathrm{i}\beta_3 \tag{1.94}$$

Of these three higher-order terms, the last term which represents the self-induced Raman effect plays the most dominant role. The first two terms can be derived by a phenomenological expansions of the non-linear refractive index. The NLS equation including the first two terms is still integrable (Kodama and Hasegawa, 1987) in this order and the soliton property is not essentially modified.

The effect of these higher-order terms on the transmission property of solitons can be obtained by evaluating the conservation laws, equations (1.75) to (1.79) including the higher-order term, h.

$$\mathrm{i}\frac{\partial}{\partial Z}\int_{-\infty}^{\infty}|q|^2 \mathrm{d}T = -\int_{-\infty}^{\infty}(hq^* - h^* q)\mathrm{d}T = 0 \tag{1.95}$$

$$\begin{aligned}\frac{\mathrm{i}}{2}\frac{\partial}{\partial Z}&\int_{-\infty}^{\infty}\left(q\frac{\partial q^*}{\partial T} - q^*\frac{\partial q}{\partial T}\right)\mathrm{d}T \\ &= -\int_{-\infty}^{\infty}\left(h\frac{\partial q^*}{\partial T} + h^*\frac{\partial q}{\partial T}\right)\mathrm{d}T \\ &= \varepsilon\sigma_3\int_{-\infty}^{\infty}\left(\frac{\partial}{\partial T}|q|^2\right)^2 \mathrm{d}T \\ &\quad - \mathrm{i}\varepsilon\beta_2\int_{-\infty}^{\infty}\left(q\frac{\partial q^*}{\partial T} - q^*\frac{\partial q}{\partial T}\right)\frac{\partial}{\partial T}|q|^2 \mathrm{d}T\end{aligned} \tag{1.96}$$

In particular, we note that these higher-order terms do not change the soliton energy, as is seen from equation (1.95). However, the momentum is modified, as shown in equation (1.96).

If we take the effect of the self-induced Raman term (the third term) and use the one-soliton solution of equation (1.64) in q of equations (1.95) and (1.96), we have

$$\frac{\mathrm{d}\eta}{\mathrm{d}Z} = 0 \tag{1.97}$$

$$\frac{\mathrm{d}\kappa}{\mathrm{d}Z} = -\frac{8}{15}\varepsilon\sigma_3\eta^4 \tag{1.98}$$

Here η is the normalised amplitude of the soliton, and κ represents the frequency of the soliton. Equation (1.98) shows that the soliton frequency decreases in proportion to the fourth power of its amplitude. If we use the original parameters, equation (1.98) reduces to

$$\frac{\mathrm{d}f}{\mathrm{d}x} = \frac{4}{15}\frac{n_2}{\lambda k''}\beta E_0^4 = \frac{2.56}{\pi^2}\frac{\lambda k''\beta}{n_2\tau_0^4} \tag{1.99}$$

Here τ_0 is the width of the soliton. In this expression, the coefficient β can be expressed in terms of the differential gain, γ_R, with respect to the frequency separation $\Delta\omega$ which designates the frequency separation between the pump and the Stokes frequencies.

$$\beta = \left|\frac{\partial\gamma_R}{\partial\Delta\omega}\right| \tag{1.100}$$

where $\gamma_R E_0^2$ gives the Raman gain per unit length of the fiber.

The fact that the central frequency of a soliton decreases in proportion to the distance of propagation implies that the group velocity decreases by the factor $\Delta v_g = (\partial v_g/\partial\omega)\Delta\omega = k''\Delta\omega/(k')^2$ (<0). Since $\Delta\omega$ is proportional to the fourth power of the soliton amplitude, the decrease in the soliton speed becomes larger for a soliton with a larger amplitude.

In the absence of the self-induced Raman process, solitons formed with the initial amplitude $A \geqslant N (N \geqslant 2)$ propagate at the same speed, with phase interference. However, in the presence of the self-induced Raman process, these N number of solitons propagate at different speeds and hence, they separate (Tai *et al.*, 1988).

1.4 Technical aspects of the design of optical soliton transmission systems

Here we discuss the technical aspects in the use of optical solitons for an all-optical transmission system.

1.4.1 *Generation of a soliton*

1.4.1.1 Pulse-width – peak power relation Solitons can be generated by a mode-locked laser (Mollenauer *et al.*, 1980) by providing a pulse with an appropriate amplitude for the given pulse width. The relation between the peak power P_0(W) of the pulse and the pulsewidth τ_0(ps) defined as the full width at half maximum of the electric field, E_0(V/m), is given by

$$(\pi n_2)^{1/2} E_0 \tau_0 = 1.76(-\lambda k'')^{1/2} \tag{1.101a}$$

or

$$\tau_0 P_0^{1/2} = 9.3 \times 10^{-2} \lambda^{3/2} (|D|S)^{1/2} \tag{1.101b}$$

where D $(= 2\pi c k''/\lambda^2)$ designates the group dispersion coefficient of the fiber in ps/nm · km, and λ is the vacuum wavelength of the light wave in μm and S is the effective cross-sectional area of the fiber in μm^2.

If the initial amplitude deviates slightly from the condition given by equation (1.101), one soliton will still be formed but its width and amplitude will differ from the initial values. If the initial power is given by $(1 + \Delta)P_0$, the soliton peak power generated in the fiber becomes $(1 + 2\Delta)P_0$ and the width becomes $\tau_0/(1 + \Delta)$, so that the width times the amplitude remains approximately constant, $P_0^{1/2}\tau_0$ as given in equation (1.101b).

If the initial power is larger than that given by equation (1.101) by a factor more than 2.25 $(= 1.5)^2$, more than one soliton are generated in the fiber. The relation between the number of solitons, N, which are generated in the fiber and the initial peak amplitude $(P_0')^{1/2}$ is given by

$$(P_0'/P_0)^{1/2} - \tfrac{1}{2} < N \leqslant (P_0'/P_0)^{1/2} + \tfrac{1}{2} \tag{1.102}$$

N solitons propagate together for a while but eventually separate due to the self-induced Raman effect.

When a mismatch exists between the soliton power (1.101) and the input power, the pulse propagates with periodic variations in its amplitude and width. The distance of this period, z_0 (km) (called the soliton period), is given by

$$z_0 = 0.96 \frac{\tau_0^2}{\lambda^2 D} \tag{1.103}$$

1.4.1.1 Minimum power requirement In order to form a soliton, a minimum power, P_m, is required for a given pulsewidth τ_0. The minimum power requirement comes from the condition of existence of at least one discrete eigenvalue in the inverse scattering problem for the given initial pulse. For sech t shape pulse, it is given by

$$P_{\mathrm{m}} \geqslant \tfrac{1}{4} P_0 \tag{1.104}$$

where P_0 is given by (1.101).

In the presence of a fiber loss (for example no Raman amplification), the soliton pulsewidth increases inversely proportional to $(P_0)^{1/2}$ as its peak power decreases as $\exp(-\gamma_{\mathrm{p}} z)$, keeping the peak-power–pulsewidth relationship of equation (1.101), where γ_{p} is the power loss rate along the fiber. Because of this, even if the initial peak power satisfies the condition given by (1.104), a soliton may not be formed if the integrated fiber loss over the distance of the soliton period z_0 is large enough that if the soliton peak power at $z = z_0$ becomes less than that given by (1.104). One can avoid this limitation by compensating the fiber loss by means of a continuous amplification.

If the initial peak power of the laser output fails to meet the condition of (1.104), a soliton can be formed by a non-adiabatic amplification, that is a sufficient amplification within a distance much shorter than the soliton period so that only the amplitude is increased without changing the pulsewidth. While, if the initial peak power satisfies the condition (1.104), but if the fiber loss over the distance of the soliton period is large enough that the soliton fails to be formed, a soliton can still be formed by compensating the fiber loss by an adiabatic amplification. The large gain erbium-doped fiber (Nakazawa *et al.*, 1989a) is appropriate for the non-adiabatic amplification, while the small gain Raman amplification (Mollenauer *et al.*, 1985) is suitable for the adiabatic amplification.

If one uses a fiber whose loss is completely compensated for by a Raman gain, a soliton can be formed even with an extremely small peak power if no requirement exists for the pulsewidth. The minimum peak power required for a formation of a soliton in such a lossless fiber is given by the condition such that the entire spectrum of the soliton stays in the anomalous dispersion region. Such a condition is given by

$$\tau_0 (P_{\mathrm{m}})^{1/2} = 9.3 \times 10^{-2} \lambda^{3/2} (|D_{\mathrm{m}}| S)^{1/2} \tag{1.105}$$

With the minimum dispersion D_{m} being approximately given by

$$D_{\mathrm{m}} \simeq -1.6 \times 10^2 \, \Delta\lambda/\lambda^2 \qquad \text{for a normal fiber}$$

and

$$D_{\mathrm{m}} \simeq -1.3 \times 10^2 \, \Delta\lambda/\lambda^2 \qquad \text{for a dispersion-shifted fiber}$$

where $\Delta\lambda$ is the spectral width in the wavelength,

$$\frac{\Delta\lambda}{\lambda} = \frac{1}{f_0 \tau_0} = \frac{n_0 c}{\lambda \tau_0}$$

Hence, equation (1.105) becomes

$$\tau_0^3 P_{\mathrm{m}} = 3 \times 10^{-3} S \tag{1.106}$$

for a normal fiber. For a lossless fiber of $S = 60\ \mu\mathrm{m}^2$, the minimum power required for a soliton with 1 ps pulsewidth becomes 180 mW, while for a 10 ps pulse, the minimum power is only 0.18 mW.

1.4.1.2 Use of modulational instability A train of solitons can be generated by the modulational instability which is induced by mixing a small amplitude light wave with a high-power continuous light wave with an wavelength separation $\Delta\lambda$ (Hasegawa, 1984a,b; Tai, *et al.*, 1986).

The modulation frequency Δf ($= c/\Delta\lambda$) should be chosen within the frequency range of modulational instability, i.e.

$$\Delta f \leqslant 8.5 \frac{P}{S} \frac{1}{(-\lambda^3 D)^{1/2}} \qquad \text{(THz)} \tag{1.107}$$

where P(W) is the power of the carrier, S ($\mu\mathrm{m}^2$) is the effective fiber cross section, D (ps/nm · km) (<0) is the group dispersion coefficient of the fiber and λ (μm) is the wavelength of the carrier. For an example of $S = 60\ \mu\mathrm{m}^2$, $\lambda = 1.5\ \mu\mathrm{m}$, $|D| = 10$ ps/nm · km, the maximum modulation frequency for the inducing modulational instability becomes $6(P(\mathrm{mW}))^{1/2}$ GHz.

The pulse train which is generated by the induced modulational instability should be taken out of the fiber and non-adiabatically attenuated in order to construct a soliton train, otherwise the pulse train recovers to an unmodulated c.w. wave after the soliton period. It is also important to phase modulate the c.w. wave around several hundred MHz to avoid stimulated Brillouin scattering.

1.4.2 *Transmission of solitons in fibers*

In order to transmit a train of solitons in a fiber for an extended distance, solitons must be amplified to compensate for the fiber loss. However, unlike a linear pulse train which is deformed by the fiber dispersion, solitons can be transmitted only by amplifications for a distance over several thousand kilometres. Two means of amplification may be considered. One is a distributed amplification by means of the stimulated Raman process and the other is a local amplification by the use of erbium-doped fiber amplifier.

If one uses a *distributed amplification* by means of Raman process in the fiber, the fiber loss can, in principle, be compensated for exactly and a soliton can be transmitted in the fiber for practically indefinite distance. However, if the Raman pump is periodically injected to the fiber, the Raman gain changes periodically along the transmission distance of solitons. This causes periodic variations of the soliton width and induces interactions with adjacent solitons. By numerical simulations it is found that the distance between two injection points of the Raman pump is given approximately by $2\pi z_0$ for a stable transmission of a soliton train with the duty cycle of 10% (Hasegawa, 1984a,b; Mollenauer *et al.*, 1986) where z_0 is the soliton period given in equation (1.103). Mollenauer and Smith (1988) have succeeded experimentally in the distortionless transmission of a soliton over a distance of 4000 km (which was later extended to 6000 km) by means of a periodic Raman amplification.

It has been shown, based on a perturbation theory, that solitons remain stable and behave adiabatically if the loss (or gain) per soliton period is sufficiently smaller than unity (Hasegawa and Kodama, 1981). The calculated as well as the experimental results presented here show remarkable stability of solitons even if the loss (or gain) per soliton period is of order unity.

If a *localised amplification* is used to compensate for the fiber loss, only the amplitude of a soliton is amplified non-adiabatically. As discussed earlier in Section 1.4.1, if the amplitude deviates from the soliton amplitude, the pulse adjusts itself to form a new soliton. For example, if the power amplitude is increased from P_0 to $(1+\Delta)P_0$ by local amplification, a new soliton is generated with the asymptotic power amplitude given by $(1+2\Delta)P_0$ and the rest of the power, $\Delta^2 P_0$, is lost to a linear dispersive wave. However it is found by Hasegawa and Kodama (1982) that if the localised amplifications are provided at a distance shorter than the soliton period, equation (1.103), the dispersive wave can be recaptured by the soliton and a transmission beyond a distance of 6000 km is feasible for a soliton train of 13% duty cycle.

Reshaping a soliton train by a localised amplification by means of an erbium doped fiber amplifier has recently been demonstrated by Nakazawa *et al.* (1989a,b) and Iwatsuki *et al.* (1989).

We note that the soliton period also corresponds to the distance that a linear pulse spreads approximately by a factor of two due to the linear dispersion. Hence, the required distance between two amplifications scales with that between two repeaters for a high bit-rate linear system where the repeater distance is limited by the group dispersion. Thus one can conclude that the merit of a soliton transmission system over a linear system lies in the replacement of repeaters by a comparable number of amplifiers which are much less expensive and reliable, particularly for a high bit-rate wavelength multiplexed system.

Although for an extended transmission local amplification requires six times more amplifiers than the distributed amplification, for a shorter transmission of less than 1000 km, the amplifier spacing can be made much longer.

1.5 Summary and conclusion

The model non-linear wave equation for a light wave envelope in a fiber has been presented and important properties of the solution have been given in the form of optical solitons. The theoretical properties of the wave envelope discussed in this chapter have been demonstrated in many experiments which will be presented in other chapters.

The excellent agreement between theory and experiments in soliton propagations and modulational instabilities is remarkable and supports the validity of the theoretical models presented here.

The current theory does not cover the effects of interactions among different modes in a fiber. This is an important subject which has been treated only for limited cases and requires future study.

References

Akhmanov, S. A., Sukhorukov, A. P. and Khokhlov, R. W. (1968) *Esp. Fiz. Nauk,* **93**, 19–70.

Anderson, A. and Lisak, M. (1984) *Opt. Lett.,* **9**, 463–490. (*Sov. Phys. – USP.*, **93**, 609–36.)

Gardner, C. S., Greene, J. M., Kruskal, M. D. and Miura, R. M. (1967) *Phys. Rev. Lett.,* **19**, 1095–7.

Gordon, J. P. (1986) *Opt. Lett.,* **11**, 662–4.

Hasegawa, A. and Brinkman, W. F. (1980) *IEEE J. Quantum Electron., QE*-**16**, 694–7.

Hasegawa, A. and Tappert, F. D. (1973a) *Appl. Phys. Lett.,* **23**, 142–4.

Hasegawa, A. and Tappert, F. D. (1973b) *Appl. Phys. Lett.,* **23**, 171–2.

Hasegawa, A. and Kodama, Y. (1981) *Proc. IEEE*, **69**, 1145–50.

Hasegawa, A. and Kodama, Y. (1982) *Opt. Lett.,* **7**, 285–7.

Hasegawa, A. (1984a) *Opt. Lett.,* **9**, 288–90.

Hasegawa, A. (1984b) *Appl. Opt.,* **23**, 3302–4.

Hermansson, B. and Yevick, D. (1984) *Opt. Commun.,* **52**, 99–102.

Iwatsuki, K., Nishi, S., Saruwatari, M. and Shimizu, M. (1989) *IOOC'89 Tech. Digest 5*, paper 20PDA-1, Kobe Japan.
Kodama, Y. and Hasegawa, A. (1987) *IEEE J. Quantum Electron.*, **23**, 510–24.
Lax, P. D. (1968) *Commun. Pure and Appl. Math.*, **21**, 467–99.
Mitschke, G. M. and Mollenauer, L. F. (1986) *Opt. Lett.*, **11**, 657–61.
Mollenauer, L. F., Stolen, R. H. and Gorden, J. P. (1980) *Phys. Rev. Lett.*, **45**, 1095–8.
Mollenauer, L. F., Stolen, R. H. and Islam, M. N. (1985) *Opt. Lett.*, **10**, 229–31.
Mollenauer, L. F., Gordon, J. P. and Islam, M. N. (1986) *IEEE J. Quantum Electron.*, **22**, 157–73.
Mollenauer, L. F. and Smith, K. (1988) *Opt. Lett.*, **13**, 675–7.
Nakazawa, M., Kimura, Y. and Suzuki, K. (1989a) *Electron. Lett.*, **25**, 199–200.
Nakazawa, M., Suzuki, K. and Kimura, Y., (1989b) *Opt. Lett.*, **14**, 1065–1067.
Satsuma, J. and Yajima, N. (1974) *Progr. Theor. Phys. (Japan) Suppl.*, **55**, 284–306.
Tai, K., Tomita, A., Jewell, J. L. and Hasegawa, A. (1986) *Appl. Phys. Lett.*, **49**, 236.
Tai, K., Hasegawa, A. and Bekki, N.(1988) *Opt. Lett.*, **13**, 392–4 and 937.
Taniuti, T. (1974) *Progr. Theor. Phys. (Japan), Suppl.*, **55**, 1–35.
Zakharov, V. E. and Shabat, A. B. (1987) *Zh. Eksp. Teor. Fiz.*, **61**, 118–34. (English trans. *Sov. Phys. JETP* **34** (1972) 62–9.)

2

Solitons in optical fibers: an experimental account

L. F. MOLLENAUER

2.1 Introduction

In optical fibers, solitons are non-dispersive light pulses based on non-linearity of the fiber's refractive index. Such fiber solitons have already found exciting use in the precisely controlled generation of ultrashort pulses, and they promise to revolutionise telecommunications. In this chapter, I shall describe those developments, and the experimental studies they have stimulated or have helped to make possible. Thus, besides the first experimental observation of fiber solitons, I shall describe the invention of the soliton laser, the discovery of a steady down-shift in the optical frequency of the soliton, or the 'soliton self-frequency shift', and the experimental study of interaction forces between solitons.

As early as 1973, Hasegawa and Tappert (1973) pointed out that 'single-mode' fibers – fibers admitting only one transverse variation in the light fields – should be able to support stable solitons. Such fibers eliminate the problems of transverse instability and multiple group velocities from the outset, and their non-linear and dispersive characteristics are stable and well-defined. The first experiments (Mollenauer *et al.*, 1980), however, had to wait a while, for two key developments of the late 1970s. The first was fibers having low loss in the wavelength region where solitons are possible, and the second was a suitable source of picosecond pulses, the mode-locked color center laser.

But the first experiments led almost immediately to further developments. For one, the manifest abilities of single mode fibers to compress and shape pulses led to the invention and development of the soliton laser (Mollenauer and Stolen, 1984). The precisely shaped pulses from that laser in turn provided the key to the further experimental work previously mentioned. Additionally, by demonstrating that there were no hidden problems in the generation of fiber solitons, the first experiments stimulated much thought about the possibility of creating an all-optical, soliton-based communications system (Hasegawa, 1983). Such ideas have been further amplified and tested by computer simulation (Hasegawa, 1984; Mollenauer *et al.*, 1986) and by real world experiment (Mollenauer *et al.*, 1985).

Many ideas about solitons were first worked out (and in some cases discovered or conceived) in highly abstract and purely mathematical terms. But the existence of simple and clear experimental results has stimulated the quest for them and has often generated more direct explanations. Such simple physical models often yield valuable insight and stimulate further invention. In the following, I shall refer to such models wherever possible.

2.2 Dispersion, non-linearity and pulse narrowing in fibers

The dispersive qualities of quartz glass and the loss per unit length of the best single-mode fibers presently available are both shown in Figure 2.1. As will be demonstrated shortly, soliton effects are possible only in the region of 'negative' group velocity dispersion ($\partial v_g/\partial\lambda < 0$). As shown by Figure 2.1, such dispersion occurs only for wavelengths greater than approximately 1.3 μm. (The net dispersion of a given fiber is also determined by the ratio of the core diameter to the wavelength, but such waveguide contributions can only push the zero of dispersion to longer wavelengths. In the so called 'normal' fiber, the net dispersion is little different from that shown in the figure.) Note that the region of negative dispersion includes the region ($\lambda \simeq 1.5\ \mu$m) of lowest energy loss (the minimum loss can be as low as 0.16 db/km). In the experiments to be described, the central frequency of the pulse has usually been at or near that loss minimum.

Dispersion alone – regardless of sign – always causes the higher and lower frequency components of a pulse to separate, and thus always serves only to broaden the pulse. Pulse narrowing and, by extension, solitons, are made possible by the fact that the medium is non-linear, that is the index of refraction is a function of the light intensity:

$$n = n_0 + n_2 I \tag{2.1}$$

For quartz glass, n_2 has the numerical value 3.2×10^{-16} cm^2/W and I is the light intensity in compatible units.

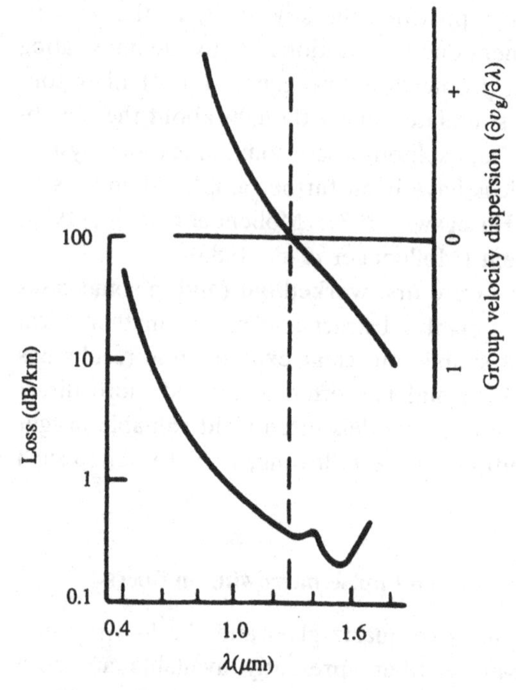

Figure 2.1. Loss of best single-mode fibers and group velocity dispersion of quartz glass as functions of wavelength.

The effect of the index non-linearity is to produce 'self-phase modulation' (Stolen and Lin, 1978). That is, a wave will experience phase retardation in direct proportion to the intensity-induced change in index and (temporarily assuming negligible pulse reshaping) to the length of fiber traversed:

$$\Delta\phi = \frac{2\pi}{\lambda} L n_2 I \tag{2.2}$$

In a pulse such as that shown in Figure 2.2(*a*), the phase retardation will be greatest at its peak. Thus there will be a crowding together and spreading apart of waves in the trailing and leading halves of the pulse, respectively, resulting in the frequency distribution shown in Figure 2.2(*b*). When such a 'chirped' pulse is acted upon by the fiber's negative group velocity dispersion, the leading half of the pulse, containing the lowered frequencies, will be retarded, while the trailing half, containing the higher frequencies, will be advanced, and the pulse will tend to collapse upon itself as shown in Figure 2.2(*c*). If the peak pulse intensity is high enough, such that the chirp is large

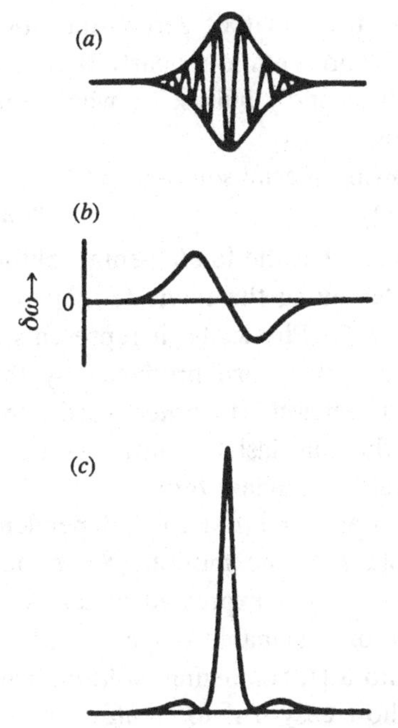

Figure 2.2. (*a*) Optical pulse that has experienced self-phase modulation, (*b*) corresponding frequency chirp and (*c*) resultant compressed pulse in a fiber with negative group velocity dispersion (see text).

compared to that produced by dispersion, the degree of pulse narrowing can be substantial.

In writing down a differential equation to describe non-linear pulse propagation in fibers, it is convenient to eliminate terms containing the central optical frequency from the wave equation. One thereby obtains an equation involving only the pulse envelope function u. The process involves expanding dispersive terms about the central frequency. When only the leading terms are kept, the result is the nonlinear Schrödinger (NLS) equation:

$$-\mathrm{i}\frac{\partial u}{\partial \xi}=\frac{1}{2}\frac{\partial^2 u}{\partial s^2}+|u|^2 u \tag{2.3}$$

where $\xi = z/z_c$ and $s = (t - z/v_g)/t_c$ represent distance of propagation along the fiber and (retarded) time measured in the soliton units z_c and t_c, respectively.

When higher order dispersive terms are unimportant, (2.3) yields a fairly accurate representation of non-linear pulse propagation in single-mode fibers. Thus, for pulses of at least a few picoseconds duration, and with central frequency well away from that of zero dispersion, the approximation is very good. The representation clearly becomes worse for very short (so called 'femtosecond') pulses, or where the dispersion varies rapidly with frequency.

The fundamental soliton is the following special solution to (2.3):

$$u(z, t) = A\,\mathrm{sech}\,(s/A)\exp(\mathrm{i}\xi/2) \tag{2.4}$$

where A is often taken to be unity. Note that the fundamental soliton does not change shape as it propagates along the fiber. (Except for the phase term, (2.4) is independent of ξ.) Physically, it represents a condition of exact cancellation between the chirp produced by the non-linearity, and that produced by dispersion. (In other words, the pulse shape and amplitude are such that the last two terms of (2.3) cancel, except for that part that generates the phase term.)

Note the area of the soliton, defined as $S = \int |u| \mathrm{d}s$, is independent of A. Thus, S is an important constant of the motion ($S = \pi$) for solitons. The stability of solitons is often best expressed in terms of area concepts. For example, any pulse of reasonable shape, for which $\pi/2 \leqslant S \leqslant 3\pi/2$, will eventually turn into a (fundamental) soliton. This is an important concept, as it shows how easy it is to create solitons, or to maintain them once they have been created.

One usually has $t_c = \tau/1.76$, where τ is the intensity FWHM (full width at half maximum) of the fundamental soliton with $A = 1$. Then one has

$$z_c = 0.322\,\frac{2\pi c}{\lambda^2}\,\frac{\tau^2}{D} \tag{2.5}$$

where λ is the vacuum wavelength, and where the dispersion parameter D reflects the change in pulse delay with change in wavelength, normalised to the fiber length. (For a 'normal' fiber, $D \simeq 15$ ps/nm · km at 1.5 μm.) Finally, the peak power corresponding to the fundamental soliton is given by the expression:

$$P_1 = 0.776\,\frac{\lambda^3}{\pi^2 c n_2}\,\frac{D}{\tau^2}\,A_{\mathrm{eff}} \tag{2.6}$$

where n_2 has the numerical value given earlier, where A_{eff} refers to the effective fiber core area, and where c is the speed of light in vacuum.

An important set of solutions to (2.3) result from an input function of the form:

$$u(0, s) = N\,\mathrm{sech}\,(s) \tag{2.7}$$

where N is an integer, the 'soliton number'. A few of those solutions are shown graphically in Figure 2.3. Note that $N = 1$ yields the

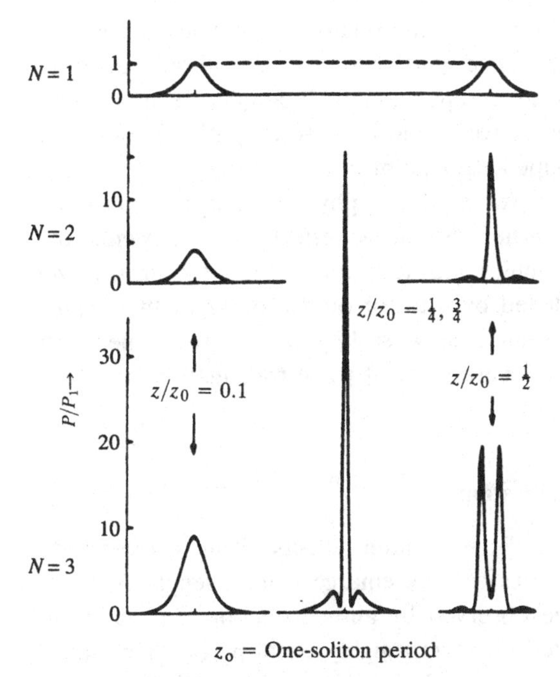

Figure 2.3. Theoretical behavior of the fundamental ($N = 1$) and two higher order solitons with propagation.

fundamental soliton, already discussed. For integers $N \geqslant 2$, sech input pulses always lead to pulse shaping that is periodic with period $\xi = \pi/2$. (The periodicity sooner or later breaks down, however, mainly because of the time dependence (Mollenauer *et al.*, 1983) of the non-linear term.) In real space, the period is

$$z_0 = \frac{\pi}{2} z_c \tag{2.8}$$

The peak input powers are, of course,

$$P_N = N^2 P_1 \tag{2.9}$$

For $N = 2$, the behavior is particularly simple: the pulse alternately narrows and broadens, achieving minimum width at the half period. For greater N, the behavior becomes more complex, but always consists of a sequence of pulse narrowings and splittings (see Figure 2.3).

Although the $N = 1$ soliton is unique, the $N = 2$ and $N = 3$ solitons shown in Figure 2.3 each represents but one member of a continuum. For example, the $N = 2$ soliton can be looked upon as a non-linear superposition of two fundamental solitons; the continuum of solutions is obtained by varying the relative amplitudes and widths of the two components. (The particular $N = 2$ soliton shown in Figure 2.3 corresponds to components with amplitude and width ratios of 3:1 and 1:3, respectively, and it is the only $N = 2$ soliton to pass through the sech shape at any point in its period.)

Finally, we note that z_0 (or z_c) has a physical interpretation independent of solitons or other non-linear effects: it represents the distance an initially minimum bandwidth pulse of width τ must travel to be dispersively broadened by a factor on the order of two. Thus, z_0 is of universal importance as a scaling factor. It is the most important and all-pervasive parameter of pulse propagation in optical fibers.

2.3 Experimental verification

To verify the predicted soliton effects, it is necessary to observe the shapes of pulses as they emerge from a length of fiber. The pulse shapes can be observed by autocorrelation. In that technique, the beam is divided into two roughly equal parts, which, after traveling separate paths, are brought together in a non-linear crystal; see Figure 2.4.

In the arrangement shown there, second harmonic light is generated only if pulses from both beams are simultaneously present in the crystal. Thus the strength of the second harmonic (registered by the photomultiplier) reflects the temporal overlap of the pulses in the two converging beams. A measurement of second harmonic intensity as a function of relative delay then yields the pulse shape in autocorrelation. As can be seen from Figure 2.3, the most varied changes in pulse shape (with changing input power) are to be observed at the output end of a fiber whose length is one-half the soliton period. The first experiments (Mollenauer *et al.*, 1980) were conducted with such a half-period fiber; Figure 2.5 summarizes the results. The autocorrelation trace labeled 'laser' describes the color center laser output (fiber input) pulses, and corresponds to $\tau = 7$ ps and an approximately sech^2 shape.

Figures 2.5(*b*)–(*e*) show the experimentally determined fiber output pulse shapes at certain critical power levels, where one sees, respectively, the expected half-period behavior of the fundamental and several higher-order solitons.

The actual length (700 m) of fiber used in this experiment agrees rather well with the half-period ($z_0/2 = 675$ m) calculated from (2.5)

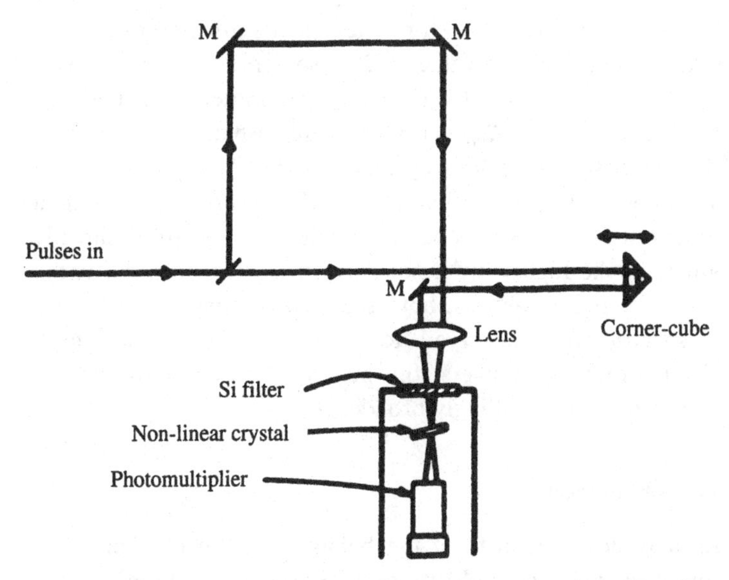

Figure 2.4. Schematic of apparatus for observing pulse shapes in autocorrelation. Variable delay is accomplished through translation of the corner cube. The silicon filter passes 1.5 μm light but keeps out visible room light.

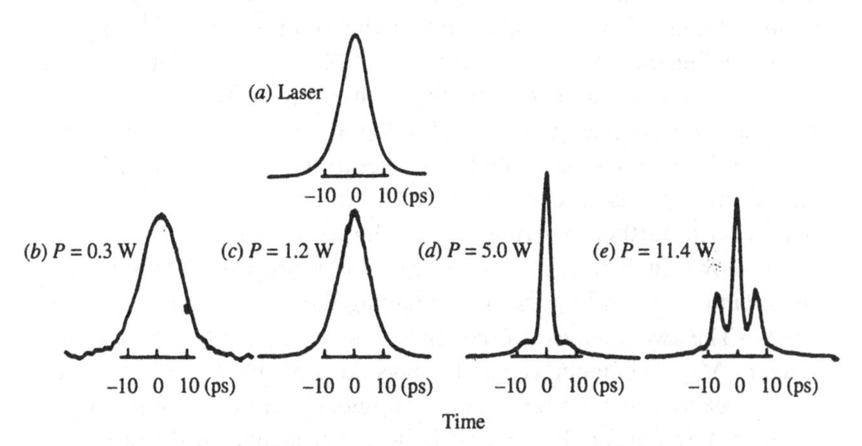

Figure 2.5. Results of experiment with a fiber whose length is one-half the soliton period. (a) Autocorrelation shape of the laser pulses launched into the fiber. (b)–(e) Autocorrelation shapes of pulses emerging from the fiber for various input powers. $P = 0.3$ W: negligible non-linear effect; only dispersive broadening is seen. $P = 5$ W: pulse narrowed to minimum width; corresponds to half-period behavior of the $N = 3$ soliton. (Note that the threefold splitting in autocorrelation corresponds to a twofold splitting of the pulse itself.)

and (2.8) for $\tau = 7$ ps and for the fiber's dispersion, $D =$ 15 ps/nm · km. The calculated value of P_1 also agrees rather well with experiment. From (2.6), $P_1 = 1.0$ W for the parameters just cited and for an effective core area $A_{\text{eff}} \approx 1 \times 10^{-6}$ cm^2, whereas the average of P/N^2 for the first three solitons yields $P_1 = 1.2$ W.

In a similar experiment (Stolen *et al.*, 1983) carried out with a full-period length of fiber, it has been possible to demonstrate directly the periodicity of the higher order solitons. In that experiment, at the critical power levels for solitons, both the pulse shapes and the pulse frequency spectra were observed to return to the input values, whereas for intermediate powers, the pulses were narrowed and the frequency spectra correspondingly broadened.

2.4 The soliton laser

As may be inferred from the behavior shown in Figure 2.3, extreme pulse narrowing should be obtained at high soliton number in a judiciously chosen length of fiber. Indeed, compression by factors as great as ×30 have been obtained experimentally (Mollenauer *et al.*, 1983). But one pays a penalty for such extreme compression: a large fraction of the pulse energy (for $N > 6$, more than half) remains in uncompressed 'wings' surrounding the central spike. The soliton laser (Mollenauer and Stolen, 1984) represents a way to obtain ultrashort pulses without having to pay that penalty. Instead, feedback from a pulse-shaping fiber enables the laser to produce minimum bandwidth, sech^2 shaped pulses of controlled width. As indicated earlier, the precisely controlled pulses from the soliton laser have been vital to further experiments on solitons.

As shown in Figure 2.6, the soliton laser consists of two cavities, the main cavity, and the cavity containing the fiber, called the control cavity. The two cavities are coupled together through their common mirror, M_6. The main cavity belongs to a synchronously pumped, mode-locked color center laser, continuously tunable over a broad band centered at $\lambda \simeq 1.5$ μm. (It is thus very similar to the laser used in the first experiments.) Without the control cavity, the laser produces mode-locked pulses greater than or equal to 8 ps width (Mollenauer *et al.*, 1982). A controlled fraction of its output pulse energy travels around the control cavity which is matched in round trip time to the main cavity. The fiber compresses the pulses and sends them back into the main cavity, thereby stimulating the color center laser to produce narrower pulses. Thus, there is a successive compression

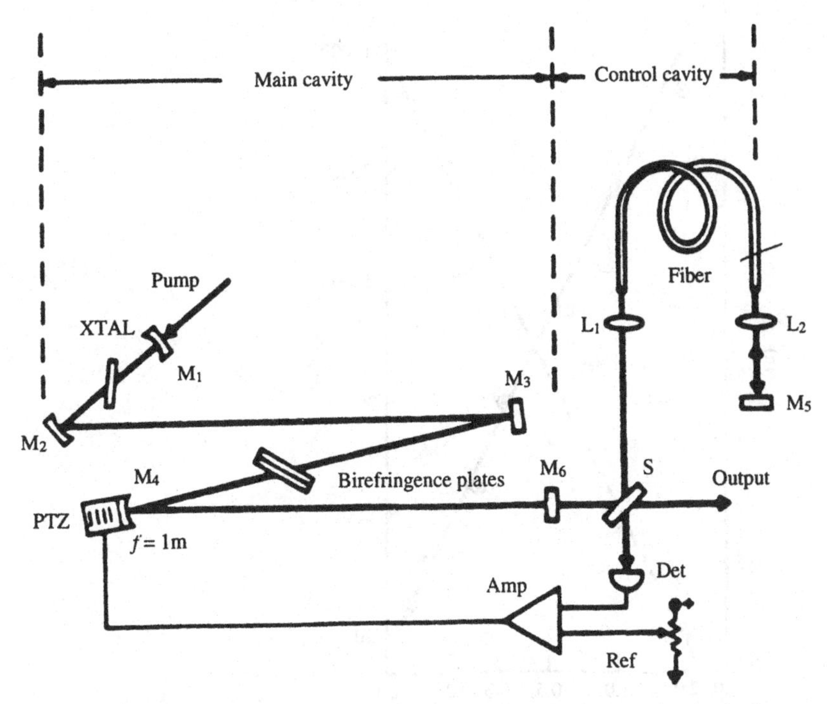

Figure 2.6. Schematic representation of the soliton laser, including stabilization loop. Beam splitter S has reflectivity R = ~30–50%. L, microscope objectives; PZT, piezoceramic translator. (This is the very latest version, which differs from that described in (Mitschke and Mollenauer, 1986a) only in that the piezoceramic translator has been removed from the control cavity to the main cavity; this reduces the translator motion required when the control fiber is long.)

until the fiber pulses become solitons, at which point the laser operation reaches a stationary state.

It was expected (Mollenauer and Stolen, 1984) that the laser would operate on an $N = 2$ soliton (see Figure 2.3). If so, and if there is a constant relation between z_0 and the control fiber length L, the width of the pulses formed by the soliton laser should scale with the square root of L. This dependence of the pulsewidths on L has indeed been confirmed experimentally in Mollenauer and Stolen (1984). That is, the data there fit the relation $L \approx z_0/2$, (see Figure 2.7), with the implication that, in the stationary state, pulses returned from the fiber have substantially the same width as those launched into it. Nevertheless, it was later learned (Mitschke and Mollenauer, 1986a) that stable solutions also exist for which L is considerably less than $z_0/2$.

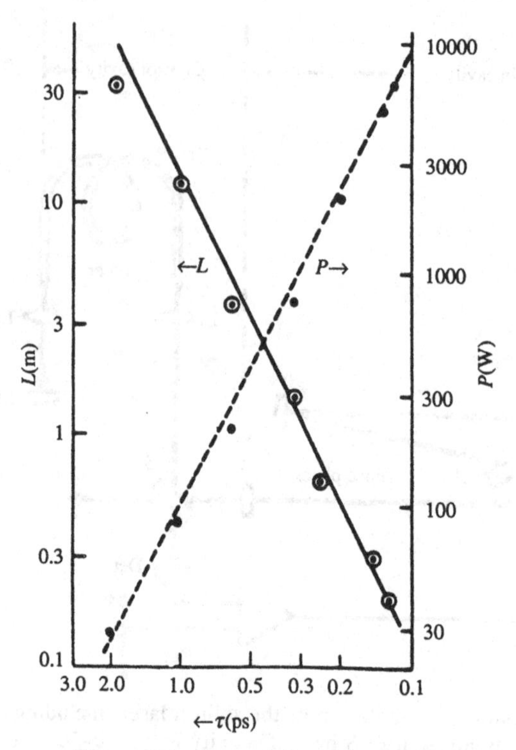

Figure 2.7. Control fiber lengths (encircled points) and peak powers at input to control fiber (full points) as functions of the obtained laser output pulsewidth. Full line: $\frac{1}{2}Z_0(\tau)$ from (2.5) and (2.8). Broken line: P_2 (τ) from (2.6) and (2.9).

Note that the system of coupled cavities allows for a high degree of independence between the laser and control cavity powers. This flexibility is vital, as the ratio of laser output to fiber soliton power is too great to be spanned by the available optical gain, as would be required in a single cavity. Successful operation of the two cavity laser requires, however, that the pulses returned from the control cavity have the correct optical phase with respect to the pulses circulating in the main cavity. The servo loop shown in Figure 2.6 maintains (Mitschke and Mollenauer, 1986a) that phase by constant correction of the relative cavity lengths against effects of vibration and thermal drift. The necessary error signal is easily obtained, based on the following. First, the power in the control cavity varies with ϕ, the round trip optical phase shift in the control cavity about as shown in Figure 2.8. Second, soliton laser action is correlated with a well

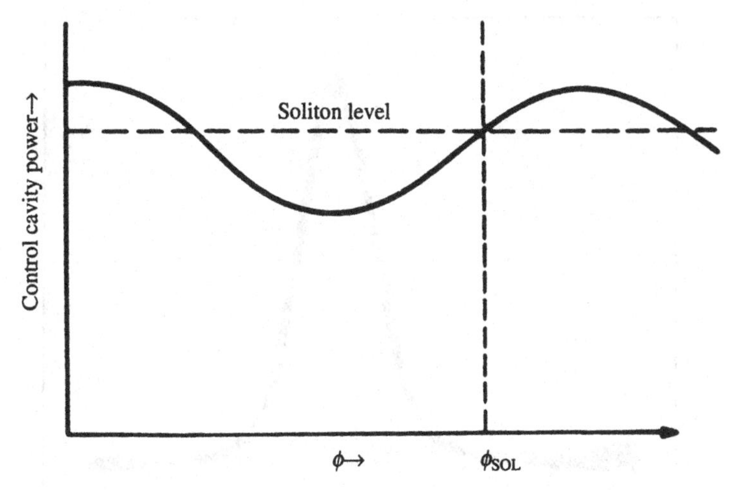

Figure 2.8. Variation of control cavity power with round trip optical phase shift ϕ.

defined level lying somewhere within the middle of the range of cavity powers.

Thus, at least in the neighborhood of the soliton level, the control cavity power is a good measure of ϕ. The appropriate error signal for the control of ϕ is then generated, simply by taking the difference between the detector signal and a reference voltage empirically set to the soliton level. The op-amp magnifies that difference and drives the piezoelectric translator of the cavity mirror M_4. Thus, assuming correct choice of signal polarity, a deviation of ϕ from the soliton value produces a change in detector signal that results in corrective displacement of M_4.

The stabilised soliton laser emits a stream of pulses that are very uniform in width and height. The noise on the pulses can be as low as approximately 1% of the full intensity. Figure 2.9 shows a time exposure of a typical autocorrelation trace as seen on an oscilloscope screen, where the sharp definition of the trace indicates the absence of serious noise.

To analyse the soliton laser more completely, conditions (input pulse shape, power) in the control fiber were accurately duplicated (Mitschke and Mollenauer, 1986a) in a second length of fiber cut from the same spool; see Figure 2.10. By gradually cutting back the length L' of this 'test' fiber, and by observing its output pulse shapes in autocorrelation, it was possible to infer what was happening in the control fiber itself.

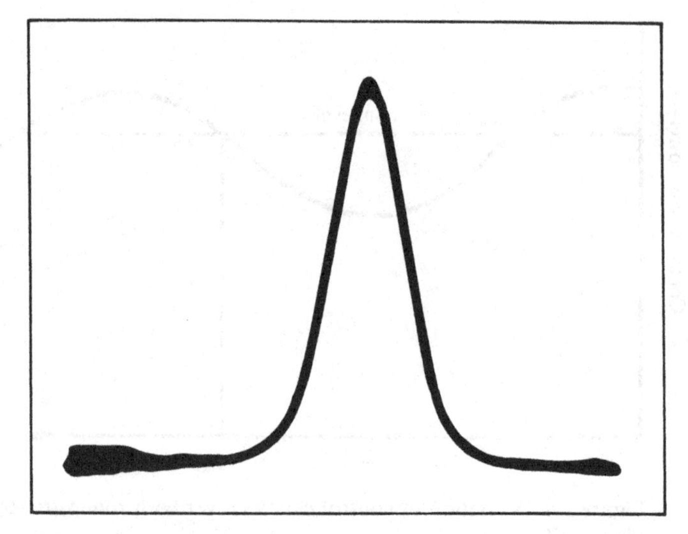

Figure 2.9. A typical autocorrelation trace of the laser output pulses. In this example, the actual FWHM of the pulses is 580 fs ($L = 1.6$ m).

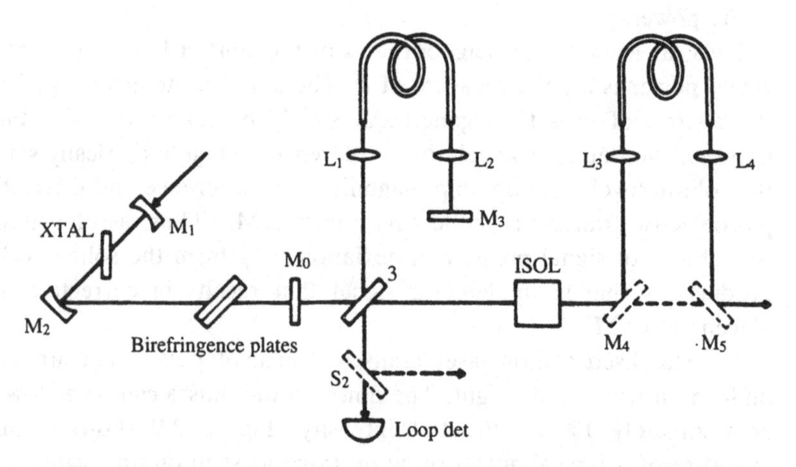

Figure 2.10. Setup to duplicate the control fiber with a 'test' fiber. Mirrors M_4 and M_5 can slide out of the beam to allow the laser output to be measured directly. (The beam from S_2 is used for direct confirmation of the datum point at $L' = 2L$. See text.)

Figure 2.11 summarises the results of a typical experiment. Note that the pulse shapes and widths make good fit to the expected behavior of sech $N = 2$ solitons. But also note that, at least in this particular example, the round trip distance in the control fiber ($2L$) is

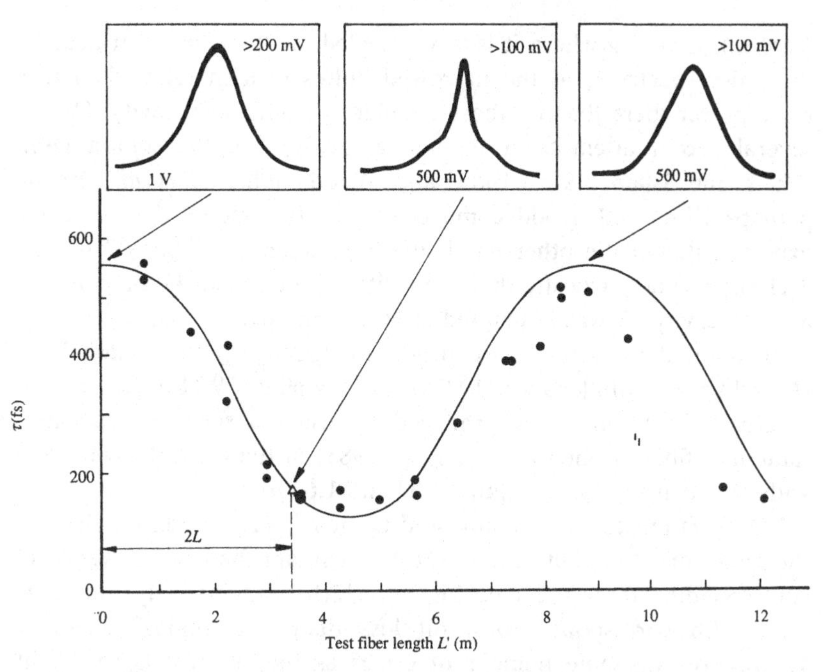

Figure 2.11. FWHM of pulses emerging from the test fiber of length L' when sech²-shaped, laser output pulses of 560 fs FWHM were launched into it. The full line is the theoretically expected value for a sech $N = 2$ soliton calculated for $D = 14.5$ ps/nm · km. The power level (in both control and test fibers) was equal to the $N = 2$ soliton power as calculated from $n_2 = 3.2 \times 10^{-16}$ cm²/W and $A_{\text{eff}} = 86\ \mu$m for 560 fs pulses. The datum point at $2L$ was found directly from the control fiber.

considerably less than the soliton period (z_0). Thus the pulses returned from the control cavity are considerably narrower than those launched into it.

This result yields additional insight into operation of the soliton laser. Without feedback from the control fiber, subpicosecond pulses will tend, with repeated round trips, gradually to broaden in time and to collapse in the frequency domain. The collapse would be brought about by dispersion-induced phase shifts and attenuation, primarily in the far wings of the pulse's frequency spectrum.

Feedback from the fiber provides the correction necessary to sustain mode locking, and thus to prevent collapse. The need for some degree of pulse narrowing is now easily understood. That is, amplitudes of the pulses returned from the control cavity are often just a few percent or less of those in the main cavity. Nevertheless, the feedback is strong where it is most needed, at the extreme fringes of the pulse spectrum, as the narrowed pulses contain relatively much more power there than do those circulating in the main cavity. Of the several recent attempts at theoretical analysis of the soliton laser (Haus and Islam, 1985; Blow and Wood, 1986; If *et al.*, 1986), perhaps Blow and Wood come closest to this model derived from experiment. On the other hand, the experiment of Figures 2.10 and 2.11 shows unequivocally that the soliton laser cannot be modeled as a single cavity, as was attempted in Haus and Islam (1985).

To date, the shortest pulse produced directly by the soliton laser (Mitschke and Mollenauer, 1987a) has $\tau = 60$ fs FWHM (see Figure 2.12(*a*)). That pulse was produced by using a special 'dispersion-flattened' fiber (Cohen and Pearson, 1983) in the control cavity, one with $D = 0$ at two wavelengths, 1.37 and 1.62 μm.

The 60 fs pulses were compressed in an $L' \simeq z_0/2$ external piece of the same fiber to about 19 fs FWHM, when launched at a power level corresponding to $N = 2$ (see Figure 2.12(b)). Incidentally, the 19 fs pulsewidth corresponds to a bit less than four optical cycles at 1.5 μm, or the same number of cycles as the shortest known light pulse (8 fs, produced by a dye laser in the visible). (Note that once again, this experiment, besides producing the shortest possible pulse, shows what the pulses must look like at one point in the control fiber itself.) The measured frequency spectrum of the 19 fs pulse just filled the space between the two $D = 0$ points of the fiber. In the mean time, the bandwidth of the 60 fs pulse is only about 25% of the gain bandwidth of the color center. Thus, one can conclude that the limit to ultrashort pulse production in the soliton laser is set primarily by the dispersion properties of the control fiber. In principle, this result could be improved upon by a more suitable fiber. It is considered possible, for example, to produce dispersion-flattened fibers with no second wavelength for which $D = 0$.

2.5 Discovery of the soliton self-frequency shift

When the experiments of Figure 2.11 were completed, the test fiber was made much longer ($L' = 392$ m, or $\sim 46 z_0$ for $\tau = 560$ fs). The object was to see how well $N = 2$ solitons could survive such a long fiber. But when pulses of P considerably greater than P_1 were launched into the fiber, the pulses displayed an unexpectedly large temporal splitting at the fiber output (see Figure 2.13).

The corresponding frequency spectrum (Figure 2.14) provided the

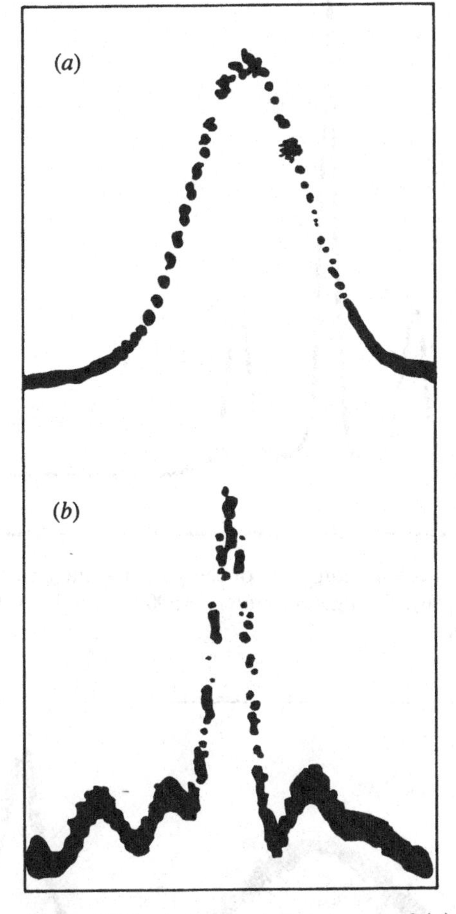

Figure 2.12. Autocorrelation traces of (*a*) 60 fs FWHM pulses produced directly by the soliton laser and (*b*) the same, but after external compression to 19 fs FWHM. Note: the regular ripples seen on the traces are not noise, but represent interference fringes between the two nearly parallel beams. Hence, the distance between bumps is the optical period (5 fs).

first clue to this strange behavior. It was clear that the pulse had split into a narrow, intense, fundamental ($N = 1$) soliton, and a weaker, non-soliton part that broadened as it traveled down the fiber. The spectrum of the weaker pulse remained more or less centered at the original laser frequency, but the soliton had undergone a strong shift to lower frequencies. The large temporal splitting could thus be explained in terms of the combined effects of group velocity dispersion and the large splitting in frequency.

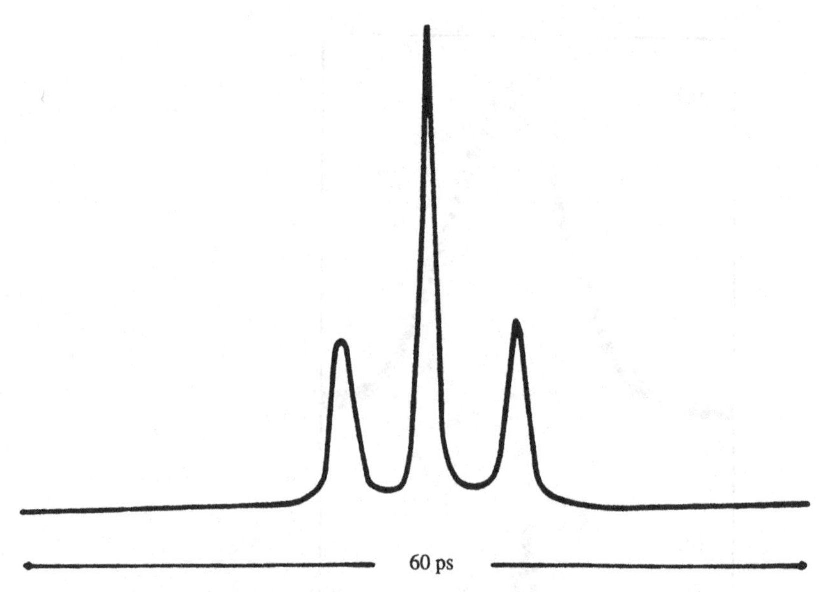

Figure 2.13. An autocorrelation trace of the pulses at the end of the test fiber ($L' = 392$ m). Parameters are: $\tau_0 \simeq 500$ fs, $p \simeq 1.5p$, full range of scan 60 ps.

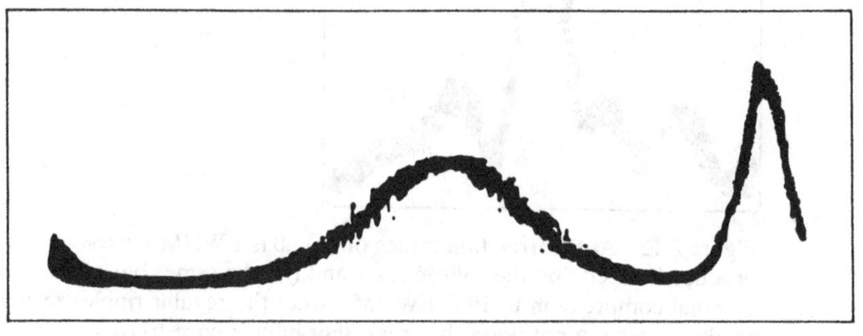

Figure 2.14. Typical spectrum of the pulses and the output end of the 392 m test fiber, under conditions similar to those of Figure 2.13. Scan width is one free spectral range of the scanning Fabry–Perot etalon (4.6 THz); frequency increases towards the right. The narrow peak near the right margin is at ν_0.

The frequency shift is made possible by the fact that in glasses, the Raman gain (Stolen *et al.*, 1984) extends right down to zero frequency difference between pump and signal (see Figure 2.15). Thus, the higher frequency components of the pulse can act as a pump to provide Raman gain for the lower frequency components, and by this means, there can be a steady flow of energy to lower frequencies.

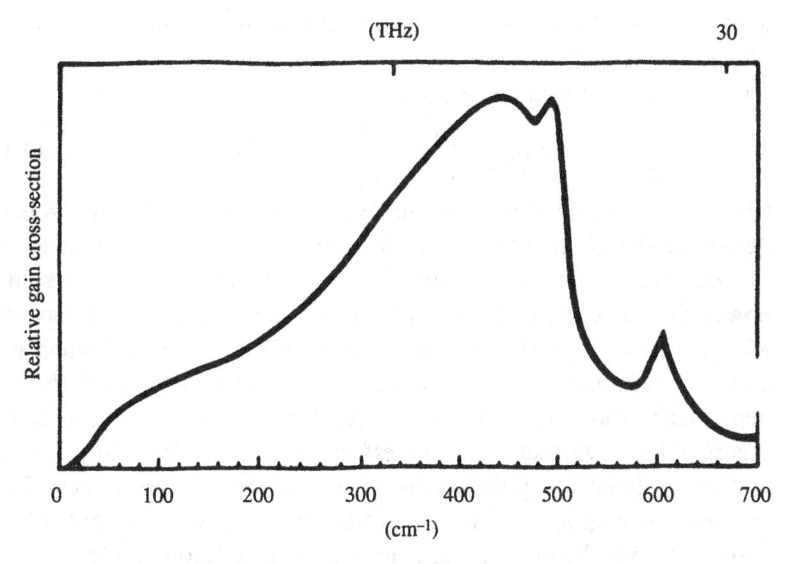

Figure 2.15. Relative Raman gain *versus* frequency difference between pump and signal (from Stolen *et al.*, (1984).

Clearly, the increasing Raman gain with increasing frequency would make the effect increase rapidly with decreasing pulsewidth (increasing pulse bandwidth). Experiment showed just such a scaling. For example, for a 260 fs soliton at output of the 392 m fiber, the down shift was 8 THz, or 4%. But for a 120 fs soliton, the shift was 20 THz (10%), and in just 52 m of fiber.

Immediately following these experiments (Mitschke and Mollenauer, 1986b), Gordon produced a proper theory (1986) of the effect. First, he noted that existence of the Raman effect requires that the non-linear term in (2.3) be modified to include a delayed response†, as follows.

$$|u|^2 u \rightarrow u(t) \int \mathrm{d}s f(s) |u(t-s)|^2 \tag{2.10}$$

where $f(s)$ is real if there are no losses other than the Raman type, $f(-|s|) = 0$ to ensure causality, and $\int f(s) \mathrm{d}s = 1$ to recover (2.3) for sufficiently short delays.

But as the modified equation (2.3) is difficult to work with directly, Gordon took its Fourier transform. The empirically known Raman

†R. H. Stolen, also of AT&T Bell Laboratories, was actually the first to note the connection between the Raman effect and delayed response of the non-linearity (Stolen, 1985).

gain (Figure 2.15) could then be introduced in a natural way, and its effects could be treated as a perturbation on the soliton. The final result is a rate of frequency shift

$$\frac{d\nu_0}{dz} = 0.0436 h(\tau)/\tau^4 \tag{2.11}$$

where $h(\tau)$, a slowly varying function of the order of unity, reflects details of the Raman spectrum. The much stronger τ^{-4} dependence can be understood as a result of the τ^{-2} dependence of the soliton power (P_1, equation (2.6)), and the τ^{-1} dependence of its spectral width, which proportionally increases both the effective Raman gain and the resulting frequency displacements. The power of the τ^{-4} dependence can perhaps be best appreciated from the log–log plot of Figure 2.16. Note that while the effect can be very large in just a few metres of fiber for pulsewidths of a few tens of femtoseconds, it becomes negligibly small for pulsewidths of several tens of picoseconds. The latter extreme is fortunate for telecommunications.

2.6 Experimental observation of interaction forces between solitons

Several papers have predicted the existence of interaction forces between copropagating solitons. In 1981, Karpman and Solov'ev (1981) analysed the interaction using a perturbation theory, and later, Gordon (1983) derived the soliton interactions directly from the exact two-soliton function. Nevertheless, in both papers, the main prediction was of a force which decreases exponentially with the initial distance of separation of equal amplitude solitons, attractive when the waves under the pulses are in phase, and repulsive for waves out of phase. The existence of such forces is of obvious significance to optical communications, and may have important implications for optical computing and other applications as well.

Recently, pulses from the soliton laser have made possible the first experimental observation (Mitschke and Mollenauer, 1987b) of such forces. The essentially isolated (separated by 10 ns) pulses from that laser were first passed through a Michelson interferometer to turn them into pulse pairs. A micrometer screw adjustment of the length of one interferometer arm allowed the pulse separation to be varied continuously from zero to many picoseconds, while piezoelectric transducer control of the same allowed for precise adjustment of the relative optical phase. The pulse pairs were then observed in autocorrelation both before and after they had passed through a length of

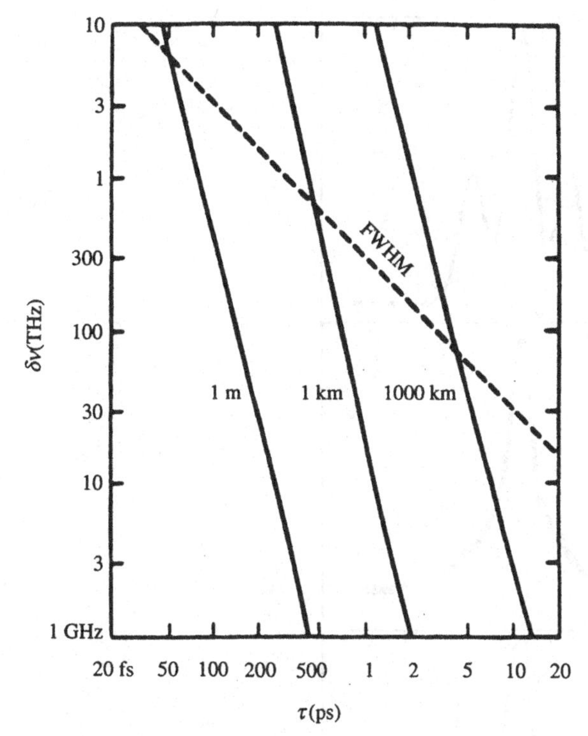

Figure 2.16. Soliton frequency shifts $\delta\nu_0$ *versus* τ, for various unit lengths of lossless sicila glass fiber having $D = 15$ ps/nm · km, and for $\lambda = 1.5\ \mu$m. The pulse bandwidth (FWHM) is shown for comparison (from Gordon (1986)).

single-mode fiber. Solitons were obtained by adjusting coupling into the fiber until the emerging pulses had exactly the same width as at input. Also, by alternately blocking the two arms of the interferometer, each member of the pulse pair could be adjusted for equal intensity as measured at the fiber output.

The first experiments were carried out in a 340 m length of polarization preserving, low loss (>0.3 dB/km) fiber. The pulsewidths were $\tau \simeq 1$ ps FWHM, which made the fiber of order ten-soliton periods long. This length would theoretically allow convergence of attractive pulses with a well-resolved initial separation (a little over 3 pulsewidths).

The results of a typical measurement are shown in Figure 2.17, for both in-phase (attractive) and opposite phase (repulsive) pulses.

Figure 2.18 is a plot of final pulse separation σ_{out} as a function of initial pulse separation σ_{in}, for both the repulsive and attractive cases.

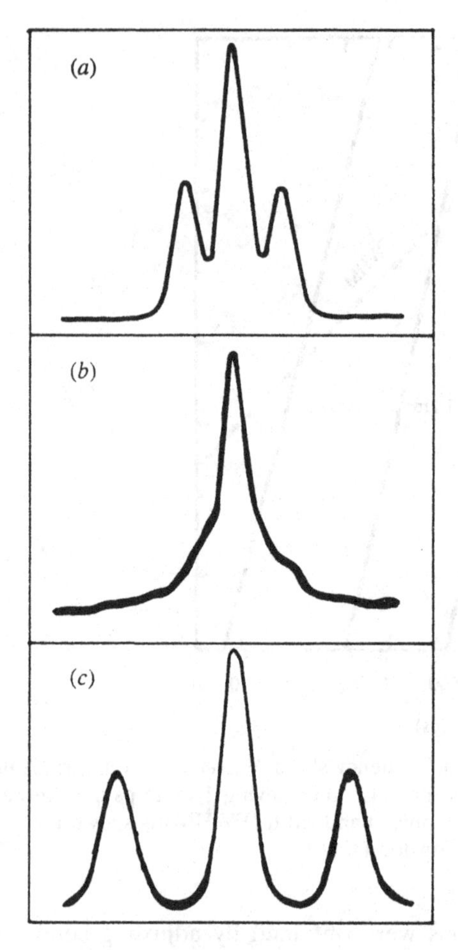

Figure 2.17. Autocorrelation traces of pairs of $\tau = 0.90$ ps pulses; (*a*) at input of 340 m fiber and (*b*), (*c*) at fiber output, for pulses in-phase (attractive) and of opposite phase (repulsive), respectively. Pulse separation at input, 2.33 ps. (All traces are to common time scale.)

The theoretical curves shown there for repulsion, and for attraction up to the first overlap, are based on equation (18) of Gordon (1983). (The cosine function given there is valid only for the attractive case ($\phi = 0$). For the repulsive case ($\phi = \pi$), that function must be replaced by the hyperbolic cosine (Gordon, private communication).) Note the close agreement between theory and experiment for the repulsive case and for the attractive case up to the point of first pulse overlap (for which $\sigma_{in} = \sigma_c$).

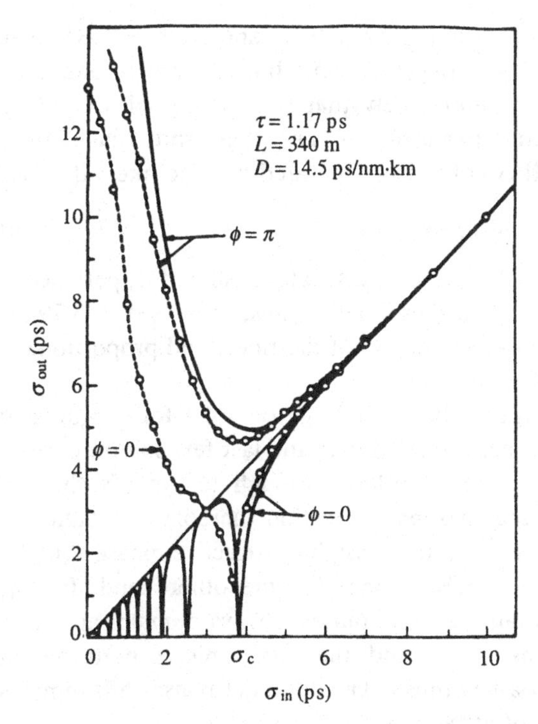

Figure 2.18. Pulse separation at output (σ_{out}) *versus* same at input (σ_{in}) of 340 fiber. Full curve, theory; broken line, experiment.

But, for the attractive branch and for smaller initial pulse separation ($\sigma_{\text{in}} < \sigma_c$), the observed behavior (again, see Figure 2.18) is different from prediction. In theory (Gordon, 1983), the two pulses reach a bound orbit around their common center, so that they periodically 'pass through each other'. (Note that this is merely a convenient manner of speaking, as the pulses are indistinguishable.) This orbital motion is reflected in the oscillatory structure of the theoretical curve, which the experimental curve clearly does not follow.

Rather, the experiment shows that the pulses never completely overlap. This is a consequence of the instability of the attractive phase (Gordon, 1983), and the fact that perturbations exist to trigger the instability. For small angles, and further assuming constant pulse separation, equation (13) in Gordon (1983) has the approximate solution

$$\phi = \phi_0 \exp(\alpha z / z_0) \tag{2.12}$$

where

$$\alpha = \pi\sqrt{2}\exp(-0.88\sigma/\tau) \tag{2.13}$$

Although it has been shown numerically (Chu and Desem, 1985) that higher order dispersion can trigger the instability, for the conditions of the experiments, it is more likely that the soliton self-frequency shift provides the effective perturbation. The phase shift caused by a common frequency shift $\Delta\nu$ of pulses separated by time interval σ is

$$\Delta\phi = 2\pi\Delta\nu\sigma \tag{2.14}$$

According to (2.11) each 1.17 ps pulse alone should experience a 3.4 GHz shift over the 340 m fiber length. Thus, for $\sigma \sim \sigma_c = 3.79$ ps, (2.14) yields $\Delta\phi \approx 0.08$ rad at the end of the fiber, and proportionally less for some fraction thereof.

When the pulses merge, α becomes large ($\alpha \approx 4.4$ for $\sigma = 0$), and the exponential growth, accumulated over the last few soliton periods of the fiber, becomes more than large enough to amplify the $\Delta\phi$ estimated above (or some reasonable fraction thereof) to a value on the order of π. In this way, the unstable attractive phase will be converted to the stable repulsive one. On the other hand, for σ_{in} considerably greater than σ_c, the pulses always remain apart, α correspondingly remains small, and the initial phase perturbation cannot grow more than a few times. Thus, the net phase shift remains well within the domain of attraction.

Once the laser was accidentally maladjusted (the polarisation axis of the control fiber was misaligned) such that the laser produced a weak (relative intensity ~2%) satellite pulse about 7 ps away from the main pulse. Then in a measurement like that of Figure 2.18, there were extra bumps in the curves near $\sigma_{\text{in}} = 7$ ps, corresponding to an approximately ±20% change in σ_{out}. This shows how even very small coherent satellites can act as emissaries of the main pulse, greatly extending the effective range of interaction. There is an obvious lesson here for telecommunications; namely, that the space between the pulses must be kept clean to avoid serious interactions. It is also clear that small deviations from the perfect pulse shape can result in large changes of the pulse interaction. The close fit of experimental results to theory (Figure 2.18) is therefore testament to the accuracy of the sech amplitude envelope shape of the soliton laser pulses.

The derivations in (Karpman and Solov'ev, 1981; Gordon, 1983) are based on the highly abstract inverse scattering theory, and other analyses (Chu and Desem, 1985; Anderson and Lisak, 1986; Blow and Doran, 1983), while differently based, are still purely mathematical. But the attraction and repulsion can be simply understood in terms of the chirp each pulse induces on the other. For the isolated, symmetrical pulse shown in Figure 2.2, the chirp is correspondingly symmetric, and (negative) dispersion can only compress the pulse. For a pulse partially overlapped by a second pulse, however, its intensity distribution, and hence its chirp, will be unbalanced. (The fact that chirp in the isolated soliton is exactly canceled out by dispersion in no way changes the argument.) If the waves under the pulses are in phase, the intensity will be increased, and the chirp *decreased*, in the region of overlap. (The chirp is proportional to the slope of intensity.) Hence, there will be a net acceleration of the pulses toward each other (attraction). By a similar argument, if the waves are out of phase, just the opposite will happen. Note that by the argument given here, the existence of a force does not require solitons. For solitons, however, the interaction is not complicated by simultaneous decay of the pulses; hence, solitons are the simplest pulses for study. Note also that a force will exist between pulses of different optical frequencies, but then, the intensities simply add, and the force will always be attractive.

2.7 Solitons in telecommunications

Conventional fiber optic telecommunications systems use but a tiny fraction of the potential information carrying capacity of optical fibers. That is, the optical signals are detected and electronically regenerated every 20–100 km before continuing along the next span of fiber. But electronic repeaters limit rates to a few Gbit/s or less per channel. Furthermore, the use of multiple channels, or wavelength multiplexing, is difficult and cumbersome, as the demultiplexing/multiplexing must be performed at each repeater. Thus, the only sensible way to achieve higher bit rates is by allowing the signals to remain strictly optical in nature.

Figure 2.19 shows a proposed all-optical system, where the fiber losses are compensated by Raman gain, and the information is transmitted as trains of solitons. To provide for the gain, c.w. pump power is injected at every distance L (the 'amplification period') along the fiber, by means of directional couplers. (The wavelength-dependent couplers provide for efficient injection of the pump power, but allow the signal pulses to continue down the main fiber with little loss.) A system might contain as many as 100 or more amplification periods.

Loss at the pump wavelength makes the Raman gain non-uniform within each amplification period. Nevertheless, the signal pulse energy fluctuations can be surprisingly small. Figure 2.20 shows gain

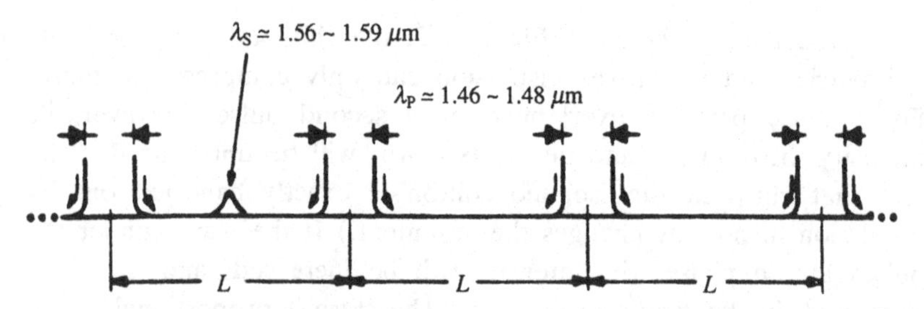

Figure 2.19. Segment of the all-optical soliton-based system. Single laser diodes are shown here at each coupler, but the required pump power, approximately 50 to 100 mW, would best be supplied by a battery of, say, a dozen lasers, each tuned to a slightly different wavelength, their outputs combined through a diffraction grating. In this way, stimulated Brillouin backscattering can be avoided. The multiplicity of pump lasers would also provide a built-in, fail-safe redundancy.

and signal energy for $L = 40$ km and low loss single mode fiber. The result of bidirectional pumping, the Raman gain is the sum of two decaying exponentials, and is adjusted (through control of the pump intensity) such that there is no net signal gain or loss over the period. Note that with the Raman gain, the signal pulse energy varies by no more than about $\pm 8\%$, whereas without it, more than 80% of the signal energy would be lost in just one period.

The stable transmission of fiber solitons over a gain compensated fiber of 10 km has already been demonstrated experimentally (Mollenauer *et al.*, 1985). Nevertheless, the question remains of just how well the solitons will stand up to the periodically varying pulse energy in the system of Figure 2.19. Here, numerical studies (Hasegawa, 1984; Mollenauer *et al.*, 1986) have provided some important answers. In (Mollenauer *et al.*, 1986), a gain/loss term $-i\Gamma u$ (where Γ corresponds to the α_{eff} of Figure 2.20, was added to the right hand side of (2.3). Solutions were then obtained on a Cray.

Figure 2.21 shows a principal result of the calculations of (Mollenauer *et al.*, 1986), the pulse distortion obtaining at the end of one amplification period, graphed as a function of the parameter L/z_0. Here the change, δS, of pulse area is used as a measure of the pulse distortion from a true soliton. (For convenience, S has been renormalised to unity.) The peak in δS, occurring at $z_0 \approx L/8$, corresponds to resonance between the soliton's phase term (see (2.4)) and the periodic pulse energy variation. Note the excellent recovery of the soliton for z_0 both long and short with respect to the resonance value.

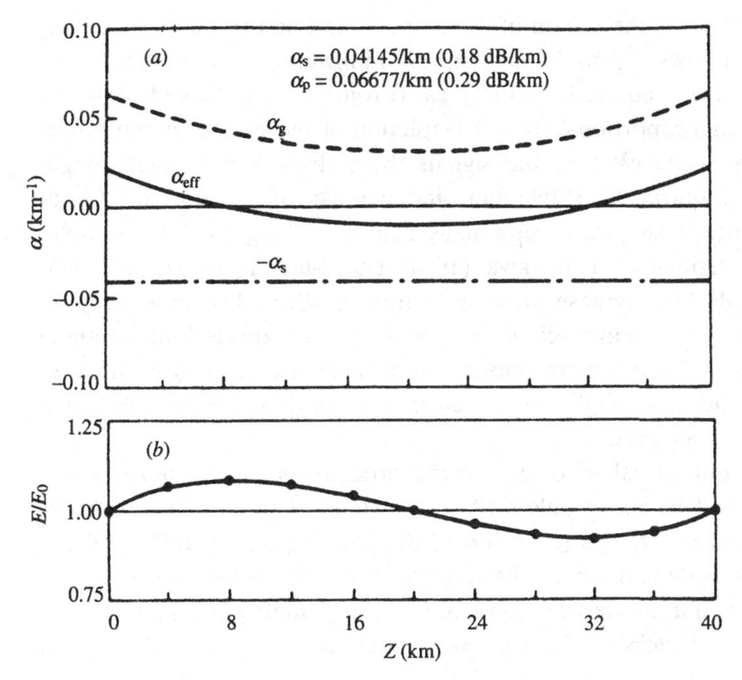

Figure 2.20. (a): Coefficients of loss ($-\alpha_s$), and Raman gain (α_g) and their algebraic sum (α_{eff}) for an amplification period $L = 40$ km. (b): The corresponding normalised pulse energy.

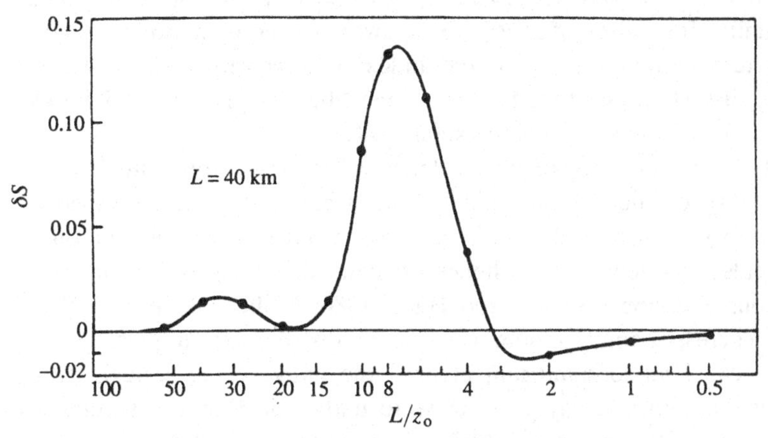

Figure 2.21. Computed change in pulse area (a measure of pulse distortion) obtaining at the end of one amplification period $L = 40$ km for a perfect soliton launched at input, *versus* the quantity L/z_0. (For other values of L, only the resonance peak height changes; its location remains the same (Mollenauer *et al.*, 1986)) Note that the region of 'large' z_0 is to the far right, while the region of 'small' z_0 is to the far left.

Nevertheless, the region of 'long' z_0 is the one of practical interest. First, it allows P_1 to be just a few milliwatts, as opposed to the greater power required for short z_0. (From (2.5), (2.6) and (2.8), P_1 is inversely proportional to z_0.) Depletion of the Raman pump power (typically ~100 mW) by the signals themselves then becomes negligible, as required for stable gain, independent of the signals. Second, the soliton pulses are exceptionally stable for long z_0. For example, the calculations of Hasegawa (1984) and Mollenauer *et al.* (1986) show negligible increase in pulse distortion after 50 or more amplification periods, and such stable behavior is expected to continue indefinitely. Furthermore, modest lumped losses, such as from couplers, should have negligible effect as long as they, too, are canceled out by Raman gain.

The required values of z_0, on the order of 40 km or more, would be obtained by using pulses of several tens of picoseconds width at low values of D; see (2.5) and (2.8). (In 'dispersion-shifted' fibers, λ_0, the wavelength where the dispersion passes through zero, can be pushed as far as desired toward the wavelength of minimum loss. Thus, in principle, D can be made arbitrarily small at that wavelength.)

The question is often raised, why not simply propagate the signals at λ_0? For one reason, that scheme would not allow for wavelength multiplexing, as λ_0 corresponds to just one wavelength. More fundamentally, for pulses of reasonable power, the combination of dispersion terms of higher order and index non-linearity leads to severe pulse distortion and broadening in long fibers (Agrawal and Potasek, 1986). Thus, the soliton is *the* stable pulse.

The maximum propagation distance will therefore be limited not by instability of the solitons, but by the accumulated effects of noise. The most serious of those effects arises from random modulation of the pulse frequencies (and hence of their velocities) by Raman spontaneous emission (Gordon and Haus, 1986; Mollenauer *et al.*, 1986). The corresponding random (Gaussian) distribution in pulse arrival times can lead to significant error if the path is long enough. (Provided the gain is always held close to unity, the Raman spontaneous emission itself increases only in direct proportion to the total path length, and remains negligibly small for many thousands of kilometres.)

In a typical arrangement, the pulses are initially spaced apart by about 10 pulsewidths (see Figure 2.22), to avoid the interaction forces discussed in the previous section. But it also helps to provide for a

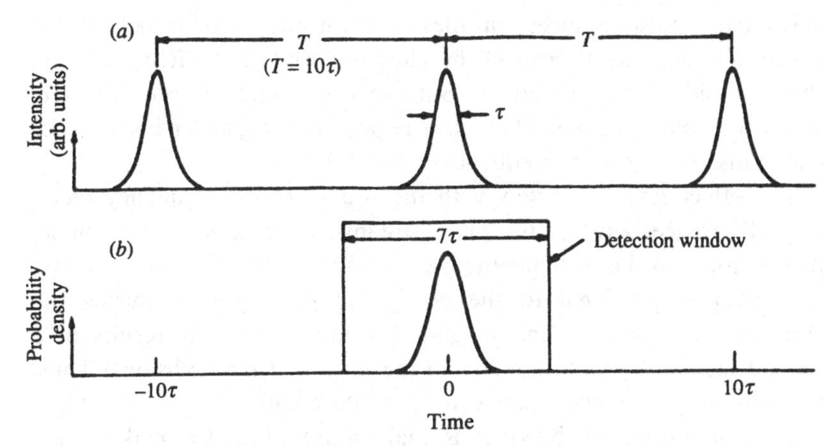

Figure 2.22. (*a*) Train of pulses in their initial relative positions. (*b*) Detection window and corresponding Gaussian probability density of pulse arrival times for an error rate of 10^{-9}.

spread in arrival times; as shown in the figure, the detection window can be nearly as large as the pulse separation.

It can be shown that for fixed fiber characteristics, and a given error rate, the bit-rate–system-length product is a constant. For example, for an assumed error rate of 10^{-9}, for $D = 2$ ps/nm · km, and for reasonable values of the other pertinent fiber parameters, the rate–length product is approximately 29 000 GHz · km. Thus, for example, for a 4 GHz bit-rate (250 ps spacing between 25 ps wide pulses), the maximum allowable distance of transmission would be $Z = 7250$ km.

In contrast to systems involving electronic repeaters, the all-optical system is highly compatible with wavelength multiplexing. By virtue of the fiber's dispersion, however, pulses at different wavelengths will have different velocities and will pass through each other. Hence, it is necessary to consider the possible effects of soliton–soliton collisions. Computer simulation has shown (Mollenauer *et al.*, 1986) that when pulses collide they modulate each other's frequencies (hence velocities), the sign of the effect depending on whether the collision takes place in a region of net gain or loss. Thus, a statistically large number of collisions can add a second source of random pulse arrival times.

The effect can be understood in terms of a cross-phase modulation, or mutually induced chirping, similar to that giving rise to the interaction forces described in the previous section. To be sure, were the collision to take place under conditions of no energy gain or loss, the chirp each pulse induces on the other in the second half of the collision would exactly cancel the chirp induced in the first half, and there would be no net effect. But the cancellation is not complete when the collision takes place in a region of net gain or loss. Hence each pulse suffers a net frequency (velocity) shift.

The effect scales inversely with the square of the frequency (velocity) difference between the pulse streams. (The effect is, of course, proportional to the distance the pulses travel during the collision. But it is also proportional to the change of pulse energy during the collision; for constant energy gain (or loss) rate, this results in a second factor of proportionality to the length of the collision.) Thus, the variance in arrival times can be made comparable to or smaller than that produced by the Raman noise, just by making the wavelength separation between adjacent channels great enough.

A few examples (Mollenauer *et al.*, 1986) of system design, with and without multiplexing, are summed up in Table 2.1. Note that the values of z_0 fall well to the right of the resonance peak in Figure 2.21. Note also that the best approach is to use the combination of many channels at a modest rate per channel. (Compare designs 2b or 2c with 1b.) In addition to producing the highest overall rate–length product, that approach puts the major burden for separating signals on simple optics, rather than on ultrafast electronics. But above all, note the tremendous overall rates made possible by the all-optical nature of the systems: in one instance, the rate is approximately 106 GHz, or about two orders of magnitude greater than possible with conventional systems!

The numbers in this table were derived as follows. The choice of D and z_0 serve to fix τ, through (2.5) and (2.8), and also P_1. The single channel bit-rate $R = 1/(10\tau)$, consistent with a spacing of 10τ between pulses. The system overall length $Z = 29\,000/R$ (see text). The total wavelength span for multiplexing is somewhat arbitrarily set at approximately 20 nm, and the algorithm for determining N, the number of channels, is detailed in (Mollenauer *et al.*, 1986). In examples 1b, 2b and c, Z is reduced, to reduce the variance in pulse arrival times caused by the Raman noise effect; this allows room for the variance caused by soliton–soliton collisions.

2.8 Conclusion

The work reviewed here demonstrates, once again, the truism that ‘physics is an experimental science’. To be sure, we have seen

Table 2.1. *Design examples of high bit-rate, soliton-based, all-optical systems $L = 40$ km and $D = 2$ ps/nm · km.*

Design No.	z_0(km)	τ(ps)	P_1(mW)	Z(km)	N	NR(GHz)
1a	30	12.3	10	3600	1	8.1
1b				2860	5	40.5
2a	100	22.6	3	6600	1	4.4
2b				5200	10	44
2c				3000	24	106

how many properties of light pulses in optical fibers, such as the existence of a stable, fundamental soliton, are described rather well by the simple nonlinear Schrödinger equation. Yet we have also seen how other important features are not. The soliton self-frequency shift, for example, came as a complete surprise. Still other features are well predicted in first approximation, but not in fine detail. Higher order solitons, for example, exhibit the expected behavior over at least a few periods, but tend eventually to break apart. The attractive force between solitons, otherwise well predicted by (2.3), has been shown unable to survive the large phase instability of close contact and the effects of self-frequency shift.

In spite of the complications, however, we have seen that there is a common and simple conceptual thread binding together almost all of the experimental results. That is, most of the effects observed so far (except for the soliton self-frequency shift) can be understood in terms of phase modulation resulting from an instantaneous index non-linearity. Thus, pulse compression and solitons result from self-phase modulation and a resultant self-chirping, while cross-phase modulation (mutually induced chirping) explains the interaction forces between solitons and the result of soliton–soliton collisions. Both the experimental results and this simple physical model give the effects a reality and an existence independent of more abstract models. Such concepts are vital to further invention and development.

Finally, the soliton laser, so important to the other experiments, itself provides another dramatic example of the need for experiment. The soliton laser began with a simple concept, but just how well that concept would work out could not possibly be foreseen without empirical test; the device is just too complex. Thus experiment has also been required for the generation of appropriate models of the soliton laser, and has been vital for the development of practical details.

Happily, here the final result has surpassed the wildest dreams of the inventor!

References

Agrawal, G. P. and Potasek, M. J. (1986) *Phys. Rev. A,* **33**, 1765.
Anderson, D. and Lisak, M. (1986) *Opt. Lett.,* **11**, 174.
Blow, K. J. and Doran, N. J. (1983) *Electron. Lett.,* **19**, 429.
Blow, K. J. and Wood, D. (1986) *IEEE J. Quantum Electron.,* **22**, 1109.
Chu, P. L. and Desem, C. (1985) *Electron. Lett.,* **21**, 228.
Cohen, L. G. and Pearson, A. D. (1983) *Proc. SPIE,* **425**, 28.
Gordon, J. P. (1986) *Opt. Lett.,* **11**, 662.
Gordon, J. P. (1983) *Opt. Lett.,* **8**, 596.
Gordon, J. P. and Haus, H. A. (1986) *Opt. Lett.,* **11**, 665.
Hasegawa, A. (1983) *Opt. Lett.,* **8**, 650.
Hasegawa A. (1984) *App. Opt.,* **23**, 3302.
Hasegawa, A. and Tappert, F. (1973) *Appl. Phys. Lett.,* **23**, 142.
Haus, H. A. and Islam, M. N. (1985) *IEEE J. Quantum Electron.,* **21**, 1172.
If, F., Christiansen, P. L., Elgin, J. N., Gibbon, J. D. and Skovgaard, O. (1986) *Opt. Commun.,* **57**, 350.
Karpman, V. I. and Solov'ev, V. V. (1981) *Physica D,* **3**, 487.
Mitschke, F. M. and Mollenauer, L. F. (1986a) *IEEE J. Quantum Electron.,* **22**, 2242 (1986a).
Mitschke, F. M. and Mollenauer, L. F. (1986b) *Opt. Lett.,* **11**, 659.
Mitschke, F. M. and Mollenauer, L. F. (1987a) *Opt. Lett.,* **12**, 407.
Mitschke, F. M. and Mollenauer, L. F. (1987b) *Opt. Lett.,* **12**, 355.
Mollenauer, L. F., Gordon, J. P. and Islam, M. N. (1986) *IEEE J. Quantum Electron.,* **22**, 157.
Mollenauer, L. F. and Stolen, R. H. (1984) *Opt. Lett.,* **9**, 13.
Mollenauer, L. F., Stolen, R. H. and Gordon, J. P. (1980) *Phys. Rev. Lett.,* **45**, 1095.
Mollenauer, L. F., Stolen, R. H., Gordon, J. P. and Tomlinson, W. J. (1983) *Opt. Lett.,* **8**, 289.
Mollenauer, L. F., Stolen, R. H. and Islam, M. N. (1985) *Opt. Lett.,* **10**, 229.
Mollenauer, L. F., Vieira, N. D. and Szeto, L. (1982) *Opt. Lett.,* **7**, 414.
Stolen, R. H. (1985) *Digest of Technical Papers, Optical Society of America Annual Meeting, October, 1985* (Washington, DC: USA) abstract WH1.
Stolen, R. H., Lee, C. and Jain, R. K. (1984) *J. Opt. Soc. Am. B,* **1**, 652.
Stolen, R. H. and Lin, C. (1978) *Phys. Rev. A,* **17**, 1448.
Stolen, R. H., Mollenauer, L. F. and Tomlinson, W. J. (1983) *Opt. Lett.,* **8**, 186.

3

All-optical long-distance soliton-based transmission systems

K. SMITH AND L. F. MOLLENAUER

3.1 Introduction

The theoretical studies of long-distance soliton transmission in optical fibres with periodic Raman gain were outlined in Chapter 2 'Solitons in Optical Fibres: An Experimental Account' by L. F. Mollenauer. From this work, it was anticipated that the overall information rates made possible by the all-optical nature of the system were potentially orders of magnitude greater than in conventional communication systems. Rate–length products were predicted to be as high as approximately 30 000 GHz · km for a single wavelength channel. In the following sections we will describe the first set of experiments designed to explore the fundamentals of such long-distance repeaterless soliton transmissions.

3.2 Experimental investigation of long-distance soliton transmission

In the following experiments, a train of soliton pulses was injected into a closed loop of fibre, the length of which corresponded to one amplification period in a real communication system. The pulses propagated around the loop many times until the net required fibre pathlength had been reached (Mollenauer and Smith, 1988). The experimental configuration is shown schematically in Figure 3.1. A 41.7 km length (L) of low-loss (0.22 dB/km at 1.6 μm) standard telecommunications single-mode fibre was closed on itself with a wavelength selective directional coupler (an all-fibre Mach–Zehnder (FMZ) interferometer). The interferometer allowed efficient coupling of the bidirectional pump waves at approximately 1.5 μm (provided

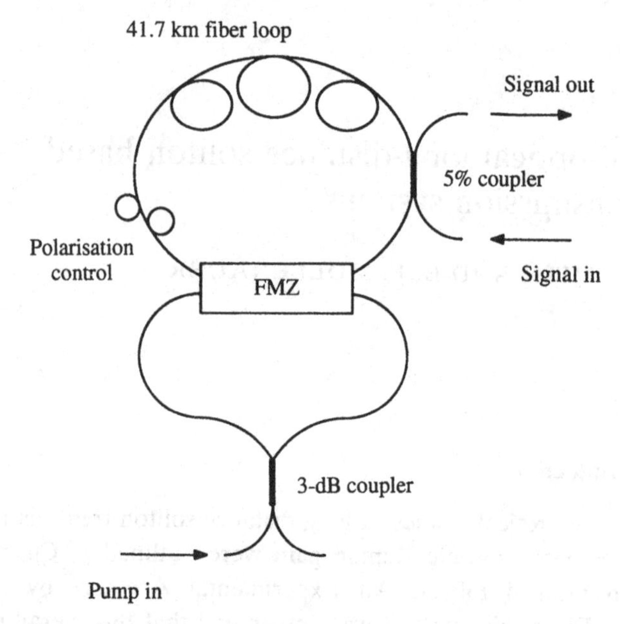

Figure 3.1. Schematic diagram of the fibre loop and input/output arrangement.

by the 3 dB coupler at the pump input) into the loop. At the same time, the signals at approximately 1.6 μm (50 ps sech^2 intensity profile pulses at a 100 MHz repetition rate from a synchronously mode-locked NaCl:OH$^-$ colour centre laser) were efficiently recirculated (<5% coupled out). The difference in the pump and signal frequencies, ~430 cm^{-1}, corresponded to the peak of the Raman gain in fused silica (Stolen *et al.*, 1984). By precise control of the c.w. Raman pump power (~300 mW from a KCl:Tl colour centre laser), the net passive propagation loss of loop configuration (~9.5 dB) could be exactly cancelled. In addition, fibre polarisation controllers were inserted into the loop in order to adjust the polarisation of the signal light entering the (slightly polarising) FMZ. In this way, the Raman gain which depends on the relative polarisation of the pump and signal, remained fairly uniform over many (>100) round-trips.

In order to avoid stimulated Brillouin scattering (SBS) of the pump light, the spectral output of the pump laser was broadened to a band at least several GHz wide. This reduced the power within a Brillouin linewidth (~40 MHz) to the level of a few milliwatts, i.e. significantly below the 8 mW SBS threshold measured for narrow linewidth excitation. A fast feedback loop controlled the laser power, such that the power fluctuations fed into the fibre were less than 0.1%. It is worth noting that the feedback loop sampled the light actually coupled into the fibre, through the use of a 0.1% fibre fusion coupler.

For the 50 ps pulses and the known fibre dispersion (group delay dispersion $D \simeq 17$ ps/nm · km) the soliton period Z_0 was calculated to be ~55 km. This makes $L/Z_0 \simeq 0.76$, which is well removed from the $L/Z_0 = 8$ distortion resonance of the soliton phase and perturbation period. Since the peak output power of the NaCl:OH^- signal laser is several orders of magnitude greater than the fundamental soliton power ($P_1 = 15$ mW), the signal input level afforded by the 5% directional coupler was more than sufficient. The signal output train leaving the loop was detected using an ultrafast diode (Bowers *et al.*, 1985) (response time 9 ps) and a computer-controlled microwave spectrum analyser. The pulse shape and duration could then be inferred from the measured pulse envelope spectrum. The time-domain measurements were complemented by simultaneous measurements of the optical spectra of the pulses. These were recorded by sending a fraction of the output to a scanning Fabry–Perot etalon.

The timing sequence of the experiment is shown in Figure 3.2. Acousto-optic modulators, controlled by a master clock, were employed as the light choppers. The pump light was turned on 200 μs (the transit time of the loop) before a 200 μs train of signal pulses was introduced into the loop. The pulse train traversed the loop the desired number of times before the pump was turned off and the signals allowed to decay. On the final round-trip (denoted by N), the spectrum analyser was triggered to measure the detected sample of the exiting pulse train. This process was repeated until the full information on the pulse profile had been obtained. Since we were observing pulse trains, the microwave spectra consisted of a sequence of harmonics of the (100 MHz) pulse repetition frequency. A single harmonic of the spectrum was recorded with each repetition of the experiment. Generally, we recorded only every fifth harmonic in order to save time accumulating data over the full 18 GHz frequency range. In all the experiments, the signal power coupled into the loop was set to that of the fundamental soliton by observing the pulse behaviour as a function of fibre distance at unity loop gain. A typical microwave spectrum of a train of pulses is shown in Figure 3.3(a) for a distance of 1002 km. The spectrum is an excellent fit to that expected theoretically for 50 ps pulses of sech^2 shape, implying that both the shape and the width of the individual pulses have been well preserved. The corresponding optical spectrum of the pulses is shown

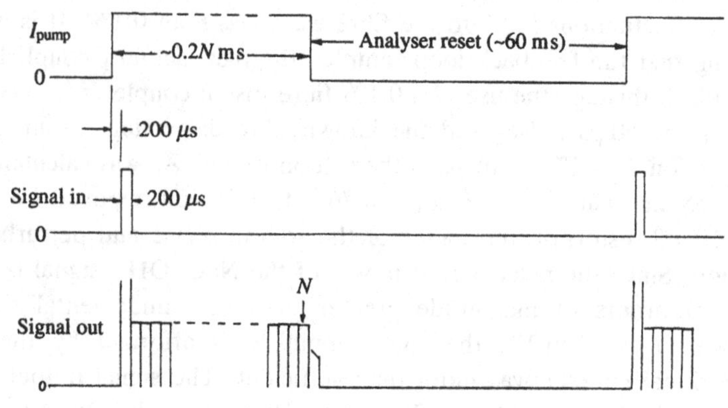

Figure 3.2. The timing sequence of the experiment.

in Figure 3.3(*b*). A time–bandwidth product of 0.32 is inferred from the measurements, which is in excellent agreement with the expected value (0.315) for a sech^2 (soliton) intensity pulse shape.

In Figure 3.4 the effective pulse duration (τ_{eff}) inferred from the measured microwave spectra is shown plotted as a function of the distance traversed for 55 ps duration input solitons. Note that τ_{eff} remains essentially constant to at least 4000 km. At greater distances, τ_{eff} is seen to increase slightly above that of the theoretical Gordon–Haus limit imposed by the random walk effects of Raman spontaneous emission (Gordon and Haus, 1986). It is therefore likely that the increase in τ_{eff} reflects a jitter in arrival times rather than a broadening of the original pulses.

3.3 Further experiments

The closed loop configuration described in the previous section constituted a unique laboratory for the study of non-linear effects over fibre optic path lengths of thousands of kilometres. We now describe briefly the results and discoveries of some of those experiments.

Precisely controlled excess (or deficient) Raman gain was easily achieved over many transits of the fibre loop. It was therefore possible to adiabatically modify the soliton energy. In this limit, where the pulse area, S (defined as the time integral of the absolute value of the pulse amplitude) is preserved, we expect that the pulse width, τ, scales inversely with the energy, E. We can, therefore, manipulate the soliton pulse duration by appropriate adjustment of the pulse

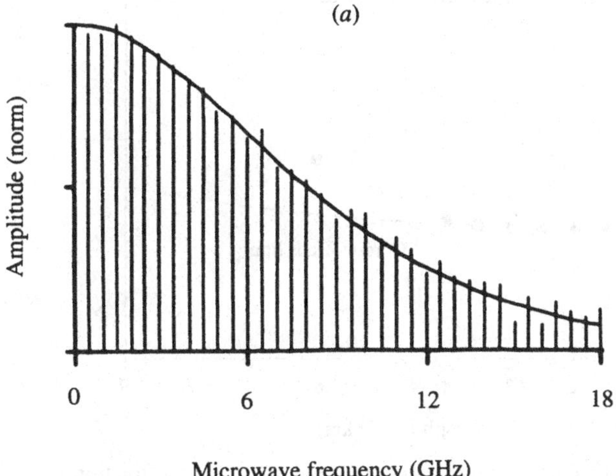

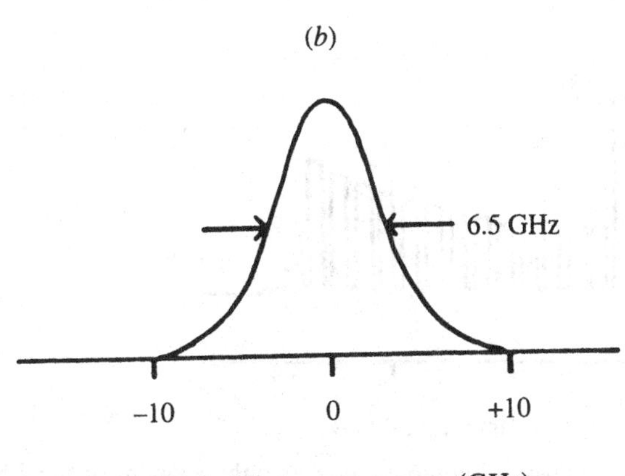

Figure 3.3. (*a*) Microwave spectrum (showing every fifth harmonic) of the intensity envelopes of 50 ps duration pulses for a distance of 1002 km with unity loop gain and (*b*) the corresponding optical spectrum.

energy (Smith and Mollenauer, 1989a). In Figure 3.5, we show a twofold adiabatic energy enhancement of the 200 μs pulse train over 13 recirculations of the loop, i.e. 542 km of fibre path. In Figure 3.6(a) we show the pulsewidth plotted as a function of the soliton energy; the full curve is of the form $\tau \propto E^{-1}$ and fits the data extremely well. We also see the direct linear dependence of the optical bandwidth, $\Delta \nu$, on the soliton energy (Figure 3.6(*b*)). Figure

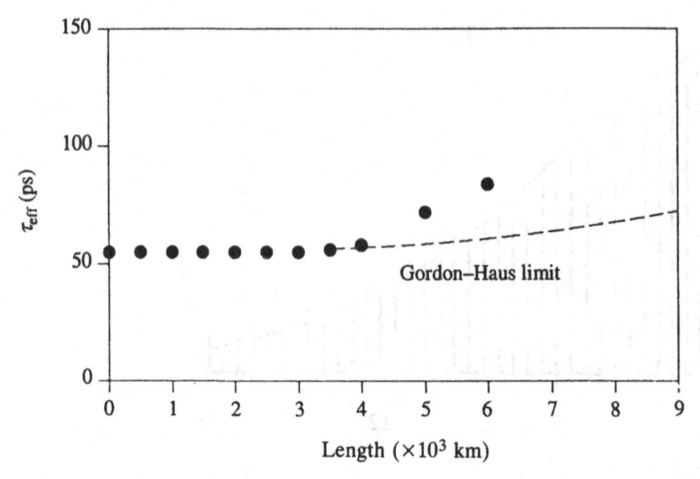

Figure 3.4. Plot of the effective pulsewidth, τ_{eff}, as a function of fibre pathlength. The broken line indicates the quantum (Gordon–Haus) limit.

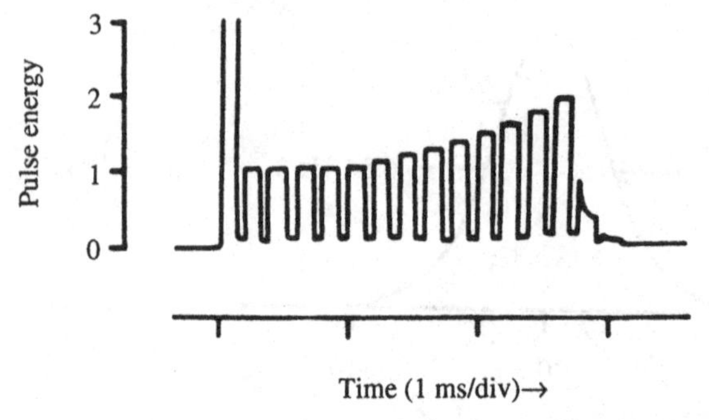

Figure 3.5. Twofold adiabatic energy enhancement of the 200 μs pulse train over 13 recirculations of the fibre loop ($\approx$ 542 km). The pulse energy is in units of the input soliton energy.

3.6(c) depicts the time–bandwidth product, $\tau\Delta\nu$, calculated from the data in Figures 3.6(a) and (b); all the points lie close to the broken line, which represents the theoretical value (0.315) for a sech^2 pulse profile. We thus confirmed the maintenance of true soliton pulses under such adiabatic energy changes.

We have also been able to study the interaction properties of solitons over many thousands of kilometres (Smith and Mollenauer, 1989b). The potential of co-propagating solitons to interact is of obvious significance to an optical communications system. In order to

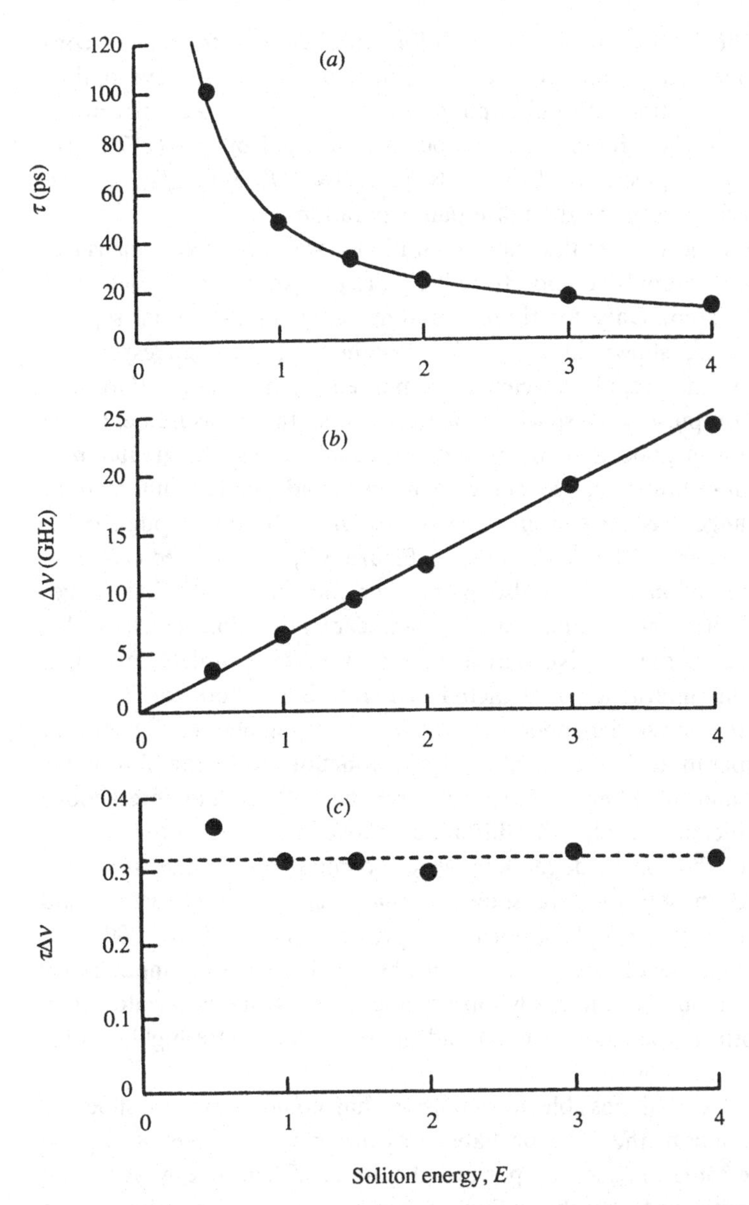

Figure 3.6. (*a*) Pulsewidths, (*b*) optical spectral widths (FWHM) and (*c*) time–bandwidth products as functions of the pulse energy (in units of the input soliton energy). All data are recorded at 1002 km.

achieve these measurements, the mode-locked signal pulse train was passed through a Michelson interferometer before entering the loop. The desired pulse separation was set by adjusting the length of one arm of the interferometer. In addition, piezoelectric transducer control of one arm facilitated precise adjustment of the relative optical phase. For a train of pulse pairs separated by T, the microwave spectrum of their intensity envelope is modulated by $|\cos(\pi fT)|$. By measuring the positions of the nulls, given by $1/T$, $3/T$, $5/T, \ldots$, we could easily determine the pulse pair separation.

For a variety of initial pair separations, we then made measurements of the repulsive and attractive pair separations as a function of fibre traversed. Data for three different values of the initial separation, T_0, are shown in Figure 3.7 together with the corresponding attractive and repulsive curves computed from theory (Gordon, 1983). For pulse pair spacings of 6τ or less, the measurements are apparently in good agreement with expected values. At greater pair separations, however, we detected a phase-independent interaction, whose range was far too great to be explained by direct pulse overlap. For example, for $T_0/\tau = 8.0$ in Figure 3.7, theory predicts negligible interaction of the pulse pairs over the entire 5000 km range, while experimental points show a significant attraction which is also insensitive to the relative optical phase. A more complete investigation of this interaction is depicted in Figure 3.8, where for $T_0/\tau > 8$ (where the phase-dependent interaction is negligible) the change in pulse separation $T - T_0$ is plotted as a function of T_0 for fibre paths of 1001 and 3004 km. For T_0/τ where phase-dependent interactions are significant, we plot the difference between the measured separation and the phase-dependent value computed for the repulsive phase. Both sets of data show alternate regions of attraction and repulsion, with peak deviations of approximately 160 ps for 3004 km. The magnitude of the effect diminishes with increased input pulse separation but is still easily observable at separations greater than 4 ns. Both the period (~1.6 ns) and the magnitude were highly reproducible.

Of the several possible mechanisms that could be responsible for this interaction, the most probable was thought to be that of dispersive wave radiation shed by perturbed solitons. Dispersive waves could collide with and thereby influence the velocities of nearby soliton pulses, in a manner similar to soliton–soliton collisions in the presence of net gain or loss. By conducting numerical simulations (Mollenauer *et al.*, 1989), we established that the modest levels of birefringence in communications grade fibre could lead to significant amounts of dispersive-wave radiation. Figure 3.9 shows one of the key simulations, where we have modelled the soliton response to a birefringence

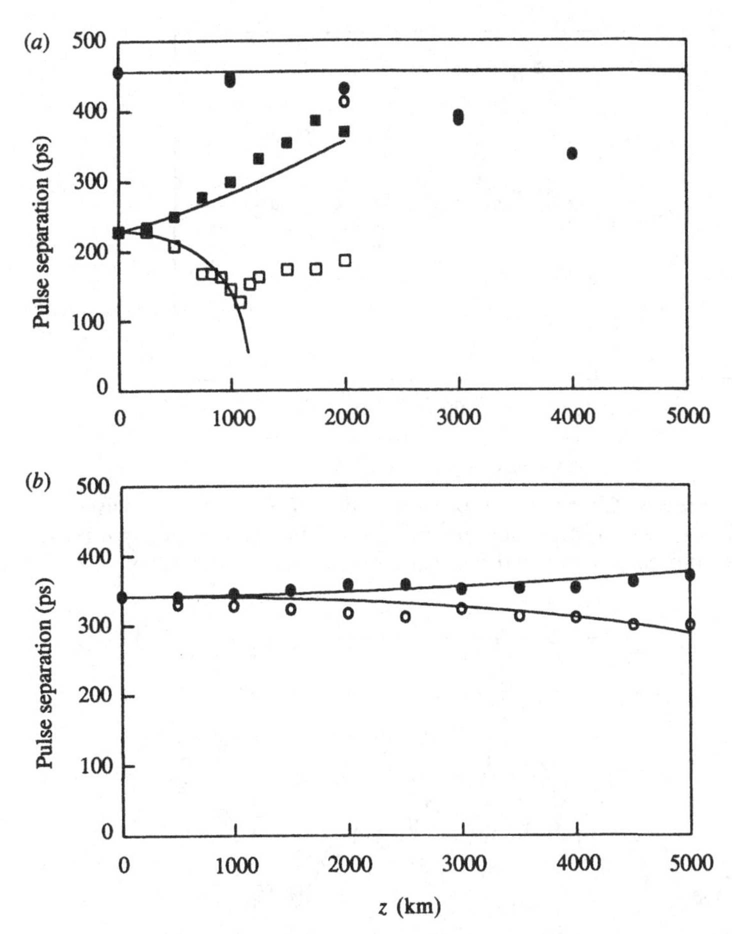

Figure 3.7. Temporal separations of 57 ps pulse pairs as a function of the fibre pathlength for (*a*) $T/\tau = 4.0$ (squares) and $T/\tau = 8.0$ (circles) and (*b*) $T/\tau = 6.0$. The filled symbols represent the attractive phase data.

of randomly varying orientation ($Z = 2000$ km, and a polarisation dispersion of 1.3 ps/km$^{1/2}$). Although the soliton avoids splitting or excessive broadening the radiation generated by polarisation dispersion is clearly illustrated. Note that the spread of the radiation is directly proportional to the fibre pathlength; the 1% amplitude point reached 100 soliton time units (corresponding to 2.84 ns) at approximately 1800 km. Further numerical studies of soliton pulse pair propagation, indicated that displacements of approximately the magnitude observed in the experiments were obtained.

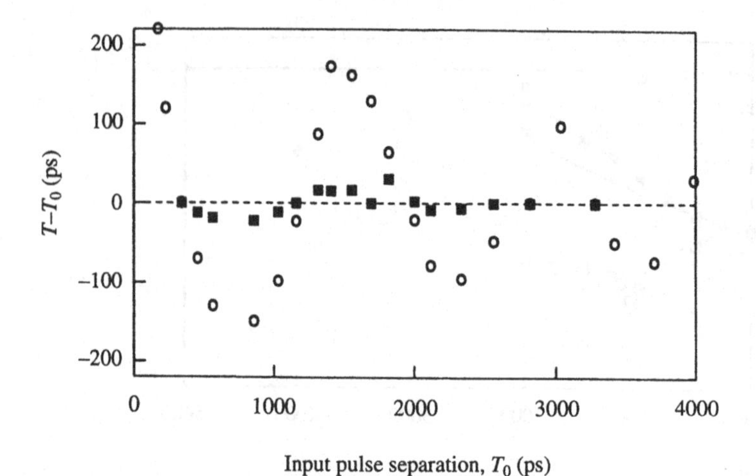

Figure 3.8. Change in the pair separation $(T-T_0)$ of 57 ps pulses due to phase-independent interaction as a function of the input pulse separation (T_0) for 1001 km (squares) and 3004 km (circles).

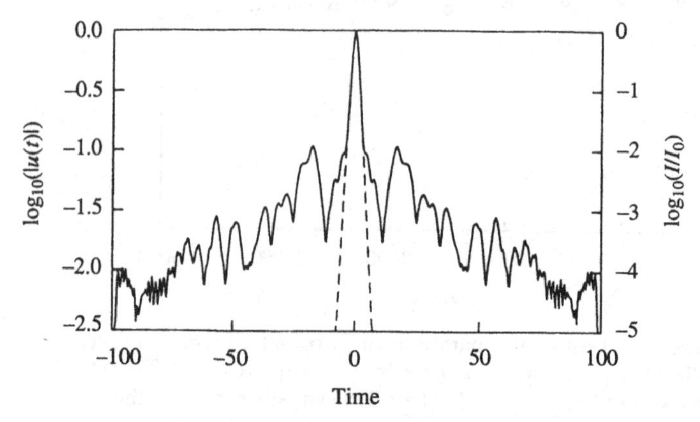

Figure 3.9. Logarithm of the normalised pulse amplitude modulus intensity as a function of time (in soliton time units, $\tau/1.76$) numerically simulated for Z = 2000 km and a polarisation dispersion parameter of 1.3 ps/km$^{1/2}$. The broken curve represents a perfect soliton.

A further interaction mechanism (Smith and Mollenauer, 1989b), that of electrostriction, has been theoretically investigated in detail by Dianov *et al.* (1990). In this case, the first light pulse propagating through the fibre stimulates a sound wave owing to the electrostriction. This induces a refractive index change which then modifies the relative time delay of the second pulse. Calculations indicated that for long propagation distances (>1000 km) the magnitude of the interaction could reach hundreds of picoseconds for initial pulse separations in the region of approximately 1 ns. It should be noted that, although further experiments are required to determine the exact contributions of the mechanisms previously outlined, both long-range interaction effects were expected to scale as D^2. The strength of the interaction should, therefore, be reduced to negligible proportions in the low dispersion fibre ($D \simeq 1$ ps/nm · km) preferred for long-distance soliton communication systems.

3.4 Conclusion

We have described some of the first experiments designed to explore the fundamentals of long-distance repeaterless soliton transmission. These experiments demonstrated the stability of solitons against the many perturbations encountered over thousands of kilometres of fibre pathlength. Furthermore, the many potential advantages of the all-optical soliton-based system have been a strong motivation for further research and development. Most recently, the advent of erbium-doped fibre amplifiers with high, polarisation independent gains (>20 dB) at pump powers of only a few tens of milliwatts has been a powerful spur to the all-optical approach. The utilisation of closely spaced ($L \simeq 25$ km) erbium amplifiers in a 75 km loop of dispersion-shifted fibre (Mollenauer *et al.*, 1990) has led to soliton transmission over more than 10 000 km with no evidence for the long-range interaction described previously. Indeed, no serious interaction was observed over distances of 9000 km until pulse separations were less than approximately 4τ. Although more experimental work is required to ascertain the far limits of the proposed system, there can be no doubt that the experiments described here point the way forward to future terahertz capacity transmission systems.

References

Bowers, J. E. and Burrus, C. A. and McCoy, R. J. (1985) *Electron. Lett.*, **21**, 812.

Dianov, E. M., Luchnikov, A. V., Pilipetskii, A. N. and Starodumov, A. N. (1990) *Opt. Lett.*, **15**, 314.

Gordon, J. P. (1983) *Opt. Lett.*, **8**, 596.

Gordon, J. P. and Haus, H. A. (1986) *Opt. Lett.*, **11**, 665.

Mollenauer, L. F., Neubelt, M. J., Evangelides, S. G., Gordon, J. P., Simpson, J. R. and Cohen, L. G. CLEO Post Deadline paper CPDP17, Anaheim, CA, May 1990.

Mollenauer, L. F. and Smith, K. (1988) *Opt. Lett.*, **13**, 675.
Mollenauer, L. F., Smith, K., Gordon, J. P. and Menyuk, C. R. (1989) *Opt. Lett.*, **14**, 1219.
Smith, K. and Mollenauer, L. F. (1989a) *Opt. Lett.*, **14**, 751.
Smith, K. and Mollenauer, L. F. (1989b) *Opt. Lett.*, **14**, 1284.
Stolen, R. H., Lee, C. and Jain, R. K. (1984) *J. Opt. Soc. Am. B*, **1**, 652.

4

Non-linear propagation effects in optical fibres: numerical studies

K. J. BLOW AND N. J. DORAN

4.1 Introduction

The study of non-linear waves has always been associated with numerical analysis since the discovery of recurrence in non-linear systems (Fermi *et al.*, 1955) and elastic soliton–soliton scattering (Zabusky and Kruskal, 1965). Since then the mathematics of non-linear wave equations has grown into the industry of inverse scattering and numerical analysis has developed a number of techniques for studying non-linear systems. Inverse scattering theory has given us much insight into integrable non-linear systems and has supplied many useful exact solutions of non-linear partial differential equations. Numerical analysis has mostly been used in the complementary field of non-integrable systems. Since most non-integrable systems of interest, in physics, are 'close' to integrable ones the combination of perturbation theory and numerical analysis provides a powerful tool for investigating such systems. In this chapter we will show through illustrative examples how simple concepts and enhanced understanding can be derived for complex non-linear problems through insight gained from numerical simulation. A good example of this is described in Section 4.6 where the concept of a 'soliton phase' emerges from numerical simulation; this is a particularly simple and useful concept enabling the design of a number of soliton switching systems.

Before we begin let us clarify the use of the term soliton. When used by mathematicians the term soliton has a precise meaning in the context of inverse scattering theory and carries with it the associated properties of a localised non-linear wave, elastic scattering amongst solitons, stability and being part of an integrable system. The term solitary wave is appropriate in a physics context where no integrable systems exist but the term soliton is usually used for simplicity. Thus, 'soliton propagation in fibres' is correctly interpreted as 'the propagation of pulses in the non-linear regime and with the correct sign of dispersion so that, in the absence of other perturbations, solitons would be observed'. In this chapter we will use the term soliton without further comment.

It is now possible to solve exactly a number of canonical equations (e.g. Dodd *et al.*, 1982) which includes the Kortveg–de Vries equation

$$\frac{\partial u}{\partial t} + u\frac{\partial u}{\partial x} + \frac{\partial^3 u}{\partial x^3} = 0 \tag{4.1}$$

the sine–Gordon equation

$$\frac{\partial^2 u}{\partial x^2} - \frac{\partial^2 u}{\partial t^2} = \sin(u) \tag{4.2}$$

and the non-linear Schrödinger (NLS) equation

$$\mathrm{i}\frac{\partial u}{\partial z} + \frac{\alpha}{2}\frac{\partial^2 u}{\partial t^2} + |u|^2 u = 0 \tag{4.3}$$

the latter is the subject of this book. The NLS equation occurs in many contexts in physics: deep water waves, plasma physics, Bose condensates, the propagation of light in optical fibres. We will discuss the results obtained from numerical analysis of the NLS equation and of various equations which are perturbed from it in the context of the non-linear optical properties of pulse transmission in optical fibres.

A detailed derivation of the NLS equation, as applied to optical fibres, can be found in Chapter 1. Here we simply note that equation (4.3) applies to a weakly guiding single-mode optical fibre. The non-linear term comes from the optical Kerr effect in which the refractive index, in silica-based fibres, is related to the optical intensity by the following equation

$$n = n_0(r) + n_2|E|^2 \tag{4.4}$$

The relationships between the normalised variables used in equation (4.3) and the real quantities are (Mollenauer *et al.*, 1980)

$$z = L\frac{|k_2|}{\tau^2} \tag{4.5}$$

$$t = \frac{T}{\tau}$$

$$|u(t)|^2 = \frac{k_0 n_2 \tau^2 P(t)}{2n_0|k''|}$$

where L, T and P are the length along the fibre, the time in the co-moving frame and the pulse peak power respectively, τ is an arbitrary time unit but is generally taken as the pulsewidth for convenience, $k'' = \partial^2 k_0/\partial\omega^2$, is the dispersion of the fundamental transverse mode propagation constant k_0. The coefficient, α, in equation (4.3), can be taken to be ± 1 according to the sign of the group delay dispersion. In order to set the scales on the physical quantities some useful numbers are: for a pulsewidth of 7 ps in a standard communication grade optical fibre at a wavelength of 1.55 μm a normalised amplitude of one corresponds to a peak power of 1 W and a normalised length of one corresponds to a fibre length of 640 m.

Let us begin with a brief review of the properties of equation (4.3) in two limits. First, at low intensities the non-linear term is negligible and we are left with the following linear wave equation

$$\mathrm{i}\frac{\partial u}{\partial z} + \frac{\alpha}{2}\frac{\partial^2 u}{\partial t^2} = 0 \tag{4.6}$$

which only includes the effect of the frequency dependence of the propagation constant. This equation can be solved by taking a Fourier transform in the time domain, solving for the evolution of the spectral components of $u(z, \omega)$ and finally taking the inverse Fourier transform to recover $u(z, t)$. The solution for an initially Gaussian pulse $(2\pi\tau^2)^{-1/4}\exp(-t^2/4\tau^2)$ is

$$u(t, z) = (2\pi)^{-1/4}\left(\tau + \frac{\mathrm{i}\alpha z}{2\tau}\right)^{-1/2}\exp\left(-\frac{t^2}{4\tau^2 + 2\mathrm{i}\alpha z}\right) \tag{4.7}$$

Several important points are contained in this result. The pulse broadens as it propagates, this property is also usually referred to as dispersion, but the broadening does not depend on the sign of α. If we write $u(z, t) = U(z, t)\exp(\mathrm{i}\phi(z, t))$ then ϕ represents the phase of the pulse. This phase factor is quadratic in time and the sign of its curvature is given by the sign of α. The derivative of the pulse phase $\mathrm{d}\phi(z, t)/\mathrm{d}t$ can be used to define a 'local' or 'instantaneous' frequency within the pulse. Since the pulse phase is quadratic in time the local frequency will be a linear function of time and this is known as a chirp. In general, pulses whose bandwidth is not limited by the transform of their intensity are referred to as chirped. The linear chirp acquired by a pulse propagating in an optical fibre depends on the sign of the dispersion which can be either positive or negative depending on the wavelength.

The second limit of equation (4.3) is obtained when $\alpha = 0$

$$\mathrm{i}\frac{\partial u}{\partial z} + |u|^2 u = 0 \tag{4.8}$$

In this limit we can solve equation (4.8) for any initial pulse shape $u_0(t) = u(t, z = 0)$ and obtain

$$u(t, z) = u_0(t)\exp(-\mathrm{i}|u_0(t)|^2 z) \tag{4.9}$$

which is known as self-phase modulation (SPM) since the phase of the pulse is modulated by its own intensity. The chirp associated with this phase is also linear near the peak of the pulse but its sign is determined by the sign of n_2. In silica-based fibres in the infrared communication wavelength range n_2 is positive or 'self-focusing' and independent of frequency.

We may already anticipate that different phenomena will occur during non-linear propagation in the presence of dispersion. These differences are fundamentally related to the different signs of the frequency chirp imposed on the pulse by the anomalous ($\alpha > 0$) or normal ($\alpha < 0$) dispersion compared to the frequency chirp imposed by the non-linearity which has a fixed sign.

In the next section we give a brief discussion of the inverse scattering method for the NLS equation and its role in numerical analysis. Following this we give an account of the best numerical methods for solving the NLS equation initial value problem and the different ways in which the numerical data can be analysed. In Section 4.4 we discuss some examples of the initial value problem for the unperturbed NLS equation. Finally, in Section 4.5, we discuss the effects of various perturbations which would exist in practice on the propagation of solitons in optical fibres.

4.2 Solitons and inverse scattering theory

We begin with a review of the solutions of the NLS equation (4.3). The simplest method is to look for propagating solutions which are separable in z and t, i.e. they have the form

$$u(z, t) = \exp(\mathrm{i}\Lambda z)T(t) \tag{4.10}$$

when equation (4.10) is substituted into the NLS equation we obtain the following differential equation for T

$$\frac{\alpha}{2}\frac{\mathrm{d}^2 T}{\mathrm{d}t^2} + T^3 - \Lambda T = 0 \tag{4.11}$$

This equation has solutions of the form sech(t) and tanh(t) depending on the sign of α. In particular, when $\alpha = +1$ the solution is

$$u(z, t) = \exp(\mathrm{i}z/2)\,\mathrm{sech}\,(t) \tag{4.12}$$

which is the single bright soliton. This solution is a self-maintaining pulse which propagates without change in shape. When $\alpha = -1$ the solution is

$$u(z, t) = \exp(\mathrm{i}z)\tanh(t) \tag{4.13}$$

which is the single dark soliton and is also a self-maintaining pulse.

If we return to the NLS equation with positive α we can also have bound states of solitons like those of equation (4.12). The simplest bound solitons occur when an initial pulse of the form

$$u(z = 0, t) = N\,\mathrm{sech}\,(t) \tag{4.14}$$

is launched where N is an integer (Satsuma and Yajima, 1974). These soliton bound states execute a periodic motion, as illustrated in Figure (4.1) for $N = 2$, where the period is $\pi/2$ for all values of N.

The only separable solutions are those given in equations (4.12) and (4.13) and this method does not give the bound states of equation (4.14). The way to make further progress was first identified by Zakharov and Shabat (1972). They used the method of inverse scattering theory to solve the initial value problem for the NLS equation. The procedure is analogous to the way in which Fourier transforms are used to solve the linear dispersion problem, equation (4.6). First the equation is transformed to obtain a description of the input pulse in terms of the eigenvalues and eigenfunctions of a scattering problem. The evolution of the eigenvalues is described by a linear system of equations. Finally, the output pulse can be reconstructed from the inverse scattering equations. This procedure results in the exact solution of the NLS equation given an initial input field. In this chapter we are mostly concerned with the effects of perturbations to the NLS equation and thus it might seem that this theory is of little use to us. In fact, the discrete eigenvalues of the direct scattering transform correspond to the 'soliton content' of the field and thus provides a useful description of the evolving field if we are prepared to calculate the eigenvalues as a function of distance. In the pure NLS equation these eigenvalues are conserved and their motion is due to the presence of the perturbation.

Let us begin by considering the pure NLS equation (4.3), with α positive and the boundary condition $|u| \to 0$, $t \to \infty$. We need to calculate the eigenvalues ζ_j of the equations:

$$\mathrm{i}\frac{\partial V_1}{\partial t} + u(t)V_2 = \zeta V_1 \tag{4.15a}$$

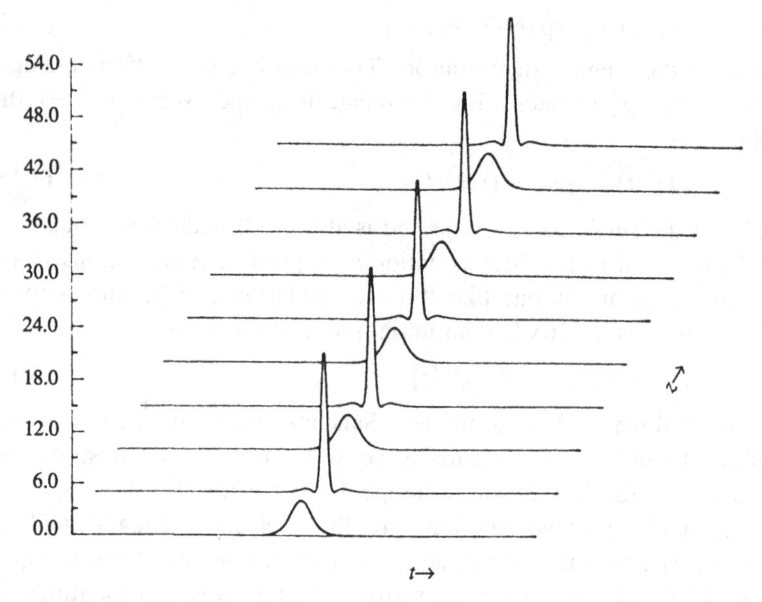

Figure 4.1. The evolution of an $N = 2$ soliton showing the periodic motion.

$$\mathrm{i}\frac{\partial V_2}{\partial t} + u^*(t)V_1 = -\zeta V_2 \tag{4.15b}$$

Zakharov and Shabat (1972) showed that the imaginary part of the bound state eigenvalues gives the energies of the solitons which are contained within the pulse and the real part of the eigenvalues gives the velocities of the solitons. The soliton energy will be less than or equal to the total pulse energy and indeed will be zero if no solitons are created. The eigenvalues are the constants of the evolutionary motion and are analogous to the frequencies in the linear dispersion problem. The number of discrete eigenvalues indicates the number of solitons in the initial pulse and we can think of any pulse as a superposition of solitons with each soliton corresponding to one of the eigenvalues together with some 'radiation energy' which disperses during propagation. Inverse scattering allows us to invert the process and determine the potential (i.e. pulse shape) from which a particular eigenvalue can be obtained. Thus soliton solutions can be obtained for any arbitrary energy. A detailed discussion of this scattering problem can be found in Satsuma and Yajima (1974).

The single bright soliton, given by equation (4.11), is the simplest of the stationary soliton solutions of the NLS equation. The most general single soliton has a velocity in the chosen co-moving frame of reference used to derive the NLS equation. This solution has a velocity parameterised by a coefficient μ and an arbitrary pulse width characterised by the parameter β. The full equation for the single soliton is given by

$$u(z,t) = \beta \exp\left(\mathrm{i}(\beta^2 - \mu^2)\frac{z}{2} + \mathrm{i}\mu t + \mathrm{i}\phi\right) \operatorname{sech}(\beta(t - t_0 + \mu z)) \tag{4.16}$$

and we have also included an arbitrary phase ϕ and initial position t_0. The parameters μ and β are the real and imaginary parts of the eigenvalue, ζ, of equation (4.15). The energy of the soliton is given by $\int |u|^2 \mathrm{d}t$ and is easily shown to be 2β for equation (4.16). Thus we have the result that the energy is related to the pulsewidth and both can be directly obtained from the imaginary part of the discrete eigenvalues. The actual soliton velocity in the laboratory frame can be obtained from the normalisations (4.5) and the transformation to the laboratory frame and is easily shown to be

$$V = \left(k_1 + k_2 \frac{\mu}{\tau}\right)^{-1} \tag{4.17}$$

We can see from this expression that μ/τ is the difference between the frequency of the pulse and the chosen centre frequency used in the slowly varying envelope approximation to define the co-moving frame. Thus, this new velocity is exactly what one would expect for a linear pulse and is not a special feature of the NLS equation.

Satsuma and Yajima (1974) studied equations (4.14) in detail and derived some exact solutions for the eigenvalues and eigenfunctions. In particular they solved the inverse scattering problem for an initial pulse of the form $2\operatorname{sech}(t)$ to obtain

$$u(z,t) = \frac{4\exp(-\mathrm{i}z/2)[\cosh(3t) + 3\exp(-4\mathrm{i}z)\cosh(t)]}{\cosh(4t) + 4\cosh(2t) + 3\cos(4z)} \tag{4.18}$$

This is also a simple scattering problem to solve numerically with $u(t)$ the potential and ζ the eigenvalues and indeed is an extremely useful method of determining the soliton content of a pulse. As an example of the use of equation (4.14) we consider the initial pulse $A\operatorname{sech}(t)$ (Satsuma and Yajima, 1974). When $A = 1$ we have a single soliton and indeed at all integer values of A the pulse is made up of that number of solitons. At arbitrary values of A the eigenvalues are purely imaginary and given by

$$\zeta_j = \mathrm{i}(A - j + \tfrac{1}{2}) \qquad j > 0, \zeta_j > 0 \tag{4.19}$$

In Figure (4.2) we plot the proportion of the total pulse energy which is contained in each individual eigenvalue (soliton) as well as the total soliton energy. It can be seen that for large A a high proportion of the energy is contained in the solitons and little 'radiation' (non-soliton) remains.

We have seen that shape-maintaining solutions exist for the simple non-perturbed non-linear pulse evolution equation for optical fibres in the anomalous dispersion regime. Any arbitrary pulse can be considered as consisting of a number of solitons, together with some dispersive radiation, the fraction of the pulse energy contained by these solitons indicates the fraction of the pulse energy that will perform soliton-type propagation. The scattering problem allows us to determine both the number and velocities of the solitons.

Another useful result obtained from inverse scattering theory is the existence of an infinite number of conserved quantities for the NLS equation. These can be expressed in the form of integrals of the field and its time derivatives. The first few can often be associated with physical quantities such as energy and momentum but the higher ones carry no (simple) physical interpretation. The first three conserved quantities of the NLS equation are

$$I_0 = \int |u|^2 \mathrm{d}t \tag{4.20}$$

$$I_1 = \int (u^* u_t - u u_t^*) \mathrm{d}t \tag{4.21}$$

$$I_2 = \int (|u|^4 + |u_t|^2) \mathrm{d}t \tag{4.22}$$

where I_0 is the energy (strictly the photon number but for a narrow band excitation the two are related by a constant factor of $\hbar\omega_0$), I_1 is the momentum and I_2 is the Lagrangian of the system.

We mentioned, at the beginning of this section, that when α is negative the NLS has dark soliton solutions. An inverse scattering theory also exists for this problem (Zakharov and Shabat, 1973) but is rather different in its structure. The most significant difference is that there are no bound states of dark solitons (Zakharov and Shabat, 1973, Blow and Doran, 1985). However, there are still an infinite number of conserved quantities and the direct scattering equations determine the dark soliton structure of an initial pulse.

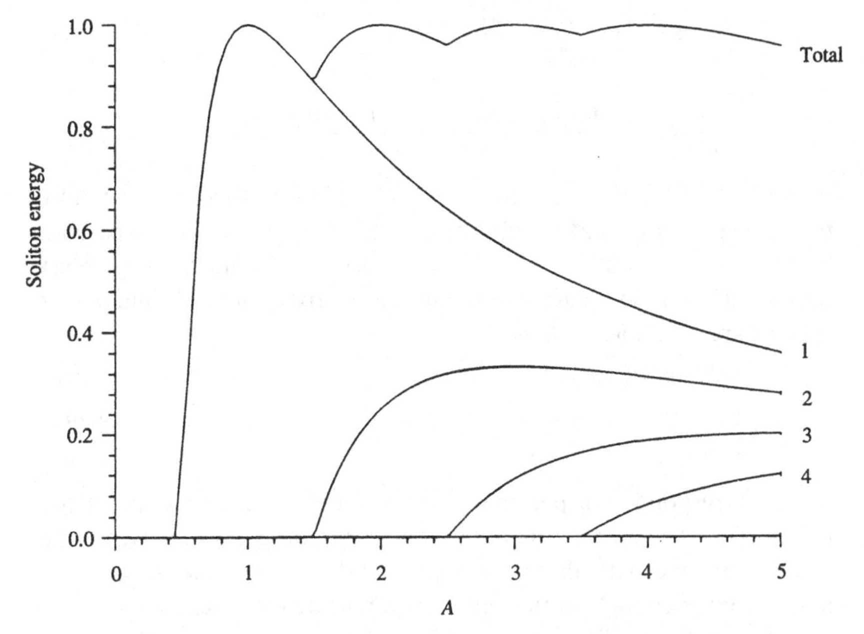

Figure 4.2. Fraction of the total pulse energy contained in the soliton eigenvalues (1–4) and the total soliton energy.

4.3 Numerical analysis

In this section we discuss some of the methods used in the numerical analysis of the NLS and equations which are perturbed from it. Numerical analysis consists of a number of stages the first of which is the formulation of the problem and this is covered in Chapter 1 by Hasegawa. The second stage is the selection and implementation of a numerical scheme for, in our case, integrating the non-linear partial differential equation. At this point we have calculated the exact, in some sense, answer to the problem but have learned nothing of the physical principles involved. The final stage, and by far the most important, consists of the representation of the results in a way which enables us to understand the physics and, hopefully, construct simple analytical models which can reproduce the main features of the results.

We begin this section with a discussion of the numerical schemes which are most useful in solving the NLS and related equations. The simplest technique is to use a finite difference approximation to the derivative terms in equation (4.3). Thus we replace

$$\frac{\partial u}{\partial z} \to \frac{u(z + \mathrm{d}z) - u(z)}{\mathrm{d}z} \tag{4.23a}$$

$$\frac{\partial^2 u}{\partial t^2} \to \frac{u(t + \mathrm{d}t) - 2u(t) + u(t - \mathrm{d}t)}{\mathrm{d}t^2} \tag{4.23b}$$

to obtain a discrete representation of the field u at times separated by the grid spacing $\mathrm{d}t$ and at subsequent points down the fibre separated by the integration step $\mathrm{d}z$. Since we must use a finite grid we must also specify the boundary conditions at the end points. A number of choices are available such as

$$u(T) = u(-T) \tag{4.24a}$$

$$u(T) = u(-T) = 0 \tag{4.24b}$$

$$u(T) = u(T - \mathrm{d}t) \tag{4.24c}$$

which correspond to a periodic grid, zero field and zero slope at the boundaries. In practice there is almost no difference between the results obtained with these boundary conditions and the grid size T must be large enough so that they do not affect the results.

Although the explicit scheme (4.23) is very simple to use, a computationally better technique is the split-step Fourier method. In this procedure equation (4.3) is replaced by two equations which are integrated alternately by the same step length, these two equations are

$$\mathrm{i}\frac{\partial u}{\partial z} + \frac{\alpha}{2}\frac{\partial^2 u}{\partial t^2} = 0 \tag{4.25a}$$

$$\mathrm{i}\frac{\partial u}{\partial z} + |u|^2 u = 0 \tag{4.25b}$$

where we now use the fact that both equations (4.25) can be integrated exactly in frequency space (4.25a) and in real space (4.25b) respectively. The transfer from the time to the frequency domain being achieved by using fast Fourier transforms which are extremely accurate and efficient computational procedures. Equations (4.25) can both be written in the form $\mathrm{i}\partial u/\partial z = Lu$ and the approximate solution of the NLS for a step length $\mathrm{d}z$ is

$$u(z + \mathrm{d}z) = \exp(\mathrm{i}L_\mathrm{a}\mathrm{d}z)\exp(\mathrm{i}L_\mathrm{b}\mathrm{d}z)u(z) \tag{4.26}$$

whereas the exact solution is

$$u(z + \mathrm{d}z) = \exp(\mathrm{i}L_\mathrm{a} + L_\mathrm{b})\mathrm{d}z\,u(z) \tag{4.27}$$

The difference between these two expressions turns out to be proportional to the commutator of the two exponents and is therefore a term of order dz^2. In practice we can use combinations of step lengths to increase the order of accuracy of the integration and this is described in more detail elsewhere (Fleck *et al.*, 1976, Feit and Fleck, 1978; Blow and Wood, 1989).

The final stage of the numerical analysis is the interpretation and representation of the results in some useful form. Let us assume that we now have sets of numerical solutions $u(z, t)$ to our problem. Simple evolutionary plots, such as those used in this chapter, usually give a good insight into the general properties of the dynamics. However, more specific and more condensed information is often required. For example, if we are interested in the potential bandwidth of a communication system then the evolution of the pulsewidth may be of interest. There are many choices for the definition of pulsewidth, the most popular being the full width at half maximum (FWHM), the root mean square width (r.m.s.) and the half energy width (HEW). The conclusions drawn from the data can depend strongly on the choice of definition. Consider applying these definitions to a pulse with a strong pedestal component. The FWHM will give a value close to the peak of the pulse whereas the r.m.s. width is dominated by the contribution from the pedestal. The HEW gives values in between these two. In some sense, the FWHM is sensitive to the soliton content of the pulse but the r.m.s. width is sensitive to the radiation. The correct choice will therefore depend on the question you are asking. This problem can be seen in more detail in the next section. This is a specific example of a general problem in numerical analysis which is that the processing of the raw numerical data can often influence the interpretation. Bearing this in mind our preference is to analyse the data in as many different ways as possible.

The final hope is that when the dynamics of the system is understood and the physics of the problem is clear then a simple model can be constructed, an example is given in Section 4.5.

In the previous section we mentioned the existence of conserved quantities of the NLS equation which took the form of integrals over the field and its derivatives. These integrals can be monitored to provide a measure of accuracy of the (bare) NLS code. The usual technique used is to check that the error in their computation scales in the correct way with the step size. The approximation used in equation (4.26) produces a local error which is $O(dz^2)$ and therefore has a global error (i.e. for a fixed length) which is $O(dz)$. The errors in the conserved quantities in this case will therefore be proportional to dz.

When we are dealing with perturbed versions of the NLS equation, which do not possess conserved quantities, checking numerical procedures becomes more difficult. One possibility is that there may be a simple equation of motion for some integrated quantity which can be checked. The classic example of this is the equation of motion of the energy in the presence of absorption and this will be discussed in Section 4.5.

4.4 Initial value problems

Although the Zakharov and Shabat (1972) formalism gives the exact solution of the NLS equation the general problem when both solitons and dispersive radiation are present involves the solution of a set of coupled integro-differential equations which essentially renders the problem intractable. Under such conditions we can solve the Zakharov and Shabat (1972) direct scattering equations (4.15) numerically and determine the soliton content of any initial pulse profile. However, there may be applications where the behaviour of the non-soliton component is important (Blow and Wood, 1986b) and then we must resort to numerical solutions of the pure NLS equation.

The use of solitons in optical communication systems has been the subject of much interest in recent years, (see, for example, Hasegawa and Tappert, 1973; Blow and Doran, 1982; Doran and Blow, 1983; Mollenauer *et al.*, 1986). It is normally assumed that the solitons are generated by lasers producing transform limited pulses. In practice this condition would not be met exactly, and we must consider the formation of solitons from non-transform-limited pulses. In this section we discuss the effects of both linear frequency chirp and random phase noise on soliton propagation. These are two particular ways in which a laser pulse might deviate from a perfect transform limit. Other initial value problems include the launching of transform limited pulses which are not exact solitons (Satsuma and Yajima, 1974) and launching many pulses which would be solitons in isolation (Blow and Doran, 1983; also Chapter 7 by Dianov *et al.*).

The direct scattering problem can be used to study this problem. In particular the distribution of solitons obtained from stochastic initial profiles can be calculated (Elgin, 1985). The symmetric and antisymmetric parts of the noise lead, to first order, to changes in the amplitude and velocity of the soliton respectively. A semiclassical solution can also be used to study the effects of a deterministic chirp on the formation of solitons (Lewis, 1985). These calculations showed that, at least perturbatively, the soliton threshold was increased and hence that only some of the total energy in the pulse ended up as a soliton. Using numerical methods we can examine the non-perturbative regime and look at the issue of whether or not solitons are created (Desem and Chu, 1986; Blow and Wood, 1986a).

We are interested in the effects of phase distortion on soliton formation and so restrict ourselves to initial pulses of the form

$$u(z = 0, t) = N \operatorname{sech}(t) \exp(\mathrm{i}\phi(t)) \tag{4.28}$$

with $N = 1, 2$, so that in the absence of phase effects the pulses would be pure solitons. For a chirped pulse $\phi(t) = \alpha t^2$ and for a pulse with noise $\phi(t)$ is a random function of t. We have used two distinct examples of random functions for the phase. For the first example (which we refer to as the random frequency model), we set

$$\frac{\mathrm{d}\phi}{\mathrm{d}t} = \beta F(t) \tag{4.29}$$

where $F(t)$ is a uniform random variable in the range $(-1.0, 1.0)$ and β is a parameter. Note that $\mathrm{d}\phi/\mathrm{d}t$ is the instantaneous frequency of the pulse. In the second example (the random walk model), we set

$$\frac{\mathrm{d}^2\phi}{\mathrm{d}t^2} = \beta F(t) \tag{4.30}$$

with $F(t)$ defined as before. Further details can be found in Blow and Wood (1989).

The results of our calculations of the eigenvalues of the Zakharov and Shabat (1972) problem (4.15) are shown in Figures 4.3 and 4.4. The magnitude of the phase, $\phi(t)$, is not easily observable, nor is it easy to obtain the instantaneous frequency along an ultrashort pulse. We have therefore chosen to characterise a particular magnitude of chirp or phase noise by the bandwidth of the optical spectrum. The particular choice of a measure of bandwidth is somewhat arbitrary. We use the HEW, as the shapes of the spectra are so strongly distorted by the chirp that the FWHM would be fairly meaningless. In all of the figures, we normalise the bandwidth to that of the unchirped pulse. In Figure 4.3 the imaginary part of the eigenvalues are plotted against the bandwidth of the chirped pulses. For $\alpha = 0.4$ the bandwidth is twice the transform limit. We can see that as the size of the chirp, and hence the bandwidth, increases the eigenvalues move toward zero, showing that the solitons are less energetic. The

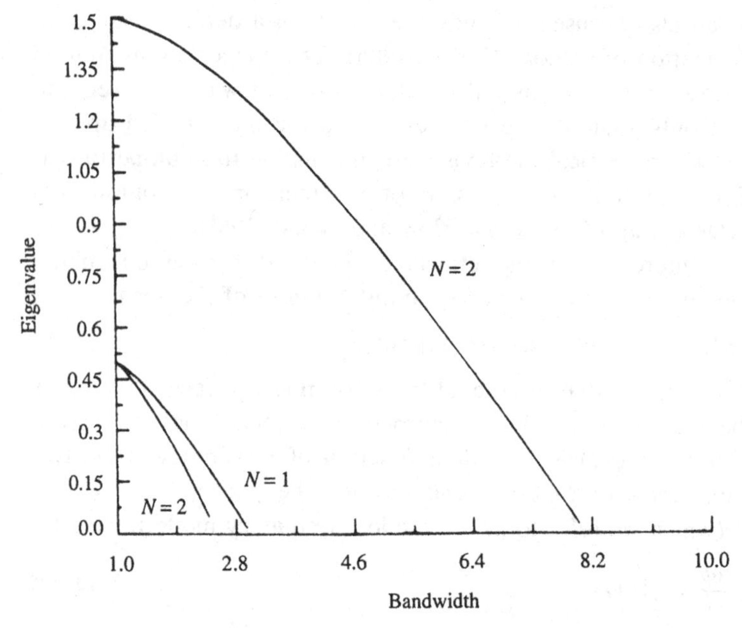

Figure 4.3. The evolution of the Zakharov–Shabat (ZS) eigenvalues for quadratically chirped pulses.

$N = 1$ soliton is destroyed for α slightly larger than 0.8 (bandwidth = 3.0). The $N = 2$ soliton loses one eigenvalue at $\alpha \simeq 0.6$, thus becoming a single soliton, and the second eigenvalue is destroyed at $\alpha \simeq 2.5$ by which point the bandwidth is eight times the transform limit.

Figures 4.4(*a*) and (*b*) are scatter plots of the eigenvalues for an $N = 1$ soliton with random frequency noise (4.4*a*) and random frequency walk noise (4.4*b*), for various values of the parameter, β, and different random sequences. Also plotted as full lines are the eigenvalues for the chirp given by the ensemble averaged phase ($t^{1/2}$ for the random frequency model and $t^{3/2}$ for the random walk frequency model). In each example, when the pulse bandwidth is less than about 1.5 times the transform limit, the eigenvalue is only slightly reduced and more than half the pulse energy remains in the soliton. This result holds for three widely different models of the phase variation across the pulse, and it seems likely that a similar conclusion could be drawn for almost any other perturbation of the phase. When the bandwidth is higher, the results are more model-dependent, but there still can be some residual soliton even when the bandwidth is three or four times the transform limit.

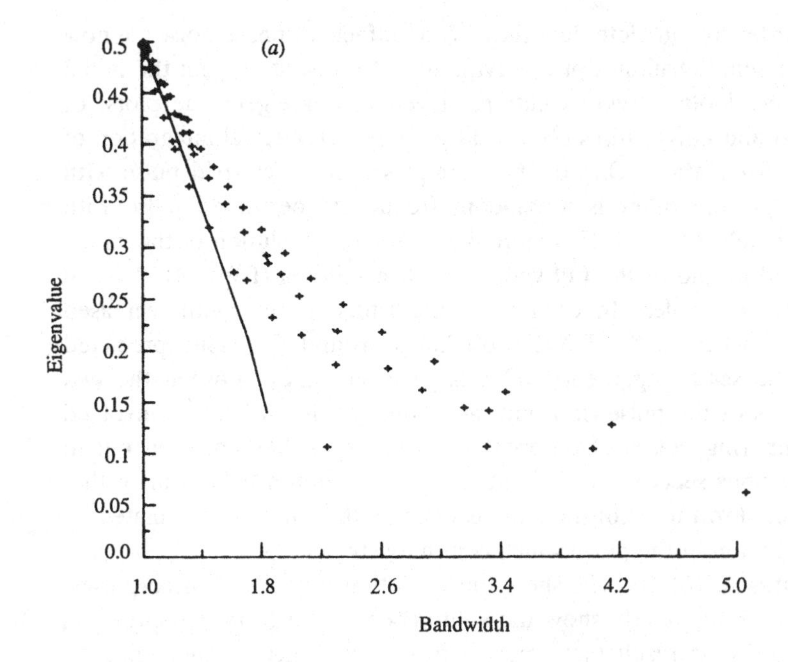

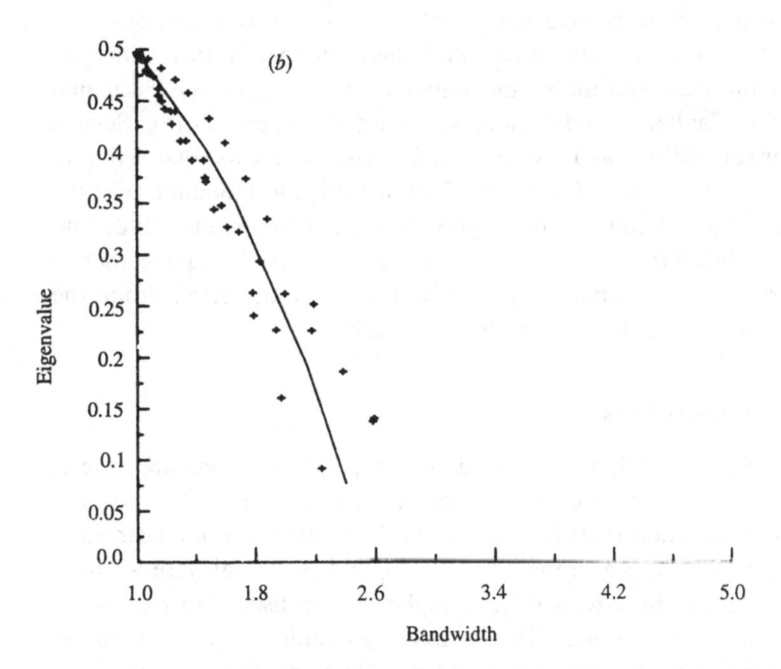

Figure 4.4. The evolution of the ZS eigenvalues for noisy pulses: (*a*) random frequency noise and (*b*) random walk frequency noise.

In order to complete our study of the effects of phase noise we now look at some examples of the evolution of the soliton from the initial field. We look at two evolutions, given in our eigenvalue plots, of chirped and noisy pulses obtained by a direct numerical integration of the NLS equation. One of the examples is for a chirped pulse with $\alpha = -0.4$, the other is a 'random frequency' perturbed pulse with initial bandwidth of 1.25. Figure 4.5 shows the evolution of the r.m.s. pulsewidth, and of the full width at half maximum (FWHM) for each of these examples. In each case, the r.m.s. pulse width increased steadily, whereas the FWHM oscillates around the value predicted from the scattering theory. This large discrepancy between the two measures of the pulsewidth indicates some of the difficulties involved in comparing pulses of different shapes. As we have pointed out in the previous section the FWHM reflects the soliton behaviour in that it settles down to a constant value whereas the r.m.s. is dominated by the radiation in the pulse which continuously disperses.

Figures 4.6(*a*) and (*b*) show the evolution of each of these pulses, and they both clearly show the soliton separating from the spreading radiation. One result that emerges from these calculations is that the transition region between the initial chirped pulse and the final asymptotic soliton is long compared with the linear dispersion distance. We can also compare the final pulse obtained from the integration of the NLS and the single soliton with the eigenvalue calculated from the Zakharov and Shabat scattering equations (4.15). Such a comparison (Blow and Wood, 1986a) clearly shows that even at $z = 20$ there is a significant contribution from the radiation near the soliton. The amplitude of the central peak oscillates as the soliton and radiation interact; these oscillations die away, with the square root of the propagation distance, as the radiation is dispersed from the central regions (Zakharov and Shabat, 1972).

4.5 Perturbations

So far we have discussed the range of solutions and effects which can be obtained for the ideal NLS in the soliton regime. Although equation (4.3) is a very good description of non-linear pulse propagation in optical fibres, it is incomplete. In a real fibre we have to account for the effects of loss, higher order dispersion and stimulated Raman scattering. The simplest example of a perturbation which we will now describe is absorption. In a real fibre the energy of an optical pulse decreases as it propagates due to various effects

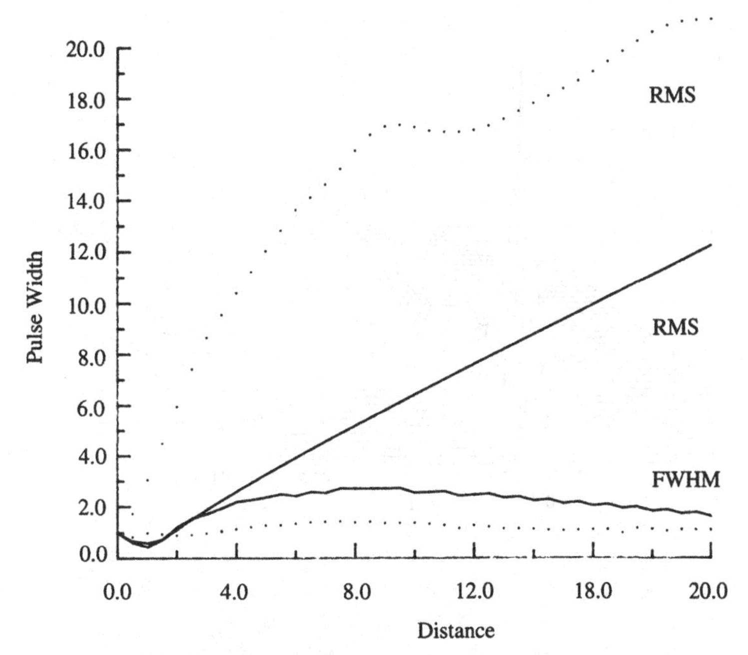

Figure 4.5. The evolution of the pulse FWHM and r.m.s. widths for noise (dotted line) and chirp (full line).

including, Rayleigh scattering from the intrinsic density fluctuations in the glass, phonon scattering in the Urbach tail and residual water vibrations. A good optical fibre has a loss of about 0.2 dB/km at 1.5 μm which corresponds to an absorption length of 15 km. The NLS can be modified (Blow and Doran, 1985) to include this effect and the equation becomes

$$i\frac{\partial u}{\partial z} + \frac{\alpha}{2}\frac{\partial^2 u}{\partial t^2} + |u|^2 u = -i\frac{\Gamma}{2}u \qquad (4.31)$$

As an example we will consider the effect of the absorption on the propagation of bright solitons. To do this we must resort to a numerical solution of equation (4.3) as we no longer have any results from inverse scattering theory to use. The specific problem we wish to address is how does the absorption affect the propagation of an initial pulse given by

$$u(z = 0, t) = A\,\mathrm{sech}\,(t) \qquad (4.32)$$

and to do this we integrate equation (4.31) numerically for a range of values of Γ and A. Either of the numerical techniques mentioned in Section 4.3 can be adapted to include the effect of loss.

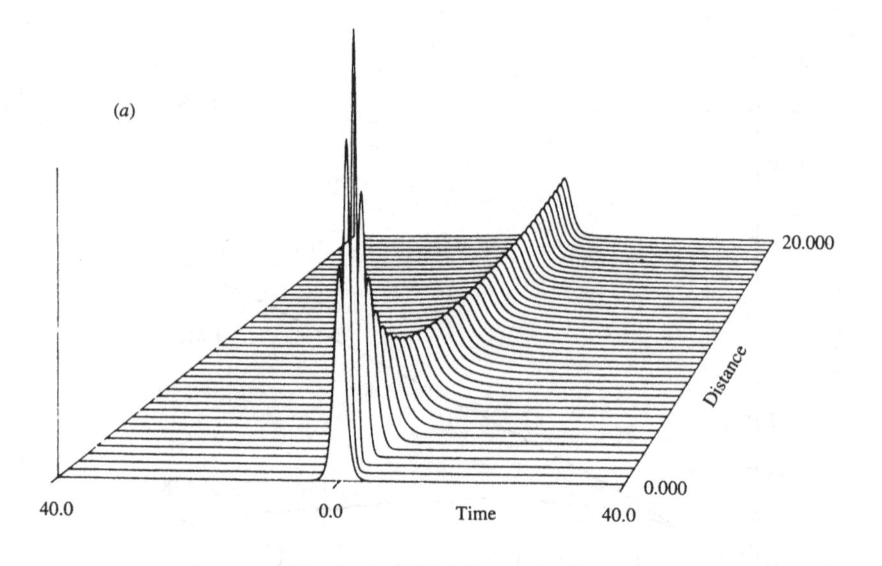

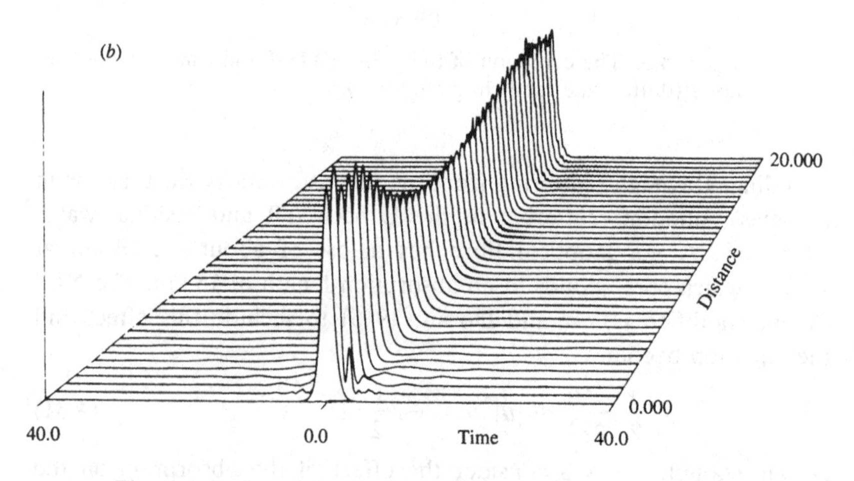

Figure 4.6. The evolution of (*a*) a chirped pulse and (*b*) a noisy pulse.

The evolution of the $N = 2$ soliton in the presence of loss is shown in Figure 4.7, the value of Γ used in this calculation was 0.02. The oscillatory motion of the $N = 2$ soliton (see Figure 4.1) is seen in the early stages of the evolution. However, the oscillation slows down and eventually freezes out at large distances. In general this is observed for all soliton states as can be seen in Figure 4.8 where we

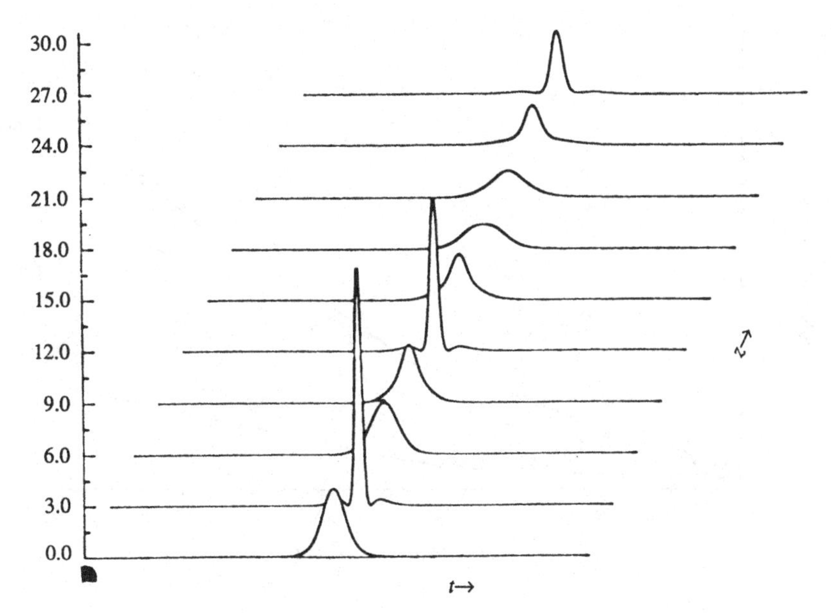

Figure 4.7. The evolution of an $N = 2$ soliton in the presence of loss.

show the evolution of the pulse FWHM for various values of A. We can already draw one important conclusion from this graph. If a communication system is operating at a given pulsewidth but for some reason needs to operate at high powers then the soliton propagation effects will, in general, be beneficial since the output pulsewidth is less than the linear result. We also note that the oscillatory motion of the pulse FWHM, characteristic of higher order solitons, persists for many soliton periods.

So can we find some general argument to enable us to understand the freezing out of the dynamics of this system. To do this we need two features of the NLS equation which we have already discussed, the conservation laws in the presence of perturbations and the scaling properties. The pulse energy, conserved by the pure NLS equation, can easily be shown to evolve according to the following equation

$$\frac{\mathrm{d}E}{\mathrm{d}z} = -\Gamma E \tag{4.33}$$

which has the solution

$$E = E_0 \exp(-\Gamma z) \tag{4.34}$$

We also note that the single soliton given in equation (4.11) has energy 2β and that all soliton solutions are subject to the scaling

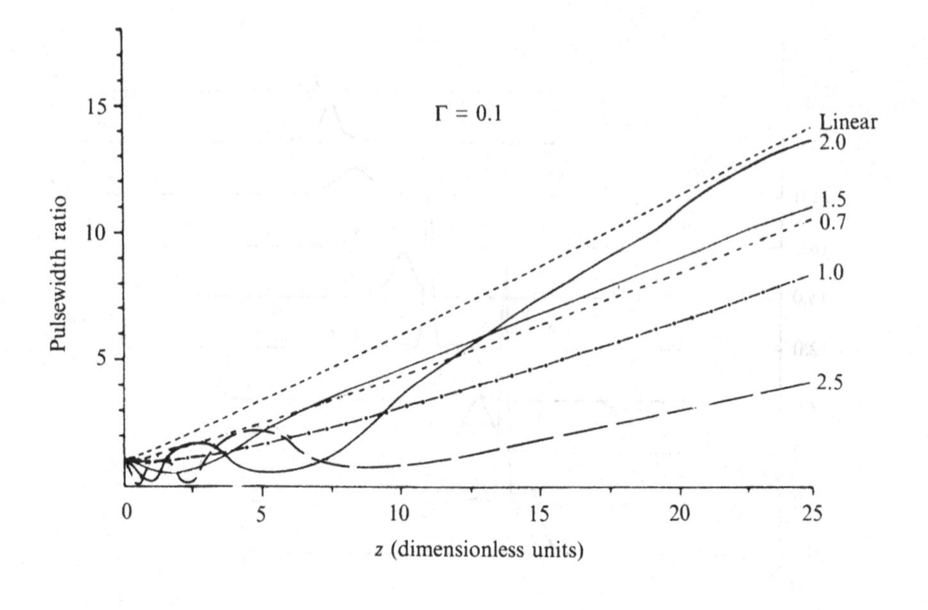

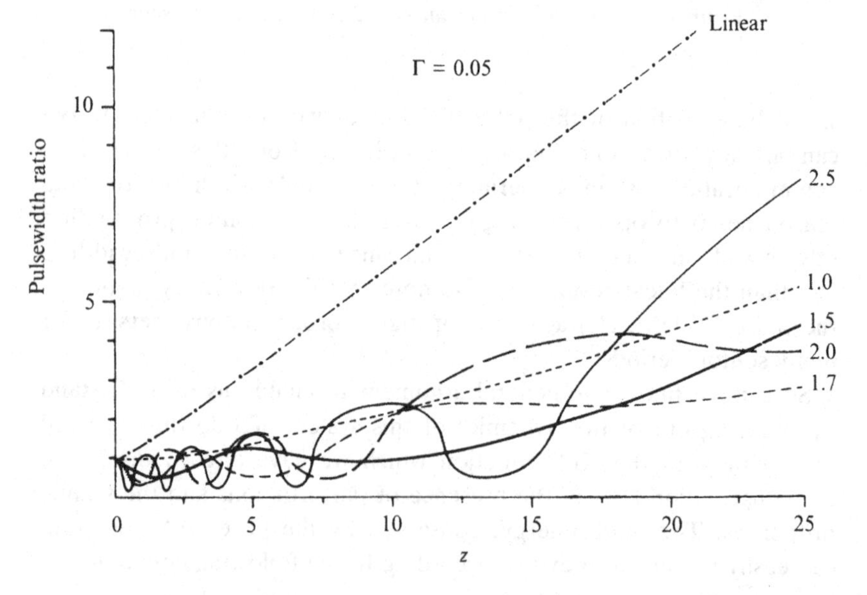

Figure 4.8. FWHM pulse widths for various values of A.

relations (4.5) which would yield a similar linear dependence on the scale parameter β. Thus, the soliton solution with $\beta = 1$ could evolve by adiabatically adjusting β to satisfy the energy equation (4.33). If we denote the $N = 2$ soliton solutions by $u_2(z, t)$ then the adiabatically connected solution would be $\beta u_2(\beta^2 z, \beta t)$ with energy 8β. Our adiabatically evolving solution in the presence of loss is finally obtained by replacing β by $\exp(-\Gamma z)$. Since the pulsewidth is β^{-1} we can see that the adiabatic solution would have a width which increases exponentially. This broadening effect can be seen clearly in the early stages of propagation in Figure 4.8 but it is clear that this situation cannot be maintained as the pulse energy would have to move outward with exponentially increasing speed. We can also see that the period of oscillation of the adiabatic solution is given by $(\pi/2)\beta^2$ which increases with propagation distance, in other words the oscillation in pulse shape must slow down and presumably freezes out when the new period becomes longer than the absorption length. This is indeed what is observed in Figure 4.8. Thus we have been able to use our knowledge of the properties and solutions of the pure NLS together with the energy equation to understand the general behaviour of solitons in the presence of loss.

If we return to the single soliton we can make one further step by using a WKB form for the total pulse phase

$$\Phi_{tot} = \int_0^z \phi(z')dz' \tag{4.35}$$

where

$$\phi(z) = -\frac{\exp(-2\Gamma z)}{2} \tag{4.36}$$

and we obtain

$$\Phi_{tot} = \frac{(1 - \exp(-2\Gamma z))}{4\Gamma} \tag{4.37}$$

Using this result and the evolution of β we can obtain an adiabatically evolving single soliton

$$u(z, t) = \exp(-\Gamma z)\exp(i(1 - \exp -2\Gamma z)/4\Gamma)\,\mathrm{sech}\,[\exp(-\Gamma z)t] \tag{4.38}$$

which is a result that can also be derived by applying perturbation theory to the direct scattering equations (Lamb, 1980).

One further technique which is useful in studying the perturbed NLS equation is to follow the Zakharov and Shabat (1972) eigenvalues during the evolution. This is easily done by solving equation (4.15) with the numerically calculated pulse $u(z, t)$. The results enable us to interpret the full solution in terms of the motion of the soliton eigenvalues. A typical evolution is shown in Figure 4.9 for the $N = 2$ eigenvalues in the presence of loss, $\Gamma = 0.1$. The eigenvalues decrease

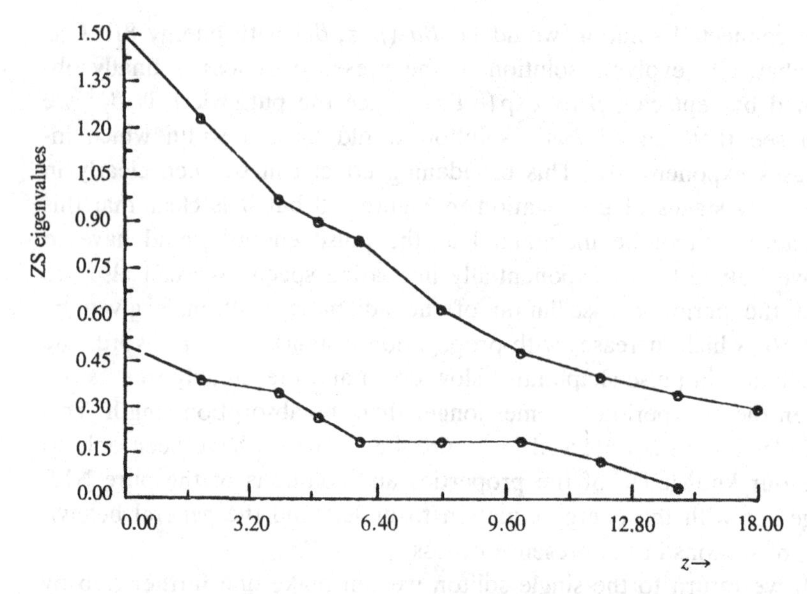

Figure 4.9. In the presence of loss the eigenvalues are no longer constants of the motion as shown here for the $N = 2$ eigenvalues.

in magnitude, characteristic of the increase in the length scale of the evolution and broadening of the pulse. At about $z = 14$ the lowest eigenvalue ceases to exist and at this point the pulse evolution loses the characteristic $N = 2$ structure.

Further details of the effects of loss on the asymptotic properties of soliton propagation can be found in Blow and Doran (1985).

So far we have been discussing the effects of absorption on solitons. In recent years a number of experiments have been done which involve gain. The energy is usually supplied by the stimulated Raman interaction which can give rise to gain over long lengths of optical fibre. In the presence of very high gain solitons can be produced (Gouveia-Neto *et al.*, 1987) and in the limit of weak gain soliton propagation has been observed over many thousands of kilometres (Mollenauer and Smith, 1988). The perturbative results obtained so far also apply to gain by changing the sign of Γ and predict the shortening of the soliton pulsewidth as it propagates. We found numerically that these predictions break down at large distances ($\Gamma L \gg 1$) for the case of loss so what happens when we include gain? In Figure 4.10 we show the evolution of the pulse FWHM for the same conditions as in Figure 4.8. Far from the perturbative result becoming less accurate at large distances it becomes increasingly more

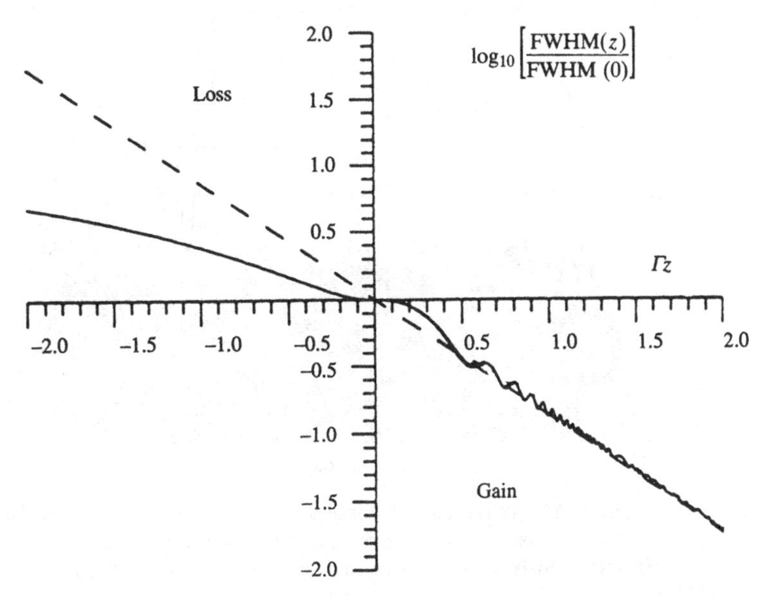

Figure 4.10. Comparison between the evolution of pulsewidth with loss (negative z) and gain (positive z). The quantity plotted is the logarithm of the pulse FWHM normalised to its initial value.

accurate at large distances. We have also plotted the equivalent calculation for loss by extending Figure 4.10 to negative values of z. The contrast between the behaviour is very marked. We can see why this is so by examining the scaling properties of solitons. We would anticipate that the perturbative result will break down when the absorption or gain length is smaller than the soliton period i.e.

$$|\Gamma| < \frac{\pi}{2}\frac{\tau^2}{k_2} \tag{4.39}$$

Since the pulsewidth τ is decreasing with propagation for gain this inequality is increasingly well satisfied with propagation. It turns out (Blow *et al.*, 1988) that the solitons are now stabilised by the gain as can be seen quite dramatically in Figure 4.11 where we show the evolution of noise. Eventually the evolution is dominated by a soliton which can have over 90% of the total energy.

4.6 Non-linear fibre devices and soliton switching

In this section we will consider the application of solitons in device configurations. In Section 4.1 we showed that the Kerr non-linearity led to self-phase modulation, equation (4.9). If we consider a

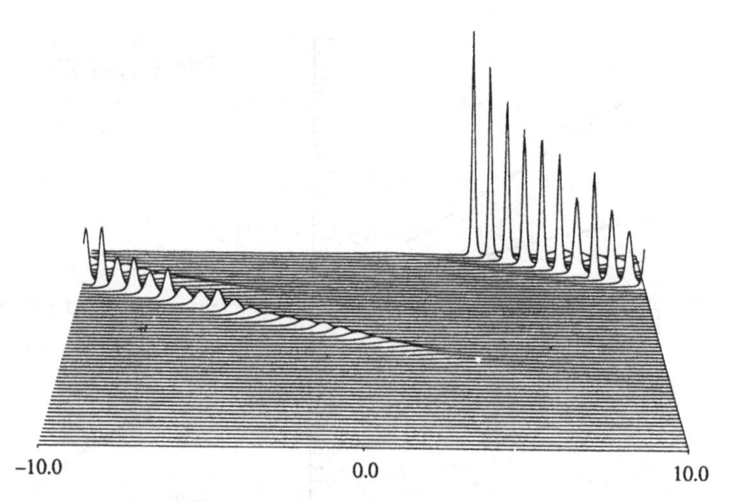

Figure 4.11. When gain is applied to this, an initial state consisting solely of noise, we see the production of solitons occurring spontaneously as a result of their strongly attracting nature.

wave of constant amplitude then we can express the non-linear phase shift in terms of the physical parameters as follows

$$\phi = k_0 n_2 L I \tag{4.40}$$

This phase shift is proportional to the length (L), intensity (I) and non-linear coefficient (n_2). In any two-mode configuration, where a coherent superposition of the fields occurs, a phase-sensitive non-linear response can be obtained if any of these three variables are different for the two modes involved. This is the principle of operation of all 'weakly' non-linear fibre devices.

These devices are often divided into two categories. Those with distributed coupling (e.g. non-linear couplers) and those with discrete coupling (e.g. Mach–Zehnder interferometers). The distinction between these types is somewhat artificial and as shown by Doran (1988) the critical switching observed in non-linear couplers can be obtained from viewing them as a concatenation of non-linear interferometers. The point is that we shall use interferometers as our example but that the role played by solitons is exactly the same in all of the devices in this general category and even the quantitative behaviour can be understood from a simple general model of the 'soliton phase'.

We begin with the all-fibre Mach–Zehnder interferometer shown in Figure 4.12. If the power splitting ratio of the couplers is α: $(1 - \alpha)$

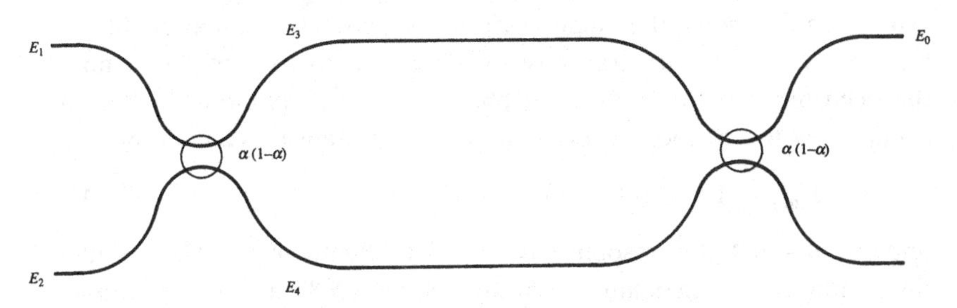

Figure 4.12. Schematic diagram of a fibre Mach–Zehnder interferometer.

then the coupling equations with inputs E_1, E_2 and outputs E_3, E_4 are:

$$E_3 = \alpha^{1/2} E_1 + \mathrm{i}(1 - \alpha)^{1/2} E_2 \tag{4.41a}$$

$$E_4 = \mathrm{i}(1 - \alpha)^{1/2} E_1 + \alpha^{1/2} E_2 \tag{4.41b}$$

When $\alpha = 0.5$ and the two arms of the interferometer are identical the device is symmetric and there will be no non-linear response. The simplest (conceptually at least) way of breaking this symmetry is to have fibres with different core areas in the two arms. This would lead to different intensities in the two arms even though the power would be the same. The output E_O with input $E_1 = E_{IN}$ as indicated in Figure 4.12 is then calculated as

$$|E_O|^2 = \frac{|E_{IN}|^2}{2}\left(1 - \cos\left[n_2 k_0 L |E_{IN}|^2\left(\frac{1}{A_1} - \frac{1}{A_2}\right)\right]\right) \tag{4.42}$$

where the A_i are the effective mode field areas in the two arms. Equation (4.42) can be written in a simplified form

$$|E_O|^2 = \frac{|E_{IN}|^2}{2}(1 - \cos\Delta\phi) \tag{4.43}$$

where $\Delta\phi$ is just the difference in the non-linear phases as given by equation (4.40). This is just the standard sinusoidal response of an interferometric device but with an intensity-dependent phase shift $\Delta\phi$.

The most practically straightforward way of breaking the symmetry in a Mach–Zehnder is to make the couplers unbalanced ($\alpha \neq \frac{1}{2}$) but leave the two arms the same. In this situation the non-linear response is

$$|E_O|^2 = |E_{IN}|^2(1 - 2\alpha(1 - \alpha)(1 + \cos[(1 - 2\alpha)n_2 k_0 L |E_{IN}|^2])) \tag{4.44}$$

This formula shows the same periodic behaviour as equation (4.43) but now the minimum transmission does not go to zero. Also note the occurrence of the difference phase factor in equation (4.44) in a similar way to equation (4.43). The minimum transmission is now

$$T_{\min} = 1 - 4\alpha(1 - \alpha) \tag{4.45}$$

and for $\alpha = 0.4$ this response is shown in Figure 4.13. The straight lines indicate the maximum transmission of 100% and the minimum transmission of 4%.

In standard silica fibres we can calculate the power required to induce a π phase shift from equation (4.40). Taking a mode area of 50 μm^2 and a wavelength of 1.5 μm gives

$$\phi = 2.7 \times PL \tag{4.46}$$

Where P is in kW and L is in metres so that about 1 kW is required for a single cycle in a 1 m long interferometer (note it is actually a phase difference which is needed so that, in practice, the power would be some multiple of this number depending on the size of α). This very high power can only be significantly reduced if the device length is increased. Note that for other devices in this category this principle still applies. Namely that a phase shift of order π as in equation (4.46) is required; the length in distributed couplers is often determined by the appropriate linear beat length and thus the opportunities for reducing the power by increasing L are reduced.

Long-phase-dependent devices are susceptible to environmental fluctuations and would be difficult to fabricate with exactly matched lengths in the arms. In order to reduce the power requirements we would need to increase the length of the arms which would lead to stability problems in the Mach–Zehnder configuration. The all-fibre Sagnac interferometer (Otsuka, 1983; Doran and Wood, 1988), illustrated in Figure 4.14, does not suffer from this effect. This is a four-port coupler with the two output ports joined together. In effect this is the same as the Mach–Zehnder previously described but now the two paths are the two routes around the loop (clockwise or anticlockwise) and the two couplers are replaced by one which is traversed twice. The symmetry can be broken in the same way as for the Mach–Zehnder interferometer; by selecting an unequal power splitting coupler. Thus the non-linear response is as given in equation (4.44) where the field E_O corresponds to the field transmitted. The lower power response (equation (4.45)) shows that linearly the configuration acts as a mirror with a reflectivity determined by α (e.g.

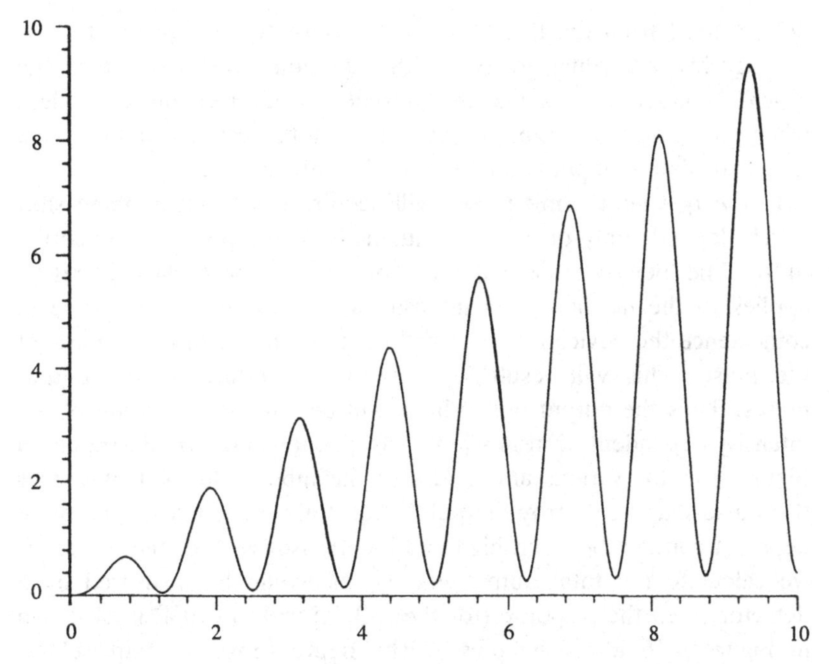

Figure 4.13. Intensity response of an unbalanced Mach–Zehnder.

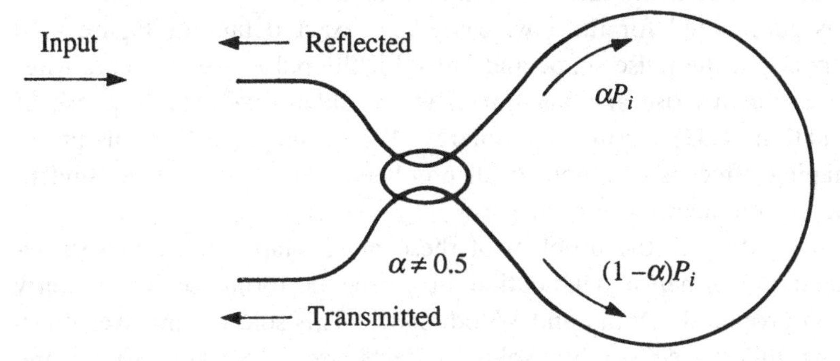

Figure 4.14. Schematic diagram of a fibre Sagnac interferometer.

for $\alpha = 0.4$ the reflectivity is 94%). For this reason the device is often called the non-linear optical loop mirror (NOLM).

The ultimate use of non-linear devices is to perform ultrafast all-optical processing and since the non-linear response in silica is of the order of femtoseconds or less silica-based devices seem to offer this possibility. We will therefore discuss the response of the devices to short pulses and take the non-linearity itself as instantaneous and include dispersion. The numerical problem is essentially the same as before apart from the fact that we now have two independent fields propagating according to the NLS equation. We noted that the Mach–Zehnder and Sagnac interferometers are formally equivalent when the pulses are short compared with the cavity length so the same calculation applies equally to both configurations.

In the quasi-c.w. limit pulses will acquire a non-linear phase shift which depends only on the local intensity of the pulse, see equation (4.9). The device response, equations (4.43) or (4.44), therefore applies to the instantaneous intensity and since the pulses arrive in coincidence the device response will be different for different parts of the pulse. This will result in pulse-shaping effects in the output pulses. Thus the output pulse shape will be altered as a result of the intensity-dependent transmission; this phenomenon is illustrated in Figure 4.15 for various amplitudes of the input field. Not only does this pulse-shaping destroy the pulses but it also seriously degrades the depth of contrast between high and low transmission of the device. If we calculate the total output energy, as would be measured by a detector, then the response (for the NOLM with $\alpha = 0.4$) is as shown in Figure 4.16 for sech^2 pulses. The figure shows a 'staircase' response for the pulse energy and results in a device with almost no contrast. The maximum transmission no longer reaches 100% which was guaranteed for the c.w. case. The exact details of Figure 4.16 depend on the pulse shape and indeed if the pulses were 'square' (i.e. instantaneous rise and fall times) we would recover the response of equation (4.42) for the total energy. We should note that this pulse-shaping effect is common to all non-linear devices which respond to the instantaneous intensity.

A solution to the problem of these pulse shape effects and subsequent performance degradation in device performance has recently been proposed (Doran and Wood, 1987). This solution involves operating the device so that soliton effects are important. So far the device response has included only the effect of self-phase modulation, ignoring the effects of dispersion necessary to observe solitons. Dispersion will be significant when the propagation lengths in the device are comparable with the dispersion lengths and this can be achieved by using short pulses and long devices.

Let us now consider the response of phase sensitive devices to solitons. Recall from equation (4.12) that the single soliton solution is

$$u(z, t) = \exp(\mathrm{i}z/2)\,\mathrm{sech}(t) \tag{4.47}$$

this shows that there is a phase proportional to the length which

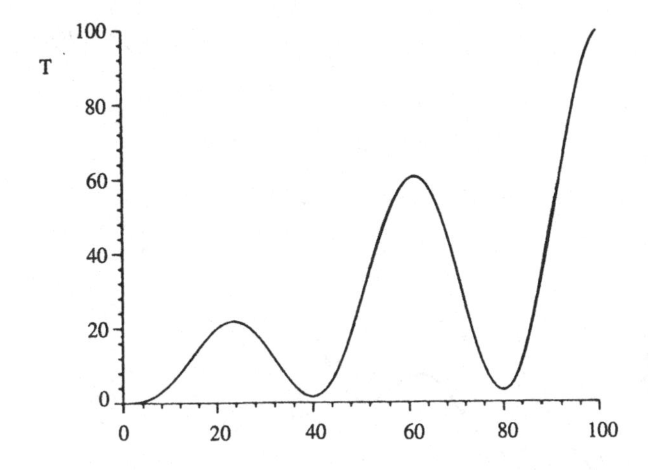

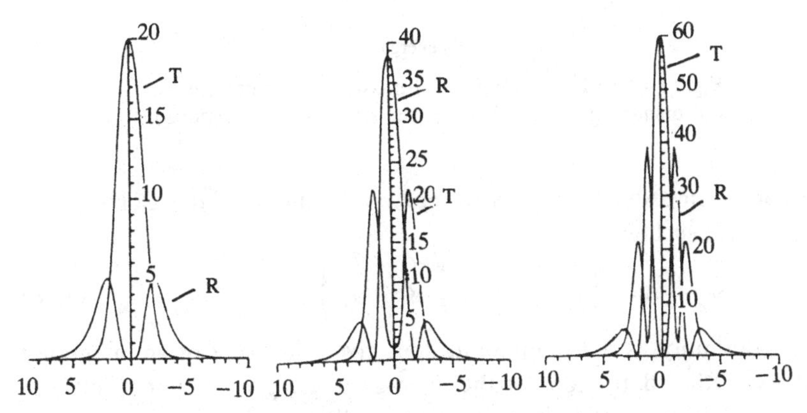

Figure 4.15. Pulse shaping occurs when a 'non-square' pulse is transmitted through a non-linear interferometer as illustrated for various values of the input intensity.

applies to the whole pulse. This is a remarkable property indicating the integrity of the soliton. What we require is such a uniform phase which applies not just to the exact single soliton but to a general pulse in the soliton regime.

Consider the general pulse $A\,\text{sech}\,(t)$, we need to know the response of the device as a function of A with the expectation that the existence of solitons will result in a uniform phase which is not only a function of distance but also of A. We can calculate numerically the propagation of such an initial pulse. Given this solution we can then calculate an average phase for this general pulse as a function of

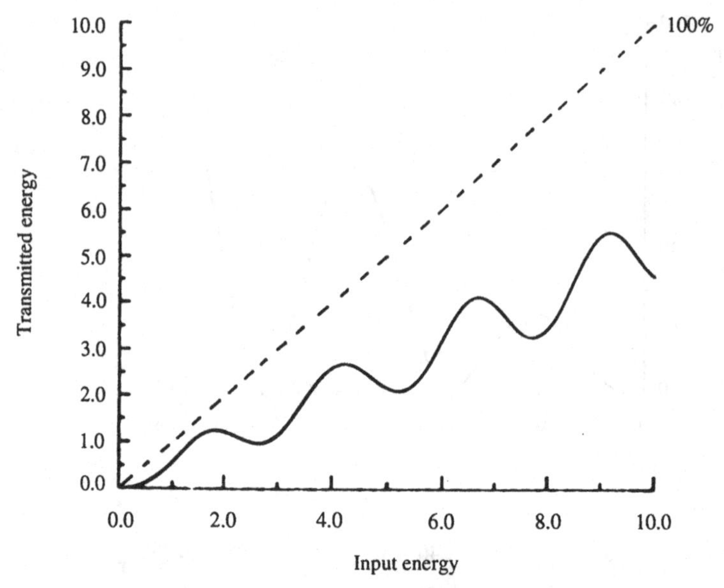

Figure 4.16. Total energy transmission in the presence of pulse-shaping effects showing the staircase-like response.

distance. In order to do this we need a definition of the global phase of a pulse such as

$$\Phi = \tan^{-1}\left(\int \operatorname{Im} u|u|^2 \Big/ \int \operatorname{Re} u|u|^2\right) \tag{4.48}$$

This definition is not unique but does give a measure of an integrated phase. Other definitions of the average phase give similar results. In Figure 4.17 we plot Φ as a function of distance for various values of A which is obtained from the numerical integration of the NLS. The linearity of the phase as a function of distance is remarkable as is the clear dependence on A.

In order to understand this phase property in more detail we will now construct a simple model. We assume that the soliton features dominate and that the uniform phase is determined by the largest eigenvalue of the Zakharov and Shabat (1972) scattering problem, see equation (4.15). Recalling equation (4.18) for the eigenvalues of $A\,\operatorname{sech}(t)$ and that the motion of the solitons is determined by exponentials of twice the eigenvalue squared (Zakharov and Shabat, 1972) then we assert that the 'soliton phase' of the pulse $A\,\operatorname{sech}(t)$ is

$$\begin{aligned} \phi_s &= 2(A - \tfrac{1}{2})^2 z \qquad & A > 0.5 \\ \phi_s &= 0 & A < 0.5 \end{aligned} \tag{4.49}$$

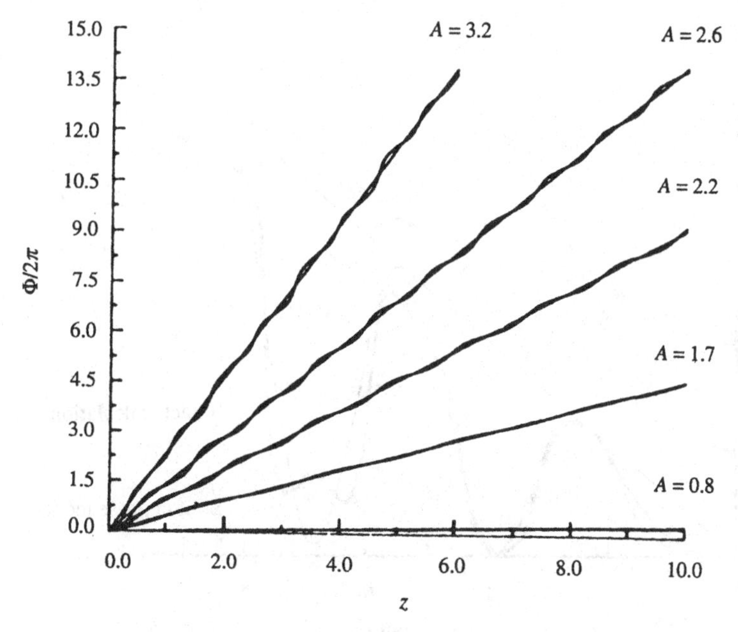

Figure 4.17. Average phase evolution for soliton pulses.

This is the linear phase plotted in Figure 4.17 at each value of A. The agreement between the model and the exact calculation is excellent. It is indeed remarkable that such a simple model is so accurately followed and further illustrates the power of the scattering transform concept. The 'soliton phase' is a general concept and can readily be applied to any of the 'weakly non-linear' devices. We shall continue with non-linear loop mirrors here but we should note that this model for example predicts the soliton equivalent of critical power in non-linear couplers with remarkable accuracy (Doran, 1988). In passing we should note that equation (4.46) is, of course, exact for the single soliton case $A = 1$.

Using equations (4.49) and (4.43) it is straightforward to derive simple response functions for the soliton interferometers. The only thing to note is that we must integrate over the pulse. The energy transmitted u_{T} of a NOLM for $A\,\mathrm{sech}\,(t)$ pulses is now quite easy to derive using the soliton phase model and is

$$u_{\mathrm{T}} = u_{\mathrm{IN}}(1 - 2\alpha(1-\alpha)(1 + \cos 2z((1-2\alpha)A^2 + (\alpha^{1/2} - (1-\alpha)^{1/2})A))) \tag{4.50}$$

In Figure 4.18 we take as an example a loop of length 2π in soliton units and display the results of an exact NLS integration for varying

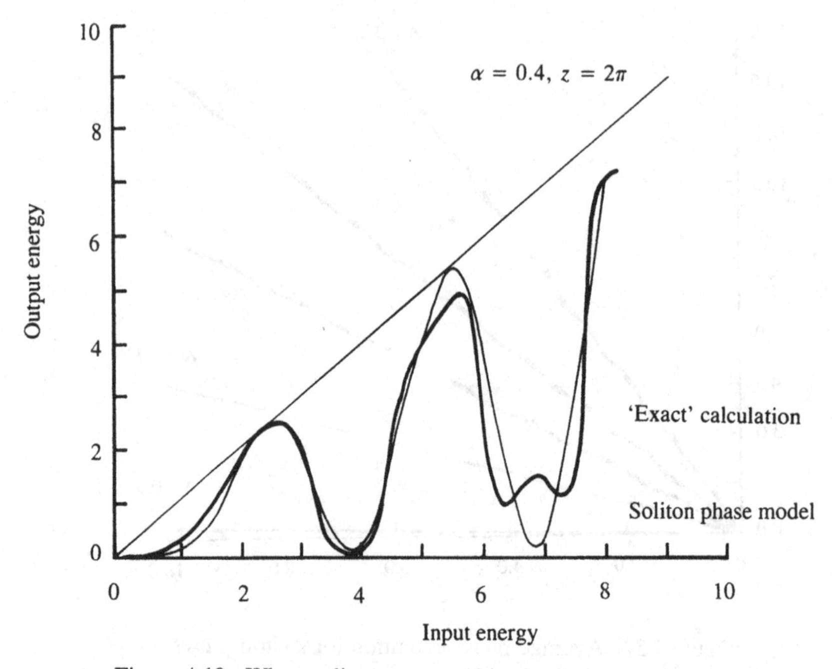

Figure 4.18. When solitons are used in the loop the response contains high contrast peaks and troughs. Also shown, for comparison, is the prediction of the soliton phase model.

input pulse energy compared with the soliton phase model (4.49). The comparison shows how reliable the simple model is in this parameter regime. Indeed the only variations occur for large launched energies where the pulse-shaping effects of the NLS equation, which we have ignored, are important. In Figure 4.19 we compare the soliton with c.w. (or square pulse) and non-soliton response of the NOLM. The restoration of the deep contrast, c.w. response obtained from incorporating soliton effects is clearly visible. These results, and those of recent experiments (Blow *et al.*, 1989), give strong support to our contention that solitons are the natural bits for all-optical processing.

4.7 Conclusion

In this chapter we have discussed the application of numerical techniques to the solution of non-linear optical pulse propagation in optical fibres. We have concentrated on the use of physical arguments and the generation of simple models to interpret the numerical data.

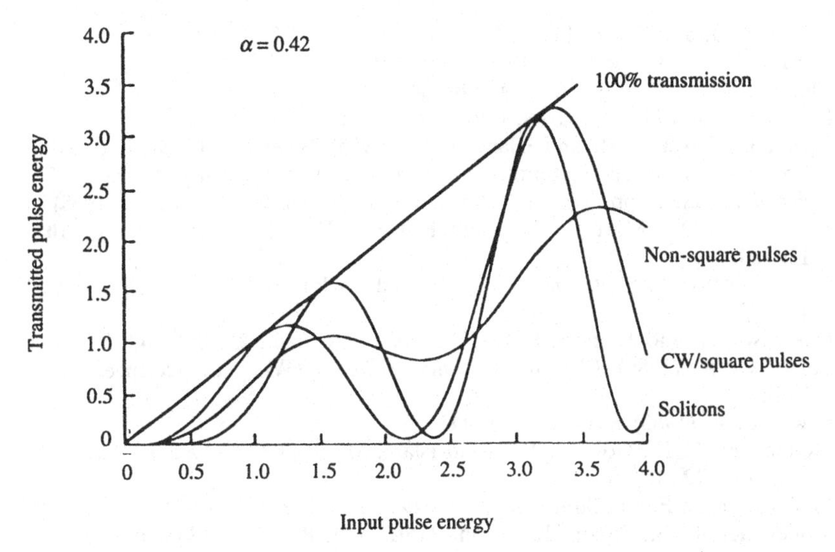

Figure 4.19. The response functions of the loop for CW/square, soliton and non-soliton operation.

Using these techniques we have been able to reach a complete understanding of the physics involved in a number of different problems and also to generate new concepts, such as the 'soliton phase', which will enable device design without the constant need for intensive numerical analysis.

References

Blow, K. J. and Doran, N. J. (1982) *Opt. Commun.*, **42**, 403.

Blow, K. J. and Doran, N. J. (1983) *Electron. Lett.*, **19**, 429–30.

Blow, K. J. and Doran, N. J. (1985) *Opt. Commun.*, **52**, 367–70.

Blow, K. J., Doran, N. J. and Nayar, B. K. (1989) *Opt. Lett.*, **14**, 754.

Blow, K. J., Doran, N. J. and Wood, D. (1988) *J. Opt. Soc. Am. B,* **5**, 381–90.

Blow, K. J. and Wood, D. (1986a) *Opt. Commun.*, **58**, 349–54.

Blow, K. J. and Wood, D. (1986b) Stability and Compression of Pulses in the Soliton Laser, *IEEE J. Quantum Electron.*, **22** 1109.

Blow, K. J. and Wood, D. (1989) *IEEE J. Quantum Electron.* (Special Issue on Ultrafast Phenomena), **25**, 2665–73.

Desem, C. and Chu, P. L. (1986), *Opt. Lett.*, **11**, 248–50.

Dodd, R. K., Eilbeck, J. C., Gibbon, J. D. and Morris, H. C. (1982) *Solitons and Nonlinear Wave Equations* (Academic Press: New York)

Doran, N. J. (1988) Solitons in Optical Fibres and Nonlinear Fibre Devices, Paper WA6 IQEC

Doran, N. J. and Blow, K. J. (1983) *IEEE J. Quantum Electron.*, **19**, 1883–8.

Doran, N. J. and Wood, D. (1987) *J. Opt. Soc. Am. B,* **4**, 1843–6.
Doran, N. J. and Wood, D. (1988) *Opt. Lett.,* **13**, 56–8.
Elgin, J. N. (1985) *Phys. Lett. A,* **110**, 441.
Feit, M. D. and Fleck, Jr J. A. (1978) *Appl. Opt.,* **17**, 3390–8.
Fermi, E., Pasta, J. R. and Ulam, S. M. (1955) Technical Report LA-1940, Los Alamos Science Laboratories. Reproduced in *Collected Works of E. Fermi* volume 2, pp. 978–88. (University of Chicago Press: Chicago, 1965).
Fleck Jr, J. A., Morris, J. R. and Feit, M. D. (1976) *Appl. Phys.,* **10**, 129–60.
Gouveia-Neto, A. S., Gomes, A. S. L. and Taylor, J. R. (1987) *Opt. Lett.,* **12**, 1035–7.
Hasegawa, A. and Tappert, F. (1973) *Appl. Phys. Lett.,* **23**, 142–44.
Lamb, G. R. (1980) *Elements of Soliton Theory* (Wiley Interscience: New York).
Lewis, Z. V. (1985) *Phys. Lett. A,* **112**, 99.
Mollenauer, L. F., Gordon, J. P. and Islam, M. N. (1986) *IEEE J. Quantum Electron.,* **22**, 157–73.
Mollenauer, L. F. and Smith, K. (1988) *Opt. Lett.,* **13**, 675–7.
Mollenauer, L. F., Stolen, R. H. and Gordon, J. P. (1980) *Phys. Rev. Lett.,* **45**, 1095–8.
Otsuka, K. (1983) *Opt. Lett.,* **8**, 471.
Satsuma, J. and Yajima, N. (1974) *Progr. Theor. Phys. Suppl.,* **55**, 284–306.
Zabusky, N. J. and Kruskal, M. D. (1965) *Phys. Rev. Lett.,* **15**, 240–3.
Zakharov, V. E. and Shabat, A. B. (1972) *Sov. Phys. – JETP,* **34**, 62–9.
Zakharov, V. E. and Shabat, A. B. (1973) *Sov. Phys. – JETP,* **37**, 823.

5

Soliton–soliton interactions

C. DESEM AND P. L. CHU

5.1 Introduction

In a communication system, it is desirable to launch the pulses close to each other so as to increase the information-carrying capacity of the fibre. But the overlap of the closely spaced solitons can lead to mutual interactions and therefore to serious performance degradation of the soliton transmission system. This has been pointed out independently by three groups (Chu and Desem, 1983a; Blow and Doran, 1983; Gordon, 1983).

Karpman and Solov'ev (1981) first considered the two-soliton interaction in their study of the non-linear Schrödinger equation (NLS) by means of single-soliton perturbation theory. Although they did not have optical fibre transmission in mind their results are applicable to fibres since the soliton propagation in optical fibres are described by the same equation. However, their method is restricted to large soliton separation only.

Our numerical investigations (Chu and Desem, 1983a) show that soliton interaction can lead to a significant reduction in the transmission rate by as much as ten times. At about the same time, Blow and Doran (1983) showed that the inclusion of fibre loss also leads to dramatic increase of soliton interactions. Gordon (1983) derived the exact solution of two counter-propagating solitons (of nearly equal amplitudes and velocities) and analysed the interaction by obtaining the approximate equations of motion corroborating the results of Karpman and Solov'ev (1981).

A considerable amount of research effort has been spent on the reduction of soliton interactions. For example the use of Gaussian-shaped pulses has been suggested (Chu and Desem, 1983b). The interaction in this case is reduced because of its steep slope but this is achieved at the expense of creating larger oscillatory tails. It has also been shown (Hermansson and Yevick, 1983; Shiojiri and Fujii, 1985; Anderson and Lisak, 1986) that introducing a phase difference between neighbouring solitons can lead to a reduction in the interaction. Incoherent interaction of solitons have also been analysed (Anderson and Lisak, 1985). The third-order dispersion of the fibre can also be used to reduce the mutual interactions (Chu and Desem, 1985a; Chu and Desem, 1987b), but it has been shown that this would result in the break-up of the bound state of the solitons. A more realistic way of reducing the interactions is to launch the adjacent pulses with unequal amplitudes (Chu and Desem, 1985b; Desem and Chu, 1987b). In this case, the solitons form a bound system and effectively maintain their initial pulse separation. The higher order non-linear effect of the fibre is also shown to break up the bound state of the solitons (Kodama and Nozaki, 1987; Kodama and Hasegawa, 1987). A detailed study of the soliton interaction in the presence of fibre loss and periodic amplification have also been presented (Desem and Chu, 1987a) where we point out that additional transmission instabilities will result due to the interactions.

5.2 Two-soliton interactions

5.2.1 *Non-linear Schrödinger equation*

An optical pulse with field envelope $q(z,\tau)$ propagating in an optical fibre whose refractive index varies with the pulse intensity may be described by (Kodama and Hasegawa, 1982)

$$\mathrm{i}\frac{\partial q}{\partial z} \pm \frac{1}{2}\frac{\partial^2 q}{\partial \tau^2} + |q|^2 q = -\mathrm{i}\Gamma q + \mathrm{i}\beta\frac{\partial^3 q}{\partial \tau^3} - \alpha_1\frac{\partial}{\partial \tau}(|q|^2 q) + \alpha_2 q\frac{\partial |q|^2}{\partial \tau} \tag{5.1}$$

where Γ is related to the fibre loss while β and α_1 represent the effect of higher order fibre dispersion and the higher order non-linear effects respectively while α_2 represents the Raman effect resulting in a self-frequency shift (Mitschke and Mollenauer, 1986). The higher order non-linear terms can be ignored unless very short pulses (femtoseconds) are considered.

In the absence of loss and the higher order terms, equation (5.1) becomes

$$\mathrm{i}\frac{\partial q}{\partial z} \pm \frac{1}{2}\frac{\partial^2 q}{\partial \tau^2} + |q|^2 q = 0 \tag{5.2}$$

and is the so-called non-linear Schrödinger (NLS) equation. This equation includes the most important terms of propagation, namely the fibre dispersion and the non-linearity. The sign of the second term depends on the dispersion of the fibre. If the sign is positive, dispersion is anomalous and the NLS equation supports bright solitons and, if it is negative, we have dark solitons. This equation has a single-soliton solution of the following form:

$$q(z,\tau) = \eta \operatorname{sech}[\eta(\tau - \xi z)] \exp[\mathrm{i}(\eta^2 - \xi^2)z/2 - \mathrm{i}\xi\tau] \tag{5.3}$$

with η as the amplitude and ξ the velocity of the soliton. The single soliton has the property that it propagates without change of shape and represents the condition of exact balance between the fibre dispersion and the non-linearity.

To calculate the evolution of pulse propagation for an arbitrary input waveform, the NLS equation needs to be solved numerically. The numerical procedure usually used is either the split-step Fourier transform method (Tappert, 1974) (or the propagating beam method (Yevick and Hermansson, 1983) which essentially involve the same computational steps of solving the linear and non-linear parts of the equation alternatively.

Zakharov and Shabat (1972) have shown that the NLS can be solved using the inverse scattering method. They show (Zakharov and Shabat, 1972, Satsuma and Yajima, 1974) that a given initial pulse, $q(0, \tau)$, evolves to form a number of solitons and a non-soliton part which decays as a dispersive tail. To obtain the steady-state soliton solutions, we first need to solve the Zakharov and Shabat (ZKS) eigenvalue equations (Zakharov and Shabat, 1972):

$$\frac{\delta v}{\delta x} + \mathrm{i}\zeta v = qu \qquad \frac{\delta u}{\delta x} - \mathrm{i}\zeta u = -q^* v \tag{5.4}$$

where $\zeta = \xi/2 + \mathrm{i}\eta/2$ are the complex eigenvalues (where ξ and η represent the velocities and the amplitudes of the solitons), and q is the initial pulse waveform. The variable x has the same meaning as τ in equation (5.2). The eigenvalues, ζ, are solved for eigenfunctions $|u|$, $|v| \to 0$ for $x \to \infty$. In general, for an arbitrary input pulse, $q(0, \tau)$, equation (5.4) needs to be evaluated numerically. Once the eigenvalues are known, the general N-soliton solution (corresponding to N discrete eigenvalues, ζ) can be constructed using an inverse transformation from $2N$ simultaneous algebraic equations (Zakharov and Shabat, 1972) or, as Gordon (1983) has shown, from just N

equations. For example a single-soliton solution, characterised by the eigenvalue $\zeta = \xi/2 + i\eta/2$ is given by equation (5.3). It can be seen that ξ and η represent the velocity and the amplitude of the soliton respectively. The interaction of two solitons is represented by the two-soliton solution of the NLS which corresponds to two discrete eigenvalues, ζ. The eigenvalues, ζ, are related to the amplitudes and velocities of the solitons. If the real part, ξ, of the eigenvalues, i.e. velocities, are equal, the solitons (for $N \geqslant 2$) are said to form a 'bound state'. This means that the solitons undergo periodic oscillations in shape, which are determined by the imaginary part of the eigenvalues (η). These solutions are also called 'breathers'. However, when all the ξs are different, i.e. no two solitons have the same velocity, they no longer form a bound state and the N-soliton solution breaks up into diverging solitons as $z \to \infty$. For larger number of solitons ($N > 2$) the explicit analytic solutions are too complicated to be useful, but the solutions can be obtained numerically.

5.2.2 *Two-soliton interactions*

The interaction between two solitons is represented by the solution of the NLS which coresponds to two discrete eigenvalues $\zeta_{1,2}$:

$$q(z, \tau) = \frac{|\alpha_1| \cosh(a_1 + i\theta_1) \exp(i\phi_2) + |\alpha_2| \cosh(a_2 + i\theta_2) \exp(i\phi_1)}{\alpha_3 \cosh(a_1) \cosh(a_2) - \alpha_4[\cosh(a_1 + a_2) - \cos(\Psi)]} \tag{5.5}$$

where

$$\phi_{1,2} = \left[\frac{(\eta_{1,2}^2 - \xi_{1,2}^2)}{2} z - \tau\xi_{1,2}\right] + (\phi_0)_{1,2}$$

$$\Psi = \phi_2 - \phi_1 + (\theta_2 - \theta_1)$$

$$a_{1,2} = \eta_{1,2}(\tau + z\xi_{1,2}) + (a_0)_{1,2}$$

$$|\alpha_{1,2}| \exp(i\theta_{1,2}) = \pm\left(\left[\frac{1}{\eta_{1,2}} - \frac{2\eta}{(\Delta\xi^2 + \eta^2)}\right] \pm i\frac{2\Delta\xi}{(\Delta\xi^2 + \eta^2)}\right)$$

$$\alpha_3 = \frac{1}{\eta_1\eta_2} \qquad \alpha_4 = \frac{2}{\eta^2 + \Delta\xi^2}$$

$$\zeta_{1,2} = \xi_{1,2}/2 + i\eta_{1,2}/2 \qquad \Delta\xi = \xi_2 - \xi_1 \qquad \eta = \eta_1 + \eta_2$$

The eigenvalues, $\zeta_{1,2}$, represent the final amplitudes and velocities of the solitons, $(a_0)_{1,2}$ the position, and $(\phi_0)_{1,2}$, the phase. They are all determined by the initial launching conditions.

This equation is different from that of Gordon (1983) since, in his case, the solution was derived assuming $\zeta_1 = v(1-a) + \mathrm{i}(1+a)$ and $\zeta_2 = -v(1+a) + \mathrm{i}(1-a)$, which represents two counter-propagating solitons with amplitudes $(1 \pm a)$ and a velocity difference of $2v$. However, equation (5.5) is of a more general form as no such assumption is made and it is valid for any combination of eigenvalues. For closely separated solitons the eigenvalues no longer give a true measure of the amplitude of the pulses and therefore we first need to evaluate the eigenvalues before constructing the solution. Thus, we need the more general form of the solution as in equation (5.5).

Of interest to us is an initial pulse waveform,

$$q(0, \tau) = \operatorname{sech}(\tau - \tau_0) + \exp(\mathrm{i}\theta)\operatorname{sech}(\tau + \tau_0) \tag{5.6}$$

which represents the launching of two soliton-like pulses into the fibre. Equation (5.6) will evolve into two solitons described by equation (5.5) and a much smaller non-soliton part which decays like a dispersive tail. The interaction of the two pulses given in equation (5.6) can therefore be analysed through the two-soliton function, equation (5.5). Given the initial separation τ_0 (in units of the effective pulsewidth, (pw), where a unity amplitude soliton has pw = 2), phase difference θ between the two pulses, the eigenvalues $\zeta_{1,2}$, a_0 and ϕ_0 can be evaluated by solving the ZKS eigenvalue equations (Zakharov and Shabat, 1972), using equation (5.6) as the initial condition. Substituting the eigenvalues thus obtained into equation (5.5) we then obtain the description of the interaction of the solitons.

Satsuma and Yajima (1974) have considered the initial condition of equation (5.6) and, using perturbation theory, have calculated the eigenvalues of the resulting solitons. They show that for $\theta \neq 0$ the real parts of the eigenvalues will be non-zero and unequal, $\xi_1 \neq \xi_2 \neq 0$. This means that the solitons move away with their respective velocities $\xi_{1,2}$ and eventually separate. For $\theta = 0$ ('in-phase' launching of solitons), $q(0, \tau)$ will evolve to form a bound state of solitons with an oscillation period given by $4\pi/(\eta_2^2 - \eta_1^2)$.

5.2.2.1 Solitons with equal phases Since for $\theta = 0$, $\xi_1 = \xi_2 = 0$ equation (5.5) can be simplified. To approximate the initial pulse waveform in equation (5.6), we set $(\phi_0)_1 = 0$, $(\phi_0)_2 = 0$ in equation (5.5). Dividing top and bottom of that equation by $\cosh(a_1)\cosh(a_2)$ and rearranging

$$q(\tau, z) = Q(\eta_1 \operatorname{sech} \eta_1(\tau + \gamma_0) \exp(i\eta_1^2 z/2) + \eta_2 \operatorname{sech} \eta_2(\tau - \gamma_0) \exp(i\eta_2^2 z/2)) \tag{5.7}$$

where

$$Q = \left(\frac{(\eta_2^2 - \eta_1^2)}{(\eta_1^2 + \eta_2^2) - 2\eta_1\eta_2[\tanh(a_1)\tanh(a_2) - \operatorname{sech}(a_1)\operatorname{sech}(a_2)\cos\Psi]}\right)$$

$$a_{1,2} = \eta_{1,2}(\tau \pm \gamma_0) \qquad \Psi = \frac{(\eta_2^2 - \eta_1^2)z}{2}$$

The right-hand side of equation (5.7) is a symmetric function when $\gamma_0 = 0$ and represents the interaction of two equal amplitude pulses. Before using equation (5.7), however, the eigenvalues $\eta_{1,2}$ need to be calculated.

The exact values of the eigenvalues for the initial pulse of equation (5.6) are obtained by solving the ZKS eigenvalue equations. However, approximate values can be found by equating the first terms of the Taylor series expansions of equation (5.6) to equation (5.7) and also using the conservation laws of the NLS equation (Zakharov and Shabat, 1972). The eigenvalues are then given by

$$\eta_{1,2} = \left(1 + \frac{2\tau_0}{\sinh(2\tau_0)} \pm \operatorname{sech}(\tau_0)\right) \tag{5.8}$$

These values are plotted in Figure 5.1 as a function of the initial separation τ_0, together with the exact values calculated from the ZKS eigenvalue equations. The maximum error incurred by the approximation is less than 2%. Since the initial pulse shape, equation (5.6), evolves into two solitons and a non-soliton part (dispersive tail), we need to estimate the amount of the initial energy that decays into the unwanted non-soliton part. The percentage of initial energy which forms the non-soliton part can be obtained from the exact eigenvalues. Less than 0.1% of initial energy of equation (5.6) is found to decay into the dispersive tail. This indicates that equation (5.6) very nearly satisfies the exact solution. In fact for $\tau_0 = 0$ and $\tau_0 \to \infty$, equation (5.6) is the exact solution at $z = 0$.

5.2.2.2 Soliton interaction and the period of oscillation The two-soliton solution, given in equation (5.7), describes the interaction of two equal amplitude pulses. However, as opposed to the case of a single soliton where the real part of the eigenvalue, η, represents the amplitude of the soliton, the eigenvalues η_1 and η_2 of the two-soliton function no longer represent the amplitudes of the individual pulses.

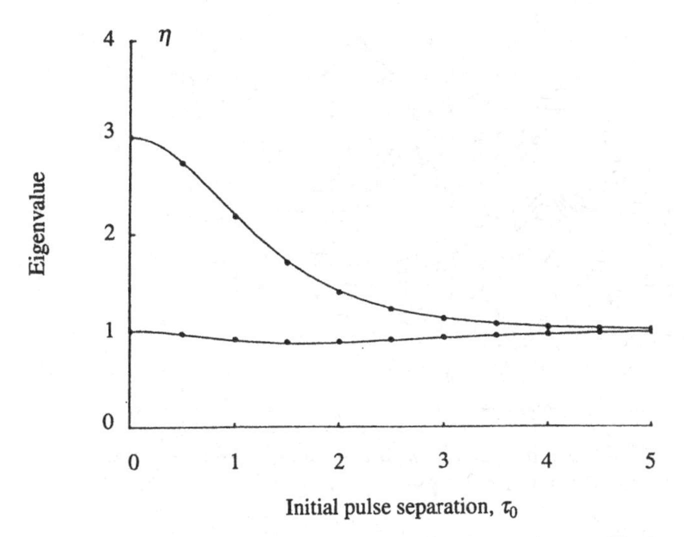

Figure 5.1. Eigenvalues, $\eta_{1,2}$, for two unity amplitude sech pulses as a function of the initial separation, τ_0. The full line is drawn using equation (5.8) and the circles are the exact values.

The two-soliton function, equation (5.7), should therefore be interpreted as two unequal amplitude solitons (η_1, η_2) bound together by the non-linear interaction. This interaction, however, manifests itself as the oscillation of two equal amplitude pulses depending on the initial condition. For example, the same eigenvalues could also represent two unequal amplitude pulses if $\gamma_0 \neq 0$ in equation (5.7). In Section 5.3, it will be shown that as a result of the higher order dispersion of the fibre, the bound state will be broken and the two solitons with amplitudes η_1 and η_2 emerge from the initial waveform of two equal amplitude pulses.

The two pulses described by equation (5.7) undergo an interaction which is periodic in z through the $\cos\Psi$ term in the expression for Q. The period is given by $4\pi/(\eta_2^2 - \eta_1^2)$. Figure 5.2 shows the propagation of two solitons with an initial separation of 3.5 pw. The two pulses initially separated by τ_0 now coalesce into one pulse at $\Psi = \pi$. Then they separate and revert to the initial state with separation τ_0, at $\Psi = 2\pi$ and so on. An approximate expression for the pulse separation as a function of the distance along the fibre can be obtained provided the two pulses are well resolved (Gordon, 1983; Karpman and Solov'ev 1981). Assuming widely separate solitons, the magnitude of equation (5.7) can be approximated by

$$q(\tau, z) = \operatorname{sech}(\tau - \Delta\tau) + \operatorname{sech}(\tau + \Delta\tau) \qquad (5.9)$$

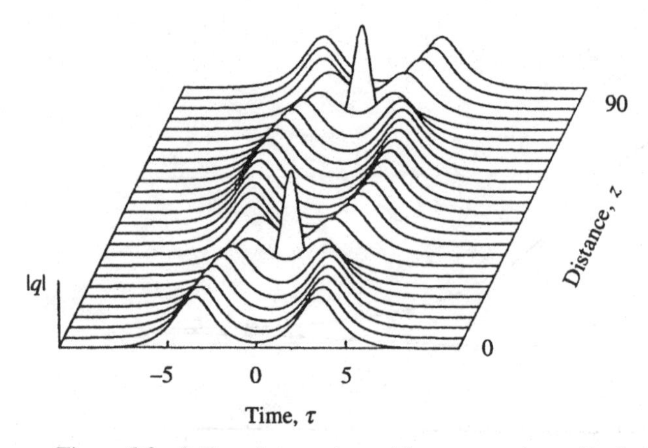

Figure 5.2. Soliton interaction with two equal amplitude pulses. Initial pulse separation = 3.5 pw.

with

$$\Delta\tau = \ln\left(\frac{2}{a}\,|\cos(az)|\right) \qquad a = 2\exp(-\tau_0)$$

and therefore the pulse separation is given by $\Delta\tau$ (Gordon, 1983; Karpman and Solov'ev, 1981). From equation (5.9), the period of oscillation is given by $z_p = \pi\exp(\tau_0)/2$ (Blow and Doran, 1983; Gordon, 1983). However, using equations (5.7) and (5.8) a more accurate expression is obtained where z_p is

$$z_p = \frac{\pi\sinh(2\tau_0)\cosh(\tau_0)}{2\tau_0 + \sinh(2\tau_0)} \tag{5.10}$$

and is plotted in Figure 5.3 together with $z_p = \pi\exp(\tau_0)/2$ and the exact values calculated from the ZKS eigenvalue equations. The period of oscillation given by $z_p = \pi\exp(\tau_0)/2$ is a good approximation for an initial separation $\tau_0 > 2$. Also shown in Figure 5.3 is the period calculated by numerically solving the NLS equation for the initial pulse waveform of equation (5.6). The period is taken as twice the initial coalescence distance of the two pulses. This gives a worst case estimate since the initial waveform takes some time to evolve into the two solitons. However, the good agreement in Figure 5.3 indicates that the initial waveform given by equation (5.6) very closely satisfies the two-soliton solution. As the initial separation is increased, the period of oscillation and therefore the distance of coalescence increases. Since the period will increase exponentially for

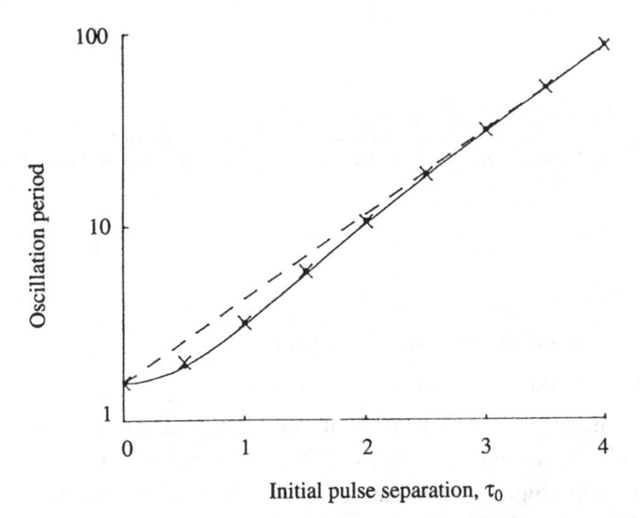

Figure 5.3. Oscillation period, z_p, for equal amplitude pulses as a function of the initial separation, τ_0. The full curve is plotted from equation (5.10). The broken curve is the exponential approximation, the circles are the values using the exact eigenvalues and the crosses are obtained by solving the NLS numerically.

large initial separation, the interaction can effectively be avoided if the solitons are widely separated.

5.2.3 *Solitons with unequal phases, ($\theta \neq 0$)*

Introducing a phase difference θ, as in equation (5.6), between the two pulses results in an unbound state where two separating solitons emerge, moving with different velocities ξ_1 and ξ_2. For $\xi_1 \neq \xi_2$, the two-soliton solution, equation (5.5), shows the interaction of two counter-propagating solitons and the initial waveform (equation (5.6)) represents the instant of collision of these two solitons. In order to observe the evolution of the two solitons, we can approximate equation (5.5) by setting $(a_0)_{1,2} = 0$, $(\phi_0)_1 = \theta_2$ and $(\phi_0)_2 = \theta_1$. Dividing top and bottom of equation (5.5) by $\cosh(a_1)\cosh(a_2)$ and rearranging we obtain

$$q(z, \tau) = Q\Big(r_1\eta_1 \operatorname{sech} a_1 \exp(\mathrm{i}(\phi_1 + \Delta\phi_1)) + r_2\eta_2 \operatorname{sech} a_2 \exp(\mathrm{i}(\phi_2 + \Delta\phi_2))\Big) \tag{5.11}$$

where

$$Q = \left(\frac{[(\eta^2 + \Delta\xi^2)(\eta^2 + \Delta\xi^2 - 4\eta_1\eta_2)]^{1/2}}{(\eta_1^2 + \eta_2^2 + \Delta\xi^2) - 2\eta_1\eta_2(\tanh(a_1)\tanh(a_2) - \operatorname{sech}(a_1)\operatorname{sech}(a_2)\cos\Psi)} \right)$$

$$\phi_{1,2} = \left[\frac{(\eta_{1,2}^2 - \xi_{1,2}^2)}{2} z - \tau\xi_{1,2} \right] + \theta_{2,1} \qquad \Psi = \phi_2 - \phi_2 + (\theta_2 - \theta_1)$$

$$a_{1,2} = \eta_{1,2}(\tau + z\xi_{1,2})$$

$$r_2 \exp(\mathrm{i}\Delta\phi_2) = \cos\theta_1 + \mathrm{i}\sin\theta_1 \tanh(a_1)$$

$$r_1 \exp(\mathrm{i}\Delta\phi_1) = \cos\theta_2 + \mathrm{i}\sin\theta_2 \tanh(a_2)$$

Equation (5.11) represents the evolution of the two solitons and Q represents the interaction between them. The two-soliton solution splits into two separate single solitons as $|z| \to \infty$, where the resultant effect of Q is to shift the relative positions of the solitons, i.e.

$$q(z \to \infty, \tau) = \eta_1 \operatorname{sech}(a_1 + \Delta a_0) \exp(\mathrm{i}(\phi_1 + \Delta\phi_1)) + \eta_2 \operatorname{sech}(a_2 - \Delta a_0) \exp(\mathrm{i}(\phi_2 + \Delta\phi_2)) \quad (5.12)$$

$$\Delta a_0 = \pm \frac{1}{2} \ln \left(\frac{\Delta\eta^2 + \Delta\xi^2}{\xi^2 + \Delta\xi^2} \right) \qquad \text{for } z \to \pm\infty$$

$$\Delta\phi_1 = \pm\theta_2 = \pm\tan^{-1}\left(\frac{-2\eta_2\Delta\xi}{\Delta\xi^2 + \eta\Delta\eta} \right)$$

$$\Delta\phi_2 = \pm\theta_1 = \pm\tan^{-1}\left(\frac{2\eta_1\Delta\xi}{\Delta\xi^2 + \eta\Delta\eta} \right) \qquad \text{for } z \to \pm\infty$$

$$\Delta\xi = (\xi_2 - \xi_1) \qquad \eta = (\eta_1 + \eta_2) \qquad \Delta\eta = (\eta_2 - \eta_1)$$

Equation (5.12) indicates that the two counter-propagating solitons come out from the collision with only a shift in their position and their phase (Zakharov and Shabat 1972, Zakharov and Shabat 1973). However, to analyse the interaction of two solitons with an initial phase difference, we need to use the two-soliton solution, equation (5.11), together with the eigenvalues which correspond to the initial pulse waveform (equation (5.6)). The eigenvalues ($\zeta_{1,2} = \xi_{1,2}/2 + \mathrm{i}\eta_{1,2}/2$), of the resulting solitons can be evaluated by using the first conservation law and equation (5.8),

$$\eta_{1,2} = \left(1 + \frac{2\tau_0 \cos\theta}{\sinh(2\tau_0)} \pm \cos(\theta/2)\operatorname{sech}(\tau_0) \right) \qquad (5.13)$$

$$\xi_{1,2} = \frac{2\sin\theta[1 - \tau_0\coth(2\tau_0)]}{\sinh(2\tau_0)}$$

$$\pm \frac{\sin(\theta/2)}{\sinh(\tau_0)}\left(1 - \frac{2\tau_0}{\sinh(2\tau_0)}\right)$$

The real and imaginary part of the eigenvalues are plotted in Figure 5.4, as a function of the initial phase difference, for three initial pulse separations. Also shown are the exact values obtained solving the ZKS eigenvalues equations. There is very good agreement between the approximate eigenvalues obtained from equation (5.13), and the exact values, especially for $\tau_0 > 2$.

As the phase difference is increased the two solitons will acquire increasing but opposite velocities as seen from the real part of the eigenvalues, ξ, in Figure 5.4, and therefore they will separate faster. For small values of θ, they exhibit an oscillatory interaction while they are slowly separating. For $\theta > 90°$ no oscillation is seen (Anderson and Lisak, 1986). For example, it can be seen that at $\theta = 180°$, the $\cos(\phi_2 - \phi_1)$ term in Q of equation (5.11) which gives rise to the oscillatory behaviour, has an argument of zero ($\phi_2 - \phi_1 = 0$), since $\eta_1 = \eta_2$. For smaller values of θ, however, the separation of the solitons as a result of the finite velocity difference, would counteract the effect of the oscillation which tries to bring the two together. The propagation of two solitons is shown in Figure 5.5 obtained from equation (5.11). The initial phase difference is $\theta = 45°$ and the separation is $\tau_0 = 3.5$ pw. An approximate expression for pulse separation, $\Delta\tau$, as a function of the distance along the fibre can be obtained (Gordon, 1983; Karpman and Solov'ev, 1981):

$$\Delta\tau = \tau_0 + \frac{1}{2}\ln\left(\frac{\cosh 2vz + \cos 2az}{2}\right) \tag{5.14}$$

$$a = 2\exp(-\tau_0)\cos(\theta/2) \qquad v = -2\exp(-\tau_0)\sin(\theta/2)$$

where τ_0 is the initial pulse separation and θ is the phase difference between the pulses.

It can be seen from equation (5.14) that introducing a phase difference, $\theta > 0$, results in the eventual separation of the pulses. Provided that the phase difference is small, the pulses also undergo oscillations before separating. Increasing the phase difference causes the solitons to separate faster. By properly phase-shifting the adjacent pulses the coalescence can therefore be avoided. Hence, this method has been proposed to reduce the soliton interaction (Hermansson and Yevick,

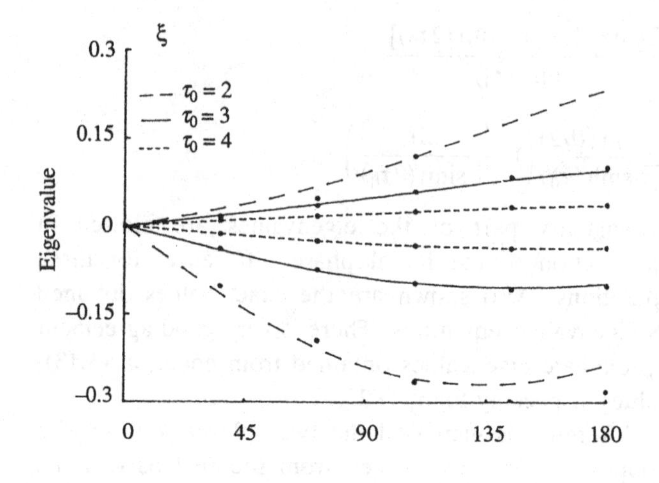

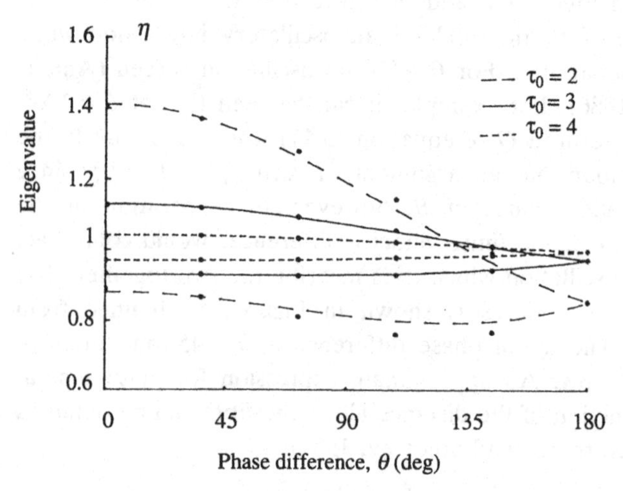

Figure 5.4. Real and imaginary parts of the eigenvalues, ($\zeta = \xi + i\eta$), as a function of the initial phase difference, θ. $\tau_0 = 2$, 3, 4.

1983; Shiojiri and Fujii, 1985; Anderson and Lisak, 1986). However, an undesirable result of this approach is that the solitons will be forced to separate.

5.2.4 *Physical interpretation*

A qualitative physical interpretation of the interaction can be given with reference to the overlap of the tails of the two pulses. For

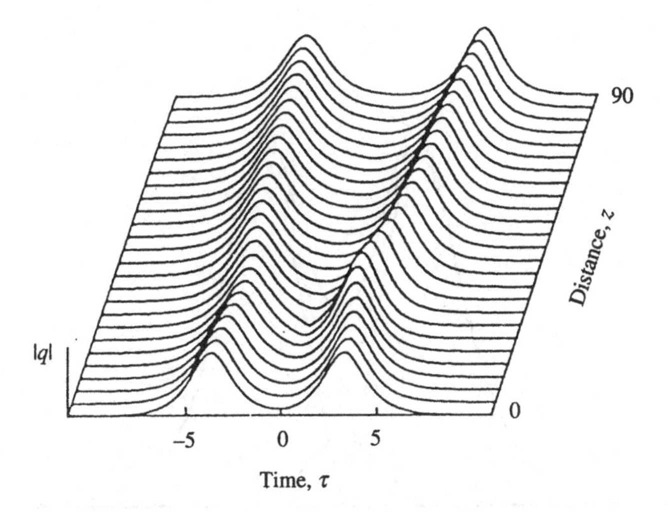

Figure 5.5. Evolution of two unity amplitude sech pulses with an initial phase difference of $\theta = \pi/4$. Initial pulse separation, $\tau_0 = 3.5$ pw.

a single pulse, self-phase modulation results in frequency chirping of the leading and trailing edges, which prevents pulse broadening due to fibre dispersion. The effect of self-phase modulation depends on the slope of the pulse envelope. Consider two sech pulses as shown in curve a of Figure 5.6. In the absence of the adjacent pulse, each propagates without change of shape. If, however, the two pulses are launched together with the same phase, then the resultant pulse waveform is given in curve b of Figure 5.6. It can be seen that the trailing edge of the first pulse and the leading edge of the second pulse are altered in shape. Since the slopes of these edges decrease, the intensity of the frequency chirping which prevents the broadening decreases. This will then result in a gradual broadening of these edges effectively bringing the two pulses closer. The increasing pulse overlap further enhances this effect. Since non-linearity is to confine the pulses, these pulses will then be forced to move together. The overall effect of the pulse overlap is therefore the attraction of the two pulses.

Consider now, two pulses with a phase difference of 180°. The resultant pulse envelope is then given in curve c of Figure 5.6. Since the slopes of the trailing edge of the first pulse and the leading edge of the second pulse are increased in magnitude, frequency chirping is enhanced. Since the enhanced chirping is more than sufficient to

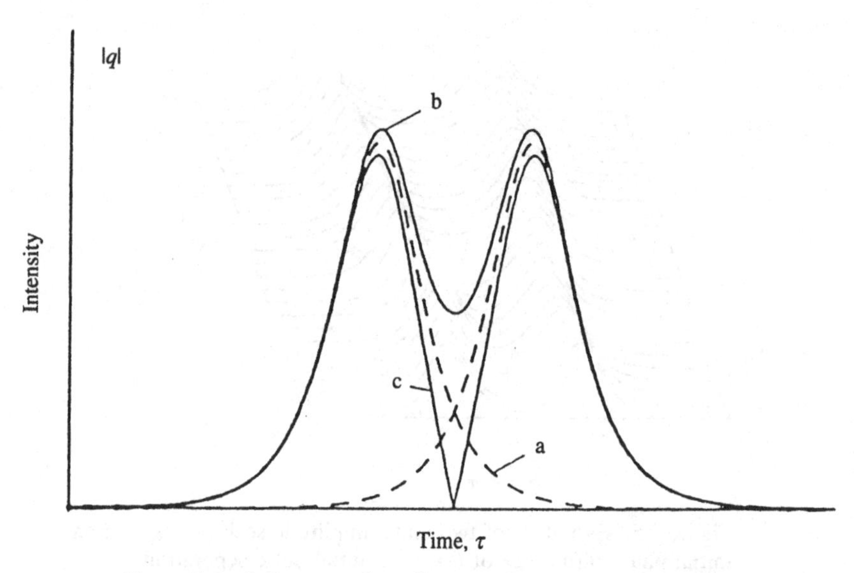

Figure 5.6. Soliton overlap. Curve *a* shows two sech pulses, *b* shows the resultant pulse waveform with two sech pulses added in phase and *c* shows the resultant pulse waveform with the two pulses out of phase by 180°.

balance the broadening due to dispersion, the two edges then separate further as opposed to the previous case, where broadening brought them closer. The two pulses are therefore effectively forced to separate.

Gordon (1983) derived the interaction force of the solitons from the equations of motion and showed that it depends sinusoidally on their phase difference:

$$q'' = -4\exp(-2q)\cos(\theta) \tag{5.15}$$

where q is the pulse separation (which is a function of z), q'' is the second derivative of q with respect to z and therefore represents the interaction force between the solitons and θ is the phase difference.

The force, q'', changes sign for $\theta > \pi/2$ and indicates that it becomes repulsive rather than attractive. This is consistent with the results which show that, for an initial phase difference $\theta > \pi/2$, the pulse separation increases monotonically.

5.2.5 *Solitons with unequal amplitudes*

Launching the adjacent pulses with different amplitudes also leads to reduced mutual interaction. This possibility was first pointed out by Gordon (1983). A single soliton propagates along the fibre with a phase change which is proportional to its amplitude squared and the distance. The phase of a larger amplitude soliton will change at a faster rate. When two solitons are launched with different amplitudes, the phase difference between them will change periodically with the distance. Therefore the force between them will change from that of attraction to that of repulsion periodically. The two pulses may then maintain their relative positions while exhibiting small oscillations.

Since we are interested in launching two unequal amplitude pulses, consider the initial pulse waveform to be

$$q(0, \tau) = \operatorname{sech}(\tau - \tau_0) + A \operatorname{sech} A(\tau + \tau_0) \tag{5.16}$$

which represents two soliton-like pulses with different amplitudes. One pulse with unity width and amplitude and the other with width and amplitude given by A. The two pulses are separated by τ_0 pw. The two pulses will evolve to form a bound state of solitons and a much smaller non-soliton part. To analyse the pulse interaction, the two-soliton function, equation (5.5) is used. Since the two pulses are initially of the same phase, the resultant eigenvalues which represent the solitons are purely imaginary ($\zeta_{1,2} = i\eta_{1,2}$). The simplified two-soliton is given as previously by equation (5.7). The parameter, γ_0, signifies the degree of asymmetry of the pulse waveform. When $\gamma_0 = 0$, equation (5.7) is symmetrical about $\tau = 0$ and describes the interaction of two equal amplitude pulses. As $|\gamma_0|$ is increased from zero, the separation of the two pulses represented by equation (5.7) increases. Their amplitudes and widths approach the values given by η_1 and η_2. However, we also need to take the effect of Q into account as it tends to shift the positions of the pulses. Before using equation (5.7), γ_0 and $\eta_{1,2}$ need to be calculated.

An empirical formula for the eigenvalues of the initial pulse waveform of equation (5.16) can be given as

$$\begin{aligned}\eta_{1,2} = &\left[\frac{A+1}{2} + \frac{2\tau_0 A^{1/2}}{\sinh(2\tau_0 A^{1/2})} + (A^{1/2} - 1)\operatorname{sech}(A\tau_0)\right] \\ &\pm \left[\frac{A-1}{2} + \operatorname{sech}(A\tau_0)\right]\end{aligned} \tag{5.17}$$

where A and τ_0 are as given in equation (5.16).

The eigenvalues are plotted in Figure 5.7 as a function of the initial pulse separation τ_0 (in units of the pulsewidth of the unity amplitude

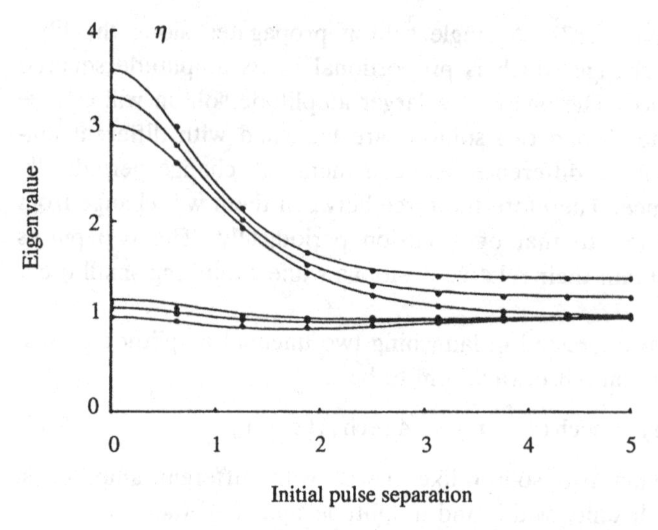

Figure 5.7. Imaginary part of the eigenvalues, $\eta_{1,2}$, for two unequal amplitude sech pulses as a function of the initial separation, τ_0. The full line is drawn using equation (5.17) and the circles are the exact values. The eigenvalues are plotted for $A = 1, 1.2, 1.4$.

pulse), together with the exact values calculated from the ZKS eigenvalue equations. The eigenvalues are shown for $A = 1$, 1.2 and 1.4. There is good agreement, especially for $\tau > 2$. The maximum error in using equation (5.17) is less than 5%. It can be seen that as the initial separation is increased the eigenvalues will be given by $\eta_2 = A$ and $\eta_1 = 1$.

In order to calculate γ_0, the eigenvalues are first evaluated. Then, using the two-soliton solution, equation (5.7), γ_0 is numerically chosen to obtain the best approximation to the initial waveform, equation (5.16). However, an approximate analytic expression can also be found by equating equation (5.7) with equation (5.16) when the pulses are well separated

$$\gamma_0 = \tau_0 - \left(1 - \frac{2\,\mathrm{sech}\,(A\tau_0\,)}{A}\right)\left(\frac{1+A}{2A}\right)\ln\left(\frac{1+A}{A-1}\right) \qquad \tau_0 > 2,\ A > 1 \quad (5.18)$$

This equation is plotted in Figure 5.8 together with the numerical values. There is good agreement between the two results for $A > 1.05$. It can be seen that γ_0 increases as the initial separation and A increase.

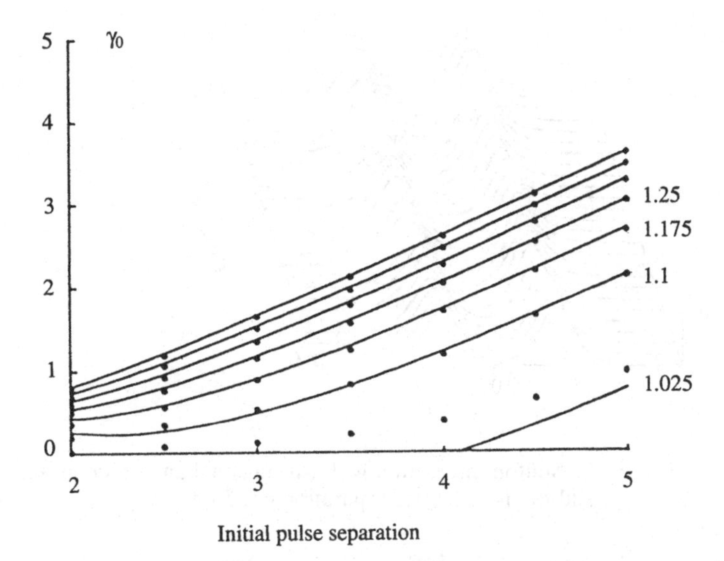

Figure 5.8. Two-soliton function parameter, γ_0, as a function of the initial pulse separation, τ_0, for different initial amplitudes A. The full line is plotted using equation (5.18) and the circles are the exact values.

5.2.5.1 Soliton interaction and pulse separation The interaction between the two solitons is represented by Q in equation (5.7). The effect of Q is to displace the positions of the solitons. At $z = 0$, the two solitons are separated by τ_0 (pulsewidths). The interaction, Q, and therefore the shift in the position of the solitons is a function of distance, z, through the $\cos\Psi$ term. It can be seen that as $|\gamma_0|$ increases by increasing A, the value of the coefficient of $\cos\Psi$ in equation (5.7) is decreased. Therefore, the dependence of Q on the spatial variable z is weakened and the interaction of the solitons is then reduced. Figure 5.9 shows the trajectories of two pulses with $A = 1.1$ and $\tau_0 = 3.5$ obtained from equation (5.7). It can be seen that the interaction is still periodic, but, at no stage do they coalesce to form one pulse as in the previous case. It should also be noted that the pulse with the larger amplitude and narrower width undergoes a larger displacement throughout the interaction. The two pulses are shifted from their initial positions, the larger pulse by a larger amount. It is also found that the pulses are out of phase at the instant of closest separation.

The period of oscillation, z_p, can be calculated by using equations (5.17) and (5.7). As the initial separation is increased, the oscillation

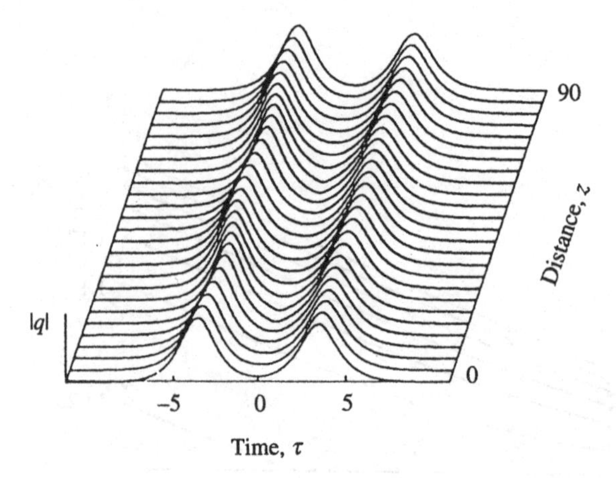

Figure 5.9. Soliton interaction with two unequal amplitude pulses, $A = 1.1$ and the initial pulse separation = 3.5 pw.

period, $z_p = 4\pi/(\eta_2^2 - \eta_1^2)$, reaches a constant value, given by $z_p = 4\pi/(A^2 - 1)$. For widely separated solitons, the oscillation period is therefore simply determined by the amplitude ratio, A. It can also be seen from equation (5.7) that the oscillation period equals the period of the relative phase change of the two solitons confirming that the force experienced by the solitons exhibits a dependence on their relative phase.

Apart from the period of oscillation, we also need to calculate how far the pulses deviate from their respective positions throughout the interaction. An approximate expression for pulse separation, $\Delta\tau$ (in units of the pulsewidth of the unity amplitude pulse), as a function of the distance along the fibre can be obtained (Blow and Doran, 1983; Karpman and Solov'ev, 1981) as

$$\Delta\tau = k \ln\left(\exp(\tau_0/k) + \frac{2}{a^2}\left[\cos 2az - 1\right]\right) \tag{5.19}$$

with

$$a = \left(4\exp(-\tau_0/k) + \left(\frac{A-1}{A+1}\right)^2\right)^{1/2} \qquad k = \frac{A+1}{4A}$$

where τ_0 is the initial pulse separation and A is the amplitude of the larger pulse as given in equation (5.16). In Figure 5.10, the pulse separation is plotted using equation (5.19) for two pulses initially separated by $\tau_0 = 3.5$ and with $A = 1$ (equal amplitude pulses), $A = 1.05$ and $A = 1.1$. As can be seen, pulse coalescence can be avoided if the two pulses are launched with different amplitudes.

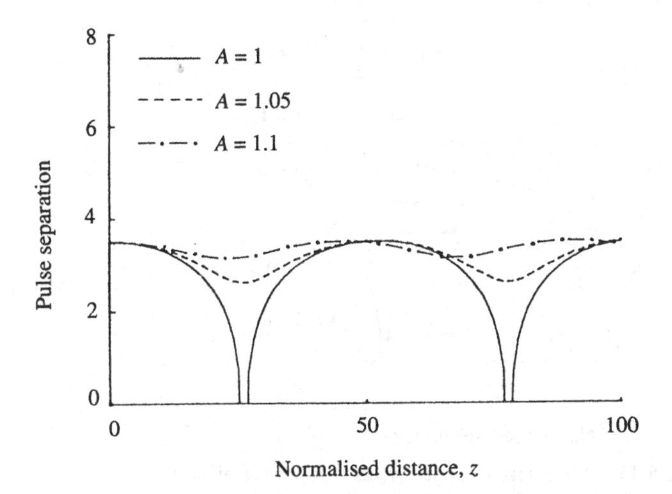

Figure 5.10. Pulse separation, $\Delta\tau$, as a function of normalised distance, z, with $A = 1$ (equal amplitude pulses, full line), $A = 1.05$ (broken line) and $A = 1.1$ (chain line).

However, we need to know the minimum separation of the pulses throughout the interaction. The maximum separation equals the initial separation, τ_0, and the minimum separation can be obtained from equation (5.19) by setting $z = \pi/2a$. Since equation (5.19) is valid for widely separated solitons, we modify the resultant expression to obtain a better approximation for the minimum separation, $\tau_{\min}$, as

$$\tau_{\min} = \tau_0 - \frac{A+1}{4A[1 + 4A\,\text{sech}\,(A\tau_0)]} \ln\left(1 + 4\left(\frac{A+1}{A-1}\right)^2 \exp\left(\frac{-4A\tau_0}{A+1}\right)\right) \tag{5.20}$$

For a given initial separation the ultimate minimum separation approaches the initial value by increasing A. For widely separated pulses only a small value of A is sufficient to prevent pulse wander. But, if the initial separation is smaller, then, a larger amplitude difference is required. The separation between the solitons can therefore be effectively maintained constant by allowing them to have unequal amplitudes. The larger the difference in amplitude the more stable the separation.

Introduction of a phase difference between the two pulses will lead to the eventual separation of the solitons as in the case of equal amplitude solitons. However, the pulse separation will be slower as

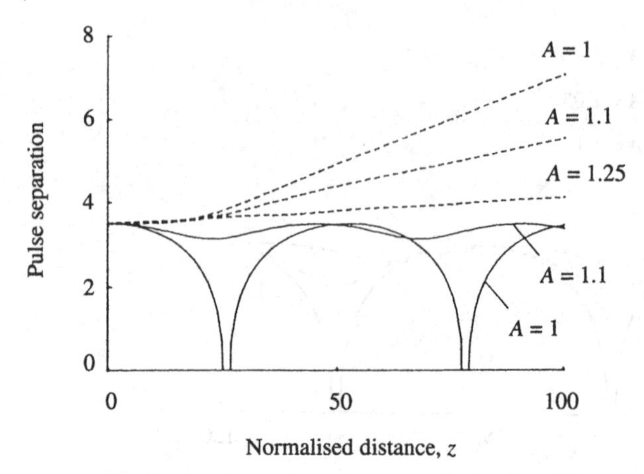

Figure 5.11. Pulse separation, $\Delta\tau$, as a function of normalised distance, z. The broken curves show the pulse separation when the initial phase difference, $\theta = \pi/2$, and the full curves when $\theta = 0$.

the amplitude difference is increased. The pulse separation is plotted in Figure 5.11.

5.3 Effect of higher-order dispersion on solitons

If the fibre is operated close to the zero-dispersion wavelength, the effect of the higher-order dispersion (here we specialise to third-order dispersion as it relates to the third-order derivative of the wavenumber) needs to be taken into account. In this case, an extra term, $i\beta\partial^3 q/\partial\tau^3$, is to be added to the right-hand side of equation (5.2).

5.3.1 *Interaction of equal amplitude pulses of equal phases*

The propagation of two equal amplitude pulses of equal phases is shown in Figures 5.12(a) and (b). Two different initial pulse separations are considered, where in Figure 5.12(b), $\tau_0 = 3.5$ and in (a), $\tau_0 = 5$. The third-order dispersion is $\beta = 0.05$. It can be seen that, when the initial pulse separation is smaller, the pulses approach closer to each other before they separate and also the rate of separation is faster.

As shown in Section 5.2, in the absence of the third-order dispersion ($\beta = 0$), the two pulses coalesce to form one pulse. However, it can be seen from Figure 5.12 that, in the presence of third-order

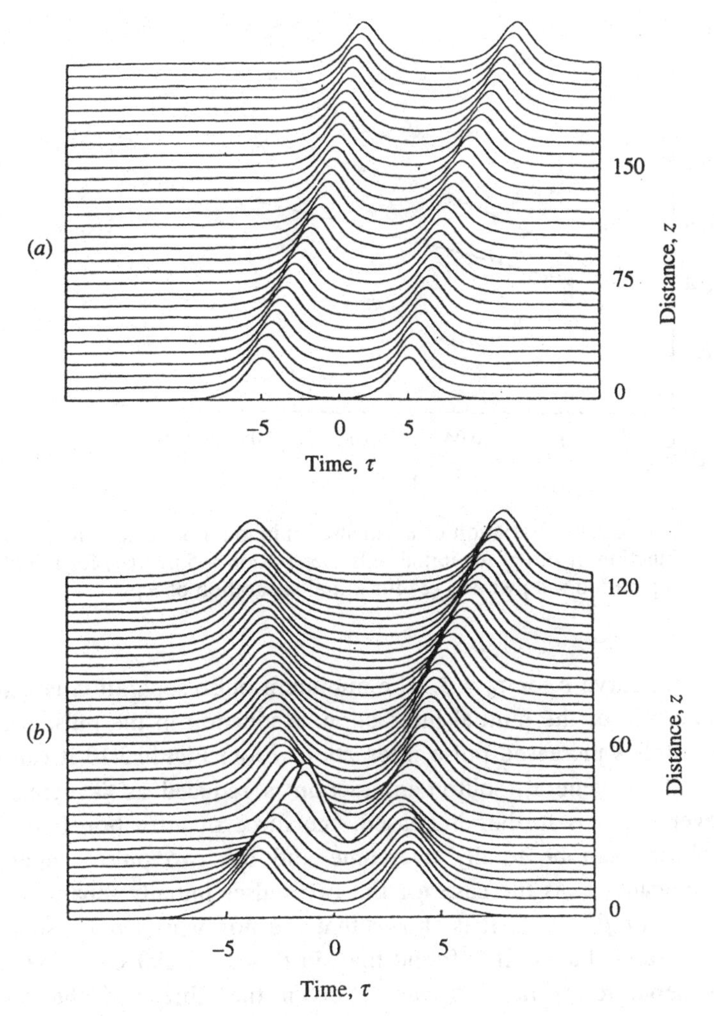

Figure 5.12. Interaction of two equal amplitude pulses in the presence of third-order dispersion, with $\beta = 0.05$: (*a*) initial separation 5 pw; and (*b*) initial separation 3.5 pw.

dispersion, the pulse coalescence is avoided. The two pulses do experience a force of attraction and approach each other, but separate afterwards. The minimum pulse separation occurs approximately at the distance of coalescence if there is no third-order dispersion ($\beta = 0$).

Figure 5.13 shows the minimum pulse separation normalised to the input pulse separation as a function of β. Three different initial pulse separations are considered, where curve *a* is for $\tau_0 = 5$, curve *b* is for

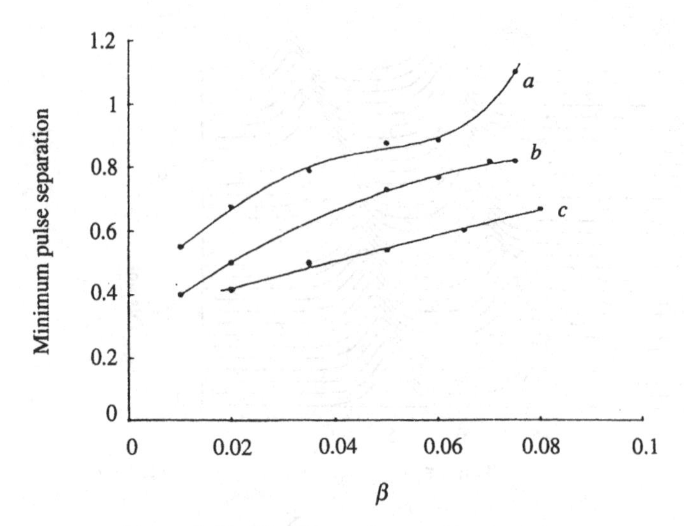

Figure 5.13. Variation of normalised minimum pulse separation as a function of β: (a) for initial pulse separation = 5 pw; (b) for initial separation = 4 pw; and (c) for initial separation of 3 pw.

$\tau_0 = 4$ and curve c for $\tau_0 = 3$. The minimum pulse separation is taken as a measure of the mutual interaction. Reduction in the pulse separation implies increased pulse interaction. From Figure 5.13 it can be seen that, in general, soliton interaction is reduced as β increases. However, as β is further increased, the force of attraction between the solitons can be totally overcome and the two pulses separate monotonically, as is the case for the two pulses initially separated by 5 pw and for $\beta > 0.07$. It is shown that the normalised pulse separation is greater than 1. It is found that, in this particular case, the two pulses separate as they propagate down the fibre and the pulse separation is always greater than the initial separation. Since the minimum separation for lower β occur approximately at the coalescence distance (when $\beta = 0$), the value plotted in Figure 5.13 is the normalised pulse separation at this distance and is greater than one. However, further increase of the third-order dispersion can lead to gross distortion of the pulses.

The mutual interaction of the solitons can therefore be reduced by the presence of the third-order dispersion of the fibre. The reduction depends on the ratio of third- to the second-order dispersion and also on the initial pulsewidth, as seen from β, as well as the initial pulse separation.

Third-order dispersion is a non-symmetric perturbation to the NLS equation and results in the break-up of the bound state of the solitons. The result is that two separating solitons with different velocities will emerge.

In the absence of any perturbation to the NLS equation, ($\beta = 0$ in equation (5.2)) the two-soliton solution for a bound state has been discussed earlier. The solution equation (5.7) corresponds to two discrete eigenvalues $\zeta = i\eta_1/2, i\eta_2/2$ and therefore to two solitons. As indicated previously, the 'equal amplitude' soliton–solution is the result of two unequal amplitude solitons bound together by the nonlinear interaction and when this interaction is broken, these two solitons re-emerge. Third-order dispersion can break up this bound state. In Figure 5.14 we plot the peak amplitudes of two pulses of initially equal amplitudes, propagating in the presence of third-order dispersion. The initial separation of the pulses is 3.5 pw and two values of β are considered, $\beta = 0.05$ (shown in Figure 5.12(*b*)) and $\beta = 0.035$. From Figure 5.14 it can be seen that the two pulses emerge with different amplitudes. The two pulses are of equal amplitudes at $z = 0$. At a distance of $z \cong 26$, which corresponds to the coalescence of the solitons there is a sudden change of the amplitudes as a result of the interaction. The two pulses then emerge with constant amplitudes which are no longer equal.

In the absence of third-order dispersion ($\beta = 0$) the two equal amplitude pulses initially separated by 3.5 pw form a bound state of two solitons with imaginary part of the eigenvalues (and therefore the amplitudes) $\eta_1 = 0.952$ and $\eta_2 = 1.072$. The straight lines in Figure 5.14 denote the values η_1, η_2. It is seen that, the two solitons, predicted by the eigenvalues, will emerge indicating that the bound state of the solitons is broken.

Since periodic boundaries are used in the numerical scheme, the presence of the non-soliton part of the waveform generated due to the third-order dispersion leads to the small fluctuations in the amplitude as seen in Figure 5.14.

As the third-order dispersion is increased, the pulses separate faster. In all of these cases considered, the pulses separate monotonically after the first coalescence distance. However, if the third-order dispersion is small, the pulse interaction cannot be overcome as quickly. Even though the pulses eventually separate, the break-up of solitons occurs at a considerably longer distance than the first coalescence distance.

The break-up of the bound state is the result of the two solitons acquiring different velocities, which can be calculated by evaluating

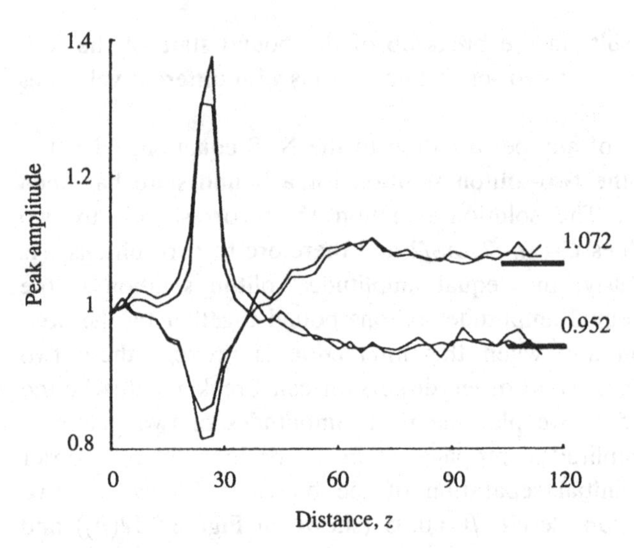

Figure 5.14. Peak amplitude variation of the solitons during propagation. The straight lines denote equal amplitude pulses with initial separation of 3.5 pw.

the eigenvalues of the pulse waveform. Even though the ZKS eigenvalue equations do no strictly apply when considering the perturbed NLS equation, it does give us useful results while the perturbation is small. Taking the pulse waveform as the initial condition to the ZKS equations, the eigenvalues ($\zeta_{1,2} = \xi_{1,2}/2 + i\eta_{1,2}/2$) are numerically calculated as a function of the propagation distance. Figure 5.15 shows the real and imaginary part of the eigenvalues (velocity and amplitude) for the two pulses as they propagate in the presence of the third-order dispersion, for $\beta = 0.035$, 0.05, 0.06, and initial separation of 3.5 pw. It can be seen from the real part of the eigenvalues, ξ (the soliton velocity), that the two solitons acquire opposite velocities. The larger the β is, the larger is the velocity difference, $\xi_2 - \xi_1$.

It is important to realise that the acquisition of the velocity is the result of the mutual interaction of the solitons. If a single soliton is considered, the velocity, ξ, will be zero, but only the position will change (Hasegawa and Kodama, 1981; Kodama and Hasegawa, 1982). The sudden change in the eigenvalues as seen in Figure 5.15 corresponds to the distance of coalescence of the solitons. The imaginary part of the eigenvalue, η (amplitude of the solitons), however, remains fairly constant except for the small change around the coalescence distance. The evaluation of the eigenvalues therefore confirms that the two solitons characterised by η_1 and η_2, which were

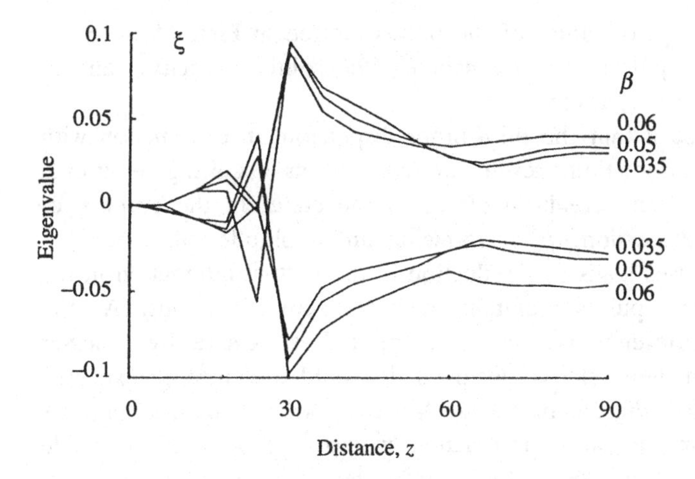

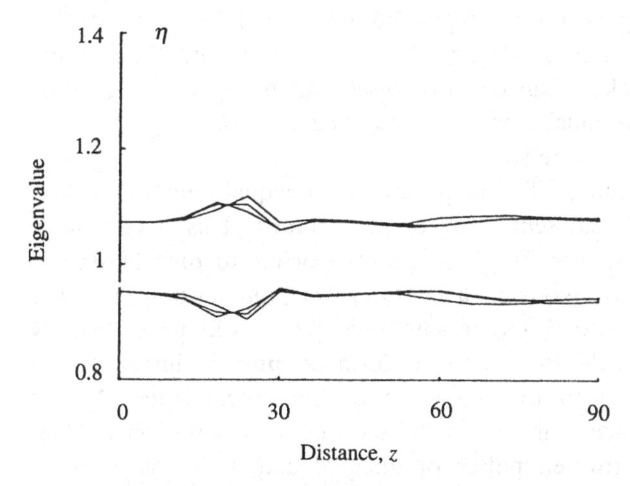

Figure 5.15. Real and imaginary parts of the eigenvalues, $\zeta_{1,2}$ ($\zeta_{1,2} = \xi_{1,2}/2 + i\eta_{1,2}/2$), as a function of distance, z. $\beta = 0.035$, 0.05, 0.06.

initially bound together, acquire different velocities, due to the third-order dispersion, and separate.

Even though we are considering the perturbed NLS equation, the use of the eigenvalues are justified while the perturbation is small, and indeed this approach gives us a good insight into understanding the effect of the third-order dispersion. The eigenvalues can also give a quantitative description of the propagation of the solitons. The imaginary part of the eigenvalues, η, in Figure 5.15 compare quite well with the peak values of the pulses plotted in Figure 5.14. Similarly the real part of the eigenvalues (the soliton velocity) can describe the pulse separation.

We have seen that the third-order dispersion, in conjunction with the mutual interaction, results in the solitons acquiring additional velocities, and the break-up of the bound state. In the absence of third-order dispersion, β, introducing an amplitude difference between the pulses leads to a reduction in the mutual interaction in the sense that the pulses maintain their respective positions. As the amplitude difference is increased, the pulses deviate by a lesser amount from their relative positions. It would therefore be expected that when the adjacent pulses are launched with unequal amplitudes the reduction in soliton interaction would lead to a more stable soliton formation when the third-order dispersion is present. Figure 5.16 shows the propagation of two unequal amplitude pulses, with an initial separation of 3.5 pw and with $A = 1.1$ and $\theta = 0$. The third-order dispersion is taken as $\beta = 0.05$. The bound state of the solitons is still broken. But the two pulses start to separate at $z \simeq 200$ as opposed to the equal amplitude case where $z \simeq 26$. Therefore the soliton formation is more stable.

However, launching adjacent pulses with unequal amplitudes does result in the eventual separation of the solitons. This is because a single pulse is shifted in its position in proportion to $\beta\eta^2$ (Hasegawa and Kodama, 1981; Kodama and Hasegawa, 1982) where η is the amplitude. Since two different amplitude pulses will be shifted by different amounts, the two pulses do not maintain their initial separation, again leading to the break up of the bound state. As the amplitude difference is increased the solitons break up sooner. Consider the two perturbed pulses of unequal amplitudes as given in equation (5.16). For $A = 1.1$, the pulses start to separate at $z \simeq 200$, for $A = 1.2$ this distance is $z \simeq 120$ and for $A = 1.3$, $z \simeq 70$. These results are obtained by solving the NLS equation with $\beta = 0.05$ and for an initial separation of 3.5 pw between the two pulses. The velocity of separation is also larger as A is increased. However, when compared with the equal amplitude case the break-up of the soliton bound state is delayed provided that the amplitude difference is not too large.

Earlier, we have seen that introducing a phase difference, θ, as in equation (5.6), between the pulses results in an unbound state where two separating solitons emerge. For large θ, the solitons experience minimal interaction. The two pulses can therefore be considered as

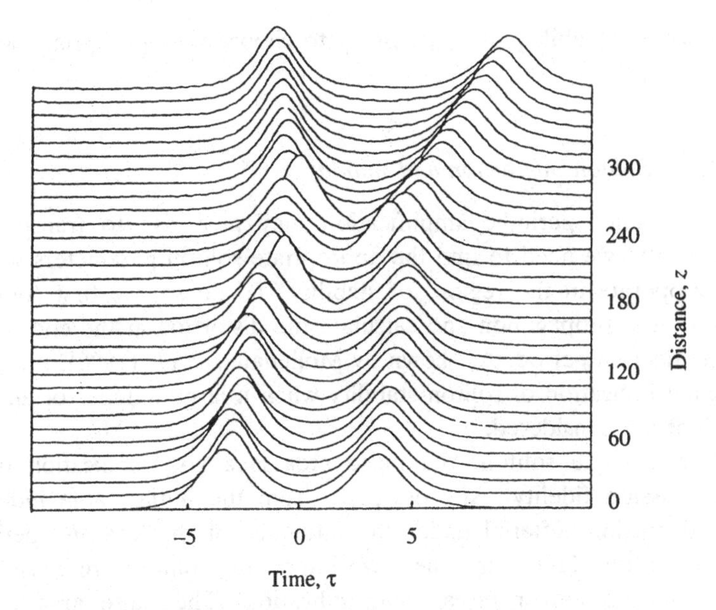

Figure 5.16. Interaction of two unequal amplitude pulses in the presence of third-order dispersion, with $\beta = 0.05$. The initial pulse separation is 3.5 pw and $A = 1.1$.

two single solitons and the third-order dispersion simply shifts their relative positions. With the decrease in the initial phase difference, the increased pulse interaction will give rise to an additional velocity increase.

5.4 Soliton interaction in the presence of periodic amplification

In practice, loss in an optical fibre is unavoidable. The fibre loss gives rise to the rapid broadening of the solitons (Hasegawa and Kodama, 1981; Kaup, 1976). To overcome this degradation, it is necessary to amplify the solitons periodically. Of the various amplification schemes proposed (Kodama and Hasegawa, 1982; Hasegawa and Kodama, 1982; Hasegawa, 1983) the use of Raman gain in the fibre is the most promising (Hasegawa, 1984; Mollenauer *et al.*, 1985; Mollenauer *et al.*, 1986). Using the Raman gain, the fibre loss can be effectively cancelled such that distortionless transmission can be achieved. In this way an all-optical high-bit-rate communication system becomes possible (Mollenauer *et al.* 1986; Mollenauer and Smith, 1988). However, soliton transmission instability may occur due to the periodic amplification (Hasegawa, 1984; Mollenauer *et al.*, 1986). To avoid the instability, it is necessary to choose the appropriate soliton parameters.

5.4.1 *Soliton interaction and stability*

Since periodic amplification may give rise to transmission instability, we need to find the appropriate soliton parameters so that they operate in the regions of stability. To do so, we first compute the soliton propagation and calculate the distortion at the end of one amplification period. As shown by Mollenauer *et al*. (1986), this gives a good indication of soliton stability when a large number of amplifications are considered.

For a single soliton, the pulse area is a good indication of its transmission fidelity. Any deviation from the initial value indicates the distortion suffered under the influence of the loss and periodic amplification. However, when two interacting solitons are considered this will no longer be a valid indication. The pulse area is not conserved even for lossless transmission. We therefore, calculate the distortion with direct reference to the two-soliton solution equation (5.7). To do so, we define a distortion parameter, D, as the measure of the deviation of the soliton shape from that given by equation (5.7):

$$D = \frac{\int_{-\infty}^{\infty} |q_N - q(\tau, z)| \mathrm{d}\tau}{\int_{-\infty}^{\infty} |q(\tau, z)| \mathrm{d}\tau} \tag{5.21}$$

$q(\tau, z)$ is given by equation (5.7) and q_N represents the emerging solitons at the end of one amplification period, calculated by solving the NLS equation with the loss coefficient and the Raman gain included. The value of z for the analytic solution is chosen such that D is minimised. This value will be approximately equal to L, the amplification period.

Figure 5.17(a) shows the distortion that will result when two equal amplitude solitons with an initial separation of 3 pw are launched for different ratio of soliton period to amplification period L/z_0 where $z_0 = \pi\sigma^2/2|k''|$ (Mollenauer *et al*., 1986) with 2σ representing the pulsewidth. Two peaks can be seen, indicating the presence of major instability for $L/z_0 \simeq 10$ and $L/z_0 \simeq 26$. Due to the interaction, the solitons change shape periodically with a period equal to $z = 4\pi/(\eta_2^2 - \eta_1^2)$ as seen from equation (5.7). For the initial condition considered here, this period corresponds to $L/z_0 \simeq 20$. This means that the amplification period L will equal the interaction period for

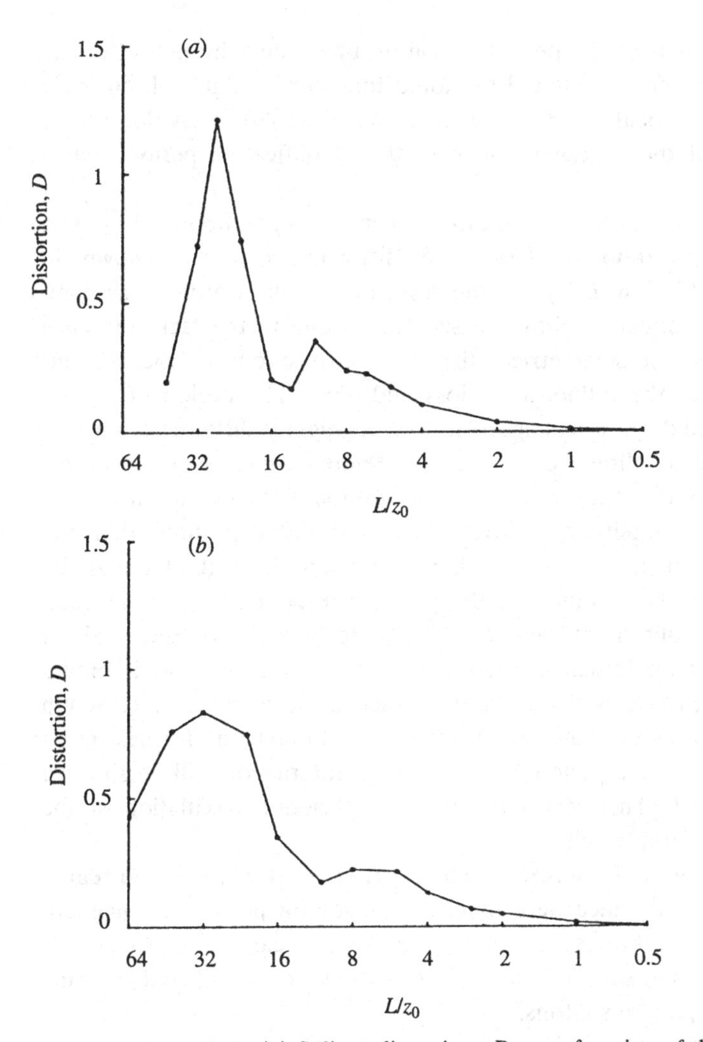

Figure 5.17. (*a*) Soliton distortion, D, as a function of the parameter L/z_0 for two equal amplitude solitons. Initial separation is 3 pw. (*b*) Soliton distortion, D, as a function of the parameter L/z_0 for two unequal amplitude solitons. Initial separation is 3 pw and the amplitude ratio is 1.1.

this value of L/z_0. Therefore the first peak represents the resonance of the period of soliton interaction with the amplification period. However, the overall phase of the solitons changes at a faster rate. This period will roughly correspond to $L/z_0 \simeq 8$, since for widely separated solitons, the two eigenvalues $\eta_1 \simeq \eta_2 \simeq 1$. Therefore, the two instability peaks are due to the resonance of the amplification period with that of the period of soliton phase and the period of the soliton interaction. It should be noted that, for a single soliton, only one instability peak exists (Mollenauer *et al.*, 1986), that due to the resonance of the soliton phase with the amplification period, which occurs at $L/z_0 \simeq 8$.

For unequal amplitude solitons with initial separation of 3 pw and an amplitude ratio of 10%, the distortion curve is shown in Figure 5.17(*b*). For $L/z_0 \gg 8$ the distortion peak is broader as compared to the equal amplitude case. This is due to the fact that since the input is not symmetrical the relative positions of the solitons change under the influence of loss and gain. The period of soliton interaction and phase change will also be slightly different from that of the equal amplitude case since the related eigenvalues are different, and therefore the positions of the instability peaks will differ.

For equal amplitude solitons, it is important to note that the instability peak at $L/z_0 \cong 8$ will be quite independent of the initial separation of the solitons but the peak corresponding to the periodicity of the soliton interactions will be to the left of the peak at $L/z_0 \cong 8$ for an initial separation greater than 2 pw, and it moves towards $L/z_0 \gg 8$ as the separation increased, since the interaction period increases with the initial separation. In fact, the furthest point that the peak corresponding to the soliton interaction will be situated is at $L/z_0 = 1$. This corresponds to the period of oscillation of the input of the form $2 \operatorname{sech} \tau$.

The main area of interest (Mollenauer, *et al.*, 1986) is in the region where $L/z_0 \ll 8$, since here much lower soliton powers are needed. In this region, however, stable transmission can be expected for closely separated solitons when soliton interaction is reduced by using different amplitude solitons.

Another measure of instability and distortion can be obtained by calculating the change of the eigenvalues which correspond to the two-soliton formation. At the end of the amplification period the resulting waveform is taken as the initial condition for the ZKS eigenvalue equation, and the eigenvalues that describes the two solitons are calculated. Figure 5.18(*a*) shows the real part of the eigenvalues which indicate the velocity of the solitons and Figure 5.18(*b*) the percentage change in the imaginary part of the eigenvalues. Since the solitons initially formed a bound system (the real part of the eigenvalues are zero), Figure 5.18(*a*) indicates that the solitons will separate when $L/z_0 > 8$ as they acquire opposite velocities. In this region therefore a bound system cannot be achieved. Figure 5.18(*a*)

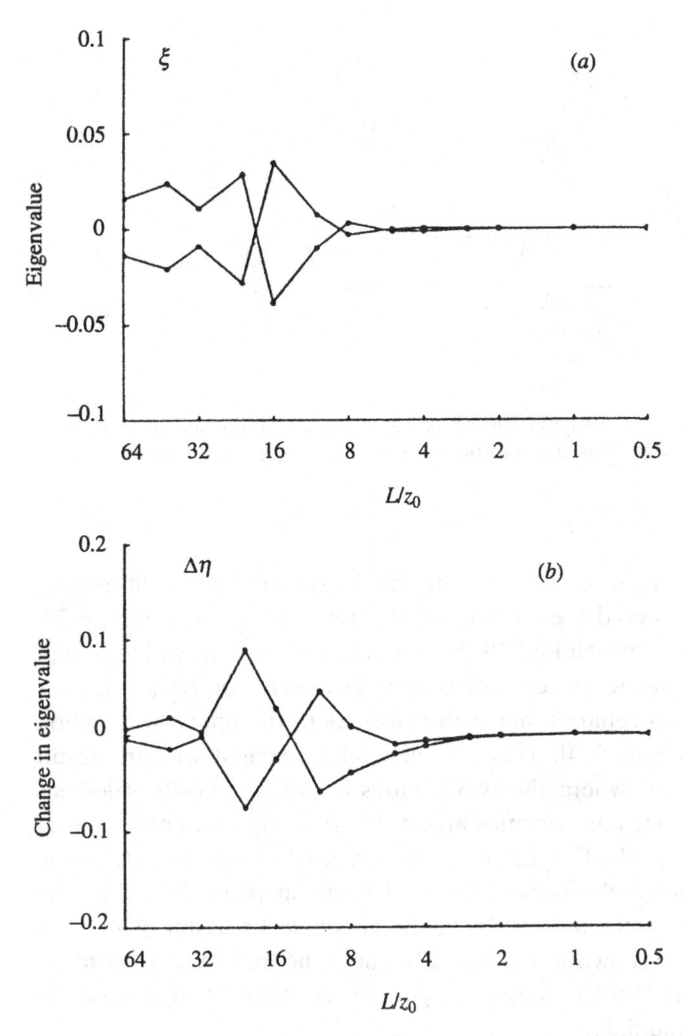

Figure 5.18. (*a*) Imaginary part of the eigenvalues, ξ, as a function of L/z_0. (*b*) Percentage change, $\Delta\eta$, in the values of the real part of the eigenvalues, η, as a function of L/z_0.

also indicate that the nature of the interaction will be different for the two instability peaks.

To have a more stable soliton system in terms of maintaining a constant pulse separation (Chu and Desem, 1985b) we choose the initial separation to be 3.6 pw (in terms of the lower amplitude soliton) and the amplitude ratio as $A = 20\%$. The resonance peak corresponding to the interaction period will occur at $L/z_0 \simeq 16$. The distortion curve will be similar to that of Figure 5.17 but will be

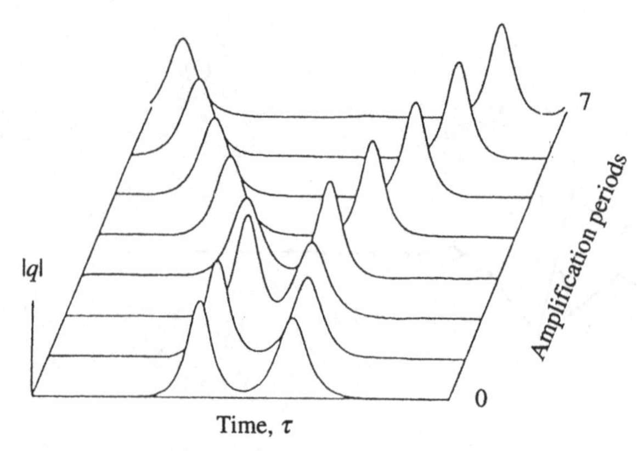

Figure 5.19. Propagation of two unequal amplitude solitons with $L/z_0 = 20$. Initial separation is 3.6 pw, and the amplitude ratio is 1.2.

broader and larger as a result of the larger amplitude difference. Figure 5.19 shows the evolution of the two solitons for $L/z_0 = 20$. The two solitons which initially formed a bound system, will separate. The two solitons do preserve their form however. For $L/z_0 = 12$, the solitons will still separate but with velocities in the opposite direction as shown in Figure 5.20. This is in general agreement with the result of Figure 5.18(*a*) where the two solitons acquires opposite velocities to each other but the velocities are of different signs on either side of the instability peak. For $L/z_0 \simeq 8$, which roughly corresponds to the resonance due to the periodicity of the soliton phase however, the instability is chaotic and results in the complete break-up of the two solitons. This is shown in Figure 5.21, where the individual solitons no longer preserve their initial shapes in contrast to the case in Figures 5.19 and 5.20.

To confirm that stable transmission can be achieved for $L/z_0 \ll 8$ we calculate the evolution of the solitons with $L/z_0 = 2$ and $L/z_0 = 1$, for fifty amplification periods which corresponds to a distance of 2000 km. For $L/z_0 = 2$, it is found that the solitons will eventually separate, but for a number of amplification periods less than 30, the deviation from the lossless case is small. For $L/z_0 = 1$, practically no distortion results. The maximum deviation of the soliton separation from that of the ideal case within the fifty links, is calculated to be less than 5%. Therefore, stable soliton transmission in the presence of loss and periodic amplification can be achieved with solitons of unequal amplitudes when $L/z_0 \ll 8$. However, this is not true for

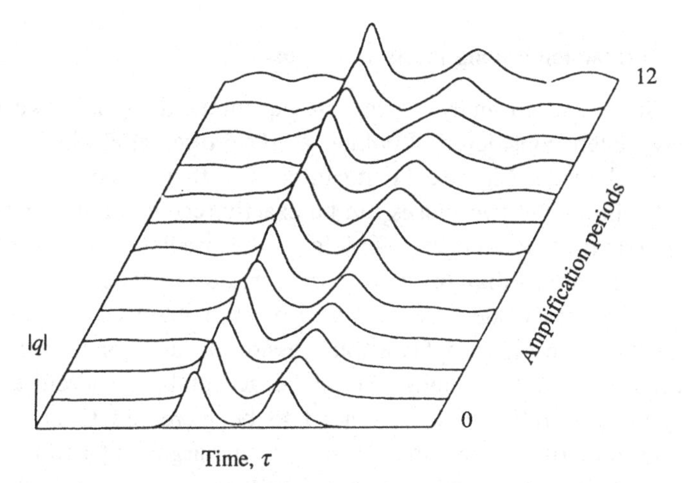

Figure 5.20. Propagation of two unequal amplitude solitons with $L/z_0 = 12$. Initial separation is 3.6 pw, and the amplitude ratio is 1.2.

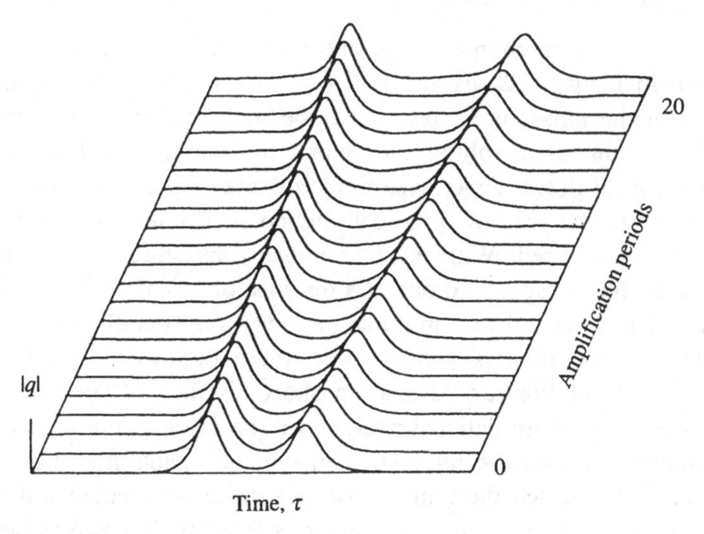

Figure 5.21. Propagation of two unequal amplitude solitons with $L/z_0 = 8$. Initial separation is 3.6 pw, and the amplitude ratio is 1.2.

equal amplitude solitons even though Figure 5.17 may indicate otherwise. The two solitons will rapidly separate just after the first pulse coalescence even though the interaction closely matches that of the lossless case prior to this break up.

5.5 Interaction among multiple solitons

So far, when analysing the mutual interaction, only two solitons have been considered. Looking at the pulse separation of two solitons as shown in Figure 5.11, it can be seen that in the case of two identical solitons the two pulses can be exactly recovered at a distance which is equal to an integral multiple of the oscillation period. One can therefore argue that the useful transmission distance can in fact be larger than the coalescence distance of the solitons. Thus, for the example shown in Figure 5.11, a transmission distance of $z = 52$, 104, 156, . . ., (in normalised units) would be acceptable. Therefore, the initial pulse separation does not need to be increased to achieve a longer distance of transmission. However, this argument is no longer true once three or more solitons are considered.

5.5.1 *Equal amplitude pulses*

The interaction of three and four equal amplitude pulses are shown in Figures 5.22(*a*) and (*b*) respectively. The initial separation between the pulses is 3.5 pw. It can be seen that in the presence of different number of solitons the interaction is different. For example two identical pulses experience total collapse at a distance of $z = 26$, whereas, in the case of three solitons no such effect is seen for the distance considered. With the four pulses considered, the two pulses in the centre collapse into one at a distance of $z \simeq 45$. The interaction will be characterised with multiple periods rather than just one period as in the case of two solitons. But, these periods cannot be fully observed from Figure 5.22 since a short distance of propagation is considered. A more detailed analysis will be given in the next section in connection with the interaction of unequal amplitude pulses. The distance over which the initial pulse waveform is repeated will therefore be different for different number of solitons. For two pulses this distance is $z \simeq 52$, but it can be seen from Figure 5.22 that for three and four solitons the pulses are not repeated at this distance.

In a communication system, since all possible pulse combinations are to be considered, the distance of transmission cannot be taken to equal the period of oscillation of two solitons. But, the distance of pulse coalescence, based on the interaction of two solitons, is a good indication of the maximum transmission distance. For a larger

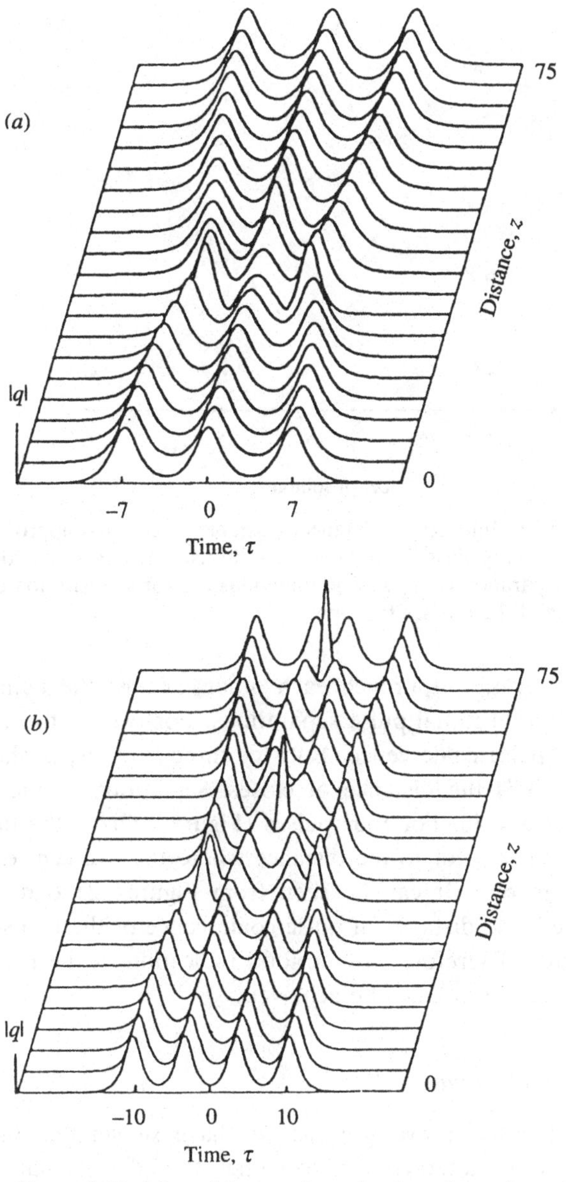

Figure 5.22. Mutual interaction of pulse of equal amplitudes and phases: (*a*) three-soliton interaction and (*b*) four solitons. The pulses are initially separated by 3.5 pw.

number of solitons, the pulses will not collapse at a shorter distance, as has been observed through the numerical studies of multiple solitons. Figure 5.23 shows the distance of the closest separations of the

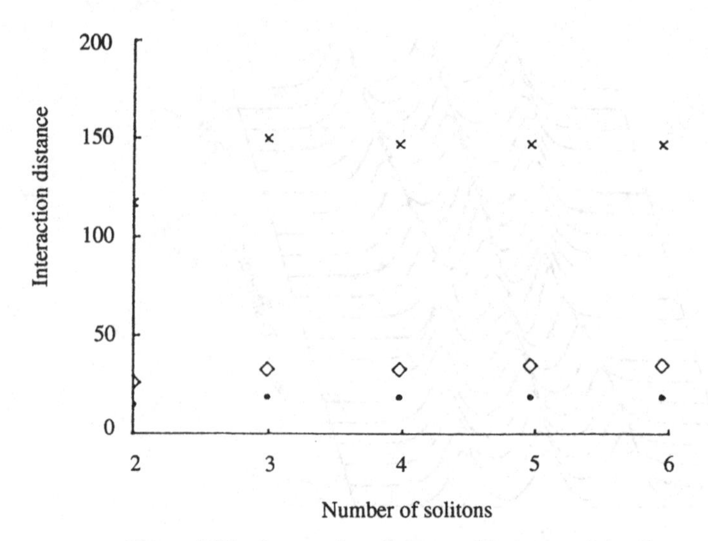

Figure 5.23. Interaction distance, (distance of the first approach of the solitons) against the number of solitons. The circles are for an initial separation of 3 pw. The diamonds are for a separation of 3.5 pw and the crosses for 5 pw.

pulses, when they first approach each other, against the number of solitons for different initial pulse separations. For two solitons, this is the distance of pulse coalescence. But, for three solitons, as shown in Figure 5.22(a), it is the distance of closest separation of the pulses which occurs at $z \simeq 30$. For four pulses (Figure 5.22(b)) the distance is again $z \simeq 30$, just prior to the distance where the two centre pulses collapse to form one. It can be seen from Figure 5.23 that, in the cases considered, the distance of pulse coalescence of the two solitons is always shorter. Therefore, it is a good indication of the maximum transmission distance that can be achieved.

5.5.2 *Unequal amplitude pulses*

On the basis of two solitons, we have shown that the best way to minimize the interaction is to launch the adjacent pulses with unequal amplitudes. Consider now that three sech-shaped pulses are launched, where the adjacent pulses are of different amplitudes and widths

$$q(0, \tau) = A_1 \operatorname{sech} A_1(\tau - 2\tau_0) + A_2 \operatorname{sech} A_2(\tau) + A_3 \operatorname{sech}(a_3\tau + 2\tau_0) \tag{5.22}$$

Only two different amplitudes need to be generated, and by alternating the amplitudes of the consecutive pulses, the adjacent pulses can always be made unequal. Therefore, given two amplitudes, v_1 and v_2, we only need to consider two cases, where $A_1 = A_3 = v_1$, $A_2 = v_2$ and $A_1 = A_3 = v_2$, $A_2 = v_1$.

The interaction of three such pulses is shown in Figure 5.24(a). $A_1 = A_3 = 1.1$ and $A_2 = 1$. The pulses are separated by 3.5 pw, $\tau_0 = 3.5$. The pulse separation as a function of the propagation distance, z, is shown in Figure 5.24(b) (note that the distance, z, considered is much longer than that of Figure 5.22(a)). It can be seen that the pulses periodically move towards the centre, but by different amount each time, and in between these times the narrower pulses exhibit small oscillations in position as in the case of two solitons. The centre pulse does not move its position due to the symmetry of the waveform. Therefore, when considering three solitons, the interaction is not necessarily minimised by simply letting the adjacent pulses have different amplitudes since they do approach each other by a significant amount.

5.5.3 *Eigenvalues and soliton interaction*

To analyse the interaction of the solitons, we calculate the eigenvalues of the initial pulse waveform. The three solitons will be characterised by three imaginary eigenvalues, $i\eta_1/2$, $i\eta_2/2$ and $i\eta_3/2$. The eigenvalues are shown in Figure 5.25(a) for the initial pulse waveform, equation (5.22), where $A_2 = 1$ and $A_1 = A_3$, for three different values, $A_1 = 1, 1.2, 1.4$. It can be seen that, as the initial separation is increased, the three eigenvalues will approach the value of the initial amplitudes of the pulses.

The interaction of the three solitons will be characterised by three oscillation periods, z_{pk}, which are related to the difference in the eigenvalues (Yevick and Hermansson, 1983), $z_{pk} = 4\pi/(\eta_i^2 - \eta_j^2)$, $i \neq j$, $i, j, k = 1, 2, 3$. The three periods are shown in Figure 5.25(b). As the initial pulse separation is increased two of the periods approach the value $4\pi/(A_1^2 - A_2^2)$ and the third increases nearly exponentially. For the case of the three solitons shown in Figure 5.24(a), the three periods calculated from the eigenvalues are $z_{p1} \simeq 42$, $z_{p2} \simeq 35$ and $z_{p3} \simeq 202$. As can be seen from Figure 5.24(b), the period, z_{p3}, represents the period of collapse of the solitons to the centre and the periods z_{p1}, z_{p2} represent the small oscillations of the position of the larger amplitude solitons.

The amount by which the pulses move from their positions, as seen

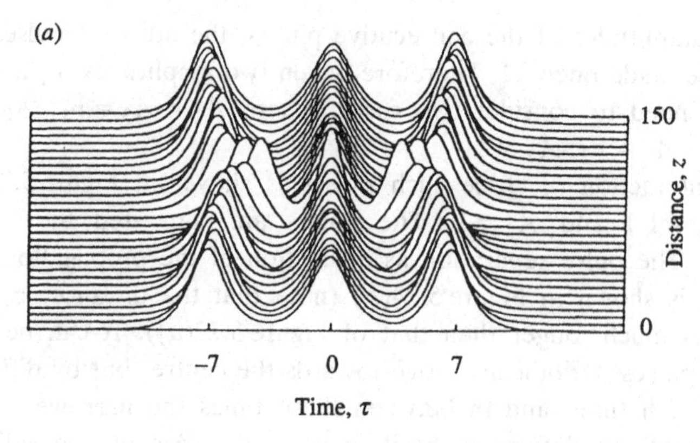

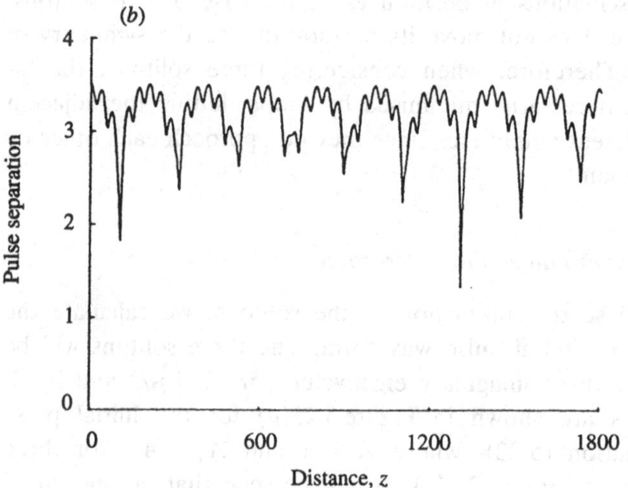

Figure 5.24. Interaction of three unequal amplitude pulses with $A_1 = A_3 = 1.1$, $A_2 = 1$. (*a*) shows the evolution of the three pulses and (*b*) shows the separation (in units of pulsewidth of the unity amplitude pulse) between the centre pulse and the side pulses as a function of the normalised distance, z.

in Figure 5.24(*b*), will also be related to the oscillation periods. A qualitative argument can be presented by simply expressing the pulse separation, τ_s, in terms of three cosine functions with periods as defined by the eigenvalues,

$$\tau_s = C_1 \cos(2\pi z/z_{p1}) + C_2 \cos(2\pi z/z_{p2}) + C_3 \cos(2\pi z/z_{p3}) + \Delta\tau \tag{5.23}$$

where, $C_1 > C_2, C_3$. The maximum separation occurs at a distance, z, where the cosine functions are all maximum, that is, $2\pi z/z_{p1} = 2\pi z/z_{p2} = 2\pi z/z_{p3} = 0, 2\pi, 4\pi \ldots$. Similarly, the minimum

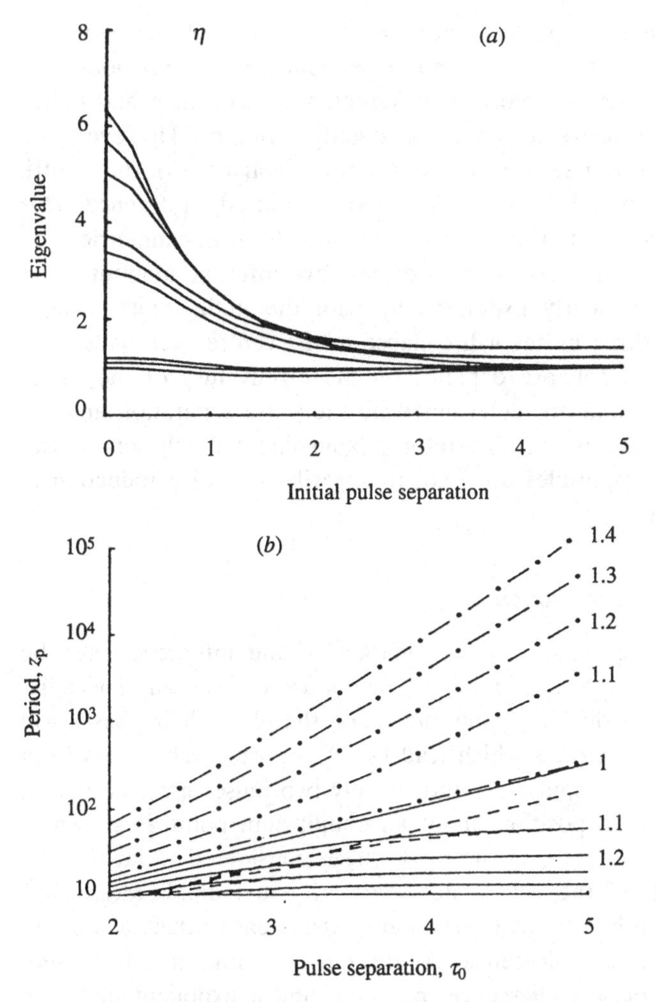

Figure 5.25. (*a*) The eigenvalues, η, for the three solitons given by equation (5.22). $A_2 = 1$, $A_1 = A_3 = 1$, 1.2, 1.4. (*b*) Period of oscillation, z_p, as a function of the initial separation with pulse amplitudes $A_2 = 1$, $A_1 = A_3 = 1$, 1.1, 1.2, 1.3, 1.4.

separation will occur when the arguments of the cosine functions are all an odd factor of π. In fact, this distance represents the total collapse of the pulses to the centre. In Figure 5.24(*b*), the maximum separation occurs at $z = 0$ but the minimum is not reached within the distance shown.

The other possible combination of the pulses constructed with two different amplitudes is obtained by letting the centre pulse have a higher amplitude than the other two, $A_1 = A_3 < A_2$ in equation (5.22). The interaction of such pulses is similar to the previous case but since the initial waveform is different, the eigenvalue and therefore the oscillation periods will be slightly different. However, the three pulses considered so far will always collapse to the centre irrespective of the initial separation or the amplitude difference. This results from the fact that the initial pulse waveform is symmetric. The pulses approach the centre periodically by different amounts (the period increases nearly exponentially with the initial pulse separation), but the three pulses will collapse to the centre eventually. The distance of total collapse depends on the eigenvalues of the initial pulse waveform and are determined by the pulse separation and the amplitudes of the pulses. Therefore, launching the adjacent pulses with different amplitudes does not necessarily lead to a reduction in pulse interaction.

5.5.4 *Physical explanation*

A simple qualitative explanation of the interaction can be given if the phase of the individual pulses are considered. The adjacent pulse are of different amplitudes and therefore their phases will change at different rates which lead to a force which alternates from being attractive to repulsive. Thus, if only two pulses are considered, they are shifted in position about an equilibrium point as shown in Section 5.2. However, with the three pulses considered so far the two on either side of the centre pulse are of equal amplitudes. They would therefore be of equal phases and attract each other, leading to the eventual pulse coalescence. As the pulse separation is increased, the period of pulse coalescence increases almost exponentially as in the case of two soliton interaction. When the pulses are of larger amplitudes (and therefore narrower widths) then the effective pulse separation is increased. The period of coalescence is therefore increased as compared to the case where the two pulses are of lower amplitudes on either side of the centre pulse.

5.5.5 *Asymmetric input pulse waveform*

In Section 5.2.4 it was shown that, apart from the eigenvalues, the two unequal amplitude pulses are characterised by an additional parameter, γ_0, which is a measure of the asymmetry of the pulse waveform. It was also shown that, with the increase of γ_0 (and therefore an increase of the asymmetry of the pulse waveform) the mutual interaction is reduced in the sense that the pulse coalescence is avoided and the minimum pulse separation is reduced. To avoid the pulse coalescence, the symmetry of the initial pulse waveform is to be avoided. The pulse waveform can always be made asymmetric if the pulses are generated using a building block of three different amplitude pulses. Here, we present some numerical solutions of the interaction of three and more solitons.

5.5.5.1 Interaction of three solitons The simplest way to construct an asymmetric set of pulses by using only three different amplitudes, A_1, A_2, A_3 is to generate the sequence of pulses as follows,

$$A_1A_2A_3A_1A_2A_3A_1 \ldots \qquad (5.24)$$

whereA_i represent sech pulses all separated by a fixed duration. For example if three consecutive pulses are launched, (i.e. three 'one's are launched: a string of digital data, . . .000111000. . .) then, there are three possible combinations, $A_1A_2A_3$ or $A_2A_3A_1$ or $A_3A_1A_2$ of forming the pulse sequence. The interaction of three such pulses are shown in Figure 5.26(*a*) and (*b*). The three amplitudes considered are $A_1 = 1$, $A_2 = 1.1$ and $A_3 = 1.2$ and the pulse separation is 3.5 pw (referred to the pulsewidth of the unity amplitude pulse). Figure 5.26(*a*) shows the interaction of three pulses arranged as $A_1A_2A_3$ and Figure 5.26(*b*) shows three pulses $A_2A_3A_1$. The pulses exhibit small oscillations in their positions similar to the case of two soliton interaction. The amount by which the pulses shift in position is different for the different combinations of the pulses. However, in both cases it can be seen that the pulse coalescence is avoided.

5.5.5.2 Interaction of four solitons and six solitons When four solitons are considered, the nature of the interaction is similar. Figures 5.27(*a*) and (*b*) show the interaction of four solitons obtained by numerically solving the NLS equation. Figure 5.27(*a*) shows four pulses arranged as $A_2A_3A_1A_2$ and Figure 5.27(*b*) $A_3A_1A_2A_3$. The pulses are initially separated by 3.5 pw. It can be seen that the pulse coalescence is avoided and the pulses maintain their relative positions.

We then look at the interaction of six pulses constructed as above, with three possible combinations $A_1A_2A_3A_1A_2A_3$, $A_2A_3A_1A_2A_3A_1$ and $A_3A_1A_2A_3A_1A_2$, shown in Figures 5.28(*a*), (*b*) and (*c*). It can be seen that for the combinations shown in Figures 5.28(*a*) and (*b*),

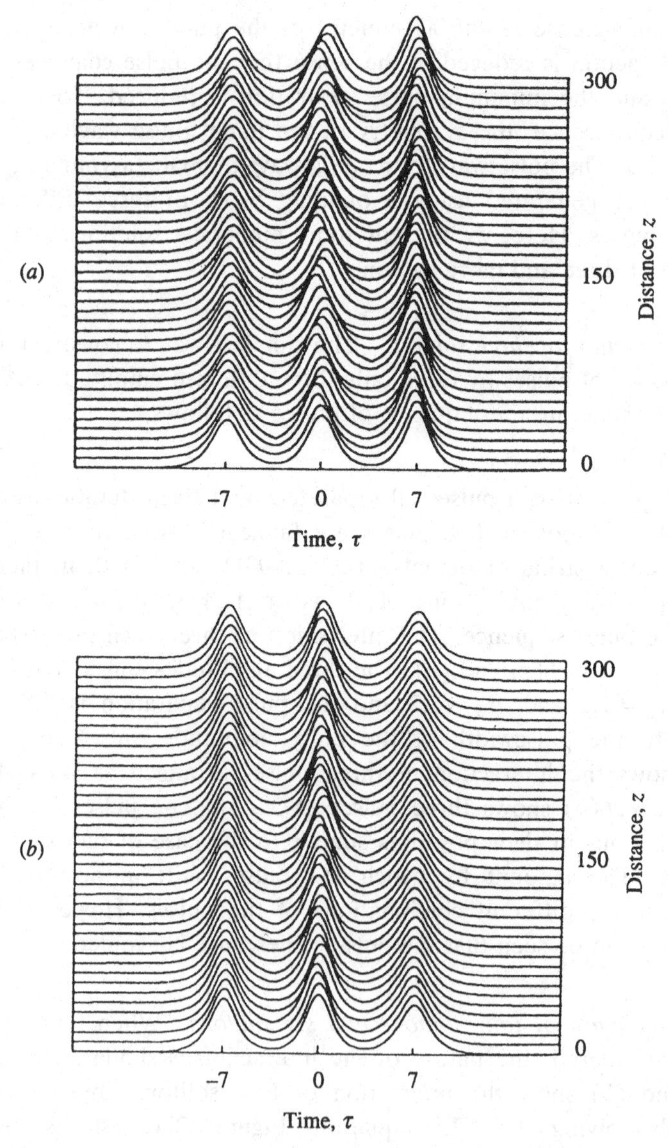

Figure 5.26. Interaction of three solitons as a function of initial separation $\tau_0 = 3.5$. (*a*) $A_1 = 1$, $A_2 = 1.1$, $A_3 = 1.2$ and (*b*) $A_1 = 1.1$, $A_2 = 1.2$, $A_3 = 1$.

the pulse coalescence is avoided. But, with the combination, $A_3A_1A_2A_3A_1A_2$, shown in Figure 5.28(*c*), the pulses do undergo a substantial amount of interaction which is clearly undesirable. However, the results presented so far indicate the possibility of redu-

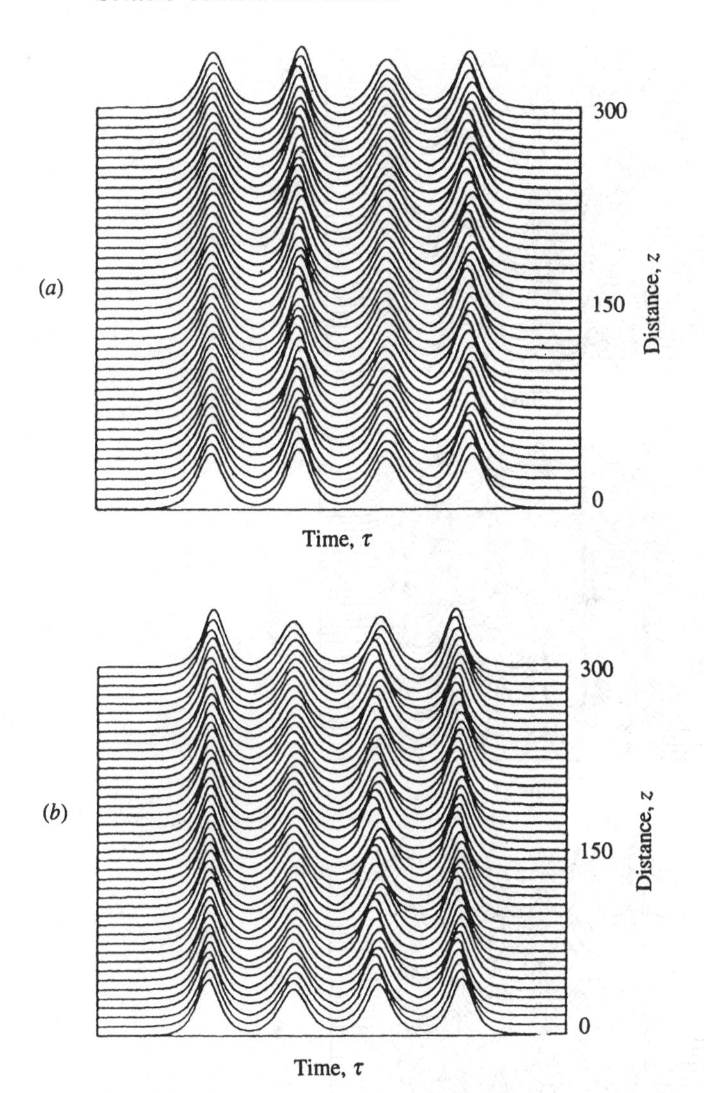

Figure 5.27. Interaction of four solitons with $\tau_0 = 3.5$ (*a*) $A_1 = 1.1$, $A_2 = 1.2$, $A_3 = 1$ and $A_4 = 1.1$ and (*b*) $A_1 = 1.2$, $A_2 = 1$, $A_3 = 1.1$ and $A_4 = 1.2$.

cing the interactions with a scheme of using only a small number of different amplitudes to generate the pulse sequence. The sequence of pulses presented here, equation (5.24), with the three amplitudes, $A = 1$, 1.1, 1.2, and separated by 3.5 pw is not the optimum. Further work, therefore, needs to be done.

Another limit in using the sequence presented in equation (5.24)

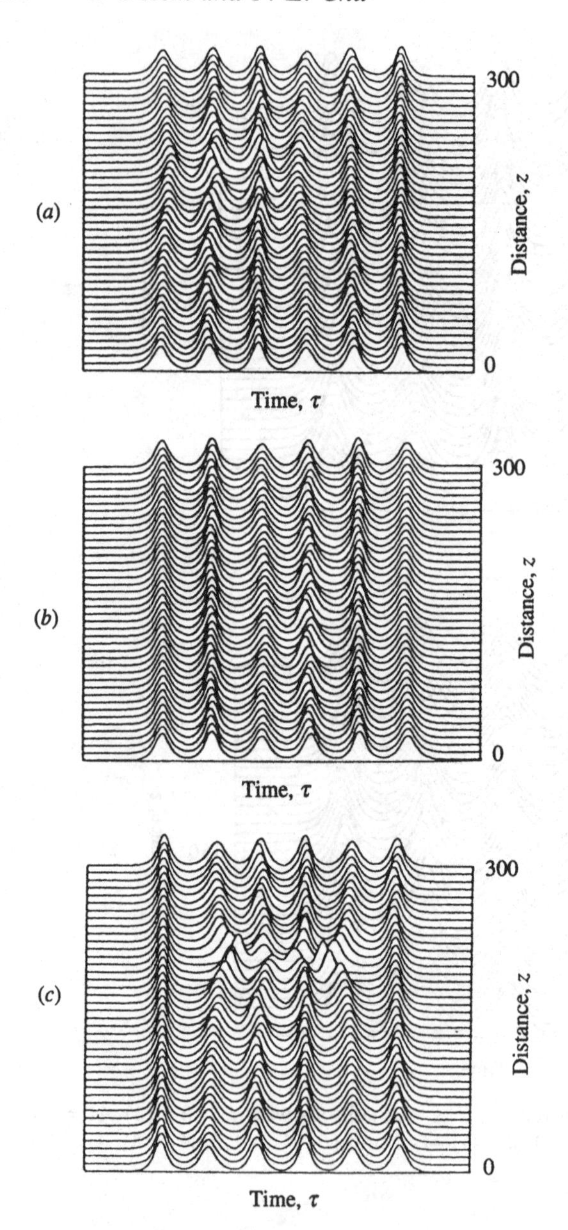

Figure 5.28. Interaction of six solitons with $\tau_0 = 3.5$. (*a*) $A_1A_2A_3A_1A_2A_3$:1,1.1,1.2,1,1.1,1.2; (*b*) $A_1A_2A_3A_1A_2A_3$:1.1,1.2,1,1.1,1.2,1; and (*c*) $A_1A_2A_3A_1A_2A_3$:1.2,1,1.1,1.2,1,1.1.

will be realised when considering the sequence of data, '. . .000010010000. . .'. That is, a string of 'zero's are transmitted, then a 'one', two 'zero's, a sequence where two equal amplitude pulses are transmitted. However, the separation is three times that of two consecutive pulses. Therefore, the coalescence distance of this sequence determines the maximum transmission distance. But again, this can be avoided by not strictly following the sequence of equation (5.24).

References

Anderson, D. and Lisak, M. (1985) *Phys. Rev. A*, **32**, 2270–4.

Anderson, D. and Lisak, M. (1986) *Opt. Lett.*, **11**, 174–6.

Blow, K. J. and Doran, N. J. (1983) *Electron. Lett.*, **19**, 429–30.

Chu, P. L. and Desem, C (1983a) *Technical Digest, IOOC'83, Tokyo, Japan*, pp. 52–3.

Chu, P. L. and Desem, C. (1983b) *Electron. Lett.*, **19**, 956–7.

Chu, P. L. and Desem, C. (1985a) *Electron. Lett.*, **21**, 228–9.

Chu, P. L. and Desem, C. (1985b) *Electron. Lett.*, **21**, 1133–4.

Desem, C. and Chu, P. L. (1987a) *Opt. Lett.*, **12**, 349–51.

Desem, C. and Chu, P. L. (1987b) IEE Proc.-J, **134**, 145–51.

Gordon, J. P. (1983) *Opt. Lett.*, **8**, 596–8.

Hasegawa, A. (1983) *Opt. Lett.*, **8**, 650–2.

Hasegawa, A (1984) *Appl. Opt.*, **23**, 3302–9.

Hasegawa, A. and Kodama, Y. (1981) *Proc. IEEE*, **69**, 1145–50.

Hasegawa, A. and Kodama, Y. (1982) *Opt. Lett.*, **7**, 285–7.

Hermansson, B. and Yevick, D. (1983) *Electron. Lett.*, **19**, 570–1.

Karpman, V. I. and Solov'ev, V. V. (1981) *Physica D*, **3**, 487–502.

Kaup, D. J. (1976) *SIAM J. Appl. Math.*, **31**, 121–33.

Kodama, Y. and Hasegawa, A. (1982) *Opt. Lett.*, **7**, 339–41.

Kodama, Y. and Hasegawa, A. (1987) *IEEE J. Quantum Electron.*, **23**, 510–24.

Kodama, Y and Nozaki, K. (1987) *Opt. Lett.*, **12**, 1038–40.

Mitschke, F. M. and Mollenauer, L. F. (1986) *Opt. Lett.*, **11**, 659–61.

Mollenauer, L. F. and Smith, K. (1988) *Opt. Lett.*, **13**, 675–77.

Mollenauer, L. F., Stolen, R. H. and Islam, M. N. (1985) *Opt. Lett.*, **10**, 229–31.

Mollenauer, L. F., Gordon, J. P. and Islam, M. N. (1986) *IEEE. J. Quantum Electron.*, **22**, 157–73.

Satsuma, J. and Yajima, N. (1974) *Progr. Theor. Phys., Suppl.*, **55**, 184–306.

Shiojiri, E. and Fujii, Y. (1985) *Appl. Opt.*, **24**, 358–60.

Tappert, F. D. (1974) *Lect. Appl. Math. Am. Math. Soc.*, **15**, 215–6.

Yevick, D. And Hermansson, B. (1983) *Optics Commun.*, **47**, 101–6.

Zakharov, V. E. and Shabat, A. B. (1972) *Sov. Phys. – JETP*, **34**, 62–9.

Zakharov, V. E. and Shabat, A. B. (1973) *Sov. Phys. – JETP*, **5**, 823–8.

6

Soliton amplification in erbium-doped fiber amplifiers and its application to soliton communication

MASATAKA NAKAZAWA

In this chapter, soliton amplification and transmission using erbium-doped fiber amplifiers are presented. First, the general features of the erbium-doped fiber amplifier are described. A new method called dynamic soliton communication is presented, in which optical solitons can be successfully amplified and transmitted over an ultralong dispersion-shifted fiber by using the dynamic range of $N = 1$–2 solitons. Multi-wavelength optical solitons at wavelengths of 1.535 and 1.552 μm have been amplified and transmitted simultaneously over 30 km with an erbium-doped fiber repeater. It is shown that there is saturation-induced cross talk between multi-channel solitons, and the cross talk (the gain decrease) is determined by the average input power in high bit-rate transmission systems.

The amplification of pico, subpico and femtosecond solitons and 6–24 GHz soliton pulse generation with erbium-doped fiber are also described, which indicates that erbium fibers are very advantageous for short pulse soliton communication. Some soliton amplification characteristics in an ultralong distributed erbium fiber amplifier are also presented. Finally, we describe 5–10 Gbit/s transmission over 400 km in a soliton communication system using the erbium amplifiers.

6.1 Introduction

Recent progress on erbium-doped fiber amplifiers (EDFA) has been very rapid since they show great potential for opening a new field in high-speed optical communication (Mears *et al.*, 1985; Poole *et al.*, 1985; Desurvire *et al.*, 1987; Snitzer *et al.*, 1988; Kimura *et al.*, 1989; Suzuki *et al.*, 1989; Nakazawa *et al.*, 1989). Their typical advantages are a high gain of more than 30 dB, low noise, wide bandwidth, polarisation insensitive gain, and high saturation output power. They are also advantageous because they operate in the 1.5 μm region where fiber loss is minimised in ordinary silica-based single-mode optical fibers. Because of these excellent characteristics, EDFAs are very useful not only for high speed communication, but also for optical soliton communication (Nakazawa *et al.*, 1990; Kubota and Nakazawa, 1990; Suzuki *et al.*, 1990).

To date, high gains of more than 30 dB have been reported using dye laser, argon ion laser and color center laser pump sources. However, for more practical applications, the pump sources should be compact and efficient excitation should be realised. It has been shown that 1.48 μm band InGaAsP laser diodes are an excellent pump source (Nakazawa *et al.*, 1989). This pumping scheme makes it possible for EDFAs to be applied practically, and it has been proven that a compact optical amplifier can be realised using laser diode pumping.

We have shown that optical solitons can be amplified successfully with EDFAs (Nakazawa *et al.*, 1990; Kubota and Nakazawa, 1990; Suzuki *et al.*, 1990). In the near future, high-speed optical soliton communications will make significant progress and offer greater flexibility through the use of in-line fiber amplifiers, especially EDFAs. The EDFA is now opening the door to the future of optical communication (Shimada, 1990).

This chapter will present the fundamental amplification characteristics of optical solitons using EDFAs and their application to soliton communications over long distances. Data transmission experiments at 5–10 Gbit/s over 400 km using 12 erbium repeaters will be described.

6.2 General features of the erbium-doped fiber amplifier

Figure 6.1 shows a block diagram of the EDFA. The pump source is a 1.48 μm Fabry–Perot type high-power laser diode (Nakazawa *et al.*, 1989), which is coupled into an erbium fiber through a WDM fiber coupler and a non-polarised optical isolator. The WDM fiber coupler passes the 1.55 μm signal and the pump beam into the erbium fiber. The non-polarised optical isolator greatly suppresses the laser oscillation of the EDFA. The spectral width of the laser diode is over 20 nm at around 1.48 μm, but all the spectral components can effectively contribute to the generation of the population inversion of

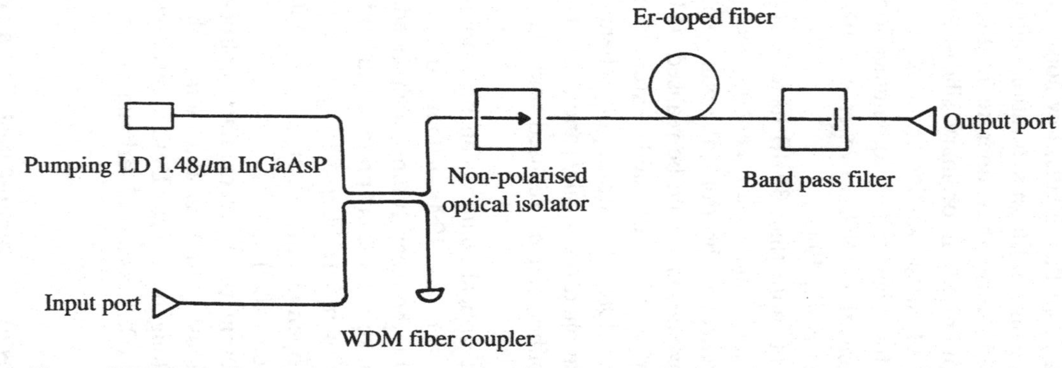

Figure 6.1. Block diagram of an erbium-doped fiber amplifier (EDFA). It consists of five main components of a laser diode pump source, a WDM coupler, a non-polarised isolator, an erbium fiber, and a narrowband filter.

the erbium ions (Kimura *et al.*, 1990). The combination of a relatively low concentration of erbium ions and a longer fiber length results in a higher gain than can be achieved with a high concentration and a shorter fiber length (Desurvire *et al.*, 1987; Kimura *et al.*, 1989). This low doping can suppress the concentration quenching which reduces the pumping efficiency. The 1.55 μm signal is fed to the input port and the amplified signal exits from the output port. A pump power of 50 mW is sufficient to obtain a gain of more than 30 dB. Optical amplifiers inevitably suffer from amplified spontaneous emission (ASE) from the EDFA. The suppression of ASE is one of the keys to the realisation of a low noise amplifier. In this case a narrowband optical filter with a passband of 1–3 nm is installed, resulting in a high signal-to-noise ratio.

EDFAs have three main applications (Shimada, 1990). The first is as a booster amplifier which is used to increase the signal power input into a transmission fiber. This amplifier is installed after the light source. For this application, a large saturation output characteristic is required. The second EDFA application is as an optical repeater which is used to amplify the attenuated signal and re-transmit the amplified signal into the next transmission fiber. High power output is also required. The third application is as a pre-amplifier to increase receiver sensitivity. This is installed in front of a photodetector, where low noise characteristics are required. EDFAs can be fully applied in these three categories.

In an EDFA, the pumping light decays along the z-axis because the pump beam is absorbed when it propagates down in the EDFA. As a result, a non-uniform population inversion occurs along the fiber, where the gain coefficient and the saturation intensity I_{sat} can be expressed as a function of z. Later, we describe some basic aspects of the EDFA.

Let the pump intensity and the signal intensity be I_{p} and I_{s}, respectively. The normalised pump intensity $\bar{I}_{\text{p}}$ and the normalised signal intensity $\bar{I}_{\text{s}}$ at a point z can be written as (Desurvire *et al.*, 1987)

$$\frac{\mathrm{d}\bar{I}_{\text{p}}}{\mathrm{d}z} = -\frac{k_{\text{s}}\bar{I}_{\text{s}} + 1}{\bar{I}_{\text{p}} + (1 + k_{\text{s}})\bar{I}_{\text{s}} + 1}\,\rho\sigma_{\text{p}}^{\text{a}}\bar{I}_{\text{p}} \tag{6.1}$$

$$\frac{\mathrm{d}\bar{I}_{\text{s}}}{\mathrm{d}z} = \frac{k_{\text{s}}\bar{I}_{\text{p}} - 1}{\bar{I}_{\text{p}} + (1 + k_{\text{s}})\bar{I}_{\text{s}} + 1}\,\rho\sigma_{\text{s}}^{\text{a}}\bar{I}_{\text{s}} \tag{6.2}$$

$$\bar{I}_{\text{p}} = I_{\text{p}}/(h\nu_{\text{p}}/\sigma_{\text{p}}^{\text{a}}\tau) \tag{6.3}$$

$$\bar{I}_s = I_s/(h\nu_s/\sigma_s^a \tau) \tag{6.4}$$

where σ_p^a and σ_s^a are the absorption cross sections at the pump wavelength and at the signal wavelength, respectively, ρ is the erbium concentration, $h\nu_p$ and $h\nu_s$ are the photon energies of the pump and signal waves, respectively, and τ is the fluorescence lifetime. We define the ratio $k_s = \sigma_s^e/\sigma_s^a$, where σ_s^e is the emission cross section at the signal wavelength.

The gain coefficient g defined as $d\bar{I}_s/dz = g\bar{I}_s$ is expressed from equation (6.2) as

$$g = \left[1 + \bar{I}_s \Bigg/ \left(\frac{\bar{I}_p + 1}{1 + k_s}\right)\right]^{-1} \frac{k_s \bar{I}_p - 1}{\bar{I}_p + 1} \rho \sigma_s^a \tag{6.5}$$

which is rewritten as

$$g = \frac{g_0}{1 + \bar{I}_s/\bar{I}_{sat}} \tag{6.6}$$

Here

$$\bar{I}_{sat} = \frac{\bar{I}_p + 1}{1 + k_s} \tag{6.7}$$

and

$$g_0 = \frac{k\bar{I}_p - 1}{\bar{I}_p + 1} \rho \sigma_s^a \tag{6.8}$$

Since $\bar{I}_{sat}$ and g_0 are functions of the pump intensity, they change with a decrease in the pump intensity along the amplifier length.

The pump power P_p and the signal power P_s are expressed as $P_p = I_p A_p$ and $P_s = I_s A_s$, respectively, where A_p and A_s are the effective core areas for the pump wave and for the signal wave. The saturation signal power P_{sat} is given by $P_{sat} = A_s(h\nu_s/\sigma_s^a \tau)\bar{I}_{sat}$.

Figure 6.2 shows the gain $G = P_s^{out}/P_s^{in}$ by solving equations (6.1) and (6.2) as a function of the input signal power P_s^{in}, where the input pump power is 59 mW. The length and the erbium concentration of the EDFA was 100 m and 92 ppm, respectively. The signal wavelength was 1.545 μm. The optical loss at the pump wavelength and at the signal wavelength were 0.13 and 0.42 dB/m, respectively. The small-signal gain G_0 was 24.5 dB and the 3 dB saturation input power P_{3dB}^{in} was −15.2 dBm. The experimental data (open circles) are also shown in this figure. For small-signal input levels the gain is constant; but for the moderate signal levels, the gain saturates with an increase in the input signal power. Gain as a function of the input pump power P_p is given in Figure 6.3, where the open circles and the

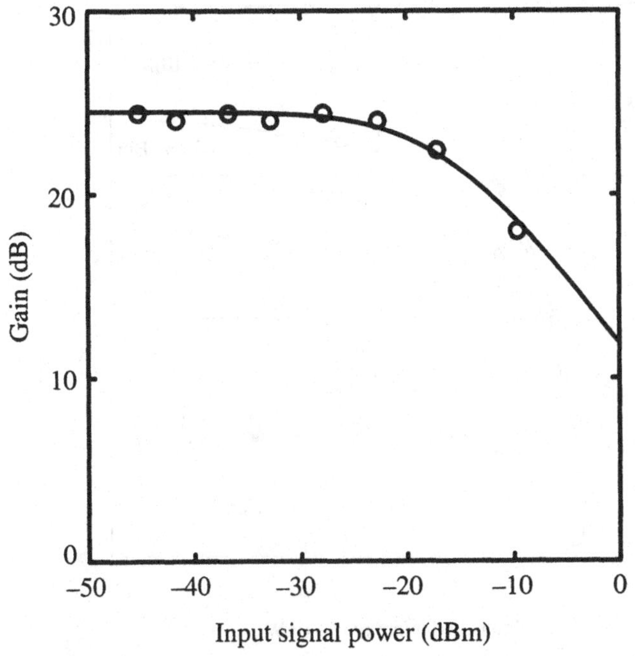

Figure 6.2. Change in gain of EDFA as a function of the input signal power. Open circles experiment; full curve, calculation.

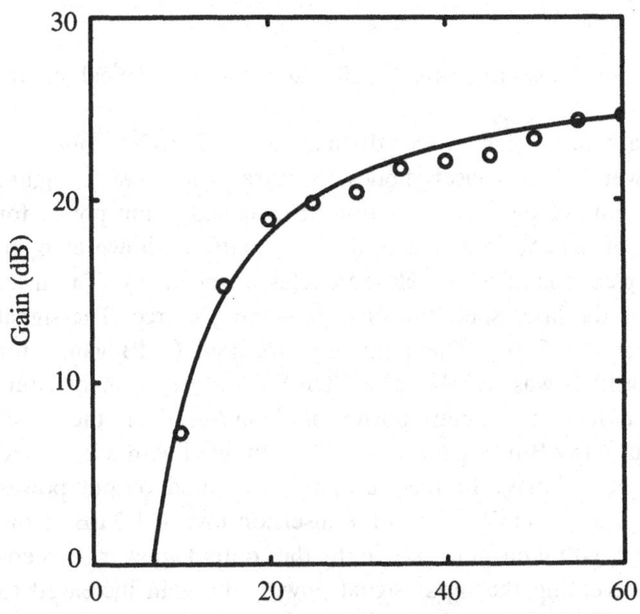

Figure 6.3. Change in gain of EDFA as a function of the input pump power. Open circles, experiment, full curve, calculation.

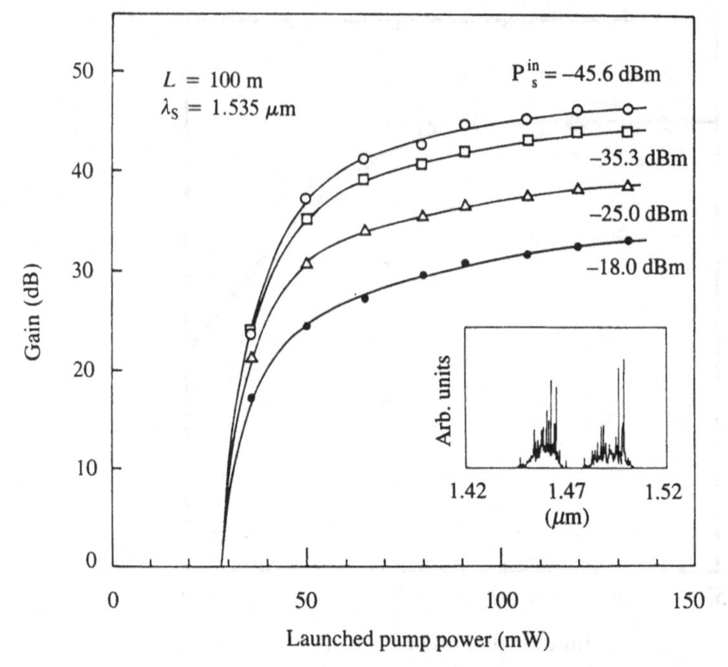

Figure 6.4. Gain *versus* EDFA pump power. The highest gain is 46.5 dB for a pump power of 130 mW. The inset is the pump spectrum.

full line represent experimental results and the theoretical curve, respectively.

Currently, a gain as high as 46.5 dB has been successfully obtained with high power 1.48 μm laser diodes (Kimura *et al.*, 1989). Figure 6.4 shows the gain change as a function of launched pump power for signal powers of −46.5, 35.3, −25.0 and −18.0 dBm, shown as open circles, rectangles, triangles and closed circles, respectively. The inset in this figure is the laser spectrum of the pumping source. The signal wavelength was 1.535 μm. The pumping threshold (0 dB gain), for the 100 m-long fiber was 28 mW. The EDFA gains are strongly saturated beyond a launched pump power of 75 mW. When the input signal power is −18 dBm, a gain of 33 dB is obtained with a launched pump power of 133 mW. In this case, the amplified output power reached as high as 32.4 mW. When the insertion loss of 1.3 dB of the optical isolator is taken into account, the output power exceeds 40 mW. By decreasing the input signal power, the gain increased to as high as 46.5 dB, and the gain coefficient is estimated at 0.76 dB/m, from the gain slope shown in Figure 6.4. The saturation output power

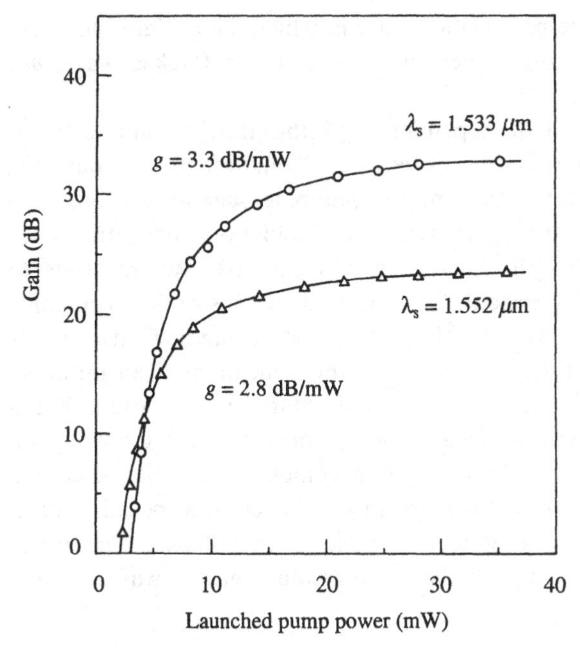

Figure 6.5. EDFA gain characteristics with a high gain coefficient. Open circles and rectangles correspond to signal wavelengths of 1.533 and 1.552 μm, respectively.

was about +11 dBm, which means that the EDFA can operate as a power amplifier for soliton amplification.

In order to realise a high-gain amplifier with low pump power, it is important that the pump intensity be confined as much as possible, so that an efficient excitation is achieved. To achieve this, special attention has been paid to structural optimisation by doping erbium ions only in the center of the fiber core, so that the high intensity part of the pump field interacts with the erbium ions, resulting in efficient pumping. EDFA gain characteristics with a high-gain coefficient thus obtained are shown in Figure 6.5. An erbium-doped fiber with a high relative refractive-index ratio was used, in which the erbium is doped only into the high-index core region. The mode field diameter, relative refractive-index difference, erbium concentration and fiber length were 4.9 μm, 1.64%, 210 ppm and 21 m, respectively. The pump threshold was as low as 2.2 mW at 1.552 μm. The gain coefficient reached as high as 3.3 dB/mW at 1.533 μm, which means that high refractive-index erbium fibers can generate high gain with a high-gain coefficient. Recently, high-gain characteristics could also be obtained by using a dual shape core (DSC) fiber in which the erbium ions were doped only into the high-index inner core region (Nakazawa *et al.*, 1990).

The pump wavelength dependence of the EDFA gain factor is briefly described here (Kimura *et al.*, 1990). When the pumping wavelength approaches 1.50 μm, the pumping scheme is close to a two-level system, resulting in inefficient population inversion. When the pumping wavelength is close to 1.40 μm, the system starts to become a three-level system, but there is no absorption (pumping line) transition from $^4I_{15/2}$ to $^4I_{13/2}$ at that wavelength. Consequently there is no gain at 1.40 μm pumping, which means that an optimum pumping wavelength has been obtained. Starting from the 1.50 μm pumping line, the gain efficiency of the erbium fiber gradually increases when the pump wavelength becomes shorter. At a certain shorter wavelength the efficiency rapidly decreases because resonanced absorption is no longer possible near 1.40 μm. The pump wavelength which gave the maximum gain was around 1.460–1.475 μm.

6.3 Optical gain of erbium-doped fiber for picosecond high-power pulses

The experimental setup for the gain measurement of an EDFA for picosecond high-power pulses is shown in Figure 6.6 (Nakazawa *et al.*, 1989b). The signal source is a color-center laser (CCL) with a repetition rate of 100 MHz. The pulsewidth is 10 ps and the wavelength is set at 1.552 μm. The peak power of the picosecond input pulse is set in the soliton power regime of a typical dispersion-shifted fiber. The low group velocity dispersion of a few ps/km · nm in dispersion-shifted fibers and the 10 ps pulsewidth give a soliton peak power of less than 100 mW. The picosecond pulse is then divided into two pulses with a beam splitter. One of these pulses is a pump pulse and the other is a probe pulse. The time delay between the pump and the probe pulses is varied with a movable corner cube. A dichroic mirror is used to couple the pump (InGaAsP laser diode) and signal beams into the erbium fiber. The signal and pump beams are coupled into a 3.5-m-long erbium-doped fiber with a core diameter of 6.9 μm, a relative refractive index difference of 0.85%, and an erbium concentration of 1370 ppm. The output pulses from the EDFA are measured with a streak camera.

The waveforms of the input pulse to the EDFA and the waveforms

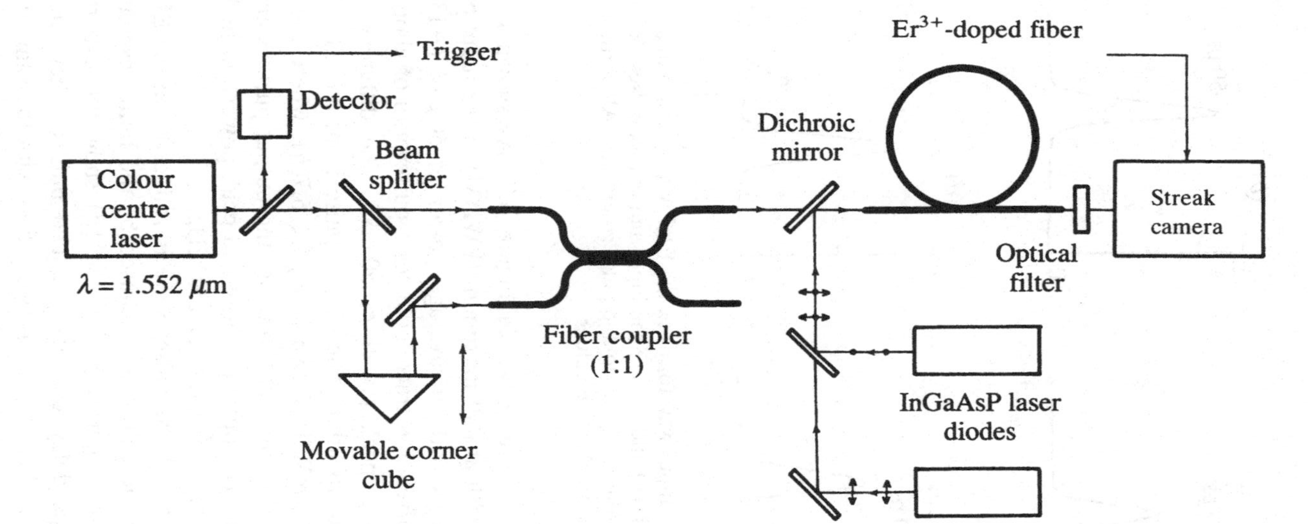

Figure 6.6. Experimental setup for the gain measurement of an EDFA for picosecond high power pulses. The pulse source is a color center laser. A time delay between the two pulses is controlled by the corner cube. The amplified pulses are measured with a streak camera.

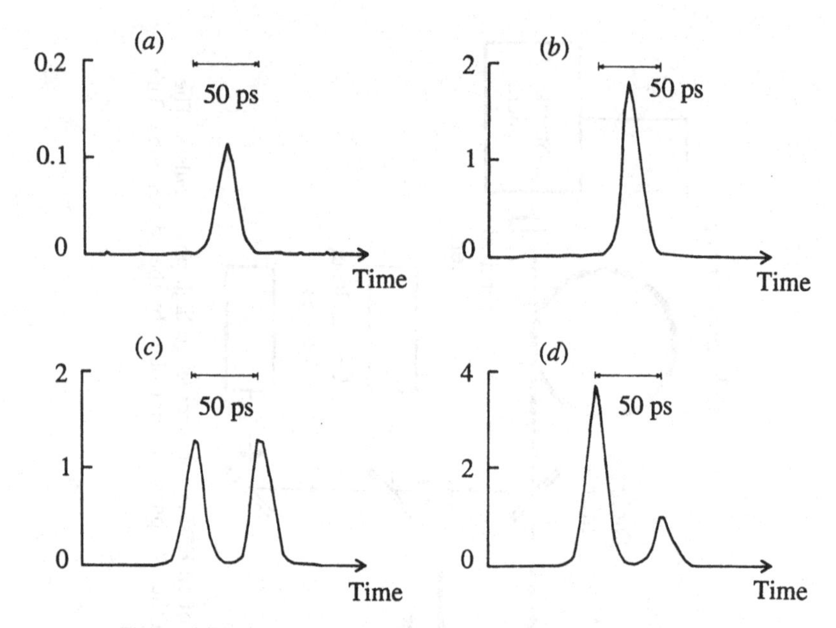

Figure 6.7. Input waveforms to the EDFA and their output waveforms. (*a*) and (*b*) are the input and output pulses. (*c*) and (*d*) are the output pulses in double pulse conditions. All vertical axes are taken with absolute values.

of the output pulse are shown in Figure 6.7. As shown in Figure 6.7(*a*), the full width at half maximum (FWHM) of the input pulses is approximately 14 ps with an average power of 50 μW (−13 dBm), where the autocorrelation trace indicates a pulsewidth of about 10 ps, corresponding to a peak power of 50 mW. The difference in the pulsewidths is due to jitter. The output pulse from the EDFA for a pump power of 90 mW is shown in Figure 6.7(*b*), where the gain is 12 dB. Figures 6.7(*a*) and (*b*) show that the output pulse is amplified with no distortion. The output energy and peak power reached 7.9 pJ and 792 mW at a repetition rate of 100 MHz.

The pump and the probe pulses were introduced into the EDFA in order to measure the gain change caused by the finite response time of the EDFA gain. Figures 6.7(*c*) and (*d*) show the experimental results for a 50 ps delay between the two pulses, when the pump power is 90 mW. In Figure 6.7(*c*), two pulses with the same intensity (50 mW peak) are coupled into the fiber. The pulse shapes and the gain of both output pulses are the same, and the gain decreases to 10.5 dB. For the single pulse shown in Figure 6.7(*b*), the gain was 12 dB. If the gain recovery time is in picoseconds, the gain of the

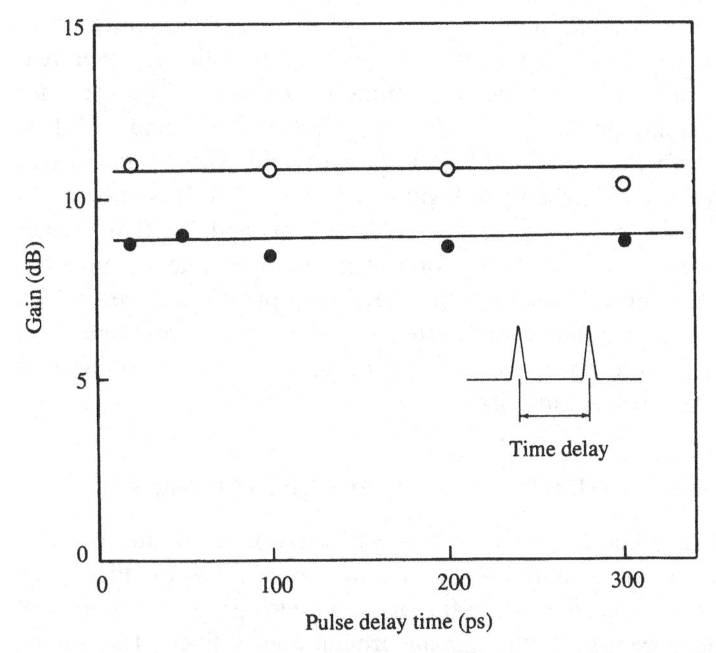

Figure 6.8. Change in EDFA gain as a function of a time delay. Open and closed circles correspond to conditions for figures (*c*) and (*d*), respectively. Pump power = 90 mW and $\lambda = 1.552\ \mu$m.

second pulse would be smaller than that of the first pulse because of the gain saturation caused by the first pulse. However, no gain decreases were observed in the second pulse. This means that the gain decrease is caused by the steady-state gain saturation in this bit-rate region. The gain decrease of 1.5 dB is due to steady-state gain saturation caused by coupling the probe pulse.

In Figure 6.7(*d*) the peak power of the front pulse is 200 mW, and that of the rear pulse is 50 mW. In this case, the gain of the front pulse is 9.2 dB and that of the rear pulse is 9.4 dB. The difference of 0.2 dB is due to the limits of measurement accuracy, and it is concluded that the gain decrease from 10.5 dB in Figure 6.7(*c*) to 9.2–9.4 dB in Figure 6.7(*d*) is also due to steady-state gain saturation caused by increasing the average input signal power. Neither pulse showed any distortion in this experiment.

The gain change of the EDFA is shown in Figure 6.8, when the time delay between the front and the rear pulses is varied from 20 to 300 ps. The open circles represent the gain of the rear pulse when the peak powers of both pulses are 50 mW, corresponding to Figure

6.7(*c*). The full circles indicate the gain of the rear pulse, corresponding to Figure 6.7(*d*). There is no change in the gain of either rear pulse as the delay time increases from 20 to 300 ps. The gains for average input powers of −10 dBm (50 + 50 mW) and −6 dBm (200 + 50 mW) are 10.9 and 9.1 dB, respectively. These results agree well with the results shown in Figures 6.7(*c*) and (*d*). It is known that the gain recovery time is of the order of 1 ms and the fluorescence decay time is 10–15 ms. This long gain recovery time is, however, effective in removing the 'pattern effect' (eye pattern change) in high bit-rate digital communication systems. If the gain recovery time were short, each pulse could be modified by the fast gain, resulting in a pattern effect in the pulse train.

6.4 Single-wavelength soliton amplification and transmission

This section explains how the EDFA is useful for non-adiabatic soliton amplification (Nakazawa *et al.*, 1989c). Figure 6.9 shows the experimental setup for single-wavelength soliton amplification and transmission through an erbium-doped fiber. The soliton source is a NaCl CCL at 1.535 μm. The repetition rate is 100 MHz. The soliton pulse is coupled to fiber A with a length of 5.6 km. Fiber A is a dispersion-shifted fiber with a zero dispersion at 1.497 μm. The fiber loss is 0.22 dB/km, the spot size W_0 is 3.6 μm, and the group velocity dispersion (GVD) at 1.535 μm is −2.7 ps/km · nm. The $N = 1$ soliton power $P_{N=1}$ is given by

$$P_{N=1} = 0.776 \frac{\lambda^3}{\pi^2 n_2 c} \frac{|D|}{\tau^2} (\pi W_0^2) \tag{6.9}$$

where λ is the soliton wavelength, n_2 is the non-linear refractive index, τ is the FWHM of the soliton, and πW_0^2 is the effective core area. In this case, $P_{N=1}$ is about 40 mW for $\tau = 9$ ps.

The EDFA pumping source is also a InGaAsP laser diode as described in Sections 6.2 and 6.3. The amplified solitons are transmitted through fiber B which is 27 km long. Fiber B is also a dispersion-shifted fiber with zero dispersion at 1.497 μm with a GVD of −2.7 ps/km · nm at 1.535 μm.

Figure 6.10 shows the gain characteristics for high-power input pulses as a function of erbium pump power. This experiment was undertaken before installing fibers A and B. For a pump power of 90 mW, the gains for three input peak powers of 63 mW (57 μW ave), 126 mW (113 μW ave), and 189 mW (170 μW ave) are 15, 13

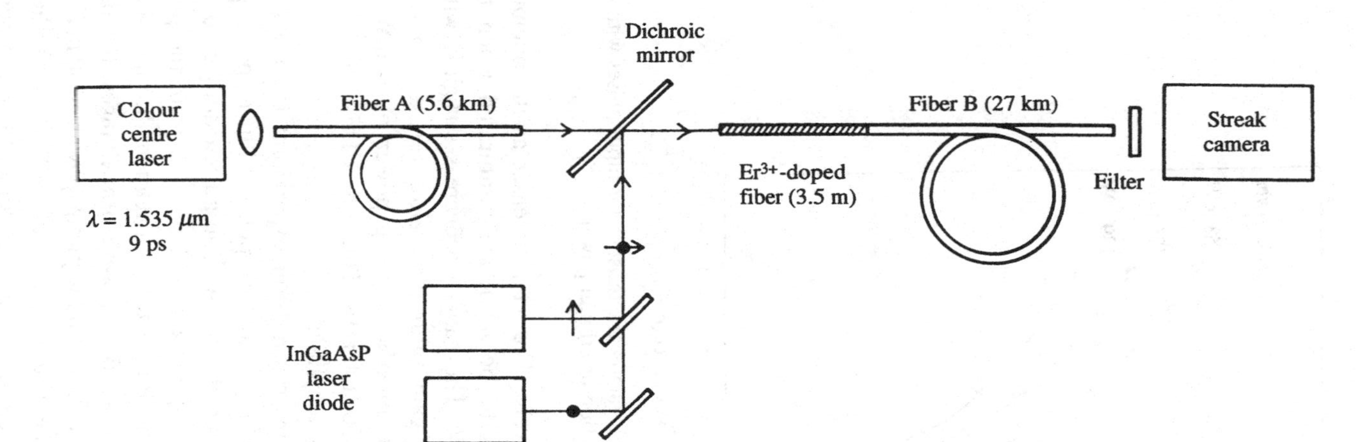

Figure 6.9. Experimental setup for single-wavelength soliton amplification and transmission through an erbium-doped fiber. The soliton source is a color center laser and the pumping source of the erbium fiber is a 1.48 μm InGaAsP laser diode.

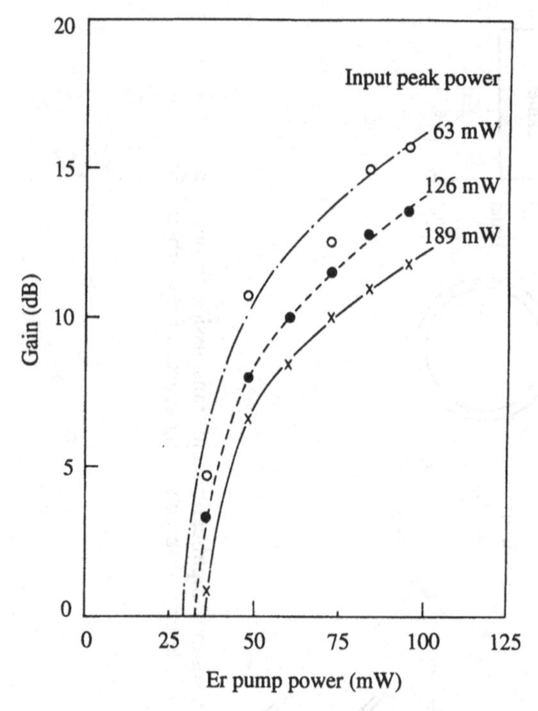

Figure 6.10. The gain characteristics for input pulses with high peak power as a function of erbium power.

and 11.8 dB, respectively. The gain increases as the pumping power increases, and begins to be moderately saturated when the pump power exceeds 50 mW. The gain also decreases slightly with higher-power inputs due to gain saturation. However, a gain larger than 10 dB is obtained for an erbium pump power of 50 mW when the soliton peak power is 50–100 mW. Thus, the EDFA can be utilised as a non-adiabatic soliton amplifier.

The result of soliton transmission experiment is shown in Figure 6.11. Figure 6.11(*a*) is a soliton pulse with a peak power of 80–100 mW at the input of fiber A. The pulsewidth for 20 accumulations is 15.9 ps. Timing jitter and a slight change in the pulse waveform will increase its width. After propagation through fiber A, the erbium amplifier and fiber B, the waveform changes into that shown in Figure 6.11(*b*) when the erbium pump power is 30 mW. Because of loss in fiber A, the coupling loss of 2 dB through the dichroic mirror, and the 0 dB gain in the erbium-doped fiber at a pump power of 30 mW, the pulse in fiber B becomes dispersive, and eventually it broadens to 35–42 ps due to the GVD of fiber B. The pulse broaden-

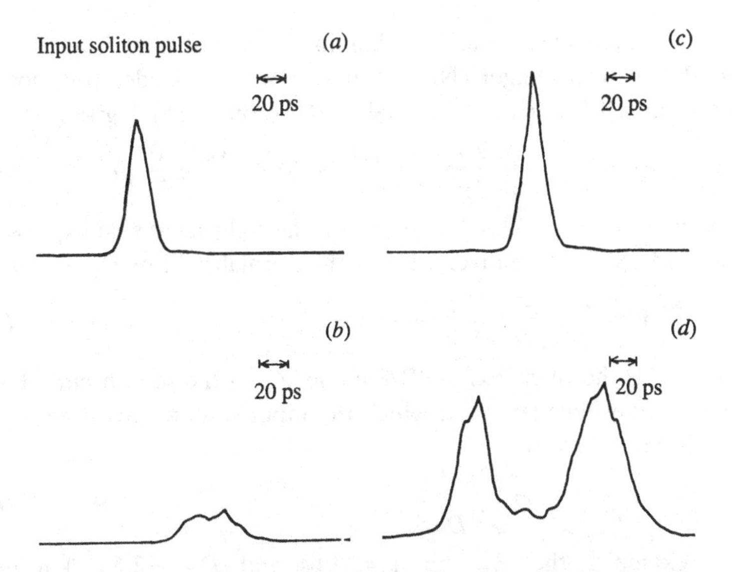

Figure 6.11. Experimental result of soliton transmission through EDFA (*a*) is the input soliton, (*b*) is the output pulse for a low gain, (*c*) is the output pulse for medium gain where the pulse width recovers its input width, and (*d*) is the output pulse for a high gain. A high order soliton was excited in (*d*).

ing due to GVD for a Gaussian pulse is given by

$$\tau_{\text{out}} = \tau\left[1 + \left\{4\ln 2\,\frac{\partial^2 k}{\partial\omega^2}\frac{l}{\tau^2}\right\}^2\right]^{1/2} \tag{6.10}$$

where τ_{out} is the FWHM of the output pulse at $z = l$, D is the GVD which satisfies $D = -(2\pi c/\lambda^2)(\partial^2 k/\partial\omega^2)$. Also, $\tau = 9$ ps, $D = -2.7$ ps/km · nm, and $l = 27$ km produces τ_{out} of 30 ps.

When the erbium pump power is increased to 43 mW, the output pulse changes to that shown in Figure 6.11(*c*), where it is seen that the output pulse almost retains the initial pulsewidth of the input soliton. Although the gain is 7–8 dB, the net gain is 5–6 dB due to a loss of 2 dB through the dichroic mirror. Solitons without wings have been obtained experimentally between gains of 6–9 dB (net gain 4–7 dB). By further increasing the pump power to 80 mW, we observe pulse splitting as shown in Figure 6.11(*d*), where the net gain is 10–12 dB.

Computational results for the experiment are given here (Nakazawa *et al.*, 1989b; Nakazawa, 1990a). We assume an $N = 1.2$ soliton at the input end of fiber A. The pulse experiences a fiber loss of 1.3 dB and a dichroic mirror loss of 2 dB, and then it is amplified by the erbium-doped fiber. We used parameters of $\tau = 12$ ps,

$D = -2.5$ ps/km · nm, and a fiber loss of 0.2 dB/km. The perturbed non-linear Schrödinger (NLS) equation which includes fiber loss and a soliton self-frequency shift (SSFS) (Gordon, 1986) is given by

$$(-i)\frac{\partial u}{\partial z} = \frac{1}{2}\frac{\partial^2 u}{\partial \tau^2} + |u|^2 u + i\Gamma u - \frac{\tau_n}{\tau_0} u \frac{\partial}{\partial \tau}|u|^2 \tag{6.11}$$

where the third and the last terms on the right-hand side express fiber loss and SSFS, respectively. Here, the normalised loss Γ is given by

$$\Gamma = \frac{2}{\pi}\left(\frac{Z_{sp}\gamma}{2 \times 10^4}\right)(\log e)^{-1} \tag{6.12}$$

where γ is the fiber loss in dB/km and Z_{sp} is the soliton period which denotes the fiber length at which the input soliton waveform repeats. Z_{sp} is expressed by

$$Z_{sp} = 0.322\frac{\pi^2 c\tau^2}{\lambda^2|D|} \tag{6.13}$$

For example, the Z_{sp} for $\tau = 12$ ps and $D = -2.5$ ps/km · nm is 23.3 km. Therefore, Γ for $\gamma = -0.2$ dB/km becomes 0.34. For SSFS, τ_n is estimated to be 5.9×10^{-15} s and $\tau_0 = \tau/1.76$, so that the coefficient τ_n/τ_0 is 8.7×10^{-4} for $\tau = 12$ ps.

Since SSFS plays an important role in explaining soliton fission in long fibers, it should be taken into account when examining soliton transmission over lengths greater than 10 km. It should be noted that SSFS appears prominently for higher order solitons and deforms the waveforms.

The waveform changes at 27 km of fiber B are shown in Figures 6.12(*a*)–(*f*) by setting net gains at 2, 3, 4, 5, 8 and 10 dBs, respectively. As observed in the experiment, single-peak solitons (*b* and *c*), single-peak solitons with small humps (*d*), and two peak solitons (*e*) clearly appeared in the simulation. These results indicate that optical solitons can be excited by using an EDFA as a lumped gain medium. In the next section, we describe how EDFAs can send solitons over long distances.

6.5 Dynamic range of the $N = 1$ solitons in a low-loss fiber

For soliton transmission, Hasegawa (1984) proposed using the gain generated by the stimulated Raman scattering to compensate for loss along the fiber. Raman gain acts as a distributed gain, so the soliton waveforms are maintained over long distances.

Here we discuss soliton transmission with lumped gain media (Nakazawa *et al.*, 1989b; Nakazawa, 1990a; Kubota and Nakazawa,

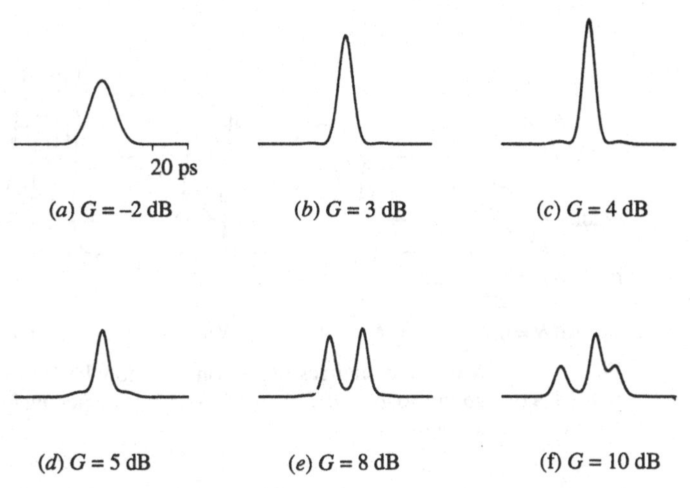

Figure 6.12. Soliton waveform changes at 27 km of fiber B. Waveforms (*a*)–(*f*) correspond to net gains of −2, 3, 4, 5, 8, and 10 dB, respectively.

1990). Instead of the Raman process, we use a new method for transmitting solitons called 'Dynamic Soliton Communication', in which we use a combination of an $N = 1.2$–2.0 soliton, the lumped gain generated by an erbium-doped fiber, and a low-loss fiber (Nakazawa *et al.*, 1990c). The key to success is to use a relatively large amplitude soliton input instead of an exact $N = 1$ soliton. Since we assume that the fiber loss is small, of the order of 0.2–0.3 dB/km, soliton narrowing occurs near the beginning of the fiber for $1 < N < 2$ solitons. Then the soliton starts to broaden since the GVD becomes dominant, because the fiber loss weakens the intensity-dependent non-linearity. In these soliton dynamics, there is a characteristic distance L_c at which the pulsewidth recovers to that of the input soliton, although the intensity is attenuated by the fiber loss. The decrease in the soliton amplitude at the fiber output can be compensated for by the lumped gain generated by the erbium-doped fiber. The non-linear phase change along the fiber, which is caused by fiber loss, should be taken into account when the total transmission distance, which includes many lumped amplifiers, exceeds 1000 km.

Figure 6.13 shows the waveform changes in the solitons in a low-loss fiber, where $\tau = 10$ ps and $D = -2.7$ ps/km · nm. Waveforms (*a*), (*b*), (*c*) and (*d*) correspond to input soliton amplitudes of $N = 0.7$, 1.0, 1.2 and 1.4, respectively. It can be seen that the width of the low amplitude soliton broadens more rapidly than the $N = 1.2$–1.4 solitons because the intensity-dependent non-linearity decreases faster.

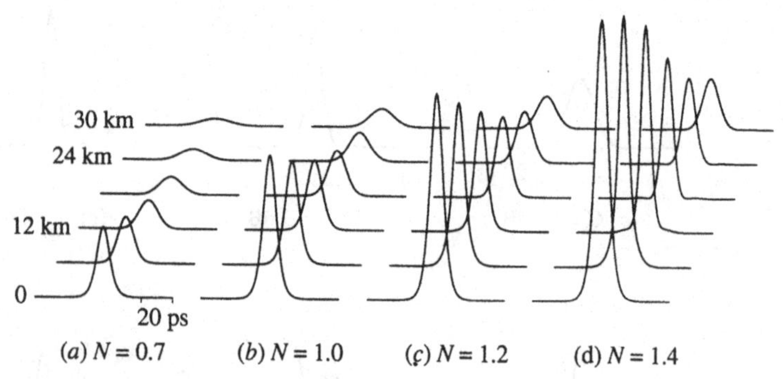

Figure 6.13. Waveform changes of solitons in a low-loss fiber. (*a*)–(*d*) correspond to $N = 0.7$, 1.0, 1.2 and 1.4, respectively. N is the input soliton amplitude.

However it should be noted that $N = 1.2$–1.4 solitons still retain their initial pulsewidth although the intensity decreases due to fiber loss. The pulsewidth and the intensity changes are shown as a function of distance in Figures 6.14(*a*) and (*b*), respectively. When N is 1.4 in (*a*), the soliton width returns to its initial value at 24 km, at which the intensity (N^2) decreases by 0.61 ($= 0.78^2$) from the input value of 1.96 ($= 1.4^2$). Therefore, the loss at peak intensity is -5.1 dB. In this Dynamic Soliton Communication system, the L_c of 24 km becomes its repeater length. Although the $N = 1$ soliton exists between $0.5 < N < 1.5$, $0.5\,\text{sech}\,(\tau)$ generates broader soliton pulses than $\text{sech}\,(\tau)$, while $1.5\,\text{sech}\,(\tau)$ generates a narrower soliton as it propagates. It should be noted that if the intensity decrease of the fiber output is less than 9 dB ($= 10\log(1.5/0.5)^2$) for an input of the $N = 1.5$ soliton, the output pulse is still considered to be a soliton. Therefore, the pulse at 24 km is still a soliton and we use it as an input pulse for the next soliton transmission by amplifying it with a 'lumped gain' such as an EDFA. In Figure 6.14(*b*) the intensity decrease is not followed by an exponential decay because of the non-linear intensity change.

6.6 Multi-wavelength soliton amplification and transmission with collision

Figure 6.15 shows the experimental setup for multi-wavelength soliton amplifications and transmission (Nakazawa *et al.*, 1989b; Nakazawa, 1990a). The soliton sources are two CCLs: one is a KCl CCL and the other is a NaCl CCL. They are pumped by a

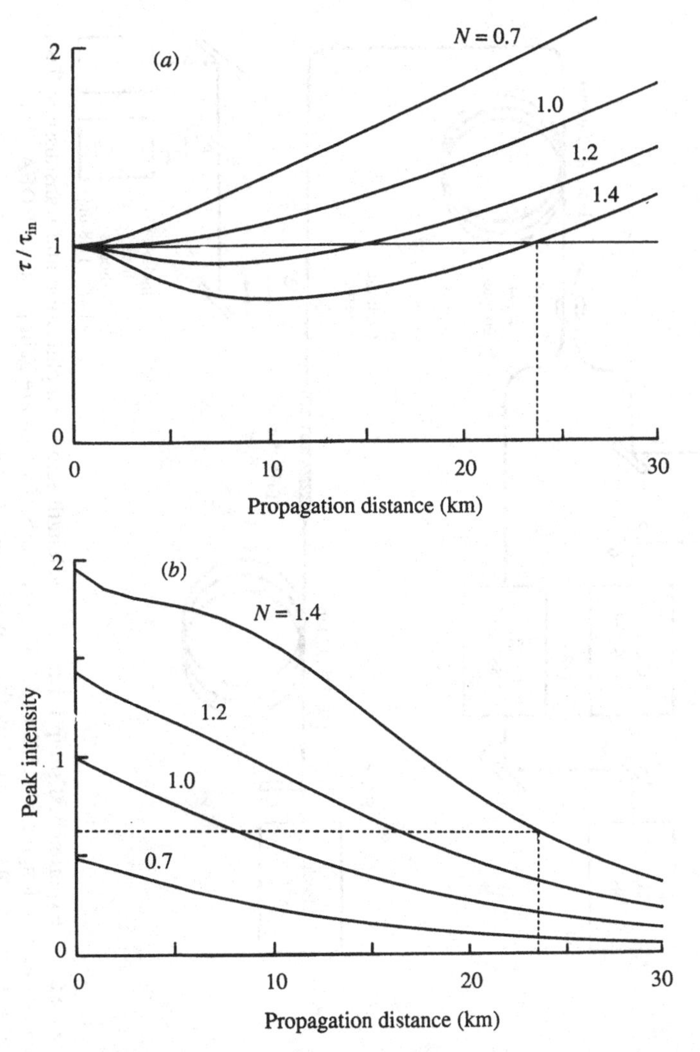

Figure 6.14. The soliton pulsewidth and the intensity changes as a function of propagation distance. (*a*) and (*b*) refer to the normalised width and the intensity, respectively, for various N numbers.

1.06 μm mode-locked YAG laser at a repetition rate of 100 MHz. One YAG laser synchronises both CCLs, otherwise the timing of the two solitons changes, causing additional jitter. One CCL wavelength is 1.535 μm and the pulsewidth is 13 ps. The other is 1.552 μm and the width is 9 ps. The delay between the two solitons is controlled by moving the corner cube. Similar to Figure 6.9, two soliton pulses are coupled into fiber A through a fiber coupler. The GVDs at 1.535 and

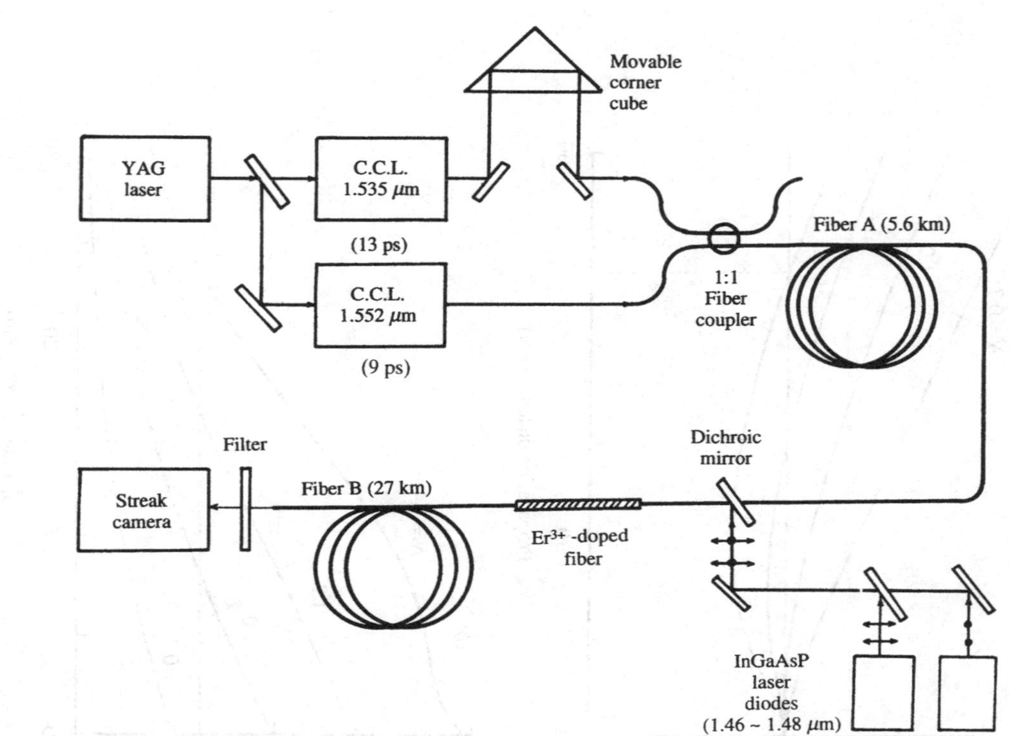

Figure 6.15. Experimental setup for multi-wavelength soliton amplification and transmission. Two solitons at wavelengths of 1.552 and 1.535 μm, both of which are gain peaks of EDFA. The collision timing was controlled by the movable corner cube.

1.552 μm are -2.7 and -3.8 ps/km · nm, respectively. The $N = 1$ soliton power for $\tau = 13$ ps at 1.535 μm is 9 mW and that for $\tau = 9$ ps at 1.552 μm is 61 mW from equation (6.9). The amplified solitons with different wavelengths are introduced into fiber B. The 1.552 μm soliton is coupled into fiber B before the 1.535 μm soliton to cause the two solitons to collide in fiber B.

Figures 6.16(*a*) and (*b*) show optical gains at 1.535 μm (front) and 1.552 μm (rear) solitons, respectively. The broken and the full curves in (*a*) indicate conditions with and without the 1.552 μm soliton, respectively. The curves in (*b*) indicate conditions with and without the 1.535 μm soliton, respectively. Gain was measured for input peak powers of 20 mW at 1.535 μm (26 μW ave) and 59 mW (53 μW ave) at 1.552 μm, both corresponding to $N = 1$ soliton power. Three delays were chosen between the two solitons to investigate how the gain of the second soliton is affected by the first soliton. Time delays of 50 ps, 200 ps and 1 ns are shown by ○, × and △, respectively. For high erbium pump powers, gain saturations are observed in each case. When there is no first soliton at 1.552 μm in Figure 6.16(*a*), the gains at 1.535 μm are 7 and 12.5 dB for erbium pump powers of 50 and 90 mW, respectively. These gains decrease by 2 dB due to the existence of the first soliton pulse at 1.552 μm. This is the gain cross talk. Although this cross talk reduces the average gain, it cannot respond at a high repetition rate, resulting in no intermodulation. In addition, the erbium pump power required for a zero net gain increases from 34 to 38 mW.

Gain changes were not observed for these three delays. This means that the gain recovery time of erbium ions is so slow that the gain reaches its average value at 10 ns repetition (100 MHz). The steady-state gain at 1.535 μm (rear soliton) depends on the 'average power' of the first soliton pulse at 1.552 μm as we mentioned in Section 6.3. The gain at the second soliton at 1.535 μm is affected by the first soliton at 1.552 μm. This may be attributed to the fact that the relatively high input power (-11 dB m ave) causes stronger saturation and therefore considerable saturation-induced cross talk occurs between 1.535 and 1.552 μm.

The gains at 1.552 μm, without the 1.535 μm soliton pulse shown in Figure 6.16(*b*) are 7 and 11 dB for erbium pump powers of 50 and 90 mW, respectively. Here, the gain decreases by about 1 dB due to the presence of the second soliton at 1.535 μm, which is smaller than that in Figure 6.16(*a*). This is because the gain is determined by the average soliton power. It is also noted that erbium pump power at

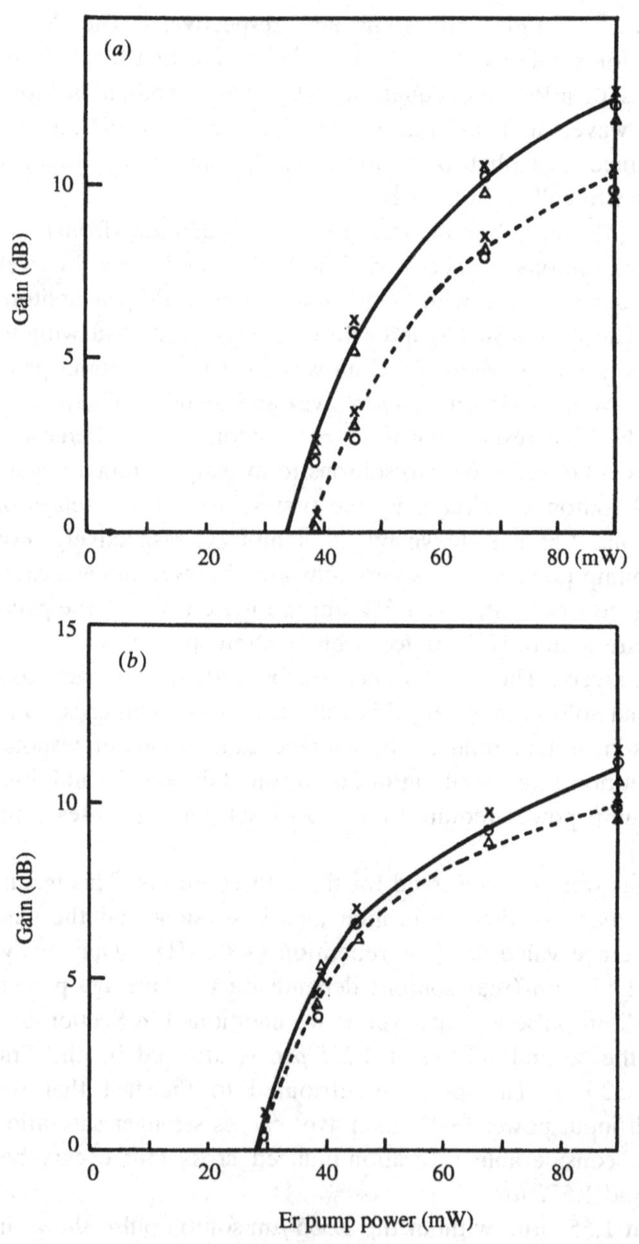

Figure 6.16. Optical gains at (*a*) 1.535 and (*b*) 1.552 μm, respectively. The gain saturation at 1.552 μm is smaller than that at 1.535 μm because the soliton energy for 1.535 μm in this case is smaller than that at 1.552 μm. In (*a*) the full curve is the gain without the 1.552 μm soliton, the broken line is with it. In (*b*) the full curve is the gain without the 1.535 μm soliton.

zero net gain for 1.535 μm is 34 mW and that for 1.552 μm is 28 mW, while the gain at 1.535 μm for 90 mW erbium pump is 12.5 dB and that at 1.552 μm is 11 dB. These results mean that the absorption at 1.535 μm is larger than that at 1.552 μm; however, fluorescence at 1.535 μm is also larger than that at 1.552 μm.

Figure 6.17 shows results of the two colored soliton transmission experiments (Nakazawa *et al.*, 1989a). The experiments were undertaken for delays between the two solitons of 50 ps, 200 ps and 1 ns. Since the wavelength of the first soliton was 1.535 μm, soliton collisions for each delay occur at 0.9, 3.7 and 18.3 km, from the input of fiber B. Although the delays at the input end of fiber B are different, each soliton waveform at the output end of fiber B was insensitive to the collision. Thus, we show the results for a 50 ps delay. The soliton waveform into the erbium fiber is shown in Figure 6.17(*a*), where the wavelength of the front soliton is 1.552 μm and the wavelength of the rear soliton is 1.535 μm. Input peak powers into the erbium-doped fiber of 59 mW (53 μW ave) at 1.552 μm and 21 mW (27 μW ave) at 1.535 μm are chosen, both corresponding to $N = 1$ soliton power. From these peak powers, the intensity between 1.552 and 1.535 μm soliton pulses should be about 3:1 in Figure 6.17(*a*). However, the radiant sensitivity of the streak camera (S_1 tube) strongly depends on wavelength in the 1.5 μm region. Therefore, the amplitude of 8 for 1.552 μm in Figure 6.17(*a*) should be read as 12 when the amplitude is compared with the 1.535 μm line.

The output waveforms for an erbium pump power of 30 mW are shown in Figure 6.17(*b1*) for 1.552 μm and (*b2*) for 1.535 μm. Since the erbium pump power is weak, the soliton cannot be amplified in the erbium fiber, resulting in pulse broadening due to the GVD of the fiber. The pulse broadening in (*b1*) (1.552 μm) is 45–52 ps and that in (*b2*) (1.535 μm) is 27–31 ps. The calculated broadening from equation (6.10) for 1.552 μm is 41 ps and that for 1.535 μm is 24 ps, both of which are slightly narrower than those in the experiment.

The waveform changes of $N = 1.5$, 2, 3 and 3.5 solitons for the two cases ($\tau = 9$, 1.535 μm and 13 ps, 1.552 μm) at 27 km of fiber B (0.22 dB/km) are shown in Figure 6.18, where Figures 6.18(*a*)–(*h*) correspond to $N = 1.5$, 2, 3 and 3.5, respectively. Figures 6.18(*a*) and (*e*) represent $N = 1.5$ (gain of 3–4 dB), and correspond to the experimental results in Figures 6.17(*c1*) and (*c2*), respectively. Figures 6.18(*c*) and (*g*) represent the $N = 3$ soliton (gain of 9–10 dB), and correspond to Figure 6.17(*d1*) and (*d2*), respectively. Quantitative agreement with the experimental results is obtained in both cases.

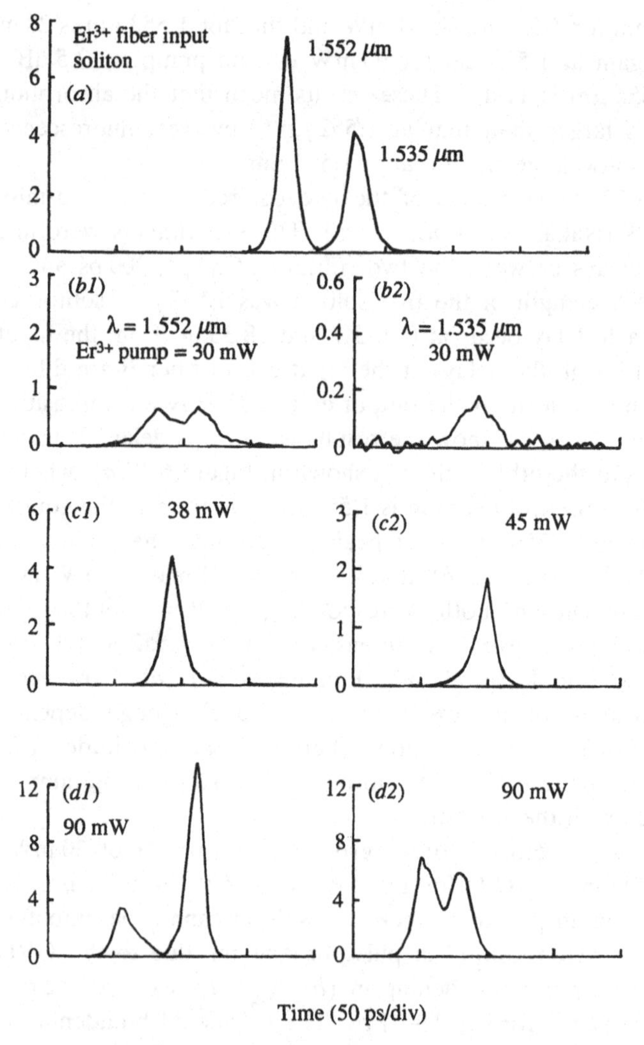

Figure 6.17. Output waveforms for two colored soliton inputs. (*a*) is the input condition, (*b1*) and (*b2*) are outputs for a low gain, (*c1*) and (*c2*) are outputs for medium gain which can reproduce the input widths, and (*d1*) and (*d2*) are outputs for a large gain.

6.7 Subpicosecond soliton amplification

When applying EDFAs to high-speed optical soliton communication systems, it is important to reveal their pico and subpicosecond response characteristics (Suzuki *et al.*, 1989b). Subpicosecond soliton pulses were generated by a combination of a CCL and a dispersion-shifted fiber. The pulsewidth of the CCL was 9 ps and the repetition

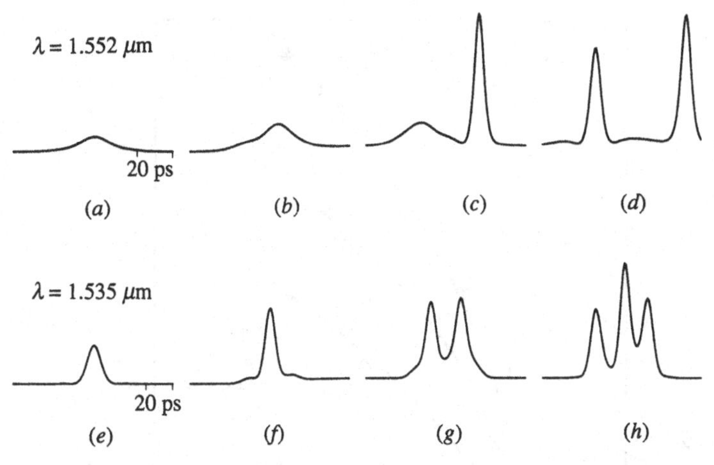

Figure 6.18. Soliton waveform changes of $N = 1.5$, 2, 3 and 3.5 solitons at 27 km of fiber B. (a)–(d) refer to $N = 1.5$, 2, 3 and 3.5 for 1.552 μm and (e)–(h) refer to those for 1.535 μm, respectively.

rate was 100 MHz. The experimental setup is the same as that shown in Figure 6.9. The $N = 24$ soliton was excited when the coupled peak power to fiber A was 22 W. Due to the excitation of high-order solitons and SSFS, the pulsewidth at the output end of fiber A was 0.7 ps, with no accompanying satellite pulses. The subpicosecond soliton thus obtained was coupled into a 3.5-m-long erbium-doped fiber through a dichroic mirror.

The pump power dependence of the gain of the EDFA for sub-picosecond pulses is shown in Figure 6.19. The open and full circles show the gain characteristics for input peak powers of 40 and 12 W, respectively. The signal wavelength was 1.552 μm and the pulsewidth was 0.9 ps. When the coupled pump power was 90 mW, gains for input peak powers of 40 and 12 W were 3.8 and 5.5 dB, respectively. The gain for an input power of 40 W is lower than that for one of 12 W, due to the gain saturation.

The triangles in Figure 6.19 show the gain for a signal wavelength of 1.535 μm. The pulsewidth was set at 0.7 ps and the input peak power was 43 W. In the present case, a gain of 2.8 dB was obtained for a pump power of 90 mW. Since a process using erbium fibers can amplify pulses with peak powers of a few tens of watts with relatively low pump power, it is superior to stimulated Raman scattering when it is used as an amplifier in soliton transmission systems.

It is found that the net gain at 1.535 μm is less than that at

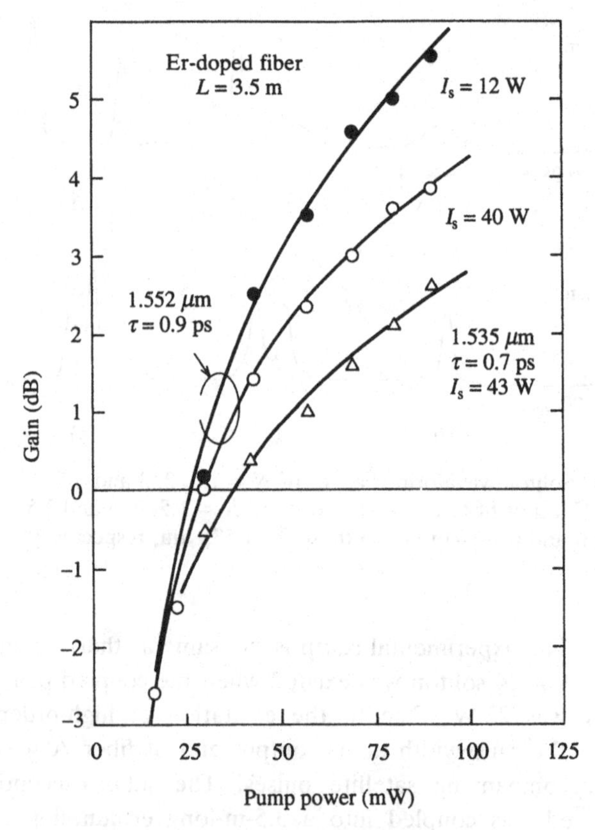

Figure 6.19. Pump power dependence of EDFA gain for subpicosecond pulses. The gain at 1.55 μm is larger than that at 1.535 μm for subpicosecond pulses.

1.552 μm. However, the fluorescence curve of an erbium-doped fiber shows that the small signal gain at 1.535 μm is larger than that at 1.552 μm. This is because a gain bandwidth of several nanometres is required to amplify subpicosecond soliton pulses. For example, assuming a hyperbolic secant-squared (sech^2) waveform, a soliton with a pulse width of 0.7 ps at 1.535 μm has a spectral bandwidth of 3.6 nm. The 3 dB gain bandwidth of an erbium-doped fiber is 5.2 nm around 1.535 μm and 14.3 nm around 1.552 μm. Since the gain bandwidth at 1.535 μm is almost equal to or less than the spectral width of the subpicosecond pulses, the effective gain for the subpicosecond soliton is small, although the narrow band gain at 1.535 μm is large. However, the gain bandwidth centred on 1.552 μm is broad enough to amplify subpicosecond solitons. Thus the gain for subpicosecond pulses in the 1.552 μm band is larger than that in the 1.535 μm band.

6.8 Femtosecond soliton amplification with adiabatic gain narrowing

We have described picosecond (9–20 ps) ~ subpicosecond (0.7–0.9 ps) soliton amplification using EDFAs, in which the soliton period was much longer than the length of the EDFA (3–5 m) and the peak power of the amplified pulse reached as high as 96 W. Recently, Zhu *et al.* (1989) reported the application of an erbium-doped fiber to the coupled-cavity mode-locking of a KCl:Tl femtosecond laser, and Ainslie *et al.* (1990) reported a few dB gain for a 200 fs soliton pulse. If the full bandwidth of approximately 30 nm can be utilised, a 100 fs pulse could be amplified. In this section, we describe the amplification characteristics of femtosecond pulses under various pumping conditions for EDFAs (Nakazawa *et al.*, 1990d).

Femtosecond infrared pulses in the 1.5 μm region are generated by using difference frequency mixing in a KTP crystal (Kurokawa and Nakazawa, 1989). Here, a synchronously pumped femtosecond dye laser pumped by a c.w. mode-locked 1.064 μm YAG laser is used. A 1.064 μm, 100 ps pulse is mixed with a 0.63 μm visible femtosecond pulse in a 3 mm KTP crystal, and the difference frequency-mixing generates a femtosecond pulse in the 1.5 μm region. To increase the 1.5 μm peak power, a cavity dumper is installed inside the dye laser cavity. Consequently, the pulse repetition rate is decreased to 3.8 MHz. By rotating a birefringent filter installed in the dye laser, tunable infrared pulses between 1.4 and 1.6 μm can be obtained. An optical filter is placed after the KTP crystal, resulting in 250 fs pulses with peak powers of 2–4 kW.

These femtosecond pulses are coupled into an EDFA through a 1.48 μm/1.55 μm WDM fiber coupler. The length of the erbium fiber is 3–5 m and the doping concentration of erbium ions is 1900 ppm. Al_2O_3 with a concentration of 500 ppm was co-doped to increase the optical gain. The zero GVD, spot size, and cut-off wavelength are 1.46, 3.5 and 1.28 μm, respectively. Since the GVD at 1.552 μm is -6.4 ps/km · nm, the $N = 1$ soliton peak power for 250 fs input is approximately 120 W, which can be easily generated by the present difference frequency-mixing technique. The pulsewidth and the spectrum of the amplified output solitons are monitored using an autocorrelator with a resolution of less than 100 fs and a spectrum analyser.

When erbium fibers about 100 m long and with a doping concentration of 100 ppm or 20 m long with a concentration of 470 ppm were used for the EDFA, the self-frequency shift become quite large and the continuum spectrum broadened from 1.55 to 1.75 μm. In this case, the spectral component of the soliton has already moved outside the EDFA gain profile, so high gain was not obtained. The key to achieving efficient femtosecond pulse amplification is to use a shorter erbium fiber with a relatively high doping concentration of between 1000 and 2000 ppm.

The gain characteristics for 266 fs input pulses are shown in Figure 6.20. The peak power of the pulse and the repetition rate are 31 W and 3.8 MHz. The average input power was 30 μW (−15.2 dBm). Figure 6.20(*a*) shows the input spectrum and (*b*) the output spectrum of the EDFA at a pump power of 20 mW. Figures 6.20(*c*) and (*d*) show input and output pulse waveforms corresponding to (*a*) and (*b*), respectively. It should be noted that one division is 206 fs in (*c*) and 323 fs in (*d*). In trace (*a*), the center wavelength of the pulse is 1.551 μm which is one of the EDFA gain peaks. The spectral width $\Delta\lambda$ is 10.3 nm and the FWHM of the pulse waveform from (*c*) is 266 fs. Hence, the pulsewidth–bandwidth product, $\Delta\tau\Delta\nu$, is 0.32–0.34, which means that the pulse is a transform-limited sech waveform. Trace (*b*) shows that the input pulse was successfully amplified and a gain of 9.2 dB was obtained for a pump power of 20 mW. The amplified spectrum is accompanied by a small pedestal in the shorter wavelength region around 1.535 μm, which is due to the existence of a gain peak at that wavelength in the EDFA. It is also noted that the spectral width of the output pulse is 9.3 nm which is a little narrower than that of the input pulse. This is attributed to the fact that the EDFA has a parabolic gain profile at around 1.552 μm, which modifies the input spectral profile when it is amplified. For convenience, an amplified spontaneous emission (ASE) profile is shown in (*b*) by a broken line. According to trace (*b*) and output photo (*d*), $\Delta\tau\Delta\nu$ is equal to approximately 0.31–0.32, which means that the pulse is a transform-limited sech pulse. When a non-transform-limited pulse with a broader spectral width and the same pulsewidth is coupled into the EDFA, the output pulse also remains at 250 fs and the spectral width is limited to 9–10 nm. These results indicate that a strong spectral shaping exists at around 1.552 μm, so that non-transform-limited pulses can be converted into transform-limited pulses by active spectral windowing.

Since the gain is 9.2 dB for an input peak power of 31 W, this produces output pulses with a peak power of about 260 W. Based on a gain coefficient of 3 dB/m (9.2 dB gain for an erbium fiber length of 3 m), the $N = 1$ soliton is already excited at around 2 m from the

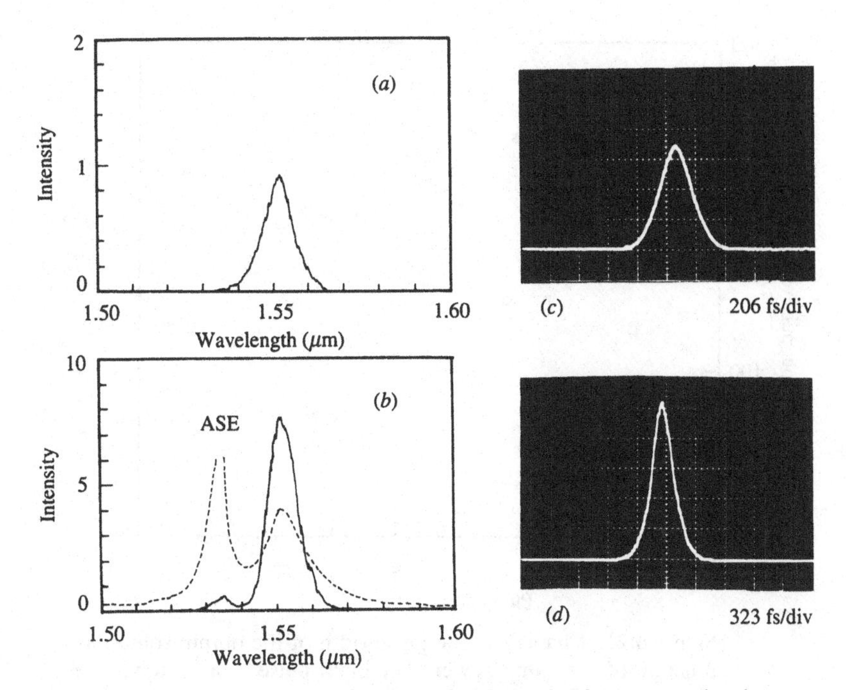

Figure 6.20. Gain characteristics of EDFA for femtosecond pulses. The input pulse width is 266 fs and the input peak power is 31 W, which is not in the soliton power regime. (*a*) and (*c*) are the input spectrum and its waveform, respectively. (*b*) and (*d*) are the corresponding output spectrum and waveform, respectively.

fiber end. When the pump power is further increased (the gain is large), SSFS appears and the carrier wavelength of the input soliton moves toward a longer wavelength region, so that the gain cannot be accurately measured.

Changes thus obtained in the pulsewidth of the output soliton as a function of the pump power are shown in Figure 6.21. As seen, the pulsewidth is linearly shortened with an increase in the pump power and the spectral width is broadened corresponding to this pulse shortening. It is important to note here that the output pulsewidth can be controlled by using adiabatic soliton narrowing in the gain medium. For pump powers smaller than 5 mW, the EDFA acts as a loss medium, which makes the output power small and means that the pulsewidth cannot be measured with the autocorrelator.

For a −7 dB m input, the gain would be roughly 4–5 dB. From the perturbation theory for soliton pulses, the soliton pulse is shortened to $1/(1+2\Delta)$ when a perturbation Δ exists. Here Δ for the gain

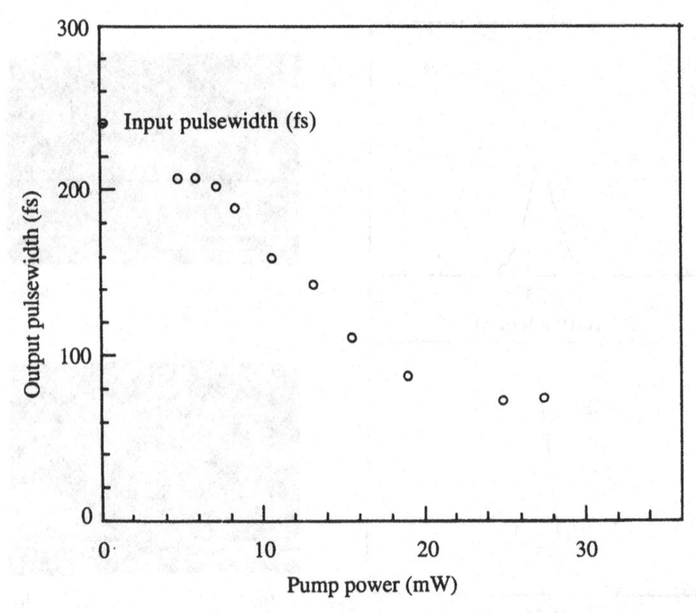

Figure 6.21. Changes in the pulsewidth of the output soliton as a function of the pump power. The input pulse is in soliton power regime. It is clearly seen that the adiabatic soliton narrowing occurs when the EDFA gain is increased.

medium is given by $2\Delta = G/(10 \log e)$. G is the power gain in dB. Therefore, when G is equal to 5 dB, 2Δ is 1.15 and hence the pulsewidth is shortened to 0.46 (= 1/2.15) that of the input pulse (240 fs × 0.46 = 110 fs) which agrees with our experimental results. Pulses of 60–70 fs result from SSFS not from adiabatic narrowing. It should be noted that soliton trapping occurs within the gain bandwidth of the EDFA when adiabatic gain narrowing occurs. However, when the pump power is increased to more than 20 mW, the soliton narrowing stops at a pulsewidth of approximately 60–80 fs, and pedestal components appear on the wing of the soliton pulse. At this stage, SSFS dominates and the center wavelength of the soliton spectrum is far from the gain bandwidth of the EDFA, resulting in no further adiabatic soliton narrowing.

6.9 GHZ soliton generation using a gain-switched distributed feedback laser diode with spectral windowing

To date, solitons have been mainly generated by high-power lasers such as CCLs and infrared dye lasers. From a practical standpoint, soliton generation from a laser diode is most desirable. It is well known, however, that the generation of transform-limited pulses from a gain-switched laser diode is very difficult since a frequency down-chirp inevitably occurs due to carrier density modulation. Even in a distributed feedback laser diode (DFB LD) transient spectral changes exist, which cause a non-linear chirp-induced dispersion penalty. This is one of an important factor limiting the high-speed transmission of the DFB laser signal in a linear scheme. The transform-limited condition is also very important in soliton transmissions and it was very difficult to generate true solitons unless spectral manipulation such as narrow-band optical optical filtering is applied to laser diodes.

In this section, we describe the generation and transmission of transform-limited solitons from a gain-switched DFB LD with spectral windowing and EDFAs (Nakazawa *et al.*, 1990e, f). The experimental setup is shown in Figure 6.22. The laser source is a bulk-type DFB LD, which is driven by a sinusoidal current modulation in the GHz region. A picosecond pulse train with a GHz repetition rate can be easily generated with this gain switching technique. Although the peak power is larger than 100 mW, the pulses are far from transform-limited. Therefore, if these pulses are sent as solitons, they do not work as solitons, resulting in broadening due to the inherent non-linear chirp.

To eliminate this disadvantage, we use a Fabry–Perot resonator (FPR) as a narrow-band spectral filter which can select only a small portion of the whole spectral component from the DFB LD. With this technique, although the transmitted average power through the Fabry–Perot resonator is reduced to 1/10 the input power, the pulse becomes a transform-limited pulse. Then, the EDFAs are used as booster amplifiers to increase the peak power up to the soliton power level.

The soliton transmission fiber is a 22-km dispersion-shifted fiber with a zero dispersion at 1.500 μm and -2.3 ps/km · nm dispersion at 1.545 μm. The $N = 1$ soliton peak power for 17 ps pulses is as small as 6 mW. Since the output from the EDFA reaches as high as 100 mW, pure solitons can be excited in the fiber.

A time-division multiplcxing technique using 3 dB(1:1) couplers is a simple way to increase the repetition rate of the original pulse train. We have described a preliminary transmission experiment on a 20 GHz soliton burst train using fiber 3 dB couplers, in which a 100 MHz repetition was converted to 20 GHz (Nakazawa *et al.*,

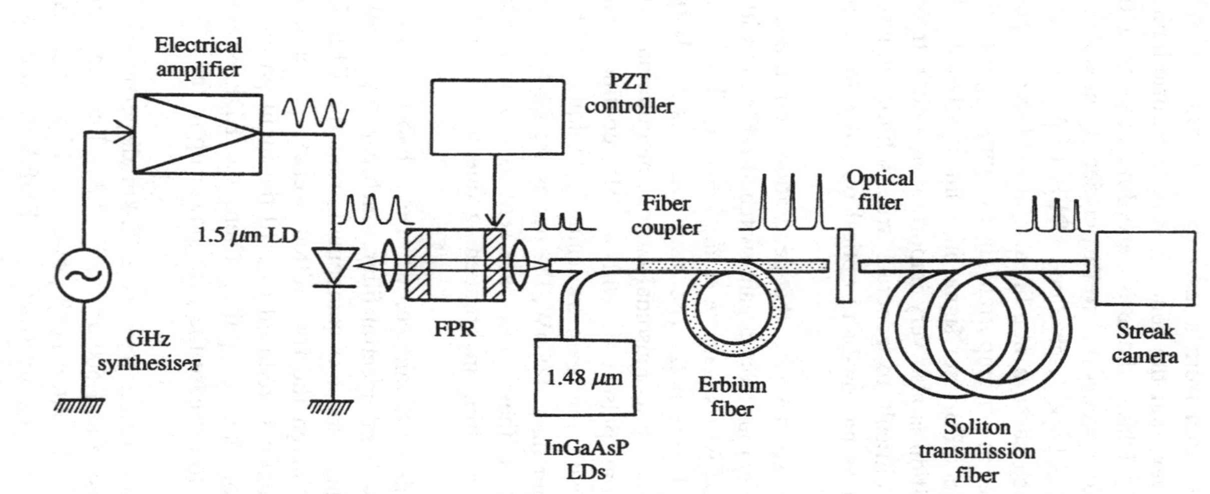

Figure 6.22. Experimental setup for generation of solitons from a gain-switched DFB laser diode with spectral windowing and EDFA.

1989d). Here, a 6 GHz soliton train from the DFB LD is successfully converted into 12 and 24 GHz soliton trains. Figures 6.23(*a1*) and (*b1*) correspond to the 12 and 24 GHz soliton trains thus obtained. For the 12 GHz pulse train, the pulsewidth broadened to 26.9 ps after a transmission of 22 km when the EDFA gain was small. This is shown in Figure 6.23(*a2*). When the EDFA gain is high, the output pulsewidth recovers to its original value as shown in Figure 6.23(*a3*), which means that a 12 GHz soliton train using a DFB LD can be completely transmitted with the EDFA. Similar to the 12 GHz results, for a low gain condition, the output pulse is no longer a soliton, since it broadened to 23.8 from 15.4 ps. This is shown is Figure 6.23(*b2*). When the gain is large as shown in Figures 6.23(*b3*), a 24 GHz soliton train can completely recover although overlapping occurs on the wings of the soliton. The average output power is about 12 mW and, therefore, the peak power at a 24 GHz repetition rate is 29 mW which is much higher than the $P_{N=1}$ soliton power of 6 mW.

The soliton narrowing caused by the excitation of high-order solitons at 6 GHz repetition is described as a function of the input peak power by changing the EDFA gain in Figure 6.24. By increasing the gain the 17 ps input solitons have been compressed to 9.6 ps. It was also possible to excite $N = 3$–4 solitons.

6.10 Soliton amplification in an ultralong distributed erbium-doped fiber amplifier

There are two types of EDFA. One is the lumped amplifier described previously and the other is a distributed amplifier. The distributed erbium amplifiers offer very attractive prospects for the realisation of ultralong lossless transmission lines. For example, they can be applied to optical soliton communication (Nakazawa *et al.*, 1990b). To achieve distributed amplifiers longer than 10 km, the doping concentration is decreased to as low as a few ppm. In this section, we describe soliton transmission over an ultralong dispersion-shifted erbium fiber with a doping concentration as low as 0.5 ppm.

Ultralong distributed EDFAs were fabricated by vapor axial deposition, which is very effective in dehydrating the core soot. Erbium ions are doped into the core soot with a solution doping technique. The glass composition is Er:GeO_2–SiO_2, the erbium concentration, estimated by absorption, is 0.5 ppm, and the fiber length is 9.4 km. The fiber has a dual shape core and the erbium ions are doped into the inner (or center) core, so that only the high intensity part of the

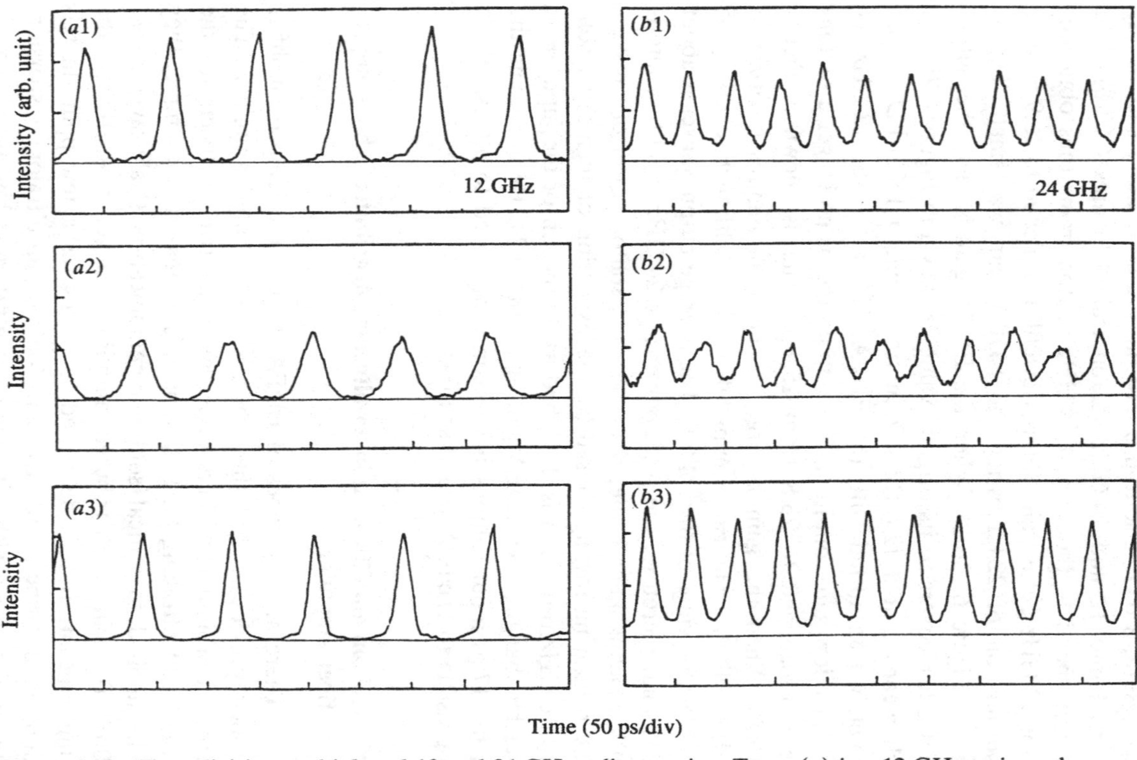

Figure 6.23. Time division multiplexed 12 and 24 GHz soliton trains. Trace (*a*) is a 12 GHz train and trace (*b*) is a 24 GHz train. (*a2*) and (*b2*) correspond to a low gain case, and (*a3*) and (*b3*) correspond to a high gain case.

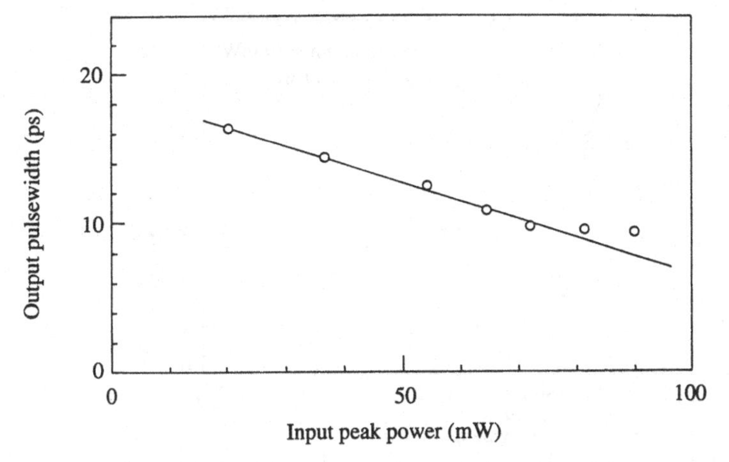

Figure 6.24. Soliton narrowing as a function of the input peak power.

pump field interacts with the erbium ions. The mode field diameter is 7.7 μm and the zero dispersion is 1.49 μm. The loss at 1.535 μm was 6.3 dB/km and that at 1.552 μm was 2.4 dB/km. The absorption at 1.48 μm was 1.3 dB/km. The contribution from OH absorption at 1.48 μm was estimated to be 0.23 dB/km by assuming a λ^{-4} loss dependency. This means that the 1.48 μm wavelength region is still an efficient pump region even in a distributed amplifier. The index profile has a Dual Shape Core (the inner, $2a_2$, and the outer, $2a_1$, cores), and the zero GVD is controlled by changing the refractive index ratios and the a_1, a_2 diameters.

The gain change as a function of signal wavelength is shown in Figure 6.25, where changes in loss measured with a wavelength tunable laser diode are also indicated by crosses. The pump power (1.48 μm InGaAsP LD) was fixed at 59 mW and the average signal power was −30 dB m. A gain as large as 23 dB was obtained at 1.535 μm. The gain bandwidth, defined by a 3 dB gain decrease, was approximately 3 nm for the 1.535 μm gain peak. The gains at 1.552 and 1.545 μm were 13.5 and 7.0 dB, where the losses were 2.5 and 1.9 dB/km, respectively. The 3 dB gain bandwidth at 1.552 μm was 9 nm.

A 20 ps soliton train was generated by a combination of a 1.533 μm gain-switched DFB LD and EDFAs as described in Section 6.9. In this case, GVD is −2.5–3 ps/km · nm, the spot size is $\pi(3.8 \times 10^{-6}\ \text{cm})^2$, and the pulsewidth is about 20 ps. Thus, $N = 1$

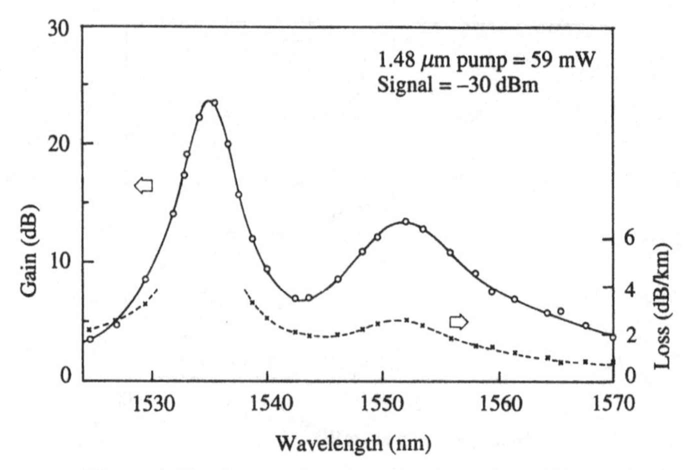

Figure 6.25. Change in gain of a dispersion-shifted, distributed EDFA as a function of signal wavelength. Open circles and crosses correspond to the gain and loss of the EDFA, respectively.

soliton peak power is approximately 8.3–10 mW. In the experiment, a pulse train with a peak power of 8.5 mW and a repetition rate of 2.5 GHz was coupled into the distributed fiber amplifier. The average input power was 0.43 mW (−3.7 dB m).

Changes in gain and output soliton width as a function of pump power at 1.48 μm are shown in Figure 6.26. Output soliton waveforms, corresponding to pump powers of 1–4 given in Figure 6.26, are shown in Figure 6.27. Figure 6.27(*a*) is an input soliton train monitored with a synchroscan streak camera. For a pump power of 9 mW (2), the fiber is still absorptive and the loss is about 8 dB and therefore, the pulsewidth was broadened to 22 ps because of the GVD, as shown in Figure 6.27(*b*). For a pump power of 14 mW (3), which is the pump threshold giving a zero net gain, the output pulsewidth has recovered to 20 ps as shown in Figure 6.27(*c*). This condition corresponds to a lossless transmission line. By increasing the pump power up to 30 mW, the gain increases to 6–8 dB and a pulse narrowing occurs down to 18.5 ps. When the pump power is further increased, the soliton narrowing is not so clear since the gain saturation is severe. Here the gain is not as high as that shown in Figure 6.26 because the coupled average power is as high as −3.7 dBm, resulting in a gain saturation. The saturation in the pulsewidth is considerable, but the amplitude saturation seems to be rather moderate. This situation varies when the input power changes. The soliton pulse waveform for a pump power of 56 mW (4) is shown in

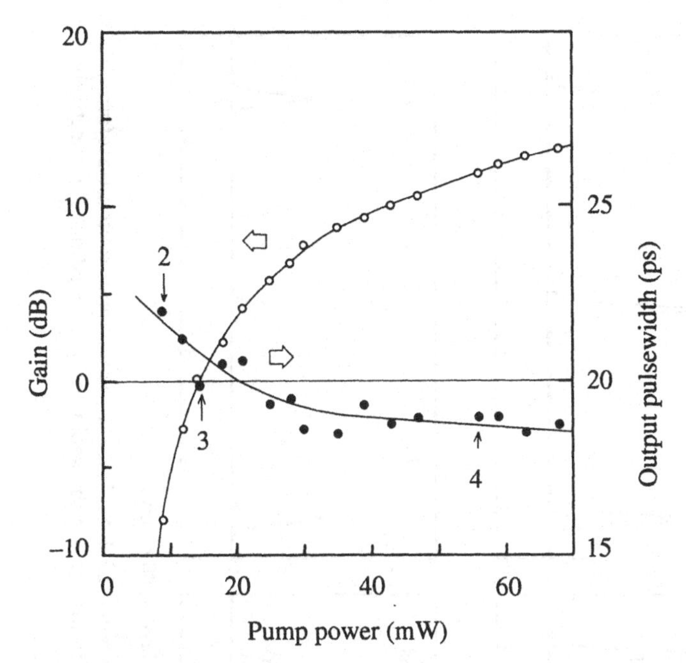

Figure 6.26. Changes in gain and output soliton width as a function of pump power at a wavelength of 1.48 μm. Open and closed circles refer to gain and output pulsewidth, respectively.

Figure 6.27(*d*), in which neither small satellite pulses nor asymmetries in pulses have been observed. They were very clean soliton pulses with a gain of about 12 dB. This gain was smaller than that given in Figure 6.27, but this is due to the difference in the input power level.

These results indicate that it is possible to amplify and transmit solitons over long distances using distributed, dispersion-shifted erbium amplifiers. Here, it is essential to use a pump source near the 1.5 μm band in order to dispatch the pump power efficiently over long distances. Only a 1.48 μm pumping scheme can achieve this.

6.11 Dynamic soliton communication at 5–10 Gbit/s

The experimental setup for soliton transmission is shown in Figure 6.28 (Nakazawa *et al.*, 1990b, Nakazawa, 1990b), in which the peak power of the output soliton pulse from EDFA 2 was set in the $N = 2$ soliton regime to extend the repeater spacing further than the L_c in Section 6.5. The laser source for soliton generation is a DFB LD as described in Section 6.9. Here, we use an Fabry–Perot resonator

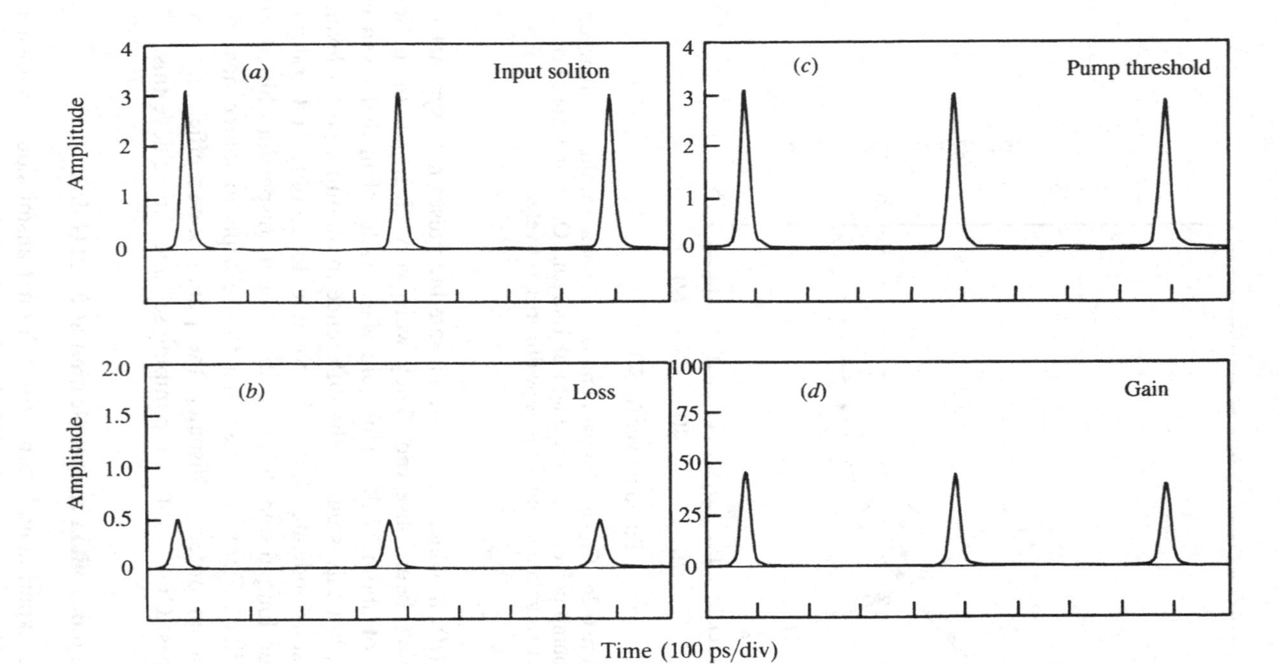

Figure 6.27. Output soliton waveforms from a distributed EDFA. Trace (*a*) is the input soliton waveform. Traces (*b*)–(*d*) correspond to output waveforms with loss, pump threshold and gain conditions, respectively.

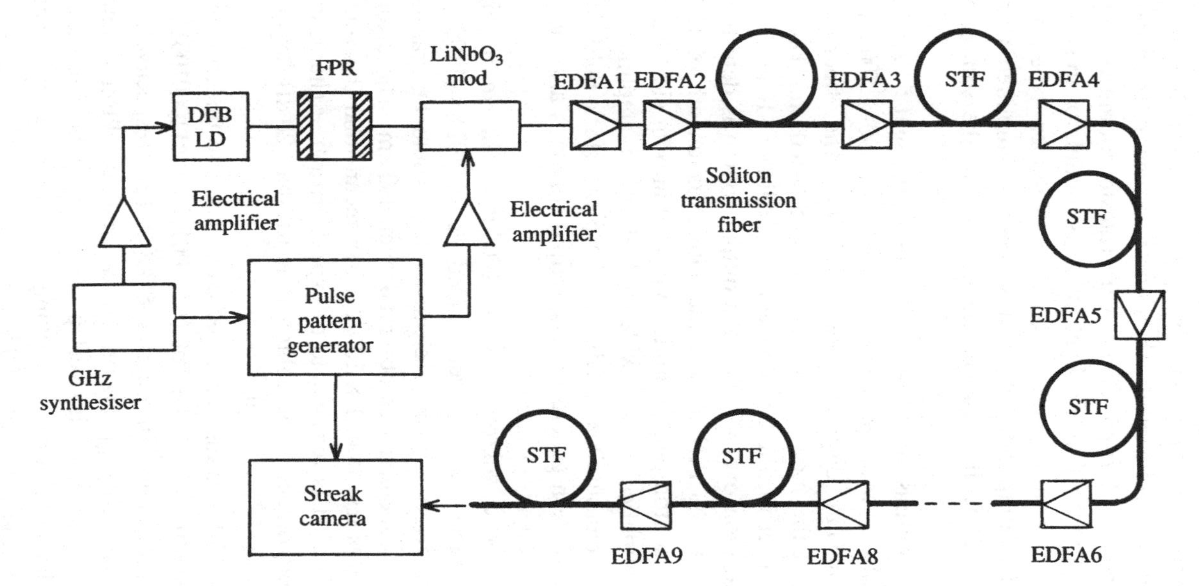

Figure 6.28. Experimental setup for coded-soliton transmission. Actual coded-soliton transmission was first demonstrated with this configuration.

with a bandwidth of 0.18 nm, a cavity spacing of about 200 μm, and a finesse of about 30. As a result, the output pulsewidth from the laser diode shortens to 20 from 27 ps. This transform-limited Gaussian pulse is coupled into a $LiNbO_3$ light intensity modulator to switch the soliton train. The modulator is a Mach–Zehnder type and the $V_\pi = 6.6$ V. The output from the modulator is amplified with two EDFAs (1 and 2) to increase the peak power in the soliton power regime.

The soliton transmission fiber (STF) consists of eight 50-km dispersion-shifted fibers with zero dispersions at 1.490–1.500 μm and −2.3–2.4 ps/km · nm dispersion at 1.545 μm. The fiber loss is 0.24 dB/km and the mode field diameter is 5.6–6.0 μm. The $N = 1$ soliton peak power for the 20 ps pulse is as small as 6–10 mW.

After propagation over 400 km, the output soliton is detected with a high-speed InGaAs PIN photodiode with a bandwidth of 15 GHz. The modulated soliton pulse train is converted into an electrical pulse train and is amplified with a wideband electrical amplifier. Then, error rates are measured at 5 Gbit/s, where the clock pulse is lead from the pulse pattern generator. The output soliton is also measured with a synchroscan streak camera.

A digitally coded soliton train at 400 km is shown in Figures 6.29(*a*)–(*c*). Traces (*a*)–(*c*) were coded as ⟨11101⟩, ⟨11001⟩ and ⟨10001⟩, respectively. As seen clean solitons could be transmitted and no pulse broadening was observed over 400 km. The error rate thus obtained was less than 1×10^{-9}, which means that this transmission is error free. It is reasonable to expect error-free characteristics because the amplitude of the soliton is more than 20 dB larger than that of ordinary IM/DD signals.

A 10 Gbit/s soliton transition experiment was also undertaken over 300 km. Transmitted data codes of ⟨0111001100⟩, ⟨1100010011⟩, and ⟨100110010⟩ are shown in Figures 6.30(*a*)–(*c*), respectively. It is clear that no distortion appears on the transmitted soliton waveforms. Although there was a small overlap on the wing between adjacent solitons, non-linear pulling or pushing was not observed since the transmission distance was only 300 km.

So far there has been no report on soliton transmission in a long single-pass fiber with optical repeaters. The present results reveal that such a transmission is possible with the dynamic soliton communication technique using erbium amplifiers and repeaters. By increasing the bit-rate, soliton transmission of several 10 Gb/s will be possible over 3000 km.

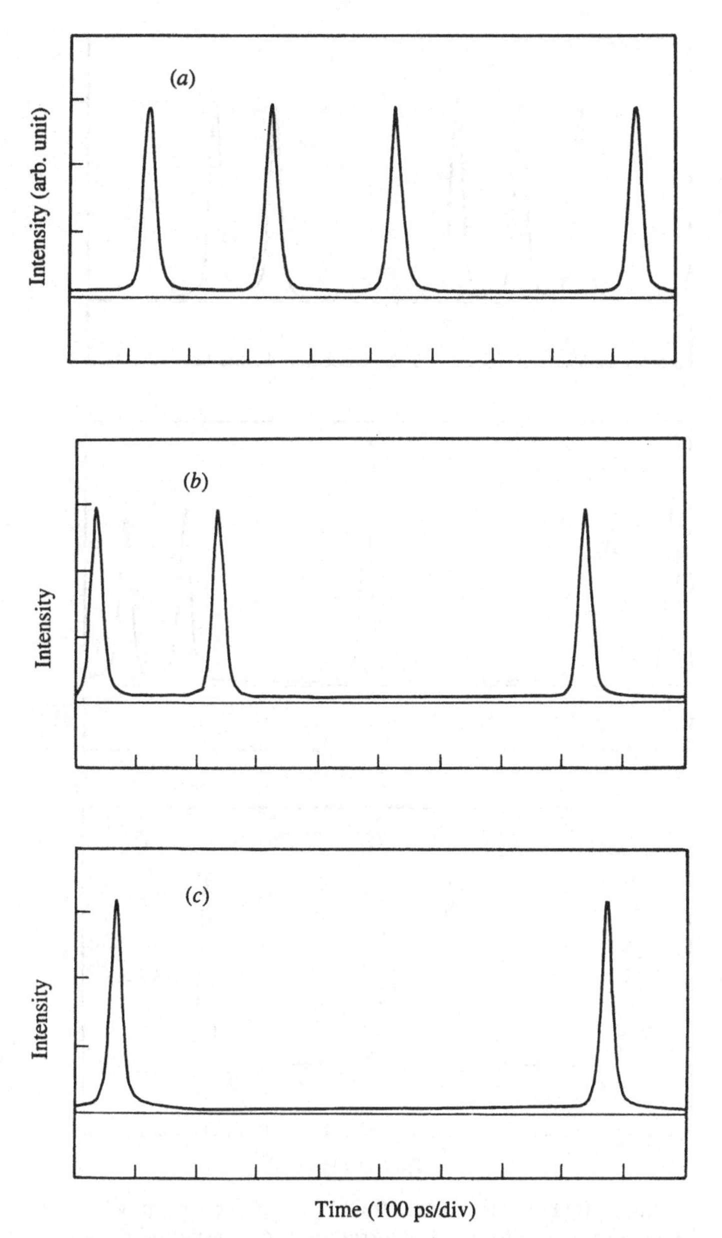

Figure 6.29. Digitally coded 5 Gbit/s soliton trains at 400 km. Waveforms (*a*), (*b*) and (*c*) correspond to transmitted data codes of ⟨11101⟩, ⟨11001⟩ and ⟨10001⟩, respectively.

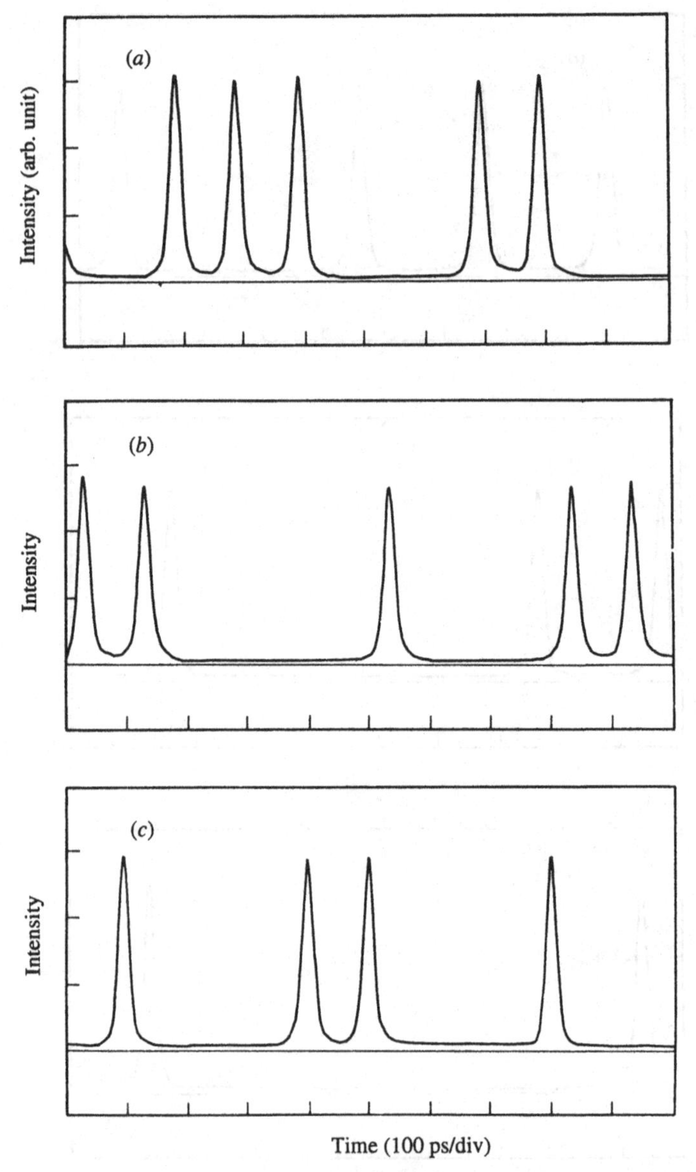

Figure 6.30. Digitally coded 10 Gbit/s soliton trains at 300 km. Waveforms (*a*), (*b*) and (*c*) correspond to transmitted data codes of ⟨0111001100⟩, ⟨1100010011⟩ and ⟨100110010⟩, respectively.

6.12 Summary

The detailed characteristics of short pulse amplification in erbium-doped fibers have been presented, and soliton amplification and transmission are mainly described. Since the erbium-doped fiber usually operates as a lumped gain medium, the dynamic range of the $N = 1\text{–}2$ soliton plays an important role in transmitting solitons over long distances when many EDFAS are used. We call this dynamic soliton communication. Multiple wavelength solitons at 1.552 and 1.535 μm are simultaneously amplified in the erbium-doped fiber and transmitted 30 km. It is shown that there is saturation-induced cross talk between the two wavelengths. This cross talk would not be a serious problem in a high bit-rate system if the average total coupled power is kept constant.

We also described the adiabatic and non-adiabatic amplification of picosecond-to-femtosecond soliton pulses, and a distributed erbium-doped soliton amplifier realised by decreasing the doping concentration to below a few ppm and by extending its length to more than 10 km. Finally, soliton transmission systems were described in which the dynamic soliton communication method was used to send solitons over long distances. Rates of 5 Gbit/s-400 km and 10 Gbit/s-300 km have been achieved with DFB LDS and EDFAs.

References

Ainslie, B. J., Blow, K. J., Gouveia-Neto, A. S., Wigley, P. G. J., Sombra, A. S. B. and Taylor, J. R. (1990) *Electron. Lett.*, **26**, 186.

Desurvire, E., Giles, C. R., Simpson, J. R. and Zyskind, J. L. (1989) *Conference on Lasers and Electro-Optics '89, Baltimore, USA PD20*. (Washington, DC: Optical Society of America).

Desurvire, E., Simpson, J. R. and Becker, P. C. (1987) *Opt. Lett.*, **12**, 888.

Gordon, J. P. (1986) *Opt. Lett.*, **11**, 662.

Hasegawa, A. (1984) *Appl. Opt.*, **23**, 3302.

Kimura, Y., Suzuki, K. and Nakazawa, M. (1989) *Electron. Lett.*, **25**, 1657.

Kimura, Y., Suzuki, K. and Nakazawa, M. (1990) *Appl. Phys. Lett.*, **56** 1611.

Kubota, H. and Nakazawa, M. (1990) *IEEE J. Quantum Electron.*, **26**, 692.

Kurokawa, K. and Nakazawa, M. (1989) *Appl. Phys. Lett.*, **55** 7.

Mears, R. J., Reekie, L., Poole, S. B. and Payne, D. N. (1985) *Electron. Lett.*, **21**, 738.

Nakazawa, M. (1990a) *Proc. SPIE*, **1171**, 328.

Nakazawa, M. (1990b) *Ultrafast Phenomena, Montrey, USA* (Optical Society of America) 232.

Nakazawa, M., Kimura, Y. and Suzuki, K. (1989a) *Appl. Phys. Lett.*, **54**, 295.

Nakazawa, M., Kimura, Y. and Suzuki, K. (1989b) *Electron. Lett.*, **25**, 199.
Nakazawa, M., Kimura, Y., Suzuki, K. and Kubota, H. (1989c) *J. Appl. Phys.*, **66**, 2803.
Nakazawa, M., Kimura, Y. and Suzuki, K. (1990b) *IEEE J. Quantum Electron.*
Nakazawa, M., Kurokawa, K., Kubota, H., Suzuki, K. and Kimura, Y. (1990d) *Appl. Phys. Lett.*, **57**, 653–5.
Nakazawa, M., Suzuki, K. and Kimura, Y. (1989d) *Opt. Lett.*, **14**, 1065.
Nakazawa, M., Suzuki, K. and Kimura, Y. (1990a) *IEEE Photonics Technol. Lett.*, **2**, 216.
Nakazawa, M., Suzuki, K., Kubota, H., Yamada, E. and Kimura, Y. (1990c) *IEEE J. Quantum Electron.*
Nakazawa, M., Suzuki, K. and Kimura, Y. (1990e) *Opt. Lett.*, **15** 588.
Nakazawa, M., Suzuki, K. and Kimura, Y. (1990f) *Opt. Lett.*, **15**, 588–90.
Poole, S. B., Payne, D. N. and Fermann, M. E. (1985) *Electron. Lett.*, **21**, 737.
Shimada, S. (1990) *Optics News*,**1**, 6.
Snitzer, E., Po, H., Hakimi, F., Tumminelli, R. and MaCollum, B. C. (1988) *Optical Fiber Communication Conference '88, Anaheim, USA, PD2.*
Suzuki, K., Y. Kimura, and Nakazawa, M. (1989a) *Appl. Phys. Lett.*, **55**, 2573.
Suzuki, K., Kimura, Y. and Nakazawa, M. (1989b) *Opt. Lett.*, **14**, 865.
Suzuki, K., Nakazawa, M., Yamada, E. and Kimura, Y. (1990) *Electron. Lett.*, **26**, 552.
Zhu, X., Kean, P. N. and Sibbett, W. (1989) *Opt. Lett.*, **104**, 1192.

7

Non-linear transformation of laser radiation and generation of Raman solitons in optical fibers

E. M. DIANOV, A. B. GRUDININ, A. M. PROKHOROV AND V. N. SERKIN

7.1 Introduction

During the thirty-year history of laser physics there have been many bright and marvellous discoveries and during this period non-linear optics has experienced a second birth. It is well known that linear electrodynamics describes field–matter interactions adequately in the case when the linear relationship between the polarisation and electrical field $P = \chi E$ is valid. At the focal point of a commercially available YAG-laser the intensity can increase up to 10^9–10^{12} W/cm^2 and to describe the phenomena occurring at such high intensity correctly it is necessary to use the general relationship $P(E)$. In particular under non-resonant conditions when the electrical field E is sufficiently less than the atomic field A_0 ($A_0 \simeq 10^{14}$–10^{16} W/cm^2) one can consider the Fourier components to be

$$P(k, \omega) = P^{(1)}(k, \omega) + P^{(2)}(k, \omega) + P^{(3)}(k, \omega) + \ldots \quad (7.1)$$

where

$$P^{(n)}(k, \omega) = \chi^{(n)}(k = k_1 + k_2 + k_3 + \ldots + k_n, \omega = \omega_1 + \omega_2 + \ldots + \omega_n)E(k_1, \omega_1) \ldots E(k_n, \omega_n)$$

$\chi^{(n)}$ is the nth order non-linear susceptibility, the magnitude of which is determined by the media in which the non-linear process takes place.

For a long time the mutual influence of laser physics on non-linear optics has been mainly restricted by the processes described by the non-linear susceptibility of the second-order $\chi^{(2)}$ which might be explained by the fact that at optical frequencies, which are far from the resonance one has $|P^{(n+1)}/P^{(n)}| \approx |E/A_0|$, and in order to study higher order non-linear effects high intensity is required. However, in a bulk medium, the interaction length of non-linear waves is short as it is restricted by diffraction divergence on one side and by the self-focusing effect from the other.

The situation has changed fundamentally with the development of low-loss optical fibers. High intensities, which might be transmitted by a fiber and long interaction length (up to a few kilometres) are both responsible for a wide range of non-linear processes, the most prominent of which are self-phase modulation, stimulated Raman scattering, stimulated Brillouin scattering, four-photon mixing, i.e. processes exploring third-order susceptibility.

The first works dealing with non-linear fiber optics brought very interesting and promising results (Stolen *et al.*, 1972, Stolen and Ippen, 1973) and led to an outburst of scientific activity in this area of quantum electronics connected mainly with methods of generation and the non-linear dynamics of ultra-short optical pulses in silica fibers. These advances have been discussed in recent review articles (Stolen, 1980a,b; Mollenauer, 1985, Lin, 1986; Dianov and Prokhorov, 1986; Akhmanov *et al.*, 1986; Gouveia-Neto *et al.*, 1988; Dianov *et al.*, 1989; Tomlinson and Stolen, 1988; Sisakyan and Shvartsburg, 1984).

At the current time there are two main directions in the study of the non-linear properties of optical fibers.

The first is the traditional one and is concerned with the generation of ultra-short pulses (USP) and the methodology of nonlinear fiber optics. In this arena a laser is used only as a source of radiation which initiates non-linear processes, a wide range of which can be used as methods for generating USPs. This approach demonstrates some competition between lasers and fibers, because both of them act as optical sources. The second direction was originated by Mollenauer and Stolen (1984) who presented the first soliton laser. They demonstrated that a fiber might help a laser manifest its abilities more fully. The aim of this branch of non-linear fiber optics is to combine a laser as the source of radiation with a fiber as the non-linear element (internal or external) of the laser cavity in order to explore the advantages of the active and guide media and to suppress the negative aspects characteristic to these elements.

Clearly both of these directions are closely related to each other since a full understanding of the physical nature of non-linear effects in optical fibers promises the successful use of fibers in laser cavities.

The aim of this chapter is the analysis of the experimental data dealing with the generation of femtosecond solitons in order to reveal the main physical effects in the transient region of non-linear pulses.

7.2 Formation of femtosecond pulses through self-action of the powerful wave packets in optical fibers: experiments

For optical frequencies the glass cubic susceptibility is associated with the electronic mechanism of non-linearity. If the field contains components with frequencies whose difference approaches the eigenfrequency of the molecular vibrations then the mixed electron–nuclear Raman non-linearity also contributes to the susceptibility.

One of the most fascinating phenomena originating from stimulated Raman scattering (SRS) in fibers is the generation of Raman solitons which had been predicted earlier by Vysloukh and Serkin (1983). The large number of theoretical and experimental works makes it possible to study, in detail, the main physical processes of femtosecond soliton formation in SRS in fibers. Great success was achieved by Dianov *et al.* (1985) and Mitschke and Mollenauer (1986) who observed, for the first time, the phenomenon of the Raman self-frequency shift of optical solitons. It has since been discovered that this phenomenon plays the most prominent role in the formation of high-contrast Raman soliton pulses in the region of the negative group velocity dispersion (GVD).

In spite of the fact that the non-linear dynamics of Raman solitons are studied experimentally in a number of scientific groups, the basic experimental set-ups are, to a large extent, identical (Figure 7.1). The radiation source initiating the SRS process in a fiber is a parametric light oscillator (Dianov *et al.*, 1985) – a color-center laser ($\lambda = 1.55\ \mu m$) (Mollenauer *et al.*, 1980), dye laser ($\lambda = 1.34\ \mu m$) (Zysset *et al.*, 1987), second harmonic of a YSGG:Cr^{3+}, Er^{3+} laser (Vodop'yanov *et al.*, 1987), YAG:Nd^{3+} laser operating at the wavelengths of $1.064\ \mu m$ (Grudinin *et al.*, 1987) or $1.32\ \mu m$ (Gouveia-Neto *et al.*, 1987a, b). The temporal characteristics of the output radiation were analysed with the help of an intensity autocorrelator. The output spectra were taken with a grating monochromator.

7.2.1 *The case when the pump wave lies in the region of the negative GVD*

Experiments on the generation of Raman solitons with the pump source emitting in the region of the negative GVD were carried

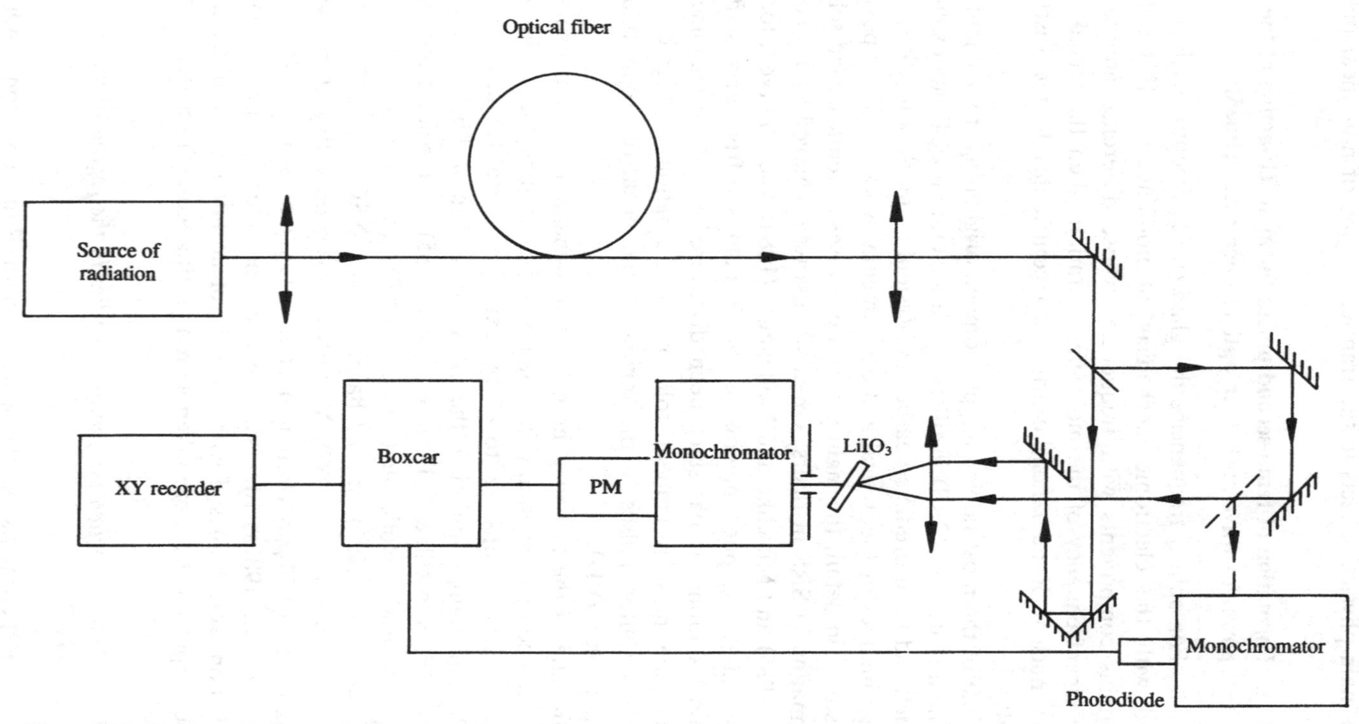

Figure 7.1. Basic experimental setup for studying the SRS process in optical fibers.

out independently by a number of scientific groups (Zysset *et al.*, 1987; Vodop'yanov *et al.*, 1987; Gouveia-Neto *et al.*, 1987a, b, c, d; Kafka and Baer, 1987) each of which used different lasers. However, in spite of the fact that the characteristics of the generated solitons are very much alike in all the experiments, the physical interpretation of the results obtained were different. That is why we thought it worthwhile at first to describe in detail the experimental results obtained by Vodop'yanov *et al.*, (1987) and then to explain them qualitatively from the point of view of simple physical considerations and quantitatively using a strong theory.

As has already been mentioned the second harmonic of a YSGG:Cr^{3+}:Er^{3+} laser was used as a pump source. The pump source emitted trains of bandwidth-limited pulses. The train duration was 150 ns, the pulse duration 150 ps, the time interval between pulses ~7 ns, the peak power at the double frequency was 200 kW. Figure 7.2 shows the output spectra of a fiber with zero chromatic dispersion wavelength $\lambda_0 = 1.34\ \mu$m and a length of 60 m at different levels of the pump power. The spectra are continuous without the prominent maxima characteristic of the SRS in the region of positive GVD. One can see Stokes and anti-Stokes components with 50 cm^{-1} shift which are due to the modulation instability of the pump wave. Results from the measurement of the temporal characteristics of the radiation are shown in Figure 7.3(*a*). The duration of the pulses generated in the Stokes wing of the spectra on the assumption of a sech-shaped pulse was 80–100 fs depending on the wavelength (see Figure 7.3(*a*)). The autocorrelation function (ACF) has intense wings with a pedestal existing for up to 100 ps. On increasing the laser pump power the contrast ratio K, i.e. the peak-to-pedestal ratio of the autocorrelation decreased and at the same time the duration of the central peak remained constant. Note that at the increase of the wavelength the ACF contrast also increased.

The measurements of the temporal structure, which were performed using a streak camera (with temporal resolution ~3 ps), were informative. Figure 7.3(*b*) shows the traces obtained from the screen of the streak camera. One can see that an increase in the pump power leads to an increase in the number of peaks in the Stokes spectral region, while the intensity of a separate peak remains unchanged. At the minimum of the possible pump power in the vicinity of 1.6 μm one or several (due to the stochastical nature of the process) peaks were generated. The measured power of a single peak was 4 ± 0.3 kW in the region of 1.65 μm which corres-

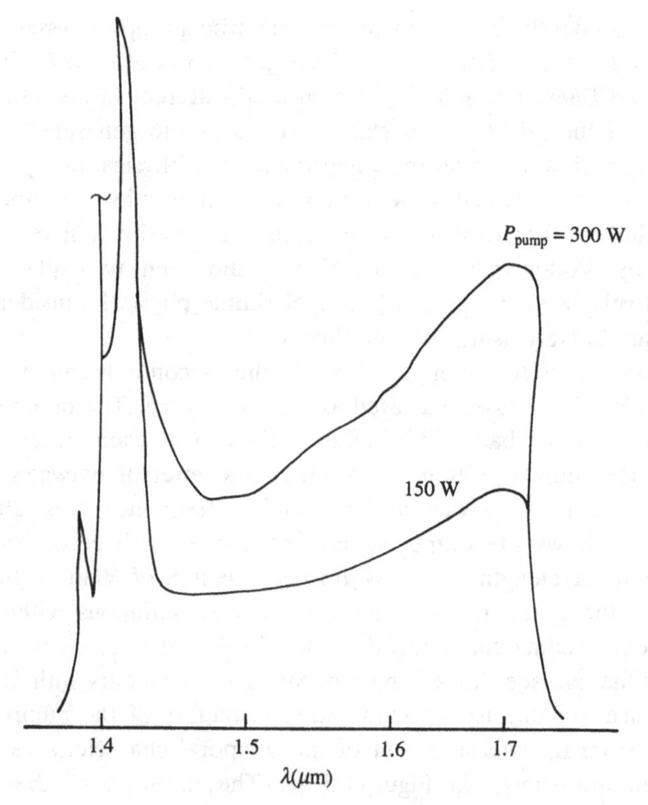

Figure 7.2. Raman spectra of a single-mode fiber with $\lambda_0 = 1.34$ μm and length of 60 m at the different levels of the pump power. Pump at 1.39 μm.

ponds to the power of a fundamental soliton with duration of 100 fs.

Let us now present a few simple estimates which clarify the physical nature of the phenomena.

First we estimate the influence of multi-soliton compression on the pulse shape at the initial propagation stage. It is known (Mollenauer *et al.*, 1983) that a multi-soliton pulse compresses at the characteristic length $Z_c = Z_0/\pi N$, Z_0 is the fundamental soliton period and N is the number of solitons. The characteristic length of modulation instability (MI) growth is

$$Z_{MI} = \frac{G_{th}}{K_{MI}} = \frac{G_{th}\lambda}{2\pi n_2 I_L}$$

where K_{MI} is the increment of MI, n_2 is the non-linear refractive index of silica glass equaling 3.2×10^{-16} cm^2/W, I_L is the peak intensity of the pump pulse, G_{th} is the MI threshold increment and equals 16. If we make Z_c equal to Z_{MI} we can find the critical number of solitons in a pump pulse, which, if exceeded, results in MI growth and the pump pulse splitting into separate spikes

$$N \simeq 2G_{th} = 25\text{–}30$$

In the present case $I_0 = 10^3$ W/cm^2 and is the intensity of the 100 ps fundamental soliton at $\lambda = 1.395\ \mu$m, $I \simeq 10^9$ W/cm^2 pump pulse intensity and $N = (I/I_0)^{1/2}$. So instead of multi-soliton compression a pump pulse tends to split into separate spikes due to MI. As for the results obtained at 1.34 μm (Zysset *et al.*, 1987) the number of solitons was about 10 and the pump pulse undergoes multi-soliton compression.

Second, from the ratio of the increases in MI and SRS one can conclude that, at the initial stage of soliton formation, the process of MI is preferable and SRS results in the appearance of spectral asymmetry. This has been confirmed by measurements (see Figure 7.2). Third, let us estimate the parameters of the solitons arising due to MI. It is known (Carpman, 1967) that MI leads to the formation of envelope solitons with amplitudes and pulse widths defined by the initial intensity of the pump wave. Since the total energy in the parametric process previously described is conserved, then the pump energy which is concentrated in a single MI period at a long distance will transform into a single soliton.

As a result we have

$$I = \frac{\pi^2}{2} I_0 \simeq \mp 5 I_0$$

$$\tau_0 = 1.76 \frac{T_{MI}}{\pi^2} = \left(\frac{0.2D\lambda^3}{In_2}\right)^{1/2}\left(\frac{1.76}{\pi^2}\right) \simeq 0.18 T_{MI}$$

In this case

$$\tau_0 = 50 \text{ fs} \qquad I_0 = 5 \times 10^9 \text{ W/cm}^2$$

Fourth, we estimate the influence of the Raman self-frequency shift (RSFS) which manifests itself through the Raman amplification of low frequency components. Using the theory of RSFS presented by Gordon (1986) which shows that

$$\frac{d\langle \nu_0 \rangle}{dz} = P\tau_0^{-4}$$

where $\langle \nu_0 \rangle$ and τ_0 are the central frequency and duration of the soliton, P is a coefficient depending on the magnitude of the GVD and wavelength. In this case $P = 0.024$ and the estimated frequency

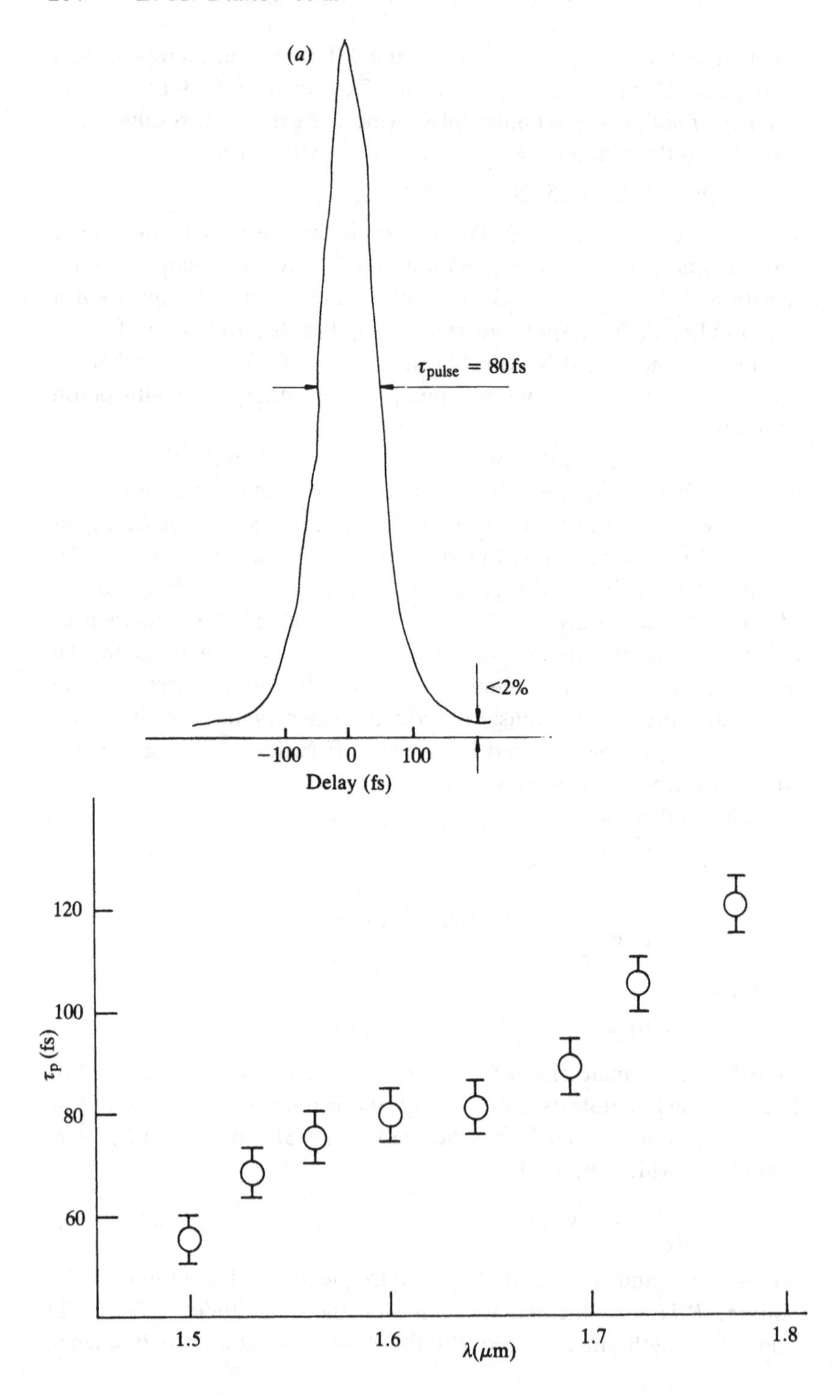
(a)
τ_{pulse} = 80 fs
<2%
−100
0
100
Delay (fs)
120
100
80
60
τ_p (fs)
1.5
1.6
1.7
1.8
λ(μm)

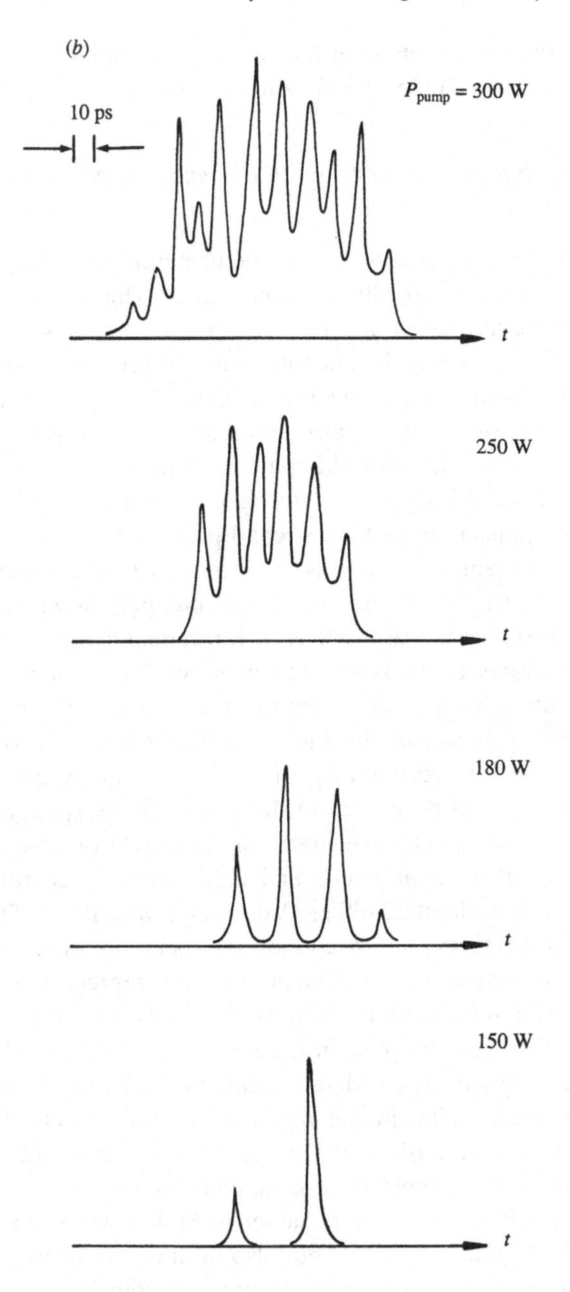

Figure 7.3. (*a*) Autocorrelation trace of the output radiation at 1.6 μm. Below, the wavelength dependence of the pulse duration near 1.6 μm. (*b*) The streak-camera results obtained at the different levels of the pump power.

shift of the 50 fs soliton is 240 nm over a 60 m length of fiber which is in reasonable agreement with the experimental results.

7.2.2 *The case when the pump wave lies in the region of positive GVD*

The YAG laser operating at its fundamental wavelength 1.064 μm is the most commercially available laser. This results in practical interest in the utilisation of such a source as a Raman soliton generator which could possibly be an alternative source compared with more expensive lasers. The main idea behind the experimental work is the fact that the Stokes pulse intensity in the region of positive GVD is comparable with the pump intensity and consequently such a source could act as a soliton generator and would be very useful in some applications of femtosecond pulses.

Basically the experimental set-up was the same as in the experiments with the YSGG:$Cr^{3+}Er^{3+}$ laser but a c.w. pumped Q-switched and mode-locked YAG laser was used as a fundamental source instead. The laser operates at 1.064 μm and provides 150 ps pulse at 700 Hz repetition rate with a peak power of approximately 600 kW. Figure 7.4 shows the spectrum of the SRS in a fiber 60 m long with zero chromatic dispersion wavelength $\lambda_0 = 1.34$ μm. The pump power is 10 kW. Note the essential difference in the nature of the spectrum before and after the zero GVD wavelength. In the region of positive GVD Stokes pulses can be clearly seen and their duration recorded by the streak camera was about 20–30 ps (Vorob'ev *et al.*, 1987). The negative GVD is characterised by a continuous spectrum as in the case YSGG laser. The dynamics of Raman spectrum represented in Figure 7.5 gives useful information which helps clarify the physical nature of the SRS. One can see that, in the region of positive GVD, successive Stokes components occur demonstrating the cascade nature of the SRS in the fiber, while, in the region of negative GVD the Stokes wing appears continuously and spreads up to 1.7 μm and all the spectral components have almost the same amplitude.

Such a qualitative difference in the behavior of Stokes waves takes place not only in the frequency domain but also in time. Autocorrelation measurements enable us to conclude that, in the region of negative GVD, there are high contrast USP with energies corresponding to those of the envelope solitons.

It is well known that a soliton is a bandwidth-limited pulse, while the bandwidth of a Stokes wing in the region $\lambda > \lambda_0$ is much wider

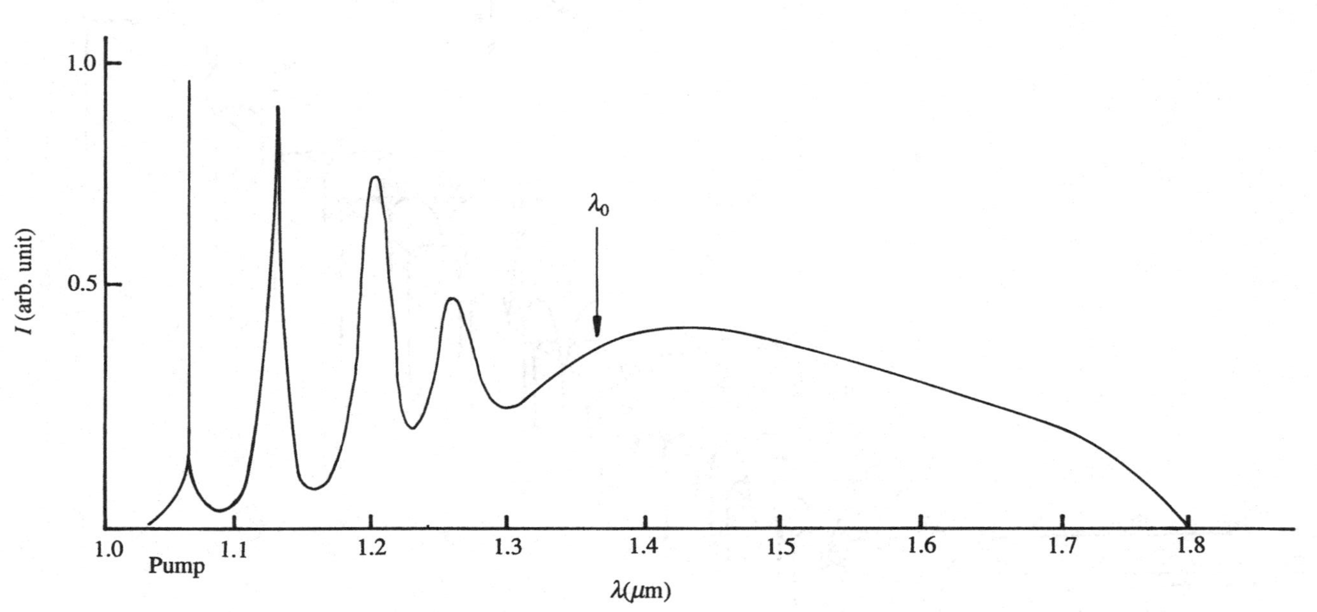

Figure 7.4. Raman spectrum in a fiber with $\lambda_0 = 1.34\ \mu m$ and length 60 m. Pump at 1.064 μm.

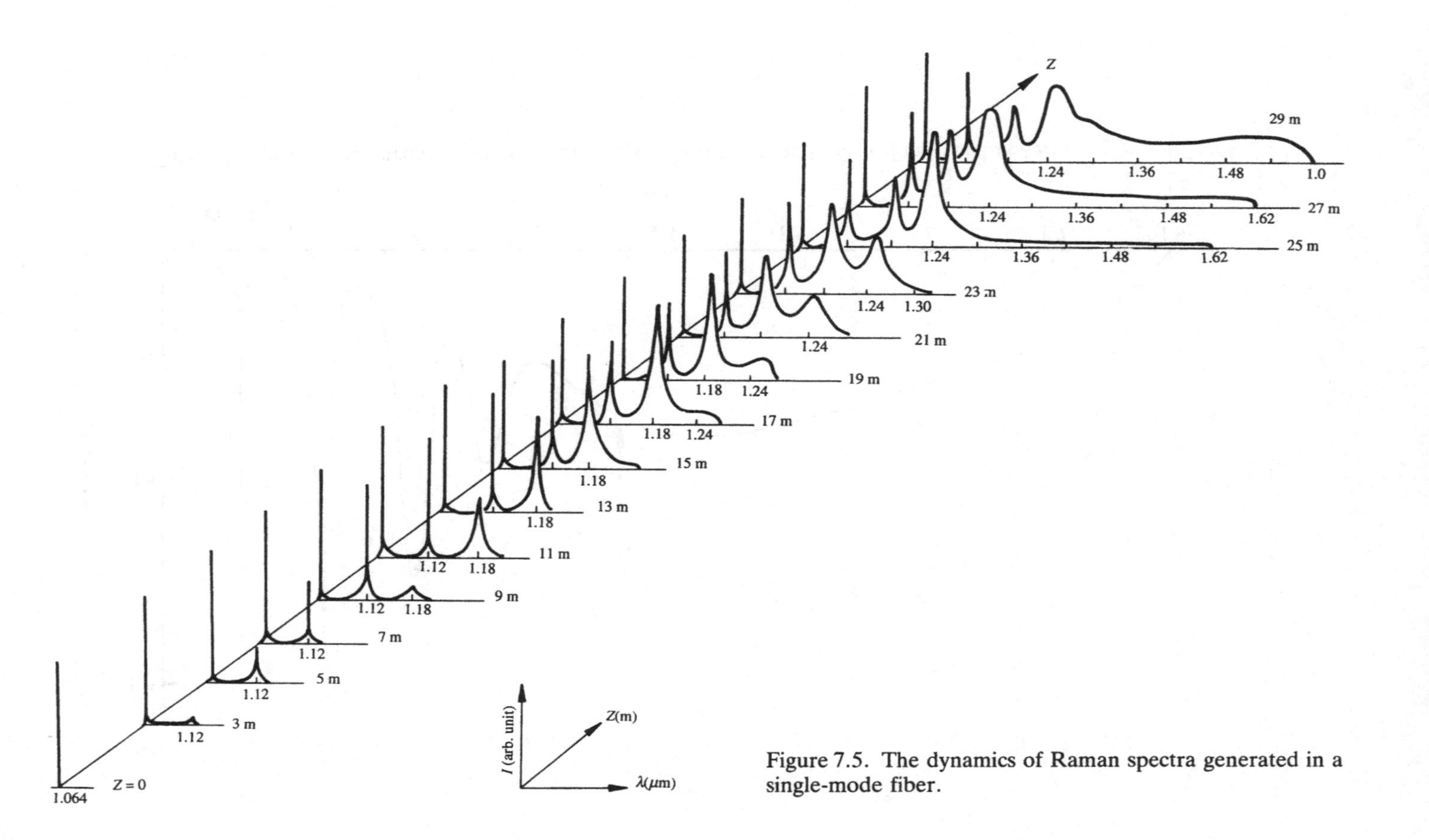

Figure 7.5. The dynamics of Raman spectra generated in a single-mode fiber.

than the bandwidth of 70 fs pulses. In order to check the soliton-like nature of these pulses interferometric correlation function (ICF) measurements were carried out and the results indicated that the duration ratio of ICF and CF at half maxima are equal to 1.09 ± 0.02 which corresponds to bandwidth-limited pulses with reasonable accuracy.

7.3 Non-linear dynamics of multi-soliton pulse propagation through single-mode optical fibers

The non-linear Schrödinger (NLS) equation, describing the propagation of picosecond solitons in single-mode fibers, predicts periodic changes in the form of the wavepacket envelope in the case of multi-soliton pulse propagation. With a decrease in pulse duration, however, higher-order dispersive and non-linear terms which do not appear in the NLS equation start to play a significant role resulting in the occurrence of a number of essentially new physical effects. Thus the reason for the attempt to exploit Raman solitons in order to check the validity of the NLS equation is quite clear.

The experimental idea was simple and is illustrated in Figure 7.6(*a*). Radiation from the first fiber, emitting Raman solitons, passes a prism after which radiation, with a spectrum which corresponds to 70 fs pulses (see Figure 7.6(*b*)), is coupled into a second dispersion-shifted fiber. The second fiber has an effective core area of 11 μm^2 and the wavelength of the zero chromatic dispersion λ_0 is 1.56 μm. Figure 7.7 shows the ACF at the input of the second fiber and the output of fibers of length 1, 2, 4 and 10 m respectively. Figure 7.8 shows the corresponding spectra. It is clear from the results that temporal and spectral splitting of the pulses has occurred. Note that the temporal separation grows with the fiber length (Figure 7.9) whereas the spectrum width is practically independent of the fiber length. In a recently published study (Mitschke and Mollenauer, 1986), similar results were interpreted as the break-up of the multi-soliton pulse into a soliton and a dispersive wave and a constant shift of the soliton's central frequency into the Stokes spectral region.

The results of our experiment are consistent with the observation that both pulses are solitons: in the first few dispersion lengths the pulsewidth does not depend on the fiber lengths within the accuracy of the measurement, but at greater lengths we have observed some pulse spreading. Such an increase in duration can be explained by the

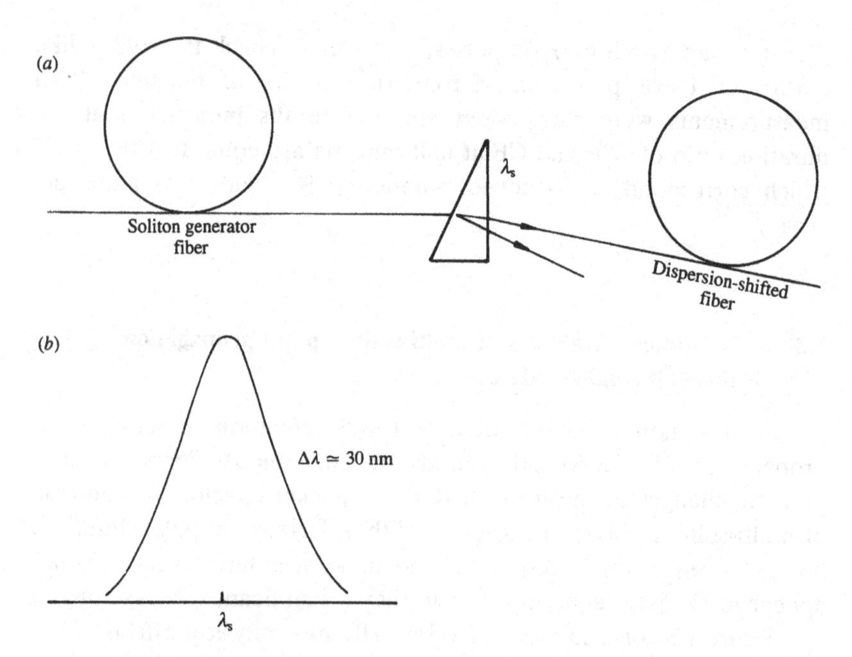

Figure 7.6. (*a*) Sketch of the experimental setup for the study of nonlinear dynamics of solitons in optical fibers. (*b*) The radiation spectrum launched into the second fiber.

energy dissipation due to the linear optical losses and by RSFS. It should be noted that the amplitude of the short-wave (and short-width) soliton was found to be less stable than the long-wave soliton.

In the case of a 500 fs pulse (Mitschke and Mollenauer, 1986) it was shown that the square of the time interval between the split components was proportional to the launched power. In our experiment this dependence resulted in some spreading of the ACF side-bands (see Figure 7.7) because the pump power was not stable enough.

The compression of the multi-soliton pulse in the initial part of the fiber enables the generation of light pulses corresponding to several periods of optical oscillations. Figure 7.10 shows the experimental results for the multi-soliton compression of femtosecond pulses. The obtained 18-fs light pulses correspond to about three optical cycles at the wavelength of 1.6 μm.

To interpret these results we first give the relevant numerical parameters. For this single-mode fiber in the vicinity of 1.58 μm the chromatic dispersion D is 1 ps/nm · km. In this case the period and power of the fundamental soliton of width $T_p = 70$ fs are

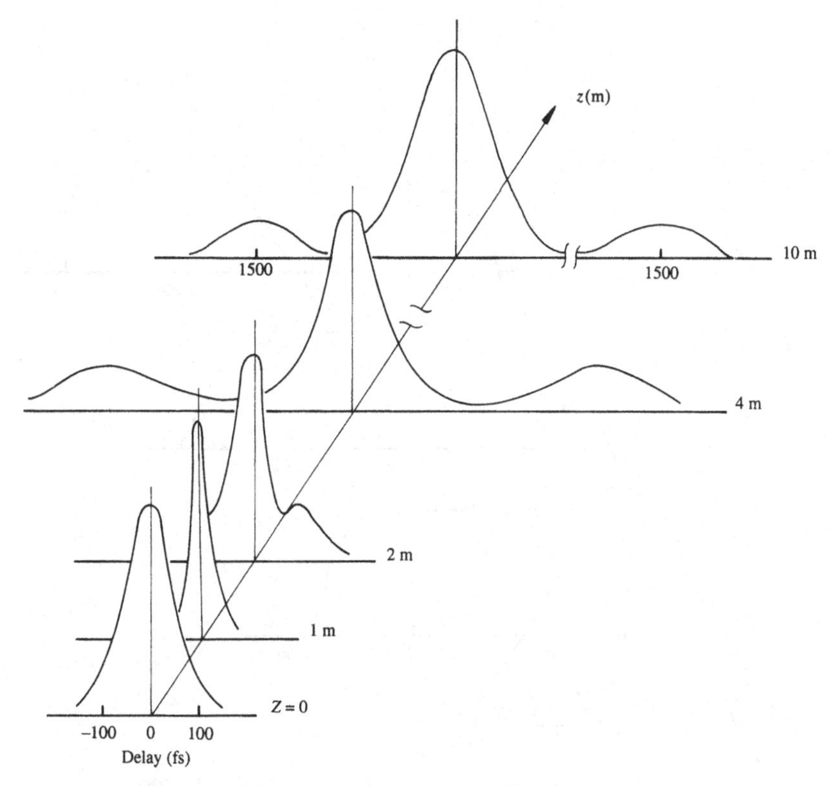

Figure 7.7. Autocorrelation traces at the input of the second fiber and the output at 1, 2, 3, 4 and 10 m length of the fiber.

$$Z_0 = \frac{\pi\tau_0^2}{2k''} = \frac{\pi c\tau_0^2}{D\lambda^2} = 1.8 \text{ m} \qquad \text{and} \qquad \frac{\lambda S_{\text{eff}}}{4n_2 Z_0} = 280 \text{ W}$$

respectively. Power $P \simeq 1.5$ kW was launched into the single-mode fiber and the corresponding solitons order is

$$N = (P/P_0)^{1/2} = 2.3$$

The dispersion lengths of the second and third order being

$$Z_{\text{disp}} = \tau_0^2/|k''| = 115 \text{ cm}$$

and

$$Z_{\text{disp}}^{(3)} = \frac{\tau_0^3}{(\partial^3 k/\partial\omega^3)} = 100 \text{ cm}$$

respectively where $\tau_0 = T_p/1.76 = 40$ fs, $k'' = \partial^2 k/\partial\omega^2 - 1.4 \times 10^{-29}$ s^2/cm and $\partial^3 k/\partial\omega^3 = 3 \times 10^{-42}$ s^3/cm and k is a propagation constant of the fundamental mode. The non-linear length $Z_{\text{nl}} = Z_{\text{disp}}/N^2$ is 23 cm. These estimates show that in order to analyse the results correctly it is necessary to take the higher-order

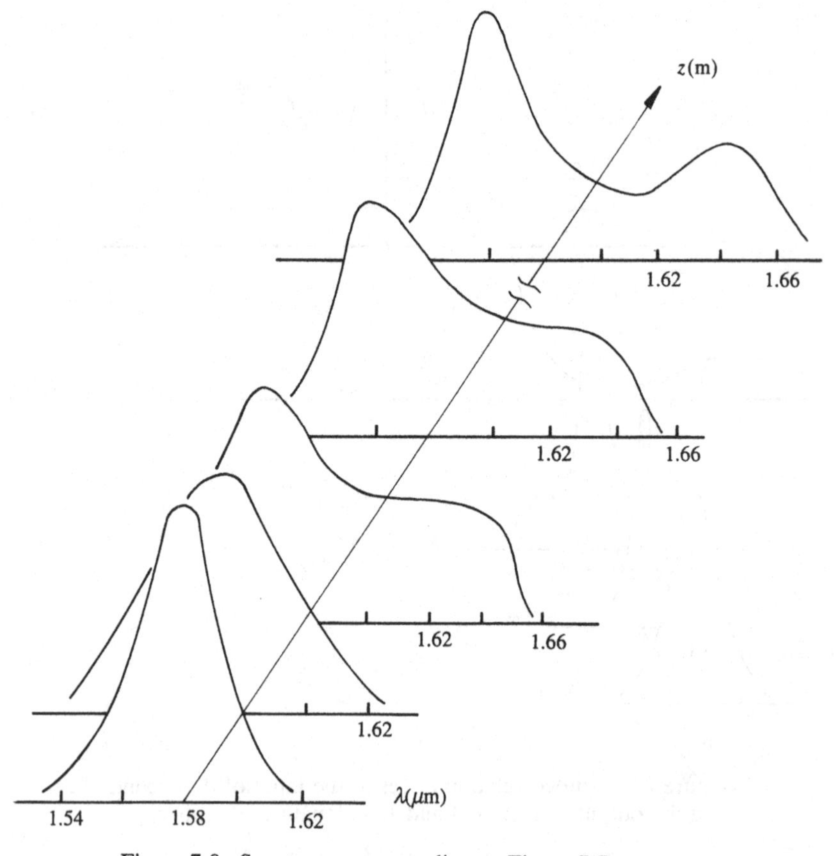

Figure 7.8. Spectra, corresponding to Figure 7.7.

dispersive and non-linear effects into consideration because their influence on the non-linear dynamics of propagation of 70 fs pulses can be compared with the influence of the low-order terms.

7.4 Non-linear dynamics of Raman solitons in optical fiber theory

7.4.1 *The model of SRS in non-linear dispersive media: the slowly varying envelope approximation*

In the semi-classical approach the electromagnetic field in non-linear dispersive media is described by Maxwell's equations while the properties of the media are governed by the quantum mechanical equations for the density matrix elements. From this set of equations one can derive the equations for the amplitudes of the molecular vibrations $\widetilde{Q}$ and the normalised population difference at resonant

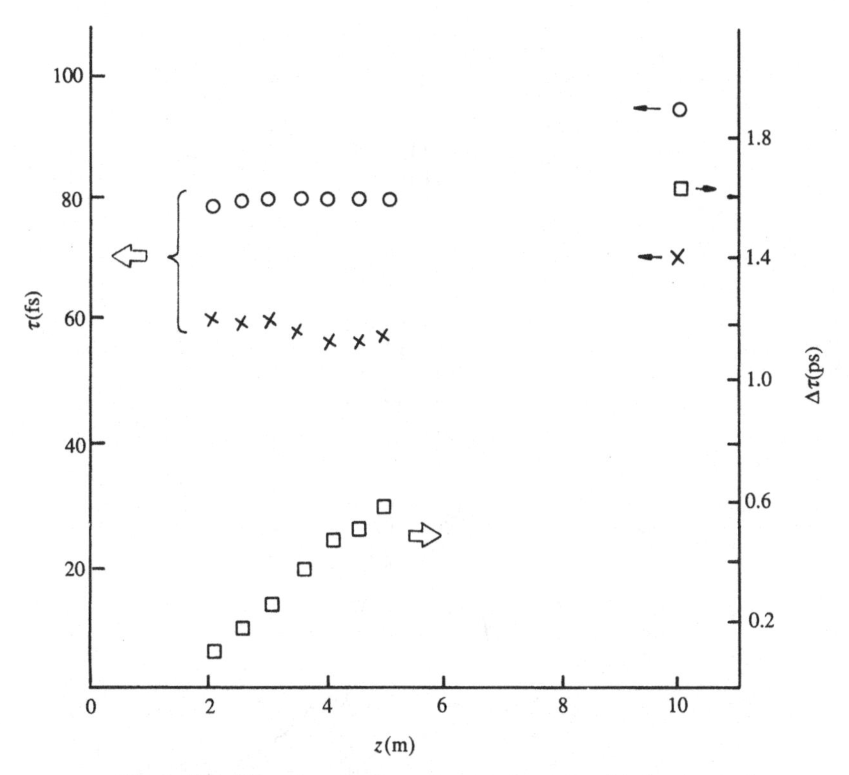

Figure 7.9. Fiber length dependence of temporal behaviour of solitons. Open squares, time interval between the color solitons; crosses, duration of 'blue' soliton and open circles, duration of the 'red' one.

levels $n = (N_1 - N_2)/N$

$$\frac{\partial^2 \varepsilon}{\partial z^2} - \frac{1}{c^2}\frac{\partial^2 e}{\partial t^2} = \frac{4\pi}{c^2}\frac{\partial^2 P^{(\text{linear})}}{\partial t^2} + \frac{4\pi}{c^2}\frac{\partial^2 P^{(\text{Kerr})}}{\partial t^2} + \frac{4\pi}{c^2} N_{\text{m}} \frac{\partial \alpha}{\partial \widetilde{Q}} \frac{\partial^2 (\widetilde{Q}\varepsilon)}{\partial t^2} \tag{7.2}$$

$$\frac{\partial^2 \widetilde{Q}}{\partial t^2} + \frac{2}{T_2}\frac{\partial \widetilde{Q}}{\partial t} + \Omega_{\text{R}}^2 \widetilde{Q} = \frac{1}{2M}\frac{\partial \alpha}{\partial \widetilde{Q}} n \langle \varepsilon \rangle^2 \tag{7.3}$$

$$\frac{\partial n}{\partial t} + \frac{n-1}{T_1} = \frac{1}{2\omega_0}\frac{\partial \alpha}{\partial \widetilde{Q}} \varepsilon^2 \frac{\partial \widetilde{Q}}{\partial t} \tag{7.4}$$

where α is the electronic polarisability of molecule, T_1 and T_2 the longitudinal and transverse times of relaxation respectively, M is an effective mass of a molecule, Ω_{R} the resonant frequency of the

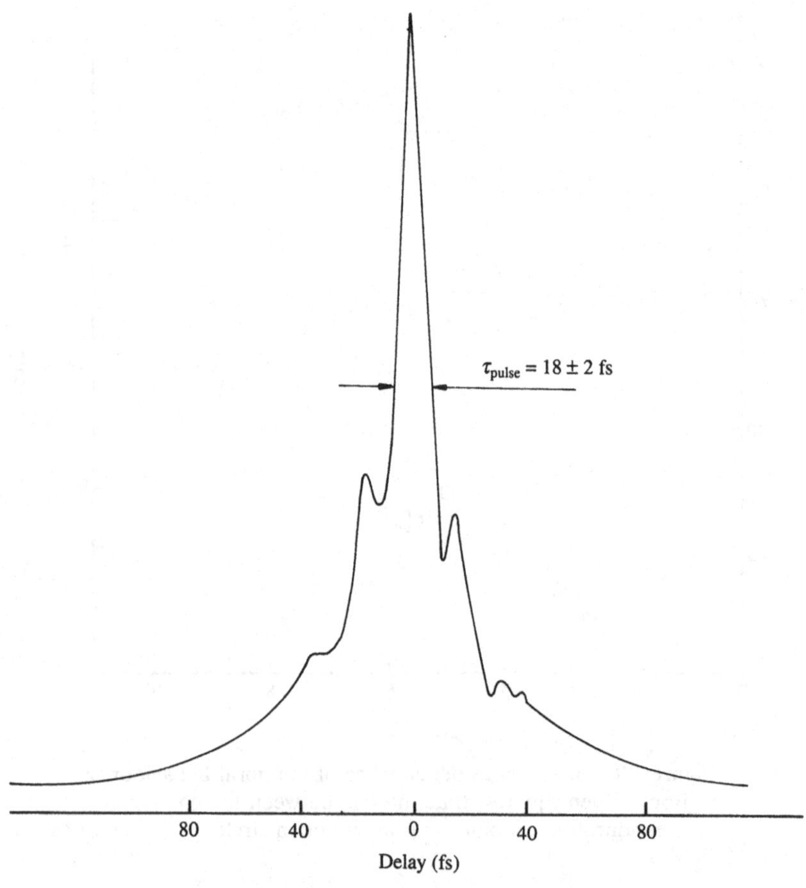

Figure 7.10. Background free autocorrelation trace of the pulse after multi-soliton compression at $\lambda_0 = 1.57\ \mu$m.

molecular oscillations, N_m the number of molecules in the unit volume. The complex amplitude of the induced polarisation in equation (7.2) is a sum of linear and non-linear parts. The linear part is caused by the linear absorption losses while the non-linear one involves two effects: the dependence of refractive index on intensity and SRS. The system equations (7.2)–(7.4) can be solved using the slowly-varying-envelope (SVE) approximation with the assumption that the spectral width of the interacting pulses is much smaller than the frequency of incident radiation (Akhmanov and Khohlov, 1964; Akhmanov, 1986):

$$\varepsilon(z, t) = \tfrac{1}{2}\boldsymbol{e}E_L(z, t)\exp i(\omega_L t - k_L z) + \tfrac{1}{2}\boldsymbol{e}E_s(z, t)\exp i(\omega_s t - k_s z) + \text{c.c.} \qquad (7.5)$$

$$\tilde{Q}(z, t) = \tfrac{1}{2}Q(z, t)\exp i(\Omega t - k_Q z) + \text{c.c.} \tag{7.6}$$

where $\boldsymbol{e}$ is the unit polarisation vector of the laser, E_L, and Stokes, E_S, waves. E_L, E_S and $Q(z, t)$ are slowly-varying functions of time compared with the optical period and the period of the molecular oscillations. In the model (equations (7.2)–(7.6)) the optical field is assumed to maintain its polarisation along the fiber length. In this interpretation the stimulated scattering process is caused by the appearance of a spatial–temporal grating of dielectric constant ε running along the medium

$$\delta\varepsilon(z, t) = 4\pi N\left(\frac{\partial\alpha}{\partial Q}\right)Q(z, t) \simeq cE_L^* E_S \exp i(\Omega t - k_Q z) + c^* E_L E_S^* \exp -i(\Omega t - k_Q z) \tag{7.7}$$

In accordance with (7.7) the scattering of the laser radiation caused by a change in the dielectric constant $\delta\varepsilon$ results in amplification of the Stokes wave (the first term in (7.7)) and attenuation of the pump wave (the second term in (7.7)). The perturbation $\delta\varepsilon$ has a non-linear Raman response time which is approximately inversely proportional to the bandwidth of spontaneous scattering $\Gamma^{-1} \simeq T_2$.

A mathematical model for SRS in glass fibers using the SVE approximation has been developed by Lugovoi (1976). In this paper the effects of phase modulation and cross-phase modulation together with first-order GVD were taken into consideration. The mathematical model of SRS in fibers in second-order dispersion theory was developed by Vysloukh and Serkin (1983). In this paper a more general case was considered when the self-action effects and Raman non-linearity were taken into account simultaneously.

Non-linear interaction of the pump and Stokes waves in the medium with third-order susceptibility can be described by the following system of dynamic equations

$$2i\left(\frac{\partial E_L}{\partial z} + \frac{1}{V_L}\frac{\partial E_L}{\partial t}\right) = -k_L''\frac{\partial^2 E_L}{\partial t^2} + \frac{k_L n_2 |E_L|^2}{n_0}E_L + \frac{2k_L n_2 |E_S|^2}{n_0}E_L - \gamma_L E_S Q \tag{7.8}$$

$$2i\left(\frac{\partial E_S}{\partial z} + \frac{1}{V}\frac{\partial E_s}{\partial t}\right) = -k_S''\frac{\partial^2 E_S}{\partial t^2} + \frac{k_S n_2 |E_2|^2}{n_0}E_S + \frac{2k_S n_2 |E_L|^2}{n_0}E_S + \gamma_S E_L Q^* \tag{7.9}$$

$$\frac{\partial Q}{\partial t} + \frac{Q}{T_2} = \gamma_Q E_L E_S^* + N_{sp}(z, \tau) \tag{7.10}$$

where

$$\gamma_{L,S} = \frac{\partial \alpha}{\partial Q} \frac{\pi N \omega_{L,S}}{c n_{L,S}} \quad \text{and} \quad \gamma_Q = \frac{\partial \alpha}{\partial Q} \frac{1}{4\pi M \Omega_R}$$

are the coefficients of non-linear coupling, $g_S (= 2T_2\gamma_Q\gamma)$ is the maximum Raman gain in the Stokes band, $\mu_{L,S}$, the coefficient of linear loss in a fiber, α_{nm} $(= \langle \chi_m^2 \chi_n^2 \rangle / \langle \chi_m \rangle)$ a geometrical factor arising due to averaging the non-linear susceptibility over the fiber's cross section, $\chi_{mn}(\mathrm{r})$ spatial functions describing the radial distribution of mode intensity in the linear approximation, n, an effective linear refractive index. $k''_{L,S} = (\partial^2 k_{L,S}/\partial \omega^2)$, is the GVD coefficient, depending in particular, on the material and waveguide dispersion of a fiber. $N_{sp}(z, \tau)$ is the source of spontaneous fluctuations at the Stokes wavelength. In the general case of an inhomogeneously broadened line with form-factor $F(\omega)$ the complex amplitude Q is related to the electromagnetic field by the expression

$$Q = \int_{-\infty}^{+\infty} q(\tau, z, \omega_0) F(\omega_0 - \Omega) \, d\omega_0 \tag{7.11}$$

where

$$q(\tau, z) = \frac{1}{T_2} \exp\left[-\left(\frac{1}{T_2} + i\Delta\right)\tau\right]$$

$$\int_{-\infty}^{\tau} E_L(z, t) E_S^*(z, t) \exp\left(\frac{1}{T_2} + i\Delta\right) t \, dt \tag{7.12}$$

here $\Delta(= \omega - \omega_0)$ is the distribution function of the SRS amplification coefficient and, ω_0 is the frequency of the vibrational transition of a separate molecule.

7.4.2 *The quasi-static regime of Raman amplification classification of the fundamental SRS regimes*

The broad lines of spontaneous scattering in silica fibers make it possible to generate femtosecond optical pulses. First, let us consider the generation of Raman pulses in the limit case when the pump pulse duration greatly exceeds the transverse relaxation time. In this case the non-linear part of the induced polarization related to the SRS quasi-statically, follows the spatial–temporal variations of the pulse envelopes at fundamental and Stokes frequencies. The dimensionless form of the initial system ((7.8)–(7.10)) transforms into

$$\mathrm{i}\,\frac{\partial \Psi_L}{\partial z} = -\frac{P_L}{2}\frac{\partial^2 \Psi_L}{\partial \tau^2} + R_{LL}|\Psi_L|^2\Psi_L + 2R_{LS}|\Psi_S|^2\Psi_L$$

$$-\,\mathrm{i}\,\frac{\omega_L}{\omega_S}\,|\Psi_S|^2\Psi_L - \mathrm{i}\mu_L\Psi_L \qquad (7.13)$$

$$\mathrm{i}\left(\frac{\partial \Psi_S}{\partial z} + \nu\frac{\partial \Psi_S}{\partial \tau}\right) = -\frac{P_S}{2}\frac{\partial^2 \Psi_S}{\partial \tau^2} + R_{SS}|\Psi_S|^2\Psi_S$$

$$+\,2R_{SL}|\Psi_L|^2\Psi_S$$

$$+\,\mathrm{i}|\Psi_L|^2\Psi_S - \mathrm{i}\mu_S\Psi_S + N_{sp}(z,\tau) \qquad (7.14)$$

$$\Psi_{L,S} = \frac{E_{L,S}}{|E_{L0}|} \qquad \tau = \frac{(t-(z/v_L))}{\tau_{L0}}$$

where Ψ_{L0} and τ_{L0} are the amplitude and duration of the input pump pulse; $Z = (Z/Z_{amp})$ where Z_{amp}, the non-linear length of the SRS amplification, is given by

$$Z_{amp} = \frac{2}{g_S\alpha_{L,S}|E_{L0}|^2} = \frac{cn_0}{4\pi g_S\alpha_{LS}I_{L0}}.$$

$$P_L = \mathrm{sign}\left(\frac{\partial^2 k_L}{\partial\omega^2}\right)\frac{Z_{amp}}{Z_{disp}}$$

$$P_S = \frac{|\partial^2 k_S/\partial\omega^2|}{|\partial^2 k_L/\partial\omega^2|}\,P_L\,\mathrm{sign}\left|\left(\frac{\partial^2 k_S}{\partial\omega^2}\right)\right.$$

where

$$R_{mn} = \frac{Z_{amp}}{Z_{NL}}\,\alpha_{mn}\,\frac{k_m}{k_n} \qquad \nu = \frac{Z_{amp}}{Z_{coh}} \qquad \text{and}$$

$$Z_{disp} = \frac{\tau_{L0}^2}{\partial^2 k_L/\partial\omega^2}$$

$$Z_{NL} = Z_{amp}\,\frac{n_0 g_S}{n_2 k_L} \qquad Z_{coh} = \frac{\tau_{L0}}{v_S^{-1} - v_L^{-1}}$$

$$Z_{sa} = (Z_{disp}Z_{nl})^{1/2}$$

are the dispersion length, non-linear length, length of group delay, and length of self-action, respectively.

The fundamental physical processes, described by the system ((7.13)–(7.14)) can be classified into active and reactive interactions.

The active one is characterised by cascade energy exchange between the waves resulting in energy transportation from one spectral region to another (for example from a region with positive GVD $((\partial^2 k/\partial\omega^2) > 0)$ to the region with negative GVD $(\partial^2 k/\partial\omega^2) < 0)$. As the refractive index is non-linear, reactive interactions of the waves cause the phase modulation at the pump pulse to be translated to the pulse at a combined frequency (cross-phase modulation) (Maker and Terhune, 1965). This also leads to the self-broadening of the pulse in the region with $(\partial^2 k/\partial\omega^2) > 0$ or to self-compression of the pulse in the region with $(\partial^2 k/\partial\omega^2) < 0$. The parameter $\xi(= Z_{\text{amp}}/Z_{\text{sa}})$, i.e. the ratio of the SRS amplification length to the self-action length, defines the relative roles of the active and reactive interactions in transient SRS (Hasegawa, 1983; Vysloukh and Serkin, 1983; Hasegawa, 1984).

There are two qualitatively different regimes of Raman pulse generation in fibers.

The first one, the so-called dispersionless regime of Raman self-action, is characterised only by the spectral broadening and linear chirp (near the top of a pulse envelope) of the original pulses. The dynamics of the temporal envelope of pump and Stokes pulses in this regime is completely defined by active SRS interactions – energy interchange between fundamental and Stokes waves. The interplay between the dispersive and non-linear effects in this regime of Raman pulse generation does not result in sufficient changes in the temporal envelopes of the interacting waves. This regime of SRS generation takes place when

$$\xi = \frac{Z_{\text{amp}}}{Z_{\text{sa}}} = \frac{G_{\text{th}}}{gI_{\text{L}}^{\max}} \frac{(k_{\text{S}}'' k n_2 I_{\text{S}}^{\max})^{1/2}}{\tau_{\text{S0}}} \ll 1 \tag{7.15}$$

In the second regime, which can be called a dispersive regime of Raman self-action, the interplay between the dispersive and non-linear effects play the main role in the process of energy transformation from the pump pulse to the Stokes one. In the spectral region of positive GVD, self-action effects cause the self-broadening of the pump and Stokes pulses and reduce the efficiency of this SRS while in the region of negative GVD self-compression effects and soliton formation occur. When $I_{\text{S}} \ll I_{\text{L}}$ and the condition $\xi \ll 1$ is valid, we have a transient region of SRS generation. However, the process of Raman amplification of a Stokes pulse leads to $\xi \to 1$. Let us define the range of critical parameters when $\xi = 1$. This can be conveniently done using dimensionless variables in the system ((7.13)–(7.14)) and taking into consideration that the non-linear parameters $R_{\text{L,S}}$ are associated with the number of solitons at the pump and the Stokes pulses $R = N^2$ and self-action length are connected with the number of solitons $Z_{sa} = (R)^{-1/2} = N^{-1}$.

It is necessary to point out that, principally, SRS, like other processes of stimulated scattering, has no threshold and can be considered to be an instability in the Stokes waves. In practice, however, there is an experimental threshold $I_S \simeq I_L$ which is defined by the condition $\exp(g_S|I_L|Z) = \exp(16\text{–}25) = \exp(G_{th})$ (Stolen, 1980a, b; Auyeung and Yariv, 1978).

The critical range of parameters might be obtained by putting the threshold amplification G_{th} equal to the amplification at the self-action length $2GZ_{sa} = 2G/(R)^{1/2} \simeq 16\text{–}25$.

So we can conclude that the various regimes of SRS are qualitatively defined by the number of solitons in the pump pulses.

7.4.3 *Transient dynamics of Raman pulses arising from spontaneous noise in region of positive GVD: the role of pulse walk-off*

In the process of SRS generation a very important role is played by the function belonging to the GVD effect which is responsible for the non-linear efficiency of the SRS through the length of the coherent amplification (Shen and Bloembergen, 1965; Carman *et al.*, 1970; Akhmanov *et al.*, 1971).

$$Z_{coh} = \tau_0 \left| \frac{1}{V_L} - \frac{1}{V_S} \right|^{-1} \tag{7.16}$$

This effect leads to the existence of a critical pump power at which the threshold Stokes intensity is achieved:

$$I_{cr} = \frac{G_{th} k_L''^{*} \Delta\omega_{nl}}{g_S(\omega_L)\tau_0} \tag{7.17}$$

In accordance with equations ((7.13)–(7.14)) the Stokes intensity is defined by

$$I_S(t, z) = I_{S0} \exp\left[g(\Omega) \int_0^z \left| E_{L0}\left(t - \frac{z}{V_L} + \frac{\xi}{\Delta V} \right) \right|^2 d\xi \right] \tag{7.18}$$

where $g(\Omega) = g(0)[1 + (\Omega T_2)^2]^{-1}$ and in the significantly non-stationary regime (4.18) can be rewritten as

$$I_S(t, z) = I_{S0} \exp\left[g(\Omega) |E_{L0}|^2 L_{eff} \right] \tag{7.19}$$

where $L_{eff} = \pi^{1/2} Z_{coh}$ for a Gaussian pulse. The total amplification increment

$$g(\Omega)|E_{\text{L0}}|^2 L_{\text{eff}} \cong \frac{g_0(\pi)^{1/2}\tau_0}{(1+\Omega^2 T_2^2)(\omega_{\text{L}}-\omega_{\text{S}})k_{\text{L}}''}\,|E_{\text{L0}}|^2 \tag{7.20}$$

indicates that moving the pump frequency towards the critical value ω_{cr} at which $k_L'' = 0$ corresponds to a situation when the maximum gain reaches ω_S. Detuning ω_L from ω_{cr} causes a reduction and the maximum gain frequency shifts towards the pump frequency. This effect was observed for the first time by Dianov *et al.*, (1986).

In the general case the SRS in a long waveguide structure can be described by an account of the randomly distributed sources of seeding noise, which represent the SRS process. Since a spontaneous scattering is a random Gaussian process then the Raman radiation fluctuates with temporal spikes with correlation times defined by the bandwidth of the spontaneous scattering.

Initially we impose pseudo-random boundary conditions for the Raman wave

$$\Psi_{\text{sp}}(z=0;\tau) = \exp\left(-\frac{\tau^2}{2\tau_{\text{sp}}}\right)[X(\tau) + \mathrm{i}Y(\tau)]$$

where $X(\tau)$, and $Y(\tau)$ are real, statistically independent random functions, having a normal distribution law with an average value of zero, τ_{sp} is the duration of the noise spikes. Autocorrelation functions of (7.18) is

$$\langle X(t+\tau)X(\tau)\rangle = \langle Y(t+\tau)Y(\tau)\rangle = \tfrac{1}{2}\sigma_k^2 \exp(-t^2/\tau_{\text{k}}^2)$$

where σ_k and τ_k are the dispersion and correlation time of the intensity fluctuations respectively, describing the scattering dynamics of spontaneous noise.

The results of the computer simulations (see Figure 7.11) showed that the formation of Raman pulses from spontaneous noise is accompanied by a change in the correlation time of the fluctuations of Raman radiation. At the first stage of SRS generation, regenerative narrowing of the spectrum plays the main role and therefore the correlation time increases monotonically with the fiber length. At the end of the linear (exponential) stage the Stokes wave there is, in fact, noise which is filtered within the bandwidth $\Delta\omega = \Delta\omega_L/(G_{th})^{1/2}$. While the pump intensity increases, approaching the threshold value of SRS, separate peaks of intensity at the Stokes wavelength are detached and are located under the top of the pump envelope. The exponential nature of the Raman amplification process in the linear regime results in identical, with respect to time, Stokes and pump waves. This effect has been called the process of SRS self-mode-locking (Isaev *et al.*, 1980).

It is easy to show that during cascade Raman generation SRS mode synchronization occurs and this effect can be used (Dianov *et al.*, 1986) for USP generation at Stokes wavelengths. Increasing the fiber length leads to an increase in the Raman gain and when the Stokes intensity reaches the order of the magnitude of the pump intensity then narrow gaps at the top of pump envelope are formed, corresponding to the fluctuation peaks in the Stokes wave (see Figure 7.11(*a*)). This point in Raman generation is very important. First, the representation of the temporal structure of the fluctuation of SRS at the pump wave results in a rapid broadening of the pump spectrum and formation within the pump spectrum of a broad and low-intensity pedestal.

Second, this point is characterised by the strong distortions in the structure of the phase modulation of the pump wave that hinders the following pulse compression in a dispersion compression scheme.

The effects of GVD and self-phase modulation (SPM) result in a partial smoothing of the fluctuation in the Stokes wave. Figure 7.11(*a*) shows the transient regime of the pump pulse with the smoothing effect. One can see that the smoothing manifests itself on the time scale $\tau = vz$. Thus with respect to noise at the Stokes wavelength detuning the group velocity acts as a spectral filter with a spectral constant $(vz)^{-1}$. Important features also arise in the spectral domain. Figure 7.11(*b*) shows the asymmetrical spectra formed by computer simulations and which have also been observed experimentally by Stolen and Johnson (1986) and Weiner *et al.* (1988). The physical nature of this effect can be explained as follows – as the energy from the leading edge of the pump pulse is transferred into the Stokes region, the rest energy of the pump pulse is responsible for the anti-Stokes shift (Schadt and Jaskorzynska, 1987; Herrmann and Mondry, 1988).

A study of the influence of cross-phase-modulation (XPM) and SPM on the SRS dynamics allows us to conclude that for a correct description of SRS in optical fibers it is necessary to take the effects of XPM and pulse walk-off into account simultaneously.

7.4.4 *Non-linear dynamics of SRS in a regime of dispersion self-action: non-linear transformation of optical solitons by means of Raman amplification*

The non-linear dynamics of USP generation is defined mainly by the spectral region in terms of the sign of the GVD where the

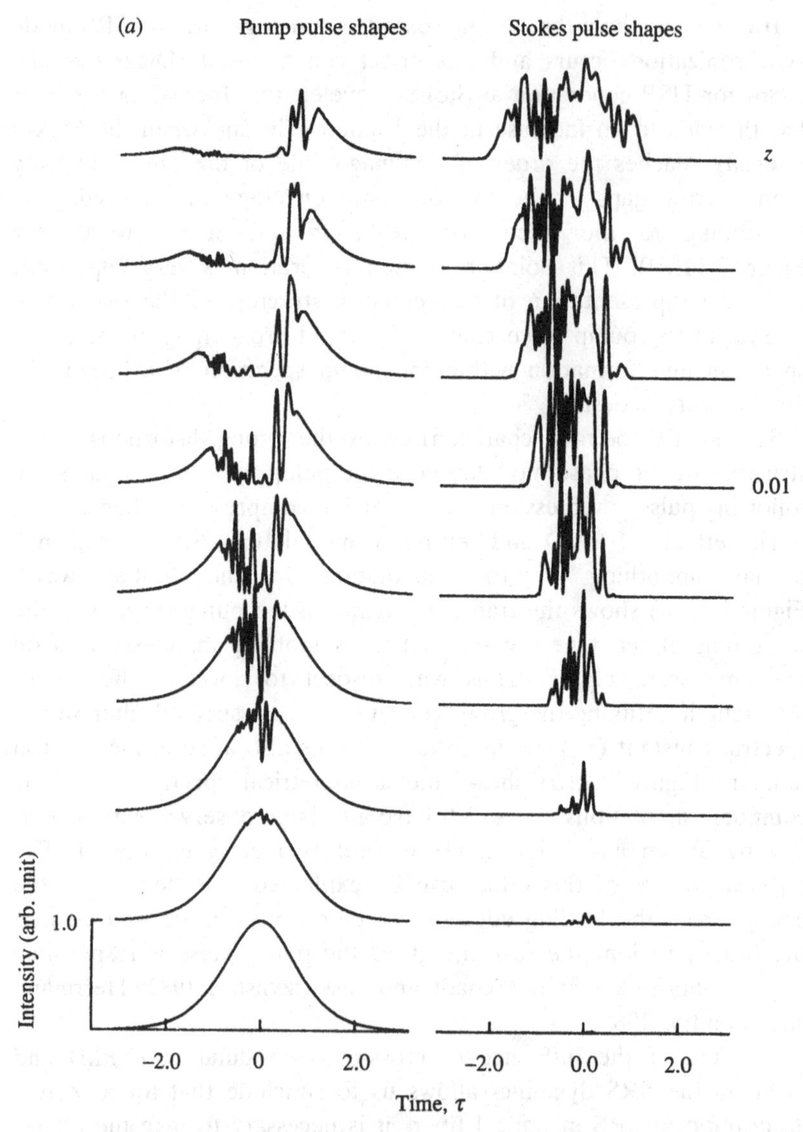

Figure 7.11. Dynamics of the formation of SRS pulses from spontaneous noise calculated in the framework of a mathematical model (7.12–7.13) at $R = 1000$, $\nu = 40$, $\Delta z = 0.002$: (*a*) the calculation of the time envelopes of the pump and SRS pulses along the fiber length; (*b*) the non-linear dynamics of the pump and SRS pulse spectra.

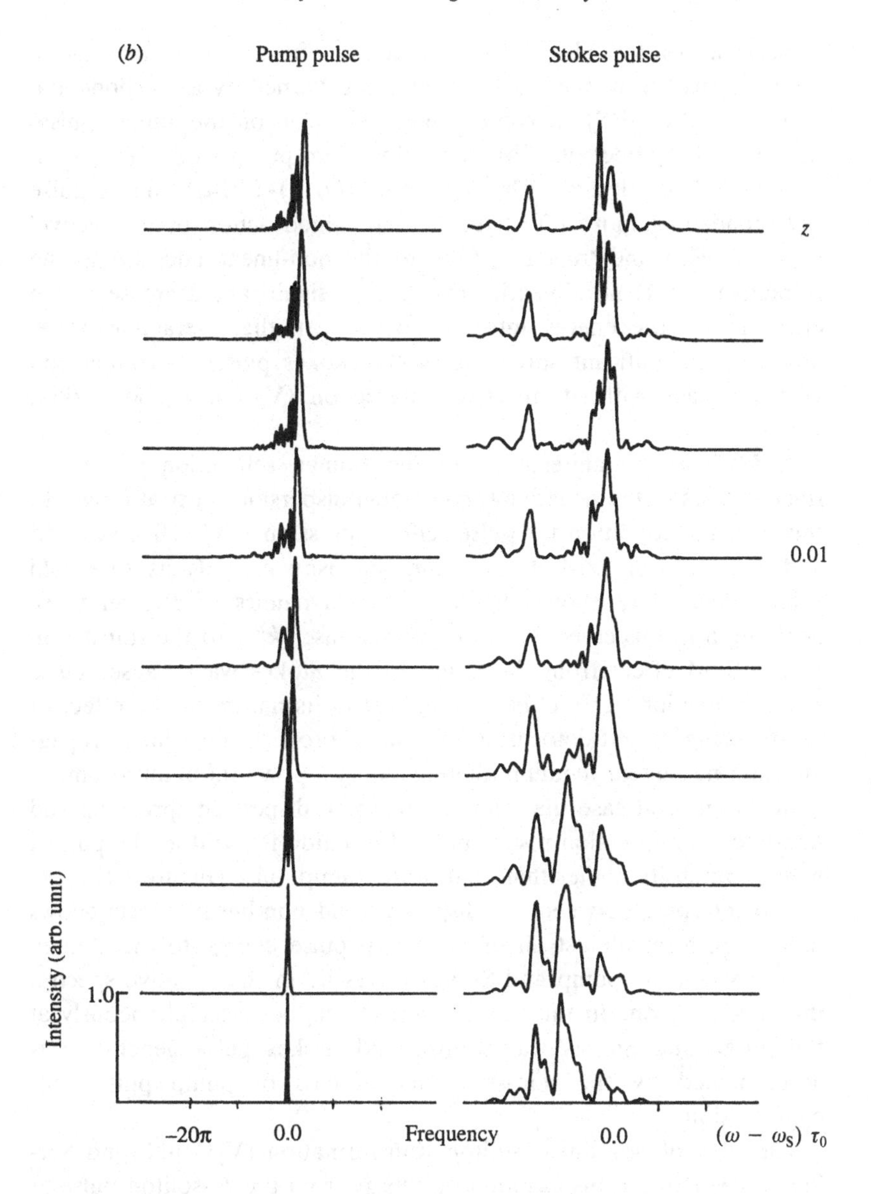

Raman pulse undergoes amplification. In the case when the pump and Stokes waves lie in the positive GVD spectral region self-action effects at pump and Stokes wavelengths have the same effect on pulse envelopes – self-action makes dispersion of the pulse-spreading process more severe. Stokes pulse formation takes place over an interaction length $Z_{sa} = (Z_{disp} Z_{nl})^{1/2}$ at which pump pulse envelope takes a rectangular shape due to the mutual actions of the non-linearity and dispersion. In the absence of self-action effects the Raman pulse duration (which, in the linear regime, is governed by the exponential nature of the SRS) is comparable with that of the pump pulse. Moreover the effect of SPM translation from pump to Stokes waves, which has been described by the model ((7.13)–(7.14)), makes pulse self-broadening more effective. The combined action of the 'active' (due to SRS) and 'reactive' (due to the non-linear addition to the refractive index) non-linearity results in a significant decrease in the efficiency of the energy interchange between the interacting waves and causes significant spreading of the Stokes pulses in comparison with the case without 'reactive' interaction (Vysloukh and Serkin, 1983).

A fascinating manifestation of the Raman self-action process in fibers occurs in the negative (anomalous) dispersion spectral range. In this case the condition for pulse self-compression $n_2 k'' < 0$ is satisfied and solitons will exist if the pump intensity exceeds its threshold value. Figure 7.12 shows the non-linear dynamics of Raman pulse generation in this case. As in the former case ($k'' > 0$) the translation of the SPM effect from the pump to the Stokes wave causes pulse self-compression. This effect is identical in its nature to the effect of self-focusing or self-defocusing of a weak probe beam which propagates in a non-linear medium illuminated by a powerful pump beam.

In the general case the interplay between dispersion spreading and non-linear compression results in the formation for soliton-like pulses, with an intensity higher than that of the pump pulse (Figure 7.12).

The interplay between the dispersive and non-linear effects allows us to transform almost all of the pump pulse energy to the Raman soliton when the pump and Stokes waves lie in the negative spectral dispersion region. In this case soliton effects occur simultaneously at the pump and Stokes wavelengths and Stokes pulse generation is accompanied by the non-stationary process of pump pulse self-compression.

The idea of non-linear soliton transformation (Vysloukh and Serkin, 1983) is the transportation of energy from the N-soliton pulse at the pump wavelength into a fundamental soliton or bound state of K-solitons pulse at the Stokes wavelength:

$$\Psi_L = N \operatorname{sech}(\tau) \rightarrow \Psi_S = \kappa_S \operatorname{sech}(\kappa_S \tau) \tag{7.21}$$

where κ denotes the form-factor. From the law of energy conservation

$$U_L + \frac{\omega_L}{\omega_S} U_S = \text{constant}$$

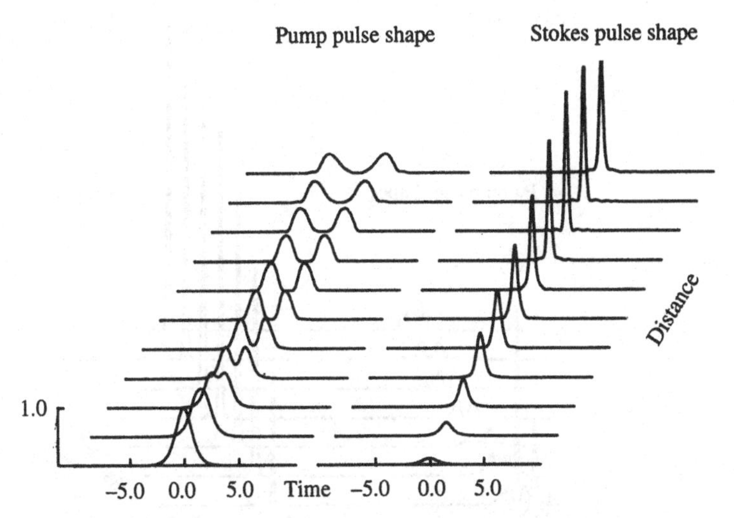

Figure 7.12. Generation dynamics of soliton-like Raman pulses excited by picosecond pump pulses with a sech-form envelope. The pump spectral range corresponds to the positive GVD, the Stokes frequency falls into the region of negative GVD. Temporal profiles of intensity are shown at various z with the step $\Delta z = \pi/40$.

$$U_{\mathrm{L}} = \int_{-\infty}^{+\infty} |\Psi_{\mathrm{L}}|^2 \mathrm{d}\tau = 2N^2 \qquad \text{and}$$

$$U_{\mathrm{S}} = \int_{-\infty}^{+\infty} |\Psi_{\mathrm{S}}|^2 \mathrm{d}\tau = 2\kappa_{\mathrm{S}} \tag{7.22}$$

we can obtain a simple estimate of the maximum compression rate $\kappa_{\mathrm{S}} = N^2$. From the condition for the formation of the bound-state soliton

$$\Psi_{\mathrm{L}} = N \operatorname{sech}(\tau) \rightarrow \Psi_{\mathrm{S}} = K\kappa_{\mathrm{S}} \operatorname{sech}(\kappa_{\mathrm{S}}\tau) \tag{7.23}$$

we can obtain an estimate of the form-factor for K-solitons $\kappa_{\mathrm{S}} = N^2/K^2$.

Figure 7.13 presents the results from computer simulations of these processes and illustrates the non-linear dynamics of the transformation of the multi-soliton pump pulse $\Psi_{\mathrm{L}} = N_{\mathrm{L}} \operatorname{sech}(\tau)$ into the fundamental Raman soliton with the envelope $\Psi_{\mathrm{S}} = \kappa_{\mathrm{S}} \operatorname{sech}(\kappa_{\mathrm{S}}\tau)$. The peculiarity of this case is that the Raman pulse is formed at the length

$$Z = \frac{\pi}{8} \frac{Z_{\mathrm{disp}}}{Z_{\mathrm{amp}}}$$

which corresponds to the maximum compression rate and zero chirp

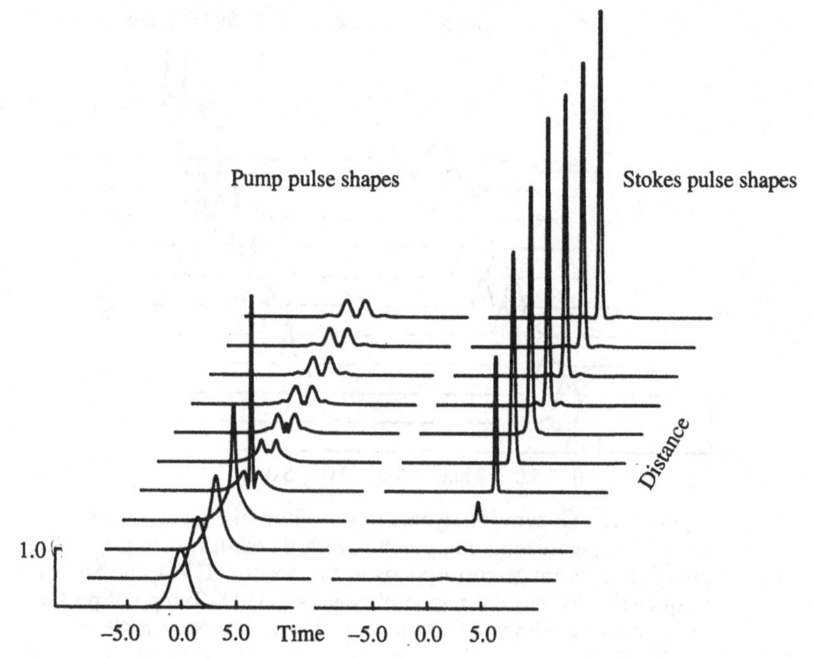

Figure 7.13. Dynamics of non-linear conversion of the bound state of $N = 3$ solitons into a single-soliton pulse at the SRS in a fiber. The intensity time profiles are shown with the interval $\Delta z = \pi/40$ along the fiber. The initial Stokes signal has the form $\psi_s(z = 0, \tau) = 0.01 \operatorname{sech} \tau$.

velocity. Finally Figure 7.14 shows the non-linear dynamics of the bound state of Raman solitons (Vysloukh and Serkin, 1984).

It should be noted that experimental verification of this description of SRS generation in the 'dispersion' regime of SPM requires a weak probe pulse at the Stokes wavelength – a 'seed pulse' in order to prevent SRS generation from spontaneous noise. An analysis of the general case of SRS generation from spontaneous noise in the region of negative GVD requires the application of Monte-Carlo statistical methods. Figure 7.15 shows the non-linear dynamics of SRS in the region of negative GVD when the pump lies in the region of positive GVD and Figure 7.16 shows the case when the pump is in the region of negative GVD. A comparison of these results leads us to the conclusion that, in the general case of SRS generation from spontaneous noise, there are soliton-like pulses (we can call them an 'outburst' of soliton noise) at the Stokes wavelength independently of the sign of the GVD for the pump wavelength. The dynamics of the

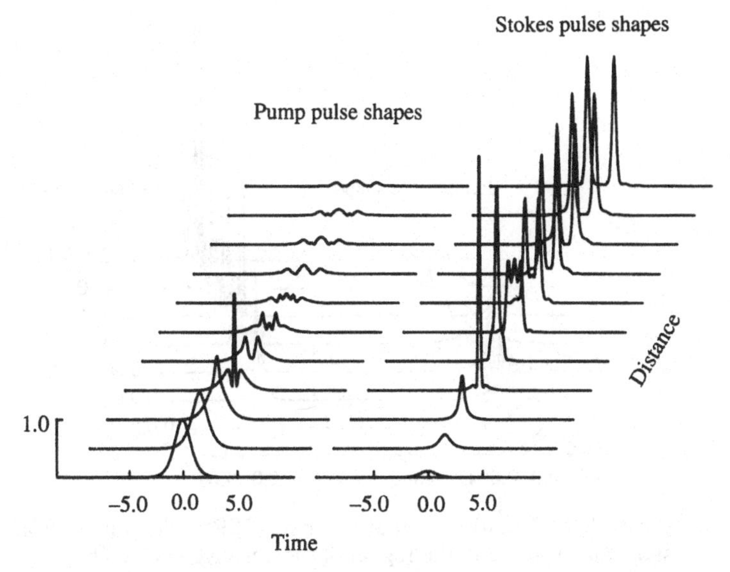

Figure 7.14. Formation of a bound state of Raman solitons and their breakdown at the Raman conversion of the $N = 3$ soliton pulse. The intensity of the initial pulse $\psi_s(z = 0, \tau) = 0.1 \operatorname{sech} \tau$.

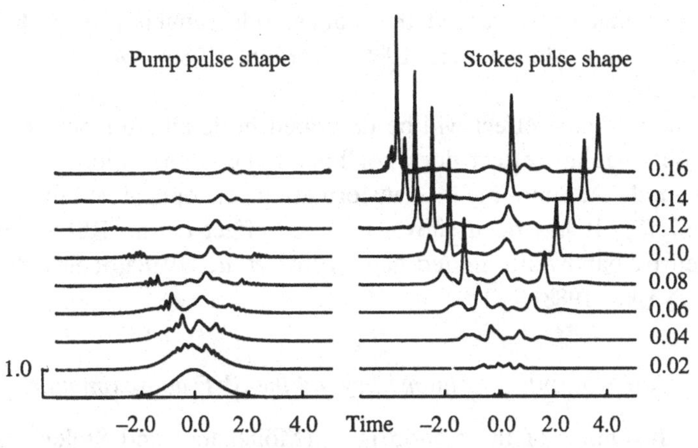

Figure 7.15. The transient dynamics of SRS soliton-like pulses, arising from spontaneous noise in region of positive GVD. Parameter $R = 100$.

solitons in these outbursts is mainly defined by the XPM effect and by the dynamics of the pump pulse envelope, which allows us to suggest various methods for distinguishing separate solitons within the 'outburst', using XPM at other wavelengths (Afanasjev and Serkin,

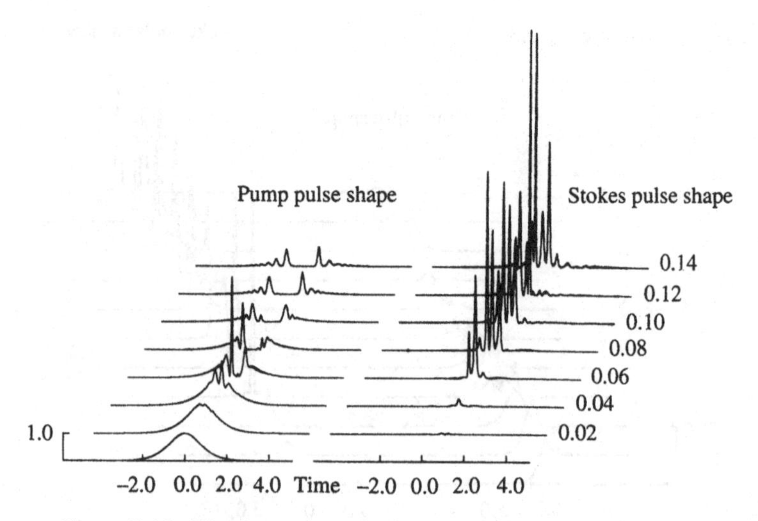

Figure 7.16. The formation of a 'burst' of SRS soliton noise from the spontaneous noise in the region of the negative GVD. The calculations are made in framework of the model (7.12–7.13) at $R = 100$, $\tau_c = 0.05$, $\varepsilon = 0.01$.

1989). A remarkable property of optical fibers is the existence of the distinctive mechanism called the Raman self-frequency shift (RSFS) (Mitschke and Mollenauer, 1986; Gordon, 1986; Dianov, *et al.*, 1985).

This remarkable effect will be described in detail later but here we point out that this effect does not hinder the compression of optical pulses and the non-linear transformation of optical solitons (see Figure 7.17). It has been shown that the effect of the RSFS can be stopped by bandwidth-limited gain (Blow *et al.*, 1988a,b; Schadt and Jaskorzynska, 1988).

7.4.5 *RSFS of optical solitons: beyond the SVE approximation*

Invention of the soliton laser (Mollenauer and Stolen, 1984) made it possible to study the non-linear dynamics of optical solitons in detail. It had been confirmed that picosecond optical solitons in fibers are real solitons for which the NLS equation is valid (Hasegawa and Tappert, 1973). In complete agreement with the inverse scattering problem (Zakharov and Shabat, 1971) the dynamics of weakly interacting solitons, with durations of about 10 ps is identical to classical particle dynamics with a twin interaction potential. However, moving into the femtosecond timescale leads to a drastic change in

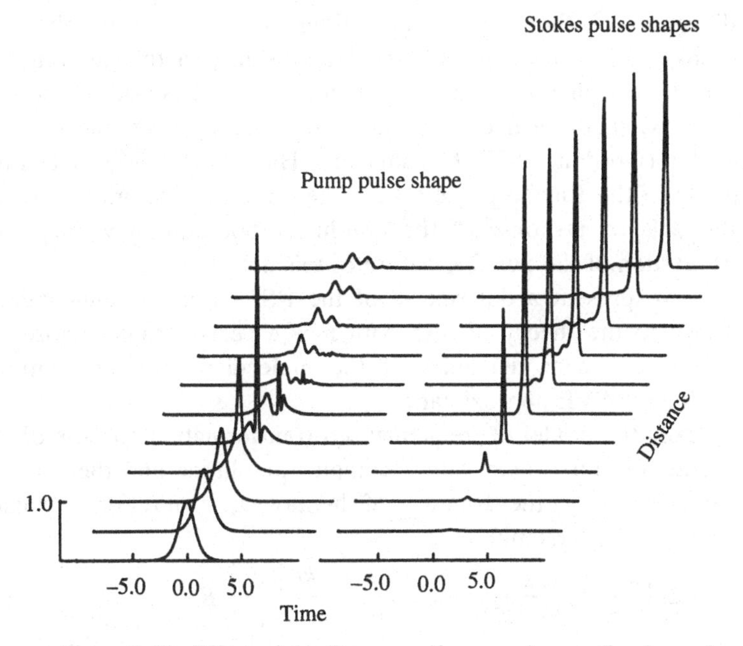

Figure 7.17. Effect of the Raman self-conversion on the dynamics of non-linear conversion of the $N = 3$ solitons into a single-soliton Raman pulse. The calculations are made in the framework of model (7.12)–(7.13). They should be compared with the results shown in Figure 7.13.

the dynamics of optical solitons. The permanent down-shift of the frequency of optical solitons, the so-called RSFS, has been found by several authors (Mitschke and Mollenauer, 1986; Gordon, 1986). This effect has been utilised for the generation of femtosecond Raman solitons by means of multi-soliton transformation (Dianov *et al.*, 1985). The RSFS effect arises in a situation in which the bandwidth of a non-linear pulse becomes as broad as the difference between the spectral components of the pulse and is comparable with the resonant frequency of the molecular vibrations. Beating at a couple of frequencies resonantly excites molecular vibrations adjusting their phases in a large volume. The phases of the stimulated vibrations are adjusted by the field of the same soliton causing Raman frequency transformation within its own spectrum: lower frequencies are amplified while higher ones are attenuated and the whole spectrum moves towards the longer wavelength region. This effect results in a change in the soliton group velocity, break up of the soliton bound state and a resultant modification of the soliton interactions that have been observed (Mitschke and Mollenauer, 1987a). Despite the fact that the physical nature of this phenomenon was qualitatively understood in the original work, theoretical descriptions had a mainly phenomenological character (Gordon, 1986; Kodama and Hasegawa, 1987). The time response of the Raman process was taken into consideration in terms of the delayed response of the non-linear polarisation without any interpretation of the possible nature of this delay.

Here we present a description of the RSFS using a semi-classical approach to the theory of SRS with new aspects, dealing rigorously with the non-linear dynamics of the molecular vibrations without exploiting the SVE approximation.

Basically the model of frequency self-transformation consists of the NLS equation for a field with a complex amplitude and the classical equation describing the dynamics of the molecular vibrations (Afanasjev *et al.*, 1990; Vysloukh *et al.*, 1989)

$$2\mathrm{i}\left(\frac{\partial E}{\partial z} + \frac{1}{V}\frac{\partial E}{\partial t}\right) = -k''\frac{\partial^2 E}{\partial t^2} + \frac{kn_2|E|^2}{n_0}E + \frac{4\pi}{c^2}N_{\mathrm{m}}\frac{\partial \alpha}{\partial \tilde{Q}}\tilde{Q}E \tag{7.24}$$

$$\frac{\partial^2 \tilde{Q}}{\partial t^2} + 2\Gamma\frac{\partial \tilde{Q}}{\partial t} + \Omega_{\mathrm{R}}^2\tilde{Q} = \frac{1}{2M}\frac{\partial \alpha}{\partial \tilde{Q}}n_0\frac{|E|^2}{2} \tag{7.25}$$

The first two terms on the right-hand side of 7.24 describe the effects associated with the GVD and non-linear refractive index due to the Kerr electronic effect. The last term takes into account the contribution of the Raman effect to the non-linear susceptibility. It is worthwhile considering this term more carefully. In accordance with (7.24) and (7.25) the powerful pump pulse initiates the appearance of additional resonances in the non-linear susceptibility of the media at frequencies $\Omega_{\mathrm{L}} + \Omega_{\mathrm{R}}$ which are the molecular eigenfrequencies. The Fourier amplitude of the force molecular vibrations Q_Ω can be defined by the expression

$$Q_\Omega = \frac{\partial \alpha}{\partial G}\frac{n_0}{4M}\frac{|E|^2\Omega}{(\Omega_{\mathrm{R}}^2 - \Omega^2 + 2\mathrm{i}\Gamma\Omega)} \tag{7.26}$$

One can conclude from (7.26) that the molecular vibrations are phased by the same broad spectrum of the pump pulse which results in the formation of additional resonances and the induced Raman dispersion. The non-linear Raman cubic susceptibility χ_{Im}, like the linear one, has a Lorenzian shape

$$\chi^{(3)}(\Omega) = \frac{\text{constant}}{(\Omega_R^2 - \Omega^2 + 2i\Gamma\Omega)} = \chi^{(3)}_{\text{Re}}(\Omega) - i\chi^{(3)}_{\text{Im}}(\Omega) \tag{7.27}$$

$$\chi^{(3)}_{\text{Re}}(\Omega) \simeq \frac{(\Omega_R^2 - \Omega^2)}{-[(\Omega_R^2 - \Omega^2)^2 + 4\Gamma^2\Omega^2]} \tag{7.28}$$

$$\chi^{(3)}_{\text{Im}}(\Omega) \simeq \frac{2\Gamma\Omega}{[(\Omega_R^2 - \Omega^2)^2 + 4\Gamma^2\Omega^2]} \tag{7.29}$$

Essentially the Stokes resonance possesses negative loss and an anomalous induced Raman dispersion. The anomalous dispersion in the region of the Raman resonance is responsible for the effect of group delay on the spectral components with maximum Raman gain. This effect of a delay in the leading edge of the Stokes pulse has been described in detail by Akhmanov *et al.* (1971), Carman *et al.* (1970) and Hellwarth *et al.* (1975). So, if the difference in the spectral components of a pulse satisfies the Raman resonance condition then the Stokes spectral component of that pulse receives an additional phase shift defined by the dispersion of $\chi^{(3)}_{\text{Re}}$ and gain.

Similarly, in the anti-Stokes region, the component ω_a acquires a phase shift defined by the normal dispersion $\chi^{(3)}(\omega_a)$ and a negative gain. It is necessary to point out that in the region of negative GVD the non-linearity of the media leads to a phase adjustment in the spectral components of the pulse while the region with $k'' > 0$, $n_2 > 0$ is characterised by phase readjustment. Therefore the process of Raman-induced frequency self-transformation is much more effective at the soliton's region of spectrum and leads to the effect of the soliton Raman self-frequency shift (Mitschke and Mollenauer, 1986; Gordon, 1986).

The process of frequency self-transformation of femtosecond pulses can be described by the next system of equations (Afanasjev *et al.*, 1989; Vysloukh *et al.*, 1989)

$$i\frac{\partial \Psi}{\partial z} = \pm\frac{1}{2}\frac{\partial^2 \Psi}{\partial \tau^2} + R|\Psi|^2\Psi + \beta RQ\Psi + O\left(\frac{T_0}{\pi\tau}\right) \tag{7.30}$$

$$\mu^2\frac{\partial^2 Q}{\partial \tau^2} + 2\gamma\mu\frac{\partial Q}{\partial \tau} + Q = |\Psi|^2 \tag{7.31}$$

In equations (7.30) and (7.31) the distance along the fiber is normalized by the dispersion length $z_{\text{disp}} = \tau_0^2/|k''|$, the running time τ by the initial pulse duration τ_0, the complex field amplitude Ψ by the input peak value $|E_0|$ and the amplitude of molecular oscillations $Q(z, \tau)$ by the value $Q_M = |\Psi^2|(2M_R^2\Omega^{-1}(\partial\alpha/\partial Q))$. The non-linearity parameter R is the ratio of the dispersion length to the non-linear length, $R = z_{disp}/z_{nl}$, where $z_{nl} = (kn_2 I_{eff})^{-1}$ is the length of the SPM, I_{eff} the effective intensity over the fiber cross section. The limits of the applicability of the NLS approximation are determined by the inequality: $T_0(R)^{1/2}/(\pi\tau_0) \ll 1$ (Golovchenko, *et al.*, 1985).

The parameter $\mu = (\tau_0\Omega_R)^{-1}$, characterising the ratio of the input pulse spectrum width to the frequency shift of the Raman resonance and parameter $\gamma = (T_2\Omega_R)^{-1}$, characterising the Q-factor of the Raman oscillator have been introduced in equation (7.31). The degree of influence of the molecular oscillations on the optical soliton pulse is determined by the parameter $\beta = g_S\gamma/(kn_2)$, where g_S is the Raman gain in the line center. From equation (7.31) is follows that the effect of the SRS self-conversion of an optical pulse occurs when the width of its spectrum becomes comparable with the resonance frequency of the molecular oscillations.

It should be noted that the SVE approximation is exploited only for an optical field. The intrinsic period of molecular vibrations in silica glass corresponding to a Stokes shift of about 440 cm^{-1} is 75 fs. This means that, for a correct description of the frequency Raman self-transformation in the femtosecond timescale, one should take into consideration the oscillatory nature of the molecular response. The physical nature results in the necessity of solving equations (7.30) and (7.31) without the SVE approximation is the low finesse of Raman oscillator, since the Q-factor of the Raman oscillator

$$Q_R = \Omega_R T_2/2 \simeq 2 \qquad (2/T_2 = \Delta\Omega_R = 250\ \text{cm}^{-1})$$

In equation (7.30) $O(T_0/\pi\tau_0)$ are terms describing higher-order non-linear and dispersive effects:

$$O\left(\frac{T_0}{\pi\tau_0}\right) = \frac{T_0}{\pi\tau_0}\left(i\beta_\alpha\frac{\partial^3\Psi}{\partial\tau^3} - iR\frac{\partial}{\partial\tau}(|\Psi|^2\Psi) - \frac{1}{2}\frac{\partial^2\Psi}{\partial z\partial\tau}\right) \qquad (7.32)$$

Here $T_0(\simeq 5\ \text{fs})$ is the optical oscillation period and the parameter β governs the effect of the higher-order dispersion.

When $\mu \ll \gamma$ the molecular response can be approximately presented in the following way

$$2\gamma\mu\frac{\partial Q}{\partial r} + Q = |\Psi|^2$$

$$Q = |\Psi|^2 - \sigma\frac{\partial}{\partial t}|\Psi|^2 \qquad (7.33)$$

where $\sigma = 2\gamma\mu = T_{rel}/\tau_0$ and is the parameter of the response delay. For silica glass we obtain $T_{rel} = 2/(T_2\Omega_R^2) \simeq 6$ fs. This approximation describes, with reasonable accuracy, the frequency shift of single-soliton pulses (Gordon, 1986; Kodama and Hasegawa, 1987).

An analysis of the 'slow' variations of the fundamental parameters of a soliton

$$\Psi(z, \tau) = \kappa(z)\,\mathrm{sech}\,[\kappa(z)(\tau + V(z)z)] \times \exp[\mathrm{i}V(z)\tau + \mathrm{i}(V^2(z) - \kappa^2(z))z/2]$$

$$V(0) = 0 \qquad \kappa(0) = 1 \tag{7.34}$$

using the SVE approximation results in a linear (in terms of distance z) decrease in the soliton velocity

$$\frac{\partial V}{\partial z} = -\frac{8}{15}\beta\sigma$$

and an increase in the amplitude of the soliton $\kappa = 1 + \beta$. In fact it means a linear decrease in the carrier frequency.

An increase in the soliton amplitude is associated with the static Raman contribution to the total non-linearity

$$\delta n = \delta n_{\mathrm{Kerr}} + \delta n_{\mathrm{Ram}}$$

In the experiment to measure n_2 this contribution is automatically taken into consideration. This is why one should renormalise the non-linearity parameter $R = R/(1 + \beta)$ (Stolen *et al.*, 1989).

In the general case the integral solution of equation (7.31) can be written using the casual Green function $H(\tau)$:

$$Q(t) = \int_0^\infty H(\tau)|\Psi(t - \tau)|^2 \mathrm{d}\tau \tag{7.35}$$

$$H(\tau) = -\frac{1}{\omega_1}\exp\left(-\frac{\gamma}{\mu}\tau\right)\sin(\omega_1\tau) \qquad \omega_1 = (1 - \gamma^2)^{1/2}/\mu \tag{7.36}$$

Note that the response function $H(\tau)$ can be determined using the real form of the Raman amplification band and the Kramers–Kronig relationship (Hellwarth *et al.*, 1975; Chraplyvy *et al.*, 1984; Golovchenko *et al.*, 1989; Stolen *et al.*, 1989). In Chraplyvy *et al.* (1984) the real part of the susceptibility was calculated using the Kramers–Kronig relationship. Recently this approach was used by Stolen *et al.* (1989) for the detailed theoretical investigation of the RSFS effect. They calculated the response function for fused silica. However, because some of the details of the Raman gain spectrum change from fiber to fiber and depend on core dopants as well as on temperature variations, the response function should be recalculated for each experiment. It was found that the main difference between the Lorentzian and silica response functions is the changing decay rate for silica for a time interval greater than 100 fs. A comparison of model

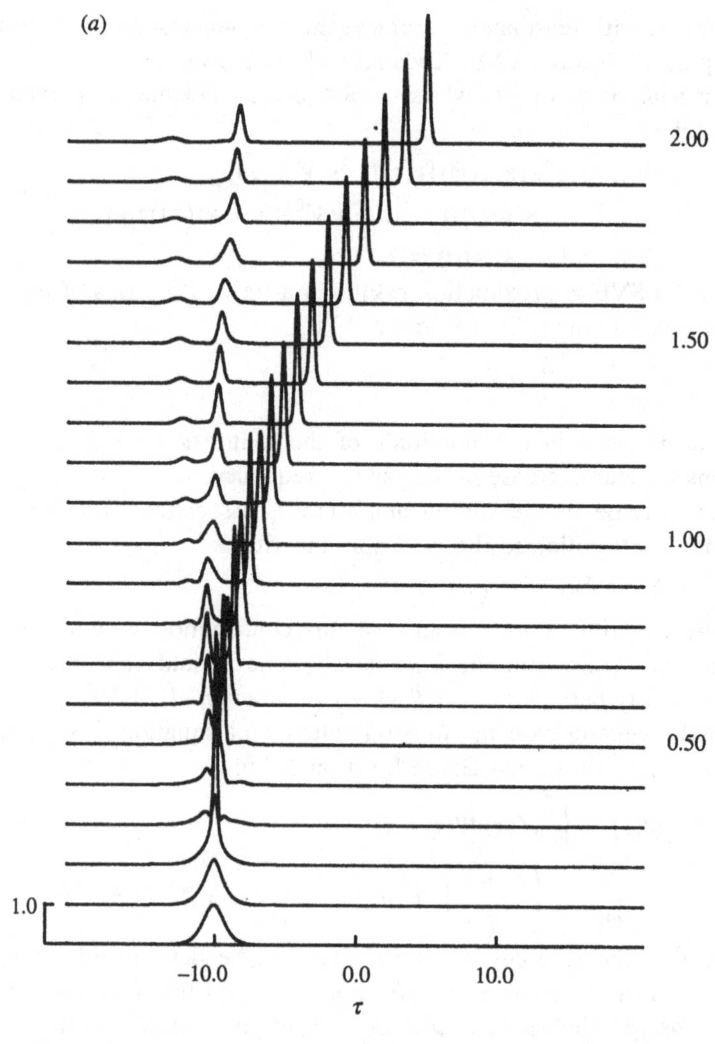

Figure 7.18. (*a*) Decay of a third-order 75 fs soliton pulse calculated in frames of a mathematical model (7.30–7.31) in the negative GVD. (*b*) The dynamics of molecular oscillation $Q(z, \tau)$.

(7.35) and that of Stolen's *et al.* (1989) allows us to choose the values of parameters $\tilde{\gamma}$ and $\tilde{\beta}$ so that the model's response function (7.36) is approximated by the real one with maximum accuracy. The approximation procedure allows us to determine the correction factors:

$$\tilde{\gamma} = \gamma k_1 \qquad \tilde{\beta} = \beta k_2 \qquad \gamma = 0.28$$
$$\beta = 0.095 \qquad k_1 = 0.85 \qquad k_2 = 2.5$$

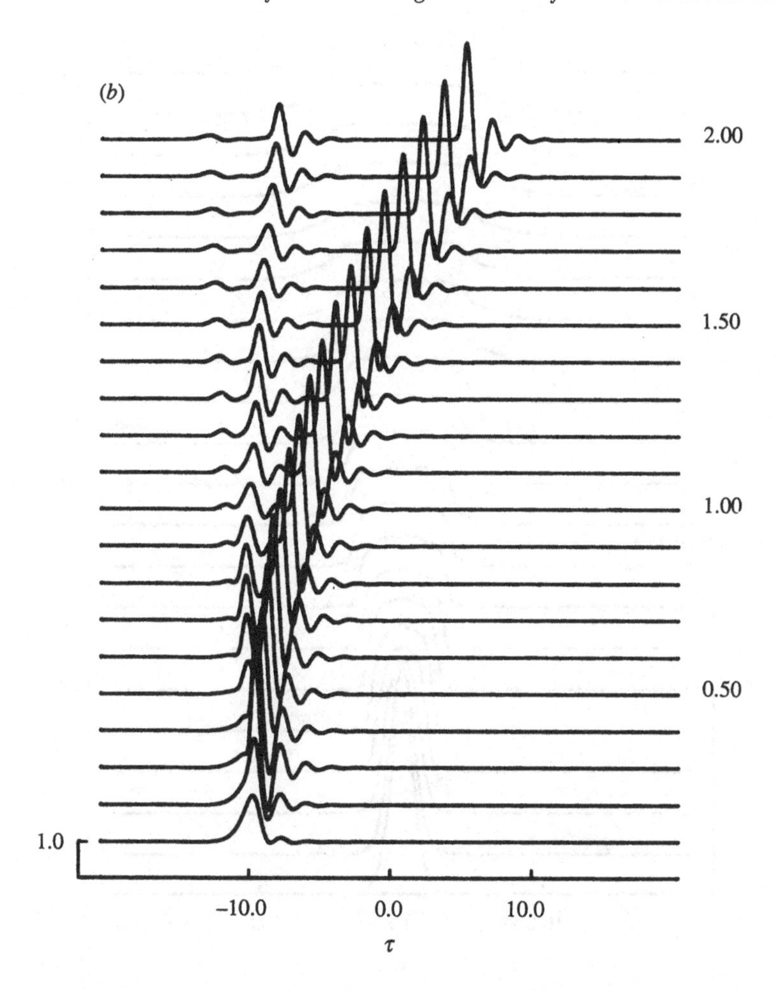

In the model ((7.30)–(7.31)) we take into consideration the effect of the finite response time of the Raman non-linearity simultaneously with the oscillation character of the Raman response. The effect of the Raman non-linearity on the higher-order solitons is to split them into their constituent parts – the phenomenon of soliton break up (Mitschke and Mollenauer, 1986; Kodama and Hasegawa, 1987; Beaud *et al.*, 1987; Grudunin *et al.*, 1987b; Serkin, 1987c, d).

Figure 7.18(*a*) shows the break up of the third-order soliton ($N = 3$) with an initial duration of 75 fs. The bound state of the three solitons breaks up into three individual solitons with their form-factors in the ratio 1:3:5 and their peak heights in agreement with those predicted by inverse scattering theory (Zakharov and Shabat,

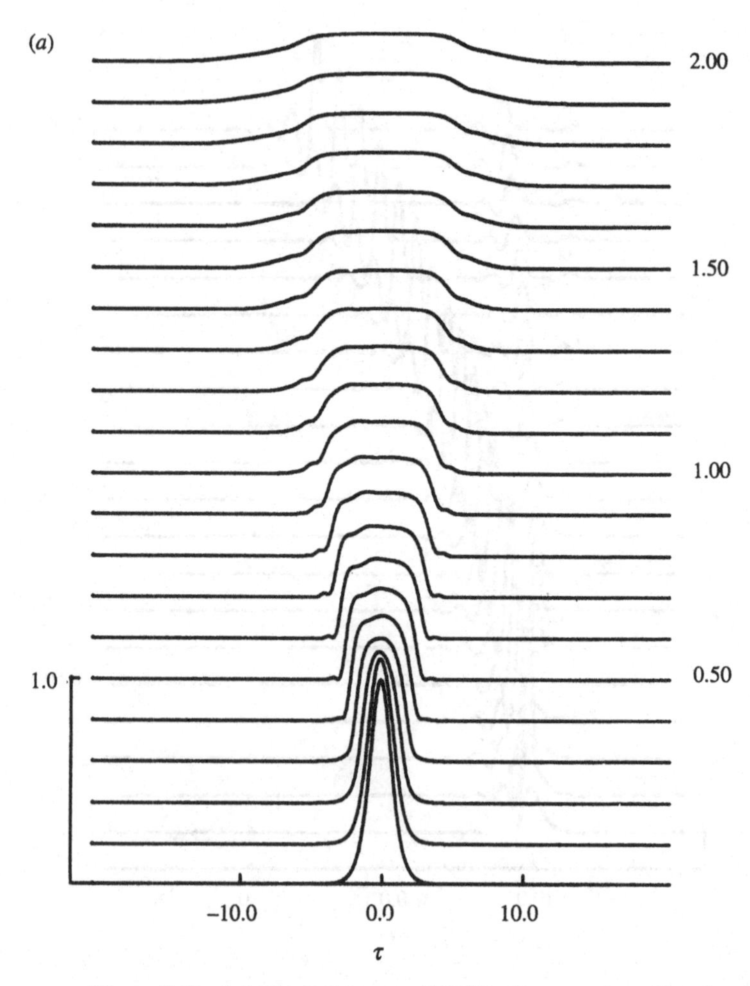

Figure 7.19. (*a*) The influence of RSFS effect on the self-action of an intense pulse ($R = 9$) in the positive GVD. (*b*) The dynamics of molecular oscillations.

1971). The transformation of the wave of molecular oscillations is also shown in this figure. The qualitative behavior of the molecular oscillation wave is similar to the decay dynamics of the time envelope. In particular from Figure 7.18(*b*), it is possible to observe the quasi-static response of the molecular oscillation wave for the soliton with the form-factor $\kappa = 1$, as well as the oscillating character of the molecular oscillation response for the solitons with form-factors $\kappa = 3$ and $\kappa = 5$. In the case considered the inertial nature of the molecular response plays a decisive role. Since the Raman addition to the

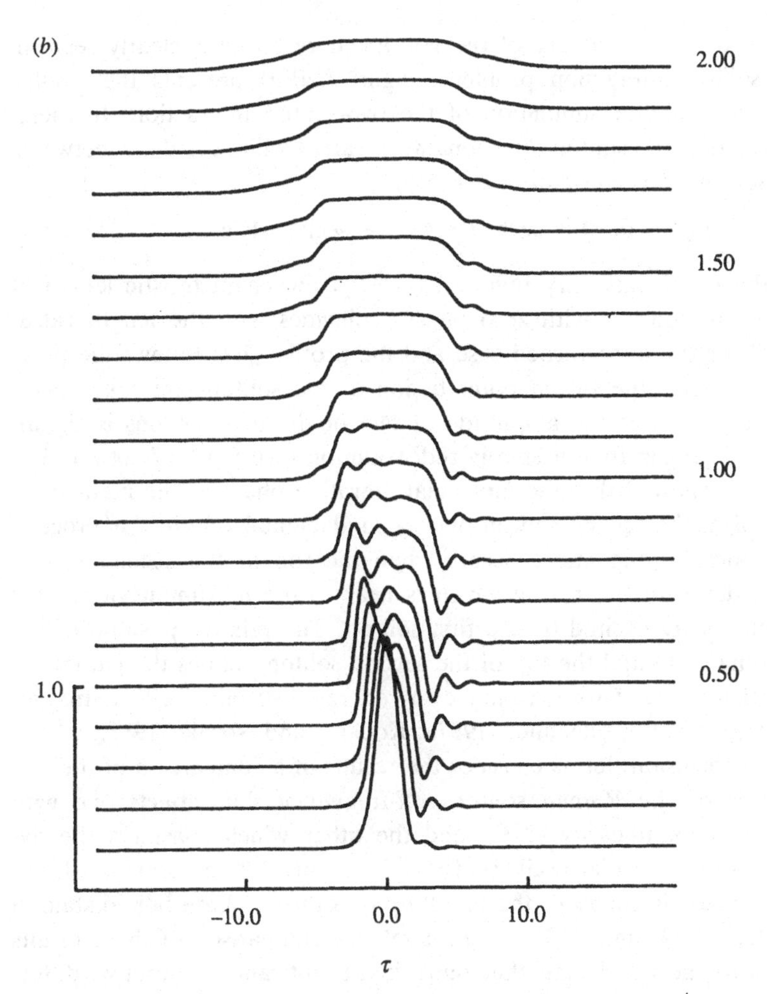

non-linearity strongly depends on the pulse duration ($\sim \tau_0^{-4}$) then the Raman delay has a different effect on each component of the multi-soliton pulse. Each component undergoes a frequency shift due to the RSFS that eliminates any degeneration in the soliton velocity and consequently a fission of the soliton bound state (Kodama and Hasegawa, 1987; Hodel and Weber, 1987; Serkin, 1987d, Tai *et al.*, 1988).

The qualitative picture of the effect of Raman non-linearity on the propagation of a pulse with the same parameters in the region of the positive GVD differs greatly from that previously considered. The results of the detailed calculation shows that the role of Raman self-action in the region of positive GVD is followed by the attenuation of the molecular oscillations and by the spreading of the wave packet (see Figures 7.19(*a*) and (*b*)).

The oscillation nature of the non-linear response is clearly seen in the soliton interaction problem. Figure 7.20(a) presents the results from a computer simulation of the two-soliton interaction. In these calculations the soliton durations were varied while the delay between the solitons was constant at $\Delta = 4$:

$$\Psi(z = 0, \tau) = \operatorname{sech}(\tau - \Delta/2) + \operatorname{sech}(\tau + \Delta/2)$$

In the quasi-stationary regime ($\tau_0 \gg T_2$) the characteristic length of fusion of solitons with zero phases coincides with the length calculated for the unperturbed case and the profile $Q(t)$ follows the pulse shape. After the fusion point fission of the soliton pair takes place and the ratio of the amplitude of two of the new solitons is significantly changed by the energy redistribution (see Figure 7.20(b)). The analysis showed that the molecular vibration phase at the moment of arrival of the second soliton plays a fundamental role in this process. The most strong effect occurs when the top of the second soliton coincides with the region which is near to the maximum vibrational amplitude Q excited by the first soliton. The relative position of the maxima of Q and the top of the second soliton defines the parameter μ which can be varyed to make the energy distributed effect stronger (Mitschke and Mollenauer, 1987a; Kodama and Nozaki, 1987).

In conclusion, let us consider the results of a comparison of the two models of the Raman soliton self-frequency shift effect: one with relaxing non-linearity (7.33) and the other which considers the dynamics of molecular oscillations (7.35). Figure 7.21 presents a calculation of the dynamics of the break up of a three-soliton bound state in models (7.33) and (7.35), respectively. A comparison of these results leads to the conclusion that there is a significant quantitative difference in the time delay of the formed Raman solitons formed on break-up. The reason for such a difference can be found if we turn to the calculation of the dynamics of molecular oscillations (see, for example, Figure 7.18(b)). As we have previously shown, the character of the interaction (attraction or repulsion) is determined by the phase of the molecular oscillations. This is why the rate of break up in model (7.35) turns out to be slower (Afanasjev *et al*., 1989).

7.4.6 *Stochastical instability of multi-soliton pulses and Raman self-transformation of optical solitons*

The phenomenon of modulation instability (MI) plays an important role in the process of Raman self-conversion of intense

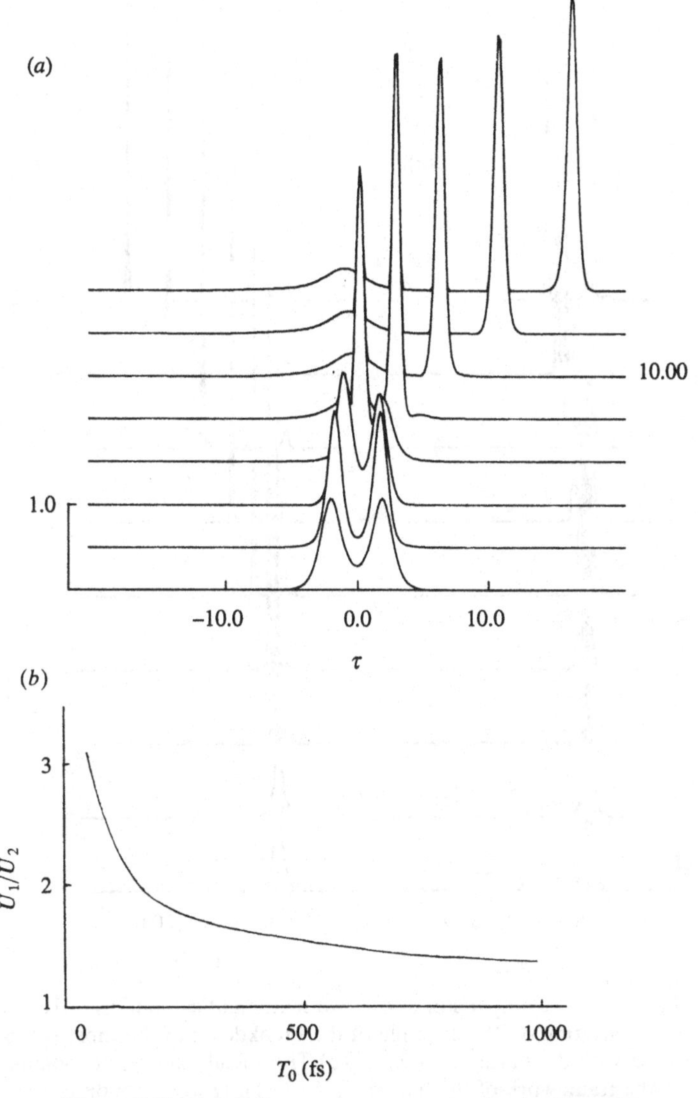

Figure 7.20. (*a*) Soliton pair propagation in the fiber. The parameters are $T_p = 75$ fs and $\Delta = 4$. (*b*) The energy redistribution between the pulses U_1/U_2 as a function of pulse duration T_p.

optical pulses in fibers.

The condition of modulation instability in a plane monochromatic wave Ψ_0 can be expressed as (Litvak and Talanov, 1967; Carpman, 1967)

$$\Omega^2 \leqslant \Omega_{cr}^2 = 4R\Psi_0^2$$

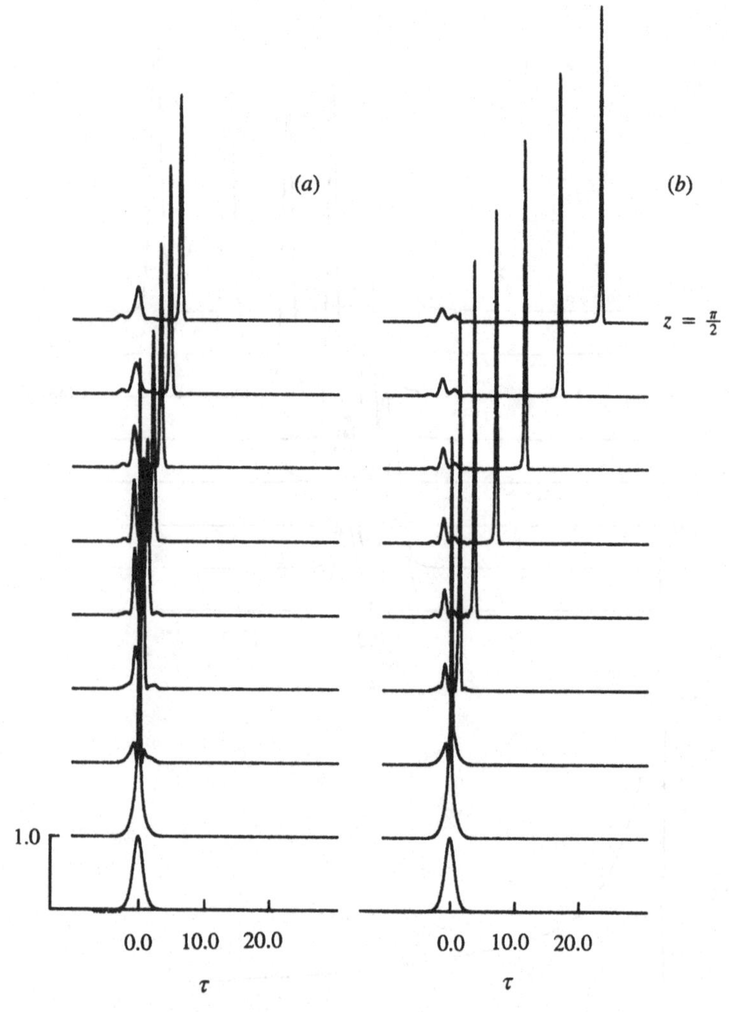

Figure 7.21. Comparison of the two mathematical models of Raman self-scattering on the example of the breakdown of the three-soliton pulse with the duration of 70 fs. (*a*) The calculation results obtained in the framework of the model (7.30–7.31); (*b*) calculation in the framework of the model (7.33).

The maximum growth rate of MI occurs at $\Omega^2_{max} = \Omega^2_c/2$ and is given by $K_{max} = \Omega^2_{cr}/4$. The transient length of MI, Z_{MI} is $2/\Omega^2_{max}$. It is easy to obtain the more general relationship for long duration pulses for which $\tau_0 \gg T_{MI}$ where $T_{MI}(= 2\pi/\Omega_{max})$ is the MI period.

The phenomenon of MI can lead to the spontaneous break up of a c.w. light wave into a periodic pulse train. So the c.w. signal is converted into a periodic pulse train whose period is $T_{\mathrm{MI}} = 2\pi/\Omega_{\max}$ (Tai *et al.*, 1986; Tai *et al.*, 1986).

It is well known that the main point of the process of multi-soliton pulses ($R = N^2 \gg 1$) self-action in optical fibers is the formation of a field singularity caused by the creation of a narrow intense peak which is located on a broad pedestal. The length of the maximum compression rate is defined by the non-linearity parameter and depends on the temporal profile of the initial wave packet (Mollenauer *et al.*, 1983; Dianov *et al.*, 1986).

In the quasi-statistical approximation ($T_{\mathrm{MI}} \ll \tau_0$) the maximal growth rate of the MI and the corresponding modulation frequency are temporally dependent and the range of MI $\Delta\Omega(= 4(R)^{1/2})$ corresponds to the maximal compression rate of pulses $\tau_0/\tau_{\min}$. The intensity fluctuations increase exponentially along the fiber with a growth rate $2K_{\max}z$ equal to $2R\Psi_0^2 z$ with its maximum value $2K_{\max}z_{\mathrm{opt}} = (R)^{1/2}\Psi_0^2$ at the distance $z = z_{\mathrm{opt}} \simeq 1/(2(R)^{1/2})$. Thus if $R = T_{\mathrm{cr}} = 25^2\text{–}30^2$ then a noise intensity of about $10^{-10}\Psi_0^2$ is sufficient for the multi-soliton pulse to break up with the critical number of solitons $N = N_{\mathrm{cr}} = 25\text{–}30$. In other words multi-soliton pulses ($N = 25\text{–}30$) are stochastically unstable in optical fibers.

Note that the physical mechanism for the formation of multi-soliton pulse stochastic instability is similar to the effect of Raman self-scattering. In fact, in pulse spectral broadening, determined mainly by the duration of the generated peak $\Delta t \simeq \frac{1}{4}N$, there is a frequency component of the pulse spectrum in the region of instability $\Omega \simeq \Omega_0$. Thus, the SPM effect can result in the formation of a self-induced seed in the spectrum, which is effectively amplified by the MI. It resembles the case in RSFS when a probe-seeding wave is amplified in the field of the pump pulse (Potasek and Agrawal, 1987; Potasek, 1987; Serkin, 1987a).

We have studied this phenomenon for a broad range of parameters, taking into account not only the regular amplitude and phase perturbations but also an amplitude-phase noise.

The results indicate that MI causes a significant reduction in the fiber compression length. For example a depth of amplitude modulation $\alpha_{\mathrm{m}} = 0.005$ results in a reduction of two in the 'non-linear focus length' at $N_{\mathrm{cr}} = 25$. Figure 7.22 shows the break up of the initial pulse into separate solitons with randomly distributed amplitude and phases – 'soliton noise'.

According to whether the self-compression of the multi-soliton pulse is a single unit ($N < N_{\mathrm{cr}}$) or the expansion of the stochastical

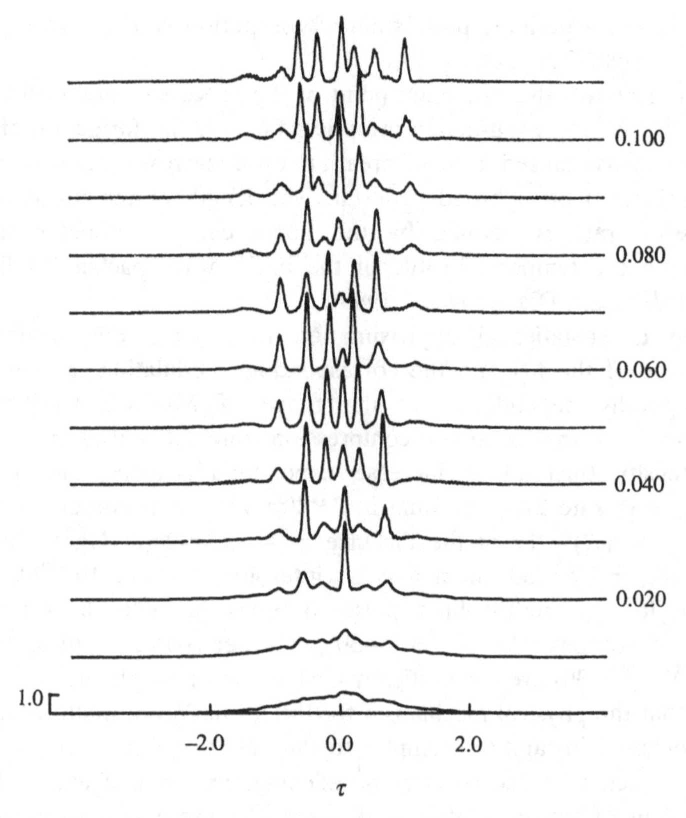

Figure 7.22. Break-up of $N = 15$ soliton pulse with randomly distributed amplitudes and phases – 'soliton noise'.

instability ($N > N_{cr}$) the pulse spectrum is significantly broadened. If the non-linear spectrum broadening during the pulse self-action is sufficient for the resonant exciting of a molecular vibrational transition then the energy redistribution process takes place and causes a Stokes wing.

Typical results from computer simulations of the RSFS dynamics of multi-soliton pulses are shown in Figures 7.23 and 7.24. Figure 7.23(*a*) presents the process of SRS transformation of the multi-soliton pulse for $N < N_{cr}$. Initially in this case the pulse acts as a single unit and the effect of pulse self-compression occurs and a field singularity arises at the point of the non-linear focus. In accordance with the theory of the RSFS effect (Gordon, 1986) the Stokes frequency shift is proportional to the peak pulse power and is inversely proportional to the fourth power of the pulse duration. Therefore the process of the N-soliton wavepacket compression is accompanied by a shift in the central peak to the tail edge of the pulse. The interaction of the shifted central peak and pedestal is non-elastic and is characterised by the energy redistribution process (see Figure 7.23(a)): the raised Stokes pulse 'absorbs' the energy from the tail edge of the pump pulse. Clearly, in this qualitative description of the pulses with different carrier frequencies the interacting pulse spectra should not overlap. Figure 7.23(b) shows the spectral dynamics for this case. One can see that, during N-soliton pulse self-compression, the Stokes wing is raised and move away from the pedestal. This fact allows us to exploit the two-pulse model for a qualitative description of the RSFS effect. The dynamics of the molecular vibration wave is interesting (see Figure 7.23(c)). The oscillating character of the Raman response results in the formation of a wave packet at the molecular-vibration carrier frequency (see Figure 7.23(c)). At the focal point, strong excitement of the optical phonons takes place and it results in the oscillating behavior of the molecular response. The non-linear discrimination of Raman solitons due to the RSFS effect might cause the appearance of additional pulses. This effect can be explained by the energy of the initial multi-soliton pulse, which is sufficient to generate sets of Raman solitons at different wavelengths (see the temporal and spectral characteristics in Figures 7.23(a) and (b)). The condition for the peak-pedestal and pedestal to move apart has been studied mathematically (Serkin, 1987c).

Another limit case occurs when $N(> N_{cr})$ is characterised by the stochastical MI at the initial stage (see Figure 7.24). As a result of the pulse break-up a random sequence of soliton-like pulses, with randomly distributed phases, amplitudes and group velocities are generated. The RSFS effect causes a non-linear discrimination of solitons not only in the time domain but also in the frequency one. In fact the process is a non-elastic interaction between the soliton with the maximum form-factor with the others. The soliton component of a pulse with the maximum form-factor $\kappa_{max} = 2N - 1$ undergoes the maximum frequency shift towards the Stokes region and consequently has the maximum group velocity. In the process of energy redistribution the Stokes wing in the initial pulse spectrum and a separate Raman soliton appear. The duration of the soliton is practically z-independent and corresponds, in order of magnitude, to the soliton with maximum form-factor κ_{max}.

So, the computer simulation results indicate that the RSFS effect

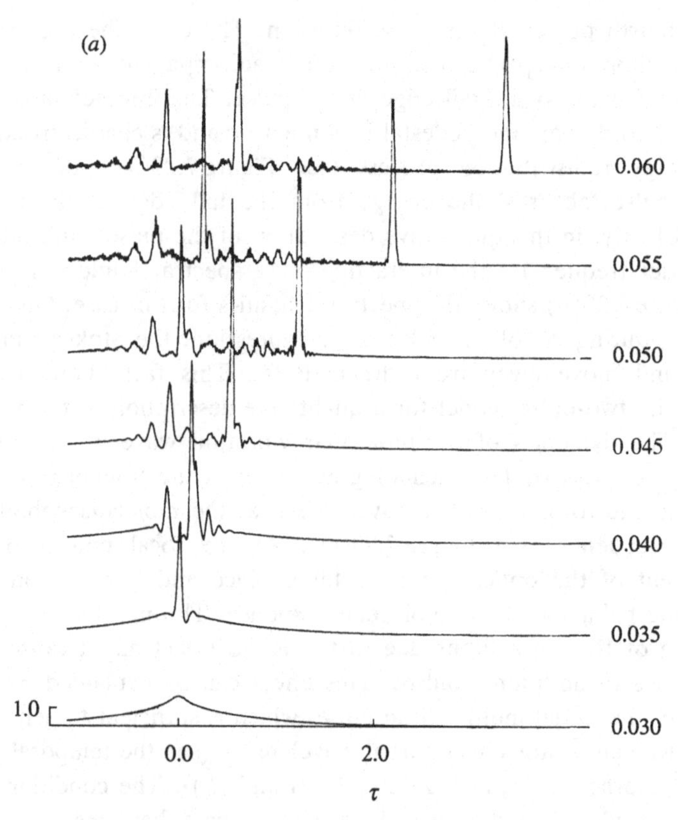

Figure 7.23. Dynamics of (*a*) the time envelope $\psi(z, \tau)$; (*b*) the spectrum and (*c*) the molecular oscillations $Q(z, \tau)$ of a multi-soliton pulse in the non-linear dispersive medium, calculated in the framework of the model (7.30–7.31) at the following parameters: $R = 225$; $T_0 = 750$ fs.

occurring in the femtosecond range results in the non-linear discrimination of solitons, that causes the appearance of soliton-like pulses with $\kappa_m = 2N - 1$ which are governed by the NLS equation and this process is accompanied by the splitting of the initial spectrum. It is very important that such non-linear transformation processes take place independently of the generation method of soliton noise in the optical fiber: through MI or SRS noise in the fiber (Beaud *et al.*, 1987; Islam *et al.*, 1989; Afanasjev and Serkin, 1989; Blow and Wood, 1989).

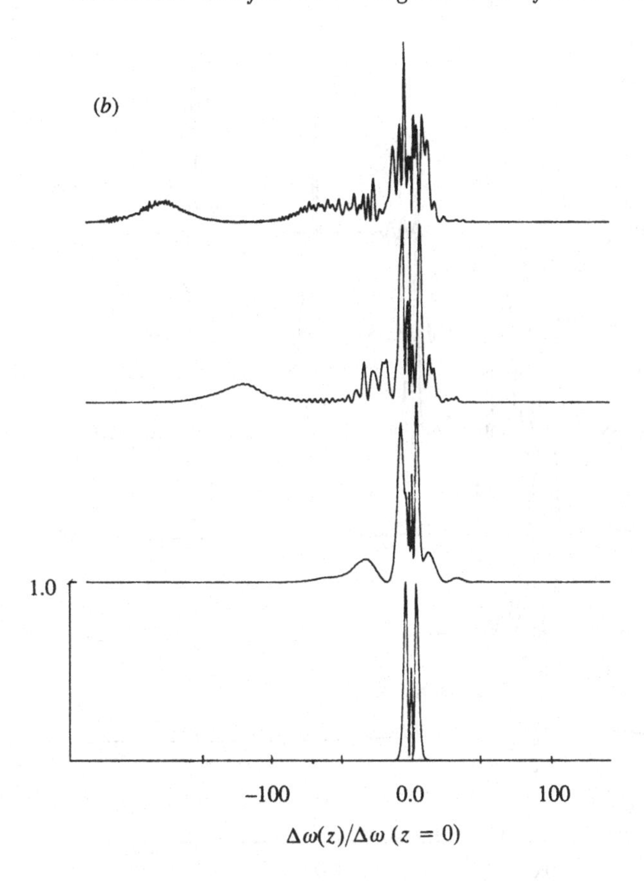

7.4.7 *Ultimate compression of optical wavepackets in non-linear dispersive media*

Generalising the method of slowly varying amplitudes for the purposes of studying the non-linear optics of femtosecond pulses is one of the most important problems in theoretical non-linear optics (Akhmanov, 1986; Kodama and Hasegawa, 1987).

Theoretical studies have also been stimulated by the fact that up until now no answer has been obtained to the key problem of the non-linear optics of femtosecond pulses: the determination of the ultimate width of a pulse that can be produced in a non-linear dispersive medium and the production of pulses with the width $\tau_{\min} \simeq T_0$, that is an optical videopulse without a high-frequency field component.

In the NLS model the propagation of an N-soliton wave packet is accompanied by the formation of a characteristic feature in the region

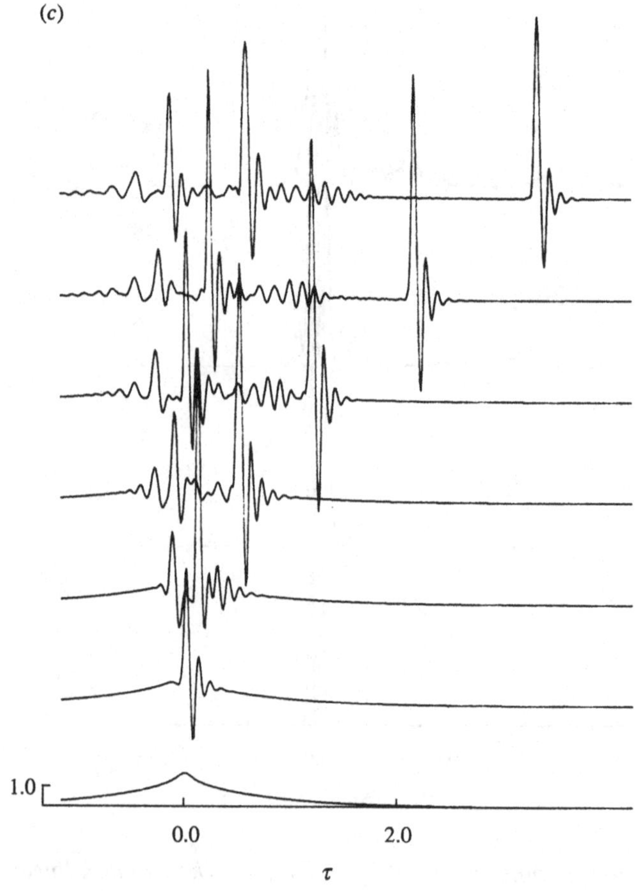

Figure 7.23. (*see page 244 for caption*)

of 'non-linear' focus, the structure of which has been investigated in detail by (Mollenauer *et al.*, 1983; Dianov *et al.*, 1986).

Note that in the framework of the NLS model there is no limitation to the pulse duration at the point of its maximum self-compression: $\tau_{\min} \simeq (4N^{-1}) \to 0$. That is why the NLS model only describes the pulse self-action correctly when the duration of the compressed pulse satisfies the condition:

$$\tau_{\min} = \frac{\tau_0}{4.1N} \gg T_0 \qquad \text{or} \qquad \frac{T_0}{\tau_0}(R)^{1/2} \ll 1 \qquad (4.37)$$

where T_0 is the period of the optical oscillations of the field.

Mathematical modeling of the powerful pulse self-action in framework of the extended NLS model allows us to predict the limit to the rate of maximum pulse self-compression and the possibility of formation and propagation in fibers of pulses with ultimate durations con-

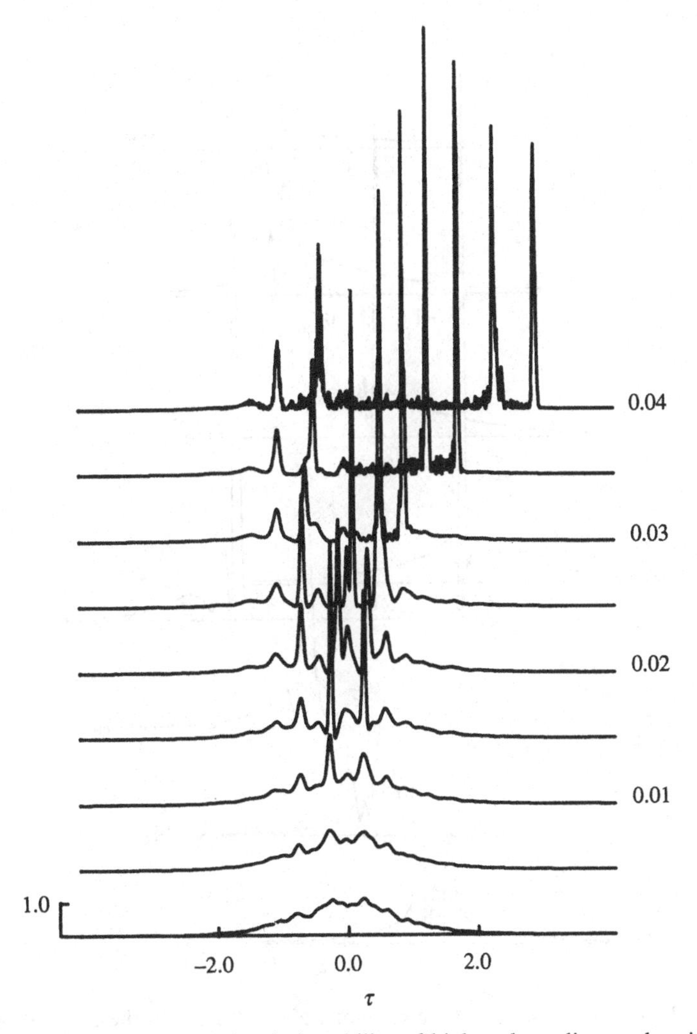

Figure 7.24. Stochastic instability of high-order soliton pulses in non-linear dispersive media.

taining only a few periods of the oscillations of the light field (Golovchenko *et al.*, 1987).

Numerical experiments enabled the determination of the critical value for the parameter $\gamma(R)^{1/2}$, associating the duration of the initial wave packet with its power, a region in which the NLS approximation is no longer valid.

Cardinal changes in the dynamics of the non-linear pulse self-compression occur in the region of the main parameters $0.1 < (T_0/\pi\tau_0)(R)^{1/2} < 0.6$ (Figure 7.25(*a*)). With an increase in the

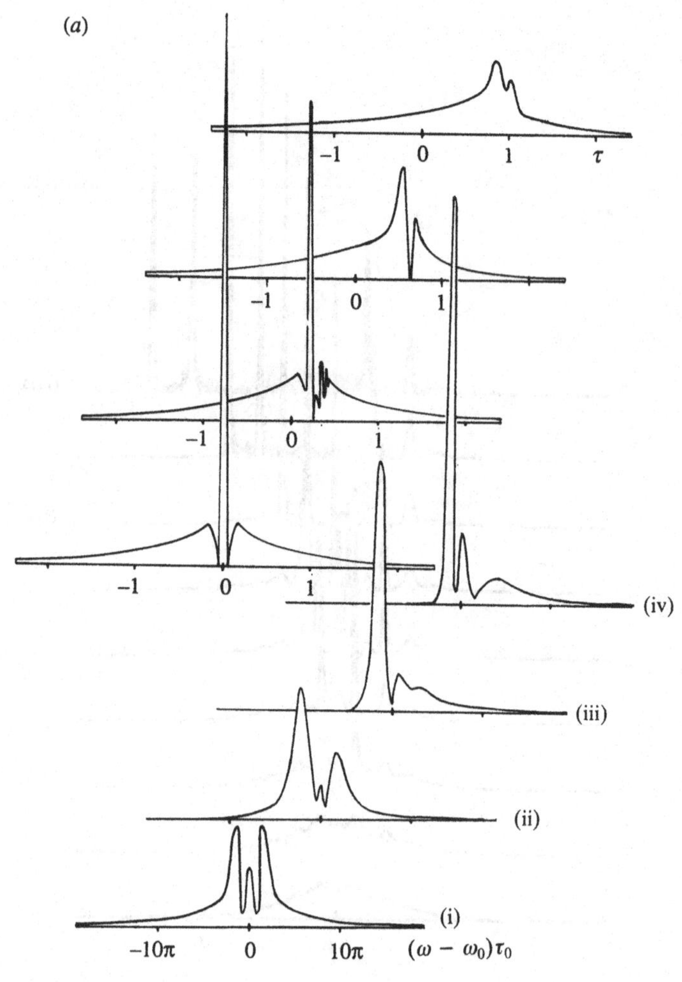

Figure 7.25. (*a*) The time envelope and spectrum of the pulse at the fiber length corresponding to the maximum self-compression for $R = 225$, $\mu_3 = 0.25$; (i) $\eta = 0$, (ii) $\eta R^{1/2} = 0.1$, (iii) $\eta R^{1/2} = 0.3$ and (iv) $\eta R^{1/2} = 0.6$.

parameter $(T_0/\pi\tau_0)(R)^{1/2}$ (which corresponds to either the shortening of the pulse duration or an increase in its power) strong asymmetry in the shape of the spectral and time envelopes of the wave packet is observed: a shift in the packet center to the envelope rear and the formation of Stokes and anti-Stokes 'wings' in the spectrum (Figure 7.25(*b*)).

The dependence of the compression at the point of maximum self-compression on the number of solitons indicates 'saturation' when the parameter $(T_0/\pi\tau_0)(R)^{1/2} \geqslant 0.3$ (Figure 7.25(*b*)).

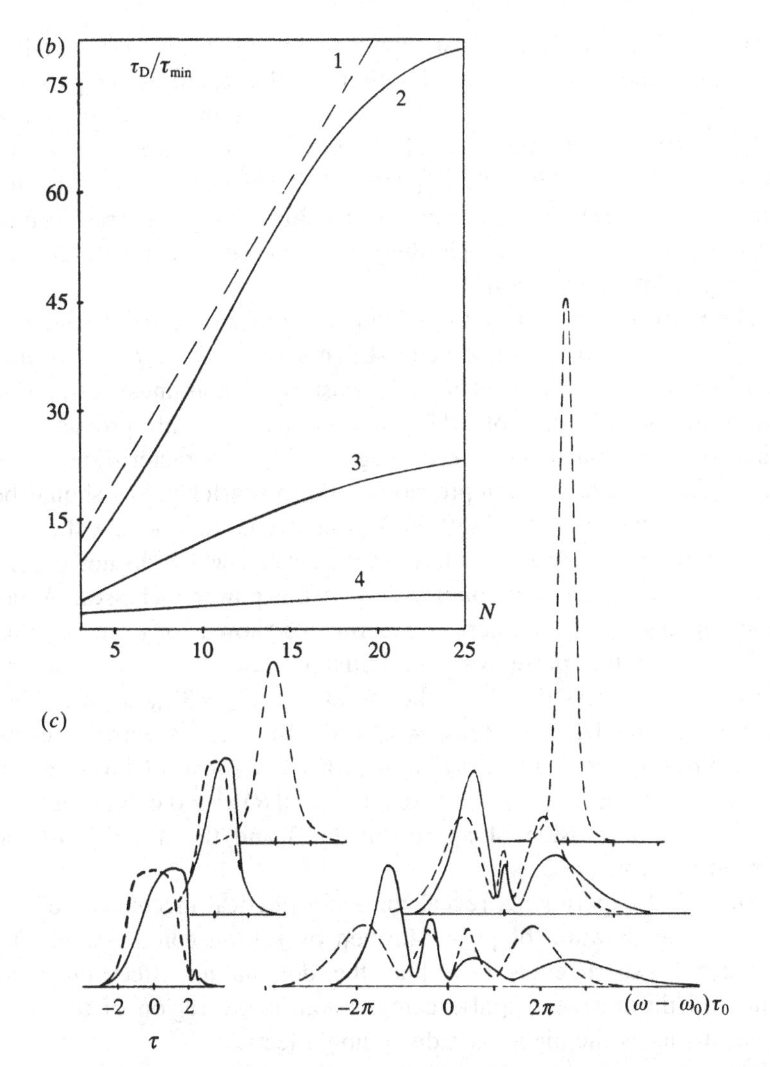

Figure 7.25 (*b*) Dependence of pulse self-compression on the number of solitons for $\eta R^{1/2} = 0$ (curve 1), $\eta R^{1/2} = 0.1$ (curve 2), $\eta R^{1/2} = 0.3$ (curve 3) and $\eta R^{1/2} = 0.6$ (curve 4). (*c*) Dynamics of the temporal envelope and of the pulse spectrum $\chi\ (z = 0, \tau) = \operatorname{sech} \tau$ in the spectral region $k_2 > 0$, calculated for $R = 225$, $\gamma R^{1/2} = 0.6$, $\gamma = 0$ (broken curves), $z = 0, 0.03, 0.06$.

It should be pointed out that the modified NSE model ((7.30)–(7.32)) allows us to understand qualitatively and quantitatively the limitation mechanisms of the pulse compression and the achievement of the ultimate duration $\tau_{min} \simeq T_0$. From a strict analysis of the results obtained one can see that this model, as well as the NLS model, at the parameters $(T_0/\pi\tau)(R)^{1/2} \ll 1$ provides results lying outside the region of applicability of the slowly-varying-amplitude method, i.e. the duration of the compressed pulse in it can turn out to be less than T_0. For example, at $N = 15$ and $(T_0/\pi\tau)(R)^{1/2} = 0.1$ the model 'predicts' the duration in the 'focus' $\tau_{min}/\tau_0 = \tau_0/60 \simeq 50 \ T_0/60 < T_0$, which is not surprising since at $(T_0/\pi\tau)(R)^{1/2} \ll 1$ model ((7.30)–(7.31)) is transformed into the NLS model, which does not impose any limitations on the degree of compression.

The most important result obtained in the numerical experiments in the framework of the modified NLS model ((7.30)–(7.32)) is that quasi-stationary wave packets can exist in a non-linear dispersive medium, the duration of which do not change with propagation. Thus, although calculations in the region of the 'characteristic feature' of the field at the self-compression of the powerful pulses should be used with care, model ((7.30)–(7.32)) allows us to evaluate the ultimate duration of the soliton-like wave packets, which do not experience any significant self-compression as the power increases. A detailed analysis of the numerical experiments shows, for example, that a $N = 25$ soliton pulse with an initial duration $\tau_0 = 14T_0$ can be 4.5-times compressed up to the duration $\tau_{min} \simeq 3T_0$ at the fiber length $z = 0.0375 \ z_{disp}(z_{disp} = \tau_0^2/k'')$. In the spectral region $\lambda = 1.5\ \mu$m and $\tau_0 = 14T_0 = 70$ fs it provides a value of the order of 1 cm. By analogy, at $N = 10$ and $(T_0/\pi\tau)(R)^{1/2} = 0.6$ we obtain a degree of compression of approximately 3 and the duration of the formed pulse $\tau_{min} \simeq 2T_0$.

Numerical experiments reveal the main physical mechanism of the limit to the duration of pulses formed by soliton compression. The conducted experiments show that the dominating effect imposing limits on the degree of pulse compression in the region of the non-linear 'focus' is the high-order dispersion effects.

Thus, the extension of the method of slowly-varying amplitudes to the problems of non-linear optics of femtosecond light pulses provides some ways for the achievement of the pulse durations $\tau_{min} \simeq$ 2–$3T_0$. The first way is utilisation of effects of the non-linear self-compression of multi-soliton pulses in the parameter region $(T_0/\pi\tau)(R)^{1/2} \ll 1$ (Figure 7.25(a), lower trace). The second way for the achievement of ultimate durations is based on the effect of decay on the soliton bound states behind the point of the non-linear 'focus'. This effect is considered in detail in the next section.

7.4.8 *Self-compression and decay of femtosecond high-order optical solitons: 'colored' solitons in fibers*

A mathematical model of the self-action process of femtosecond light pulses in non-linear dispersive media is based on the modified NLS equation:

$$\mathrm{i}\frac{\partial \Psi}{\partial z} = \frac{1}{2}\frac{\partial^2 \Psi}{\partial \tau^2} + R\delta n_{\mathrm{eff}}\Psi - \mathrm{i}\eta R\frac{\partial}{\partial \tau}(\delta n \Psi_{\mathrm{eff}})$$

$$+ \mathrm{i}\mu_3 \frac{1}{6}\frac{\partial^3 \Psi}{\partial \tau^3} - \frac{1}{2}\eta\frac{\partial^2 \Psi}{\partial z \partial \tau} \tag{7.38}$$

$$\eta = \frac{T_0}{\pi \tau_0}$$

The electronic Kerr effect as well as the mixed electron–nuclear non-linearity of Raman type contributes to the non-linearity of the effective refractive index δn_{eff}:

$$\delta n_{\mathrm{eff}} = \delta n_{\mathrm{Kerr}} + \delta n_{\mathrm{Ram}} \tag{7.39}$$

$$T^{\mathrm{rel}}_{\mathrm{Kerr}}\frac{\partial \delta n_{\mathrm{Kerr}}}{\partial \tau} + \delta n_{\mathrm{Kerr}} = |\Psi|^2 \tag{7.40}$$

$$\mu^2\frac{\partial^2 \delta n_{\mathrm{Ram}}}{\partial \tau^2} + 2\gamma\mu\frac{\partial \delta n_{\mathrm{Ram}}}{\partial \tau} + \delta n_{\mathrm{Ram}} = \beta|\Psi|^2 \tag{7.41}$$

A number of new effects occur in the framework of this extended model. The most interesting of them is the effect of the structural instability of higher-order Schrödinger solitons, causing the break up of such solitons into the 'colored' solitons – stationary (or quasi-stationary) wave packets with the average frequency shifted into the 'red' and 'blue' spectral regions (Serkin, 1987b).

The physical picture of the break up of high-order solitons into color solitons is complicated. That is why we consider first the solution of this problem in the framework of simpler models of self-action obtained by setting the parameters β, μ_3 or η equal to zero. Then moving onto the example of the break up of femtosecond soliton pulses in fibers with 'shifted' dispersion (Grudinin *et al.*, 1987b) we consider the general solution of this problem. Such a possibility in theory ignoring this or that physical mechanism means the role of each can be clarified individually as well as discovering which physical effects dominate the experiments (Mitschke and Mollenauer, 1986; Grudinin *et al.*, 1987a,b; Mitschke and Mollenauer, 1987a; Beaud *et al.*, 1987; Gouveia-Neto *et al.*, 1988a,b; da Silva *et al.*, 1988).

Let us consider in more detail the process of self-compression and break-up of femtosecond wave packets taking into account the non-linear polarisation dispersion (self-steepening) in the framework of the following model (Anderson and Lisak, 1983):

$$\mathrm{i}\frac{\partial \Psi}{\partial z} = \frac{1}{2}\frac{\partial^2 \Psi}{\partial \tau^2} + R|\Psi|^2\Psi - \mathrm{i}\eta R\frac{\partial}{\partial \tau}(|\Psi|^2\Psi) \tag{7.42}$$

Equation (7.42) is a totally integrable model and the mechanism of soliton break up can be understood by using inverse scattering methods. The analysis carried out in Okhuma *et al.*, (1987) and Golovchenko *et al.* (1985, 1986) shows that, in the modified model considered, the existence of soliton-like solutions in the region of the positive and negative GVD is possible. The characteristic feature of such soliton-like solutions is the presence of frequency modulation in the pulse, the form of which is determined directly by the functional form of the envelope:

$$\Psi(z, \tau) = \kappa \operatorname{sech}(\kappa R^{1/2}\tau)\exp\left[-\frac{\mathrm{i}R\kappa^2 z}{2} + \frac{\mathrm{i}3\eta R\kappa}{2R^{1/2}}\operatorname{th}(\kappa R^{1/2}\tau) - \frac{\mathrm{i}3(\eta R)^2 z\kappa^4}{8}\operatorname{sech}^4(\kappa R^{1/2}\tau)\right] \tag{7.43}$$

$$z \ll \frac{4}{3}(\eta^2 R^3 \kappa^7)^{-1}$$

Thus, we can say that in model (7.43) there are soliton solutions with the frequency

$$\omega = \frac{3}{2}\frac{T_0}{\pi\tau_0}R^2|\Psi(\tau = \tau_{\max})|^2$$

differing from the frequency of the initial wave packet.

A detailed analysis of the amplitude-phase structure of the pulse arising at self-action with consideration of the non-linear polarization dispersion is given in Golovchenko *et al.* (1986). Let us summarise the main results.

The presence of an additional term in equation (7.42) results in a dependence of the group velocity on the intensity so that the apparent non-linear addition to the group velocity causes an increase in the group delay of the pulse peak and an associated asymmetry of the pulse envelope followed by the steepening of the rear of the pulse and an enhancement of the anti-Stokes frequency components. So, the effect of self-steepening breaks the degeneracy of the speed of the constitutent solitons, and, as a result, they separate from each other.

In the experiment by Grudinin *et al.* (1987b) the non-linear dynamics of soliton bound states has been studied. The solitons had form-factors $\kappa_1 = 2\ R^{1/2} - 1$ and $\kappa_2 = 2\ R^{1/2} - 3$ that corresponded to the non-linear parameter $R = 5$. For the parameters of the experiment $T_p = 70$ fs ($\tau_0 = T_p/1.76$) and $T_0 = 5$ fs the calculated time interval between the two forming pulses is about $\Delta t = 3\tau_0$ at $z = 2z_0 = \pi$.

Let us estimate the influence of non-linearity delay related to the non-linearity of electronic δn_{Kerr} and Raman δn_{Ram}. The simplest model taking into consideration both effects can basically be presented in the form (7.38) by setting parameters $\eta = 0$ and $\mu_3 = 0$. If $\tau_0 \ll T^{\text{rel}}_{\text{Ram}} = 2\gamma\mu$ and $\mu^2 \ll 2\mu\gamma$ equations (7.40)–(7.41) might be reduced to a single equation with an effective response time $T^{\text{rel}}_{\text{eff}}$:

$$\delta n_{\text{Kerr}} = |\Psi|^2 - T^{\text{rel}}_{\text{Kerr}} \frac{\partial|\Psi|^2}{\partial\tau} \tag{7.44}$$

$$\delta n_{\text{Ram}} = \beta|\Psi|^2 - \beta T^{\text{rel}}_{\text{Ram}} \frac{\partial|\Psi|^2}{\partial\tau} \tag{7.45}$$

$$\delta n_{\text{eff}} = |\Psi|^2 (1 + \beta) - (T^{\text{rel}}_{\text{Kerr}} + \beta T^{\text{rel}}_{\text{Ram}}) \frac{\partial|\Psi|^2}{\partial\tau} \tag{7.46}$$

Note that $R = R/(1 + \beta)$. The effect of the delayed non-linear response on soliton break up can be studied by including a term proportional to the effective relaxation time $T^{\text{rel}}_{\text{eff}} = T^{\text{rel}}_{\text{Kerr}} + \beta T^{\text{rel}}_{\text{Ram}}$. However, for pulse width $\tau_0 \simeq 100$ fs, a more accurate approach should be used taking into account the oscillating time-dependent response of the Raman non-linearity (7.41). Numerical results show that in the framework of model (7.38)–(7.41) Raman self-scattering was one of the main effects in the experiments (Grudinin *et al.*, 1987b). Figure 7.26 shows the break up of higher-order solitons ($N = R^{1/2} = 5$) induced by Raman response non-linearity and pulse spectrum evolution. The other higher-order effects are ignored here to isolate the features associated with the Raman non-linearity. The time interval between the two forming pulses is about $\Delta t = 14\tau_0$ at $z = 2z_0 = \pi$.

Concluding the detailed discussion of the various effects arising from the self-action of powerful multi-soliton pulses we would like to pay attention to the fundamental process which limits the maximum compression rate of pulses – higher-order dispersion effects. Let us consider the self-action of wave packets using NLS equation with a third-order dispersion term.

In the region of minimum loss ($\lambda = 1.55\ \mu$m) the dispersion effects are small. Parameter μ_3 in (7.38) does not exceed 10^{-2} when $\tau_0 \simeq 10^{-12}$ s. In this pulse duration range the higher order dispersion effects are described in detail by the perturbation theory. However as the carrier frequency approaches the point of zero GVD there is a growth in the influence of the higher-order term and a sufficient

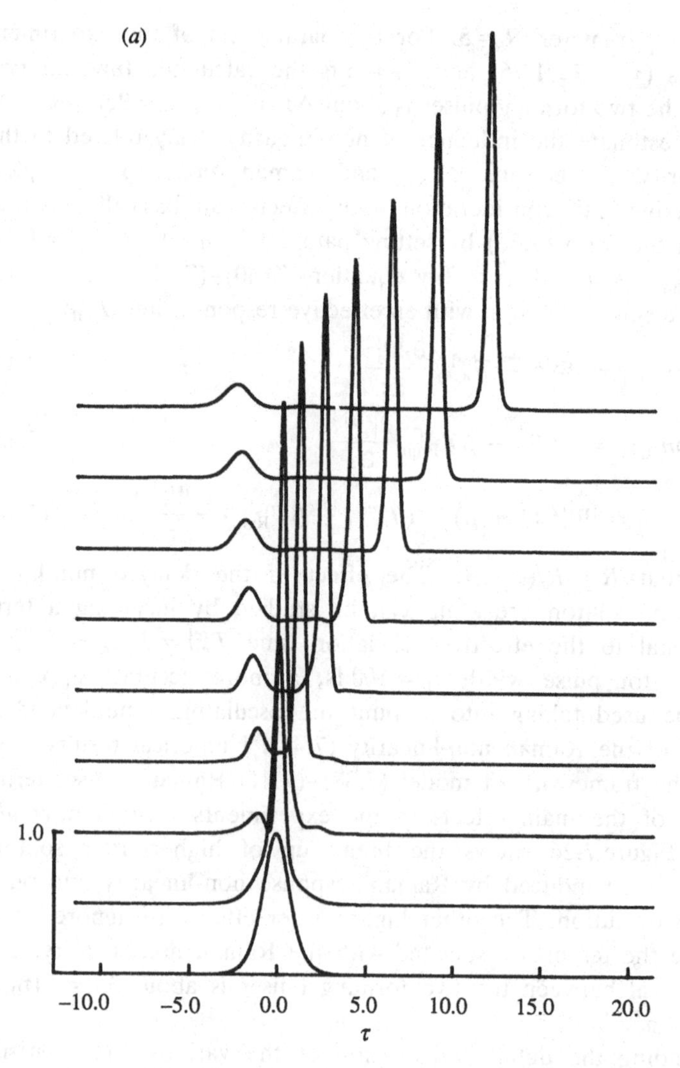

Figure 7.26. The effect of Raman self-conversion on the dynamics of the breakdown of a soliton bound state at $\sqrt{R} = 5$ and $\tau_0 = 40$ fs. (*a*) Dynamics of the time envelope; and (*b*) spectrum breakdown.

change in the pulse non-linear dynamics. It is well known that the higher-order effects are responsible for the fragmentation of the initial pulse and for the broadening of the wave-packet envelope. The higher-order dispersion terms play an especially significant role in experiments with dispersion-shifted fibers. Figure 7.27(*a*) and (*b*) show the calculated non-linear spectral and temporal dynamics of a

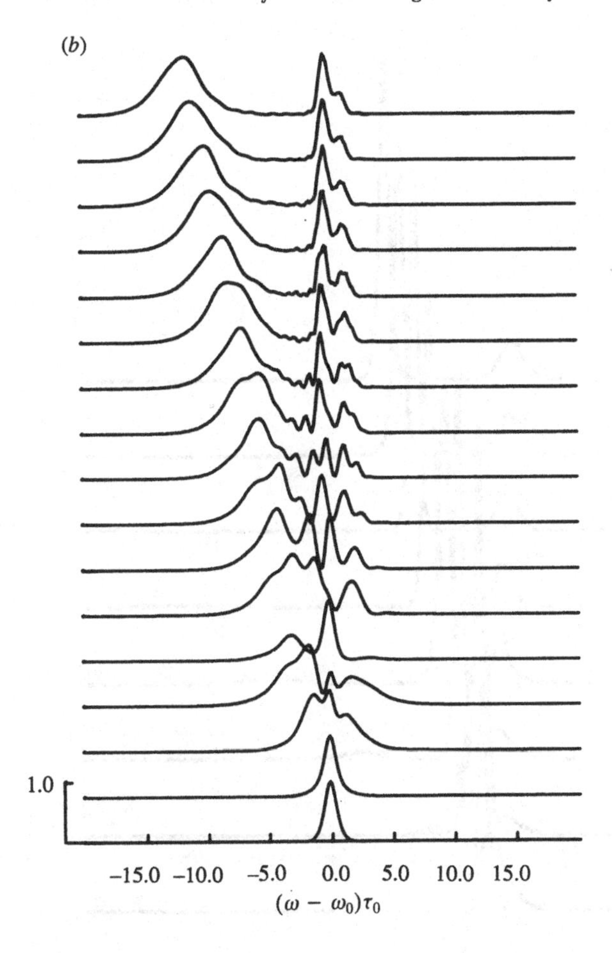

wavepacket corresponding to one of the experiments (Grudinin *et al.*, 1987b). The comparison between the theoretical and experimental results indicates that third-order linear dispersion causes the strong spectral asymmetry: the amplitudes of the Stokes components are stronger and the whole spectrum shifts towards the Stokes region. In the anti-Stokes region there is an additional wing corresponding to $\frac{1}{2}\mu_3$ (Wai *et al.*, 1986, 1987).

The result of the spectrum asymmetrisation and splitting is a decay of the soliton bound state that is illustrated in Figure 7.27(a) and (b). The time interval between the two forming pulses is about $\Delta t \simeq 7\tau_0$ at $z = 2z_0 = \pi$ for pulsewidth $\tau_0 = 70$ fs and $\mu_3 = 0.2$. So this effect is one of the dominating effects in the experimental situation discussed.

The results of the calculations taking into consideration all the higher-order linear and non-linear effects ((7.38)–(7.41)) are shown

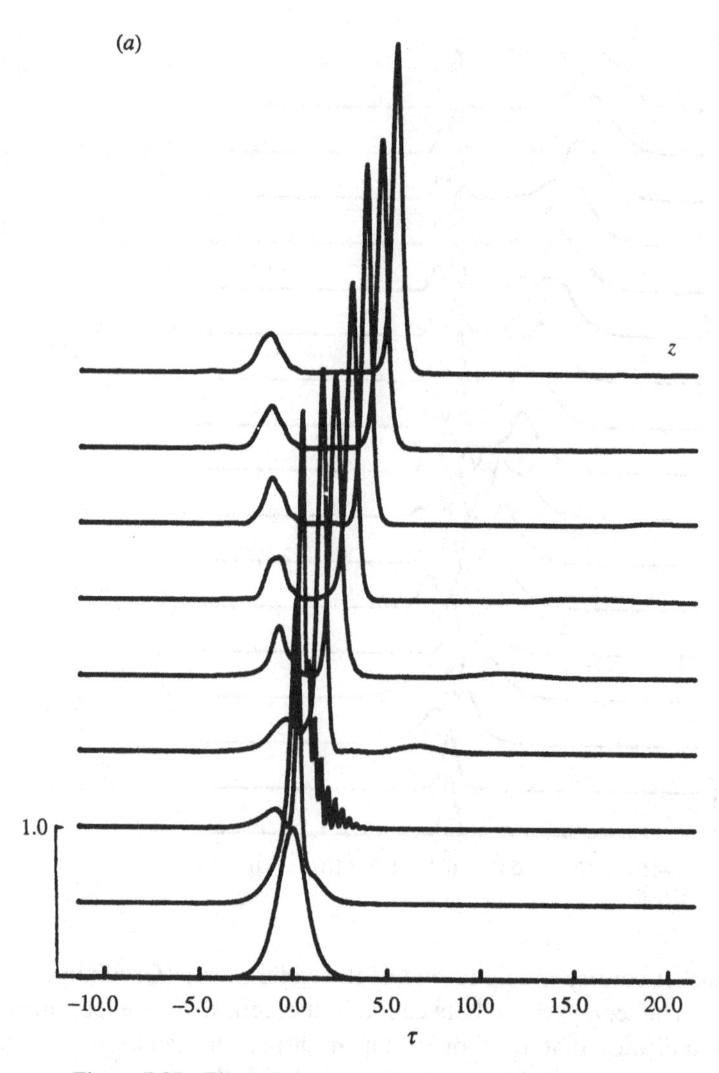

Figure 7.27. The influence of the third-order dispersion on the dynamics of a two-soliton pulse: (*a*) evolution of the time envelope in different cross-sections of the fiber ($\Delta z = \pi/8$); and (*b*) spectrum evolution ($\Delta z = \pi/16$). The calculations are made at the parameters $\sqrt{R} = 5$ and $\beta = 0.2$.

in Figures 7.28 and 7.29. The calculations shown in Figure 7.28 describe the experiments by Grudinin *et al.* (1987b) and those shown in Figure 7.29 those by Mitschke and Mollenauer (1986).

Thus we have presented a detailed analysis of the self-action processes of the femtosecond pulses using various modifications of the

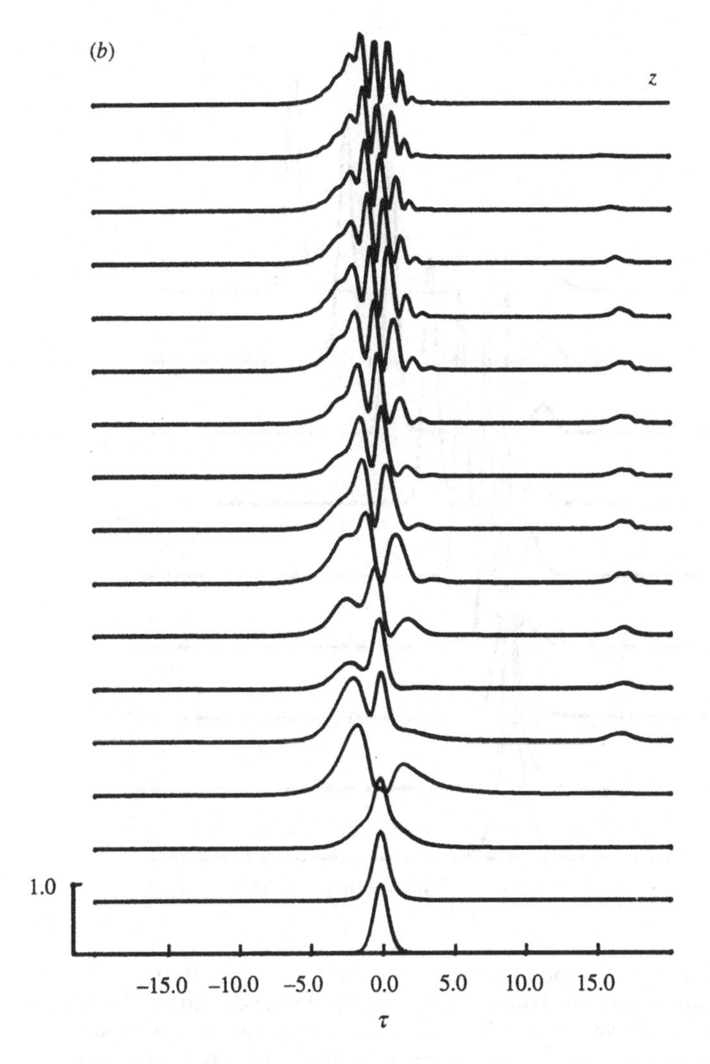

general model ((7.38)–(7.41)): a NLS model with a non-linearity dispersion, a NLS model with a non-linearity delay and a NLS model with a higher-order linear dispersion. The characteristic effect arising in all of the models is a soliton bound-state fission and generation of colored solitons – a 'long-living' (in comparison with the dispersion length) non-linear wave packets with a continuously shifting carrier frequency.

It has been shown (Mitschke and Mollenauer, 1987b; Grudinin *et al.*, 1987a) that the possibility of extreme compression and decay of high-order solitons into 'colored' ones can be realised experimentally using an auxiliary dispersion-shifted fiber. A comparison of different

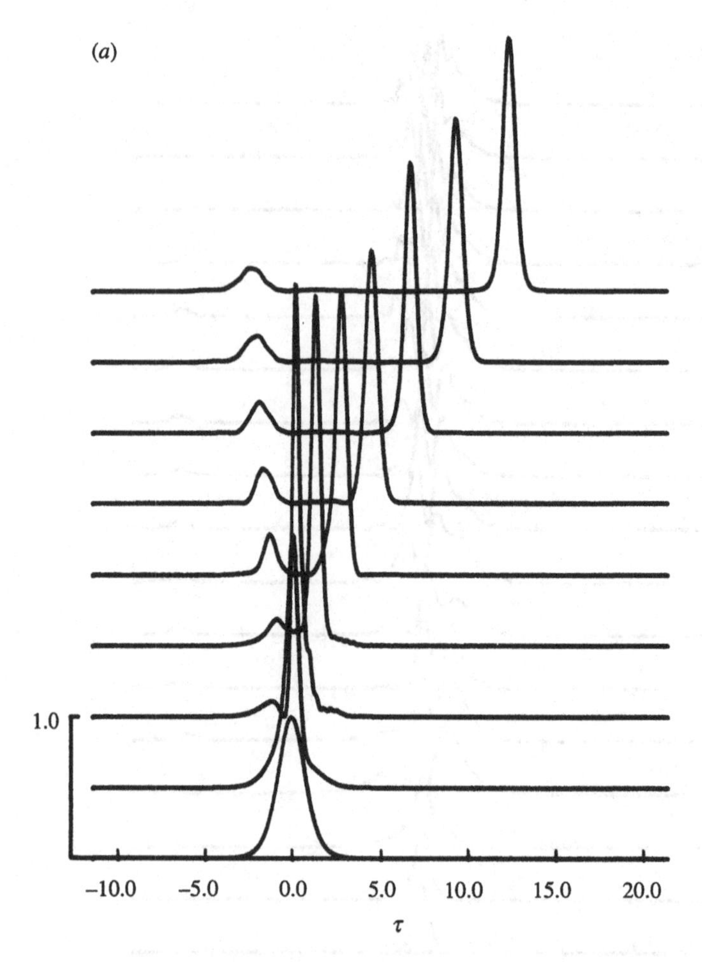

Figure 7.28. The influence of the self-steepening effect, third-order dispersion and Raman self-conversion on the dynamics of a bound state of two solitons. The calculated third-order dispersion and Raman self-conversion are made in the framework of the model (7.42) at $\sqrt{R} = 5$ and $\tau_0 = 40$ fs: (*a*) the breakdown of a bound state; and (*b*) spectrum evolution along the fiber.

models of soliton decay effect indicates that in these experiments the main effects are non-linear delay and third-order linear dispersion.

For pulsewidths $\tau_0 \sim 75$ fs, corresponding to oscillations of the molecular wave it is necessary to include all the higher-order terms in model ((7.38)–(7.41)), since all the effects of higher-order dispersion, self-steepening, Raman self-frequency shift and delayed response of electronic Kerr nonlinearity become non-negligible.

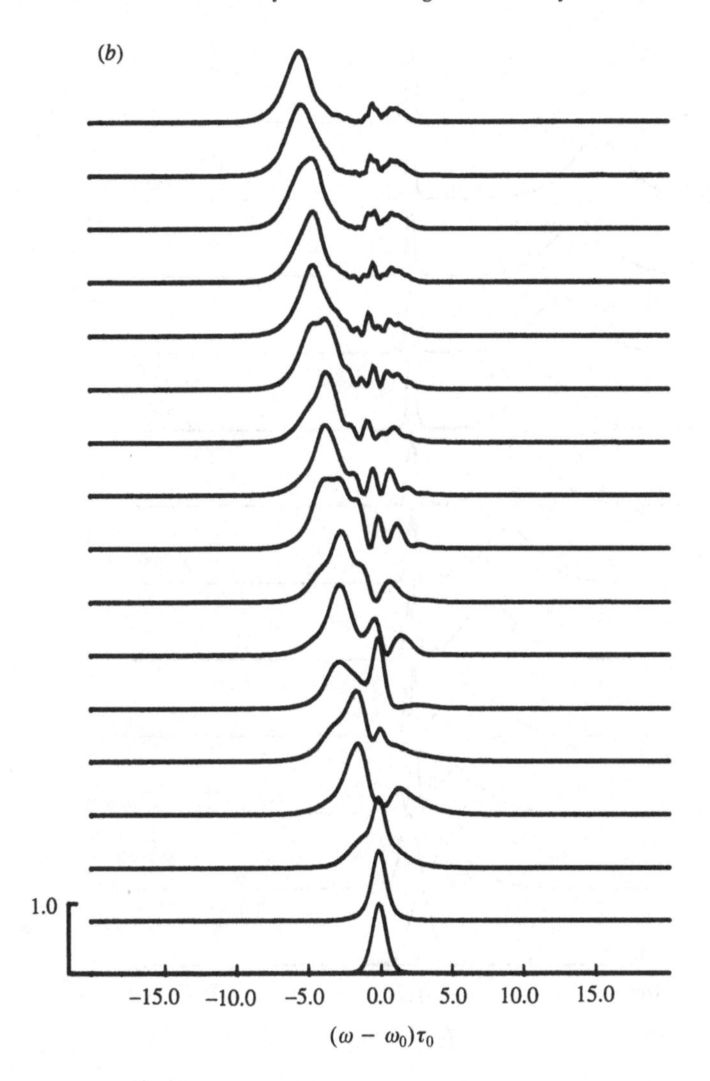

7.5 Conclusion

Nowadays the generation of optical solitons at SRS in fibers is widely used in different laboratories. The main advantage of this method is the simplicity and ingenuity of physical experiment. The possibility of obtaining Raman optical solitons in any optical laboratory equipped, for example, with a mode-locked Nd:YAG laser, makes this method very promising. Since experimental realisation of this method is accessible even for students, it is possible to discuss the

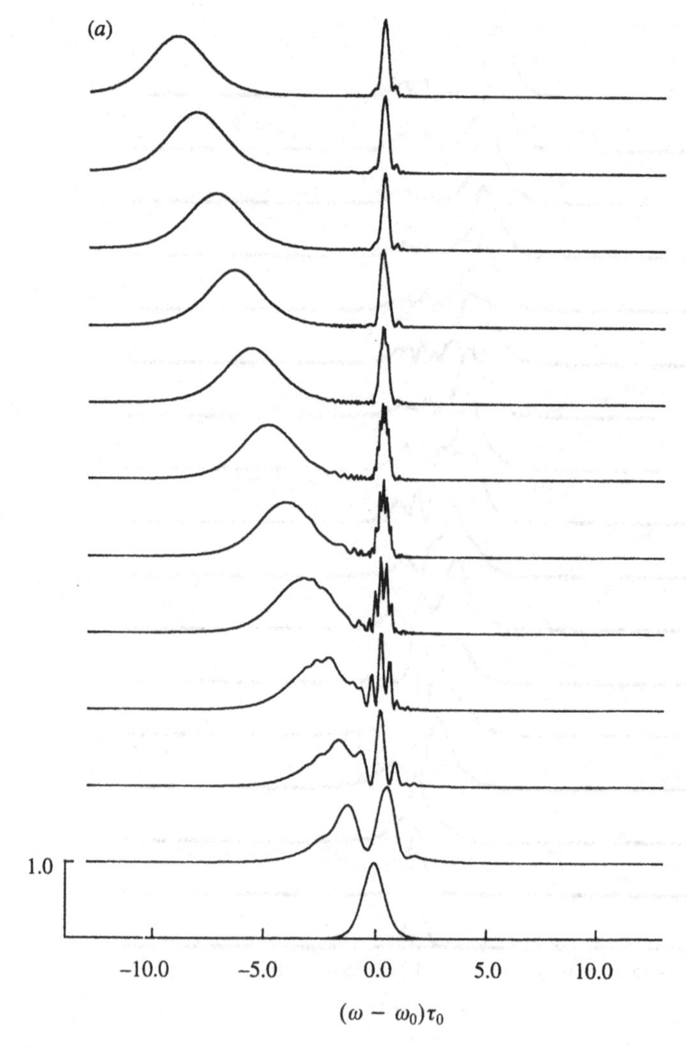

Figure 7.29. Numerical modelling of Mitschke and Mollenauer (1986) experiment on discovery of self-frequency shift in the frame of model (7.38–7.41).

corresponding problems during a practical course of optics in universities. Although the idea of the generation of Raman solitons in fibers is evident, as yet the theoretical consideration of this problem is far from being complete.

In so short a chapter we have not been able to cite a detailed bibliography. Those interested can easily find this informaion in recently published books including Hasegawa (1989), Agrawal (1989),

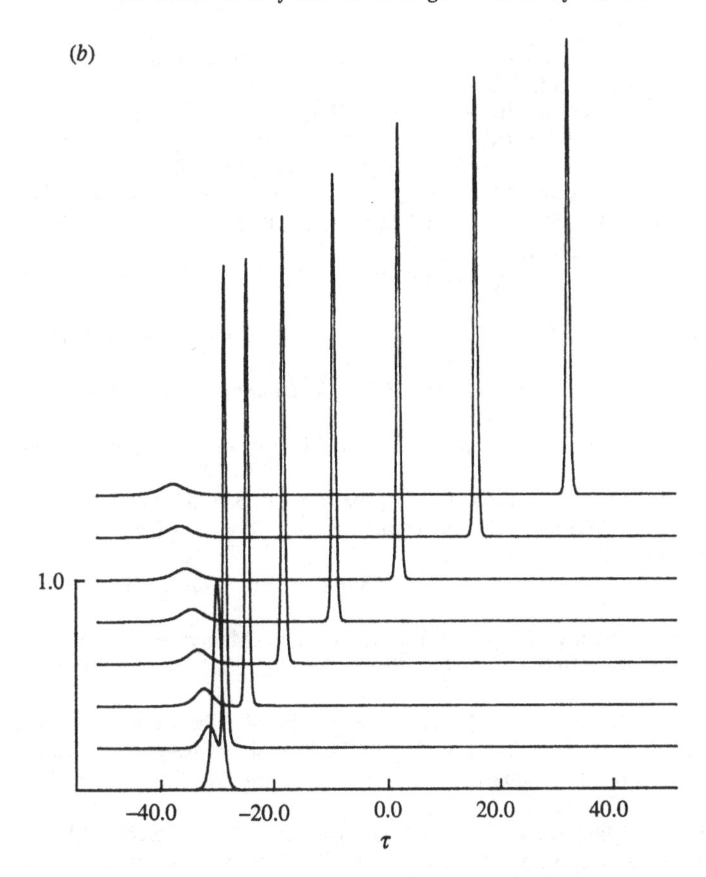

Akhmanov *et al*. (1988), Suchorukov (1988), Abdullaev *et al*. 1987, and *Ultrafast Phenomena V* edited by Fleming and Siegman (1986).

References

Abdullaev, F. Kh., Damanian, S. A. and Khabibullaev, P. K. (1987) in Baryachtar, Fan, Tashkent (eds.) *Optical Solitons* (in Russian).

Afanasjev, V. V., Dianov, E. M. and Serkin, V. N. (1989) Raman self-action of femtosecond solitons in fibers. In press.

Afanasjev, V. V. and Serkin, V. N. (1989) *Izv. AN SSSR, ser. fiz.*, **53**, 1590–8. (*Bull. Acad. Sci. USSR, Phys. Ser.*)

Afanasjev, V. V, Vysloukh, V. A. and Serkin, V. N. (1990) Opt. Lett., in press.

Agrawal, G. P. (1989) in P. L. Liao and P. L. Kelley *Nonlinear fiber optics*. (Academic Press: New York).

Akhmanov, S. A. (1986). *Usp. Fiz. Nauk*, **149**, 361–90. (*Sov. Phys. USP.*)

Akhmanov, S. A., Drabovich, K. N., Suchorukov, A. P. and Chirkin, A. S. (1971) *Zh. Eksp. Teor. Fiz.*, **58**, 485–93. (*Sov. Phys. – JETP*, **32** (1971) 266–73).

Akhmanov, S. A., Khohlov, R. V. Problems of nonlinear optics. (1964), Moscow, AN USSR, VINITI.

Akhmanov, S. A., Vysloukh, V. A. and Chirkin, A. S. (1988) *Optics of Femtosecond Laser Pulses* (Moscow: Nauka) (in Russian).

Anderson, D. and Lisak, M. (1983) *Phys. Rev. A*, **27**, 1393–8.

Auyeung, A. and Yariv A. (1978) *IEEE J. Quantum Electron.*, **14**, 347–352.

Beaud, P., Hodel, W., Zysset, B. and Weber, H. P. (1987) *IEEE J. Quantum Electron.*, **23** 1938–46.

Blow, K. J., Doran, N. J. and Wood, D. (1988a) *J. Opt. Soc. Am. B*, **5**, 381–90.

Blow, K. J., Doran, N. J. and Wood, D. (1988b) *J. Opt. Soc. Am. B*, **5**, 1301–04.

Blow, K. J. and Wood, D. (1989) *IEEE J. Quantum Electron.*, **QE25**, 2665–73.

Carman, R. L., Shimizu, F., Wang, C. S. and Bloembergen, N. (1970) *Phys. Rev. A*, **2**, 60–72.

Carpman, V. I. (1967) *Pisma Zh. Eksp. Teor. Fiz.*, **6**, 829–32. (*JETP Lett.*)

Chraplyvy, A. R., Marcuse, D. and Henry, P. S. (1984) *J. Lightwave Technol.*, **2**, 6–10.

Dianov, E. M. and Prokhorov, A. M. (1986) *Usp. Fiz. Nauk*, **148**, 289–321. (Sov. Phys. - USP.)

Dianov, E. M., Ivanov, L. M., Karasik, A. Ya., Mamyshev, P. V. and Prokhorov, A. M. (1986) *Zh. Eksp. Teor. Fiz.*, **91**, 2031–38 (*Sov. Phys. - JETP.*)

Dianov, E. M., Karasik, A. Ya., Mamyshev, P. V., Prokhorov, A. M., Stel'makh, M. F. & Fomichev, A. A. (1985). Stimulated Raman conversion of multisoliton pulses in quartz optical fibers. *Pisma Zh. Eksp. Teor. Fiz.*, **41**, 294–7. (*JETP Lett.*)

Dianov, E. M., Mamyshev, P. V., Prokhorov, A. M. and Serkin, V. N. (1989) in V. S. Letokhov, V. Shank, Y. R. Shen and H. Walter (eds) *Lasers Science and Technology* (Harwood: New York).

Dianov, E. M., Nikonova, Z. S., Prokhorov, A. M. and Serkin, V. N. (1986) *Pisma Zh. Tech. Fiz*, **12**, 756–60. (*Sov. Tech. Phys. Lett.*)

Dianov, E. M., Prokhorov, A. M. and Serkin, V. N. (1986) *Opt. Lett.*, **11**, 168–70.

Fleming, G. R. and Siegman, A. E. (eds.) (1986) *Ultrafast Phenomena*, volume 5 (Berlin: Springer).

Golovchenko, E. A., Dianov, E. M., Prokhorov, A. M. and Serkin, V. N. (1985) *Pisma Zh. Eksp. Teor. Fiz.*, **42**, 74–7. (*JETP Lett.*)

Golovchenko, E. A., Dianov, E. M., Prokhorov, A. M. and Serkin, V. N. (1986) *Dokl. AN SSSR*, 288, 851–6 (*Sov. Phys. Dokl.*, **31**, 494–7.)

Golovchenko, E. A., Dianov, E. M., Pilipetskii, A. N., Prokhorov, A. M. and Serkin, V. N. (1987) *Pisma Zh. Eksp. Teor. Fiz.*, **45**, 73–6. (*JETP Lett*, 45 (1987), 91.)

Golovchenko, E. A., Dianov, E. M., Karasik, A. Ya., Mamyshev, P. V., Pilipetskii, A. N. and Prokhorov, A. M. (1989) *Kvantovaya Electron.*, **16**, 592–4. (*Sov. J. Quantum Electron.*)

Gordon, J. P. (1986) *Opt. Lett.*, **11**, 662–4.

Gouveia-Neto, A. S., Gomes, A. S. L. and Taylor, J. R. (1987a) *Electron. Lett.*, **23**, 537–8.

Gouveia-Neto, A. S., Gomes, A. S. L. and Taylor, J. R. (1987b) *Opt. Lett.*, **12**, 1035–7.
Gouveia-Neto, A. S., Gomes, A. S. L., Taylor, J. R., Ainslie, B. J. and Cgaig, S. P. (1987c) *Electron. Lett.*, **23**, 1034–5.
Gouveia-Neto, A. S., Gomes, A. S. L. and Taylor, J. R. (1987d) *Opt. Lett.*, **12**, 927–9.
Gouveia-Neto, A. S., Gomes, A. S. L. and Taylor, J. R. (1987e) *IEEE J. Quantum Electron.*, **23**, 1183–98.
Gouveia-Neto, A. S., Gomes, A. S. L. and Taylor, J. R. (1988a) *IEEE J. Quantum Electron.*, **24**, 332–40.
Gouveia-Neto, A. S., Faldon, M. E. and Taylor, J. R. (1988b) *Opt. Lett.*, **13**, 1029–31.
Grudinin, A. B., Dianov, E. M., Korobkin, D. V., Prokhorov, A. M., Serkin, V. N. and Khaidarov, D. V. (1987a) *Pisma Zh. Eksp. Teor. Fiz.*, **45**, 211–13 (*JETP Lett.*)
Grudinin, A. B., Dianov, E. M., Korobkin, D. V., Prokhorov, A. M., Serkin, V. N. and Khaidarov, D. V. (1987b) *Pisma Zh. Eksp. Teor. Fiz.*, **46**, 175–7. (*JETP Lett.*, **46**, 221–5.)
Hasegawa, A. (1983) *Opt. Lett.*, **8**, 650–2.
Hasegawa, A. (1984) *Appl. Opt.*, **23**, 3302–09.
Hasegawa, A. (1989) *Optical solitons in fibers. Springer Tracts in Modern Physics*, volume 116 (Springer: Berlin).
Hasegawa, A. and Tappert, F. (1973) *Appl. Phys. Lett.*, **23**, 142–4.
Hellwarth, R. W., Cherlow, J. and Yang T.-T. (1975) *Phys. Rev. B*, **11**, 964–7.
Herrmann, J. and Mondry, J. (1988) *J. Mod. Opt.*, **35**, 1919–32.
Hodel, W. and Weber, H. P. (1987) *Opt. Lett.*, **12**, 924–6.
Hook, A., Anderson, D. and Lisak, M. (1988) *Opt. Lett.*, **13**, 1114–16.
Isaev, S. K., Kornienko, L. S., Kravtsov, V. N. and Serkin, V. N. (1980) *Zh. Eksp. Teor. Fiz.*, **79**, 1239–56. (*Sov. Phys. JETP.*)
Islam, M. N., Sucha, G., Bar-Joseph, I., Wegener, M., Gordon, G. P. and Chemla, D. S. (1989) *Opt. Lett.*, **14**, 370–2.
Kafka, J. D. and Baer, T. (1987) *Opt. Lett.*, **12**, 181–3.
Kodama, Y. and Hasegawa, A. (1987) *IEEE J. Quantum Electron.*, **23**, 510–24.
Kodama, Y. and Nozaki, K. (1987) *Opt. Lett.*, **12**, 1038–40.
Lin, C. (1986) *J. Lightwave Technol.*, **4**, 1103–15.
Litvak, A. G. and Talanov, V. I. (1967) *Izv. vuz Radiophys.*, **10**, 539–51.
Lugovoi, V. N. (1976) *Zh. Eksp. Teor. Fiz.*, **71**, 1307–19. (*Sov. Phys. JETP*, **44**, 638–89.)
Maker, P. D. and Terhune, R. W. (1965) *Phys. Rev. A*, **137**, 801–18.
Mitschke, F. M. and Mollenauer, L. F. (1986) *Opt. Lett.*, **11**, 659–61.
Mitschke, F. M. and Mollenauer, L. F. (1987a) *Opt. Lett.*, **12**, 355–7.
Mitschke, F. M. and Mollenauer, L. F. (1987b) *Opt. Lett.*, **12**, 407–09.
Mollenauer, L. F. (1985) *Phil. Trans. Roy. Soc. London A*, **315**, 437–50.
Mollenauer, L. F. and Stolen, R. H. (1984) *Opt. Lett.*, **9**, 13–15.
Mollenauer, L. F., Stolen, R. H. and Gordon, J. P. (1980) *Phys. Rev. Lett.*, **45**, 1095–8.
Mollenauer, L. F., Stolen, R. H., Gordon, J. P. and Tomlinson, W. J. (1983) *Opt. Lett.*, **8**, 289–91.

Okhuma K., Ichikawa Y. H. and Abe Y. (1987) *Opt. Lett.*, **12**, 516–18.
Potasek, M. J. (1987) *Opt. Lett.*, **12**, 717–19.
Potasek, M. J. and Agrawal, G. P. (1987) *Phys. Rev. A*, **36**, 3862–7.
Schadt, D. and Jaskorzynska, B. (1987) *J. Opt. Soc. Am. B*, **4**, 856–62.
Schadt, D. and Jaskorzynska, B. (1988) *J. Opt. Soc. Am. B*, **5**, 2374–8.
Serkin, V. N. (1987a) *Krakie Soobsh. Fiz. FIAN*, **6**, 30–2. (*FIAN Short Commun.*)
Serkin, V. N. (1987b) *Pisma Zh. Tekh. Fiz.*, **13**, 772–5. (*Sov. Tech. Phys. Lett.*, **13**, 320–1.)
Serkin, V. N. (1987c) *Pisma Zh. Tekh. Fiz.*, **13**, 878–82. (*Sov. Tech. Phys. Lett.*, **13**, 366–7.)
Serkin, V. N. (1987d) *Kratie Soobsh. Fiz. FIAN*, **6**, 33–35. (*FIAN Short Commun.*)
Shen, Y. and Bloembergen, N. (1965) *Phys. Rev. A*, **137**, 1787–1805.
da Silva, V. L., Gomes, A. S. L. and Taylor, J. R. (1988) *Opt. Commun.*, **66**, 231–4.
Stolen, R. H. (1980a) *Fibre and Integrated Optics*, **3**, 21–51.
Stolen, R. H. (1980b) *IEEE Proc.*, **68**, 1232–6.
Stolen, R. H., Ippen, E. P. and Tynes A. R. (1972) *Appl. Phys. Lett.*, **20**, 62–4.
Stolen, R. H. and Ippen, E. P. (1973) *Appl. Phys. Lett.*, **22**, 276–8.
Stolen, R. H. and Johnson, A. M. (1986) *IEEE J. Quantum Electronics*, **22**, 2154–60.
Stolen, R. H., Gordon J. P., Tomlinson W. J. and Haus H. A. (1989) *JOSA B*, **6**, 1159–66.
Suchorukov, A. P. (1988) *Nonlinear Wave Interaction in Optics and Radiophysics* (Nauka Moscow) (in Russian).
Sisakyan, I. N. and Shvartsburg, A. B. (1984) *Kvantovaya Electron.*, **11**, 1703–21. (*Sov. J. Quantum Electron.*, **14**, 1146–57.)
Tai, K., Hasegawa, A. and Bekki, N. (1988) *Opt. Lett.*, **13**, 392–4. and also errata *Opt. Lett.*, **13**, 937.
Tai, K., Hasegawa, A. and Tomita, A. (1986) *Phys. Rev. Lett.*, **56**, 135–8.
Tai, K., Tomita, A., Jewell, J. L. and Hasegawa A. (1986) *Appl. Phys. Lett.*, **49**, 236–8.
Tomlinson, W. J. and Stolen, R. H. (1988) *IEEE Commun. Mag.*, **26**, 36–44.
Vodop'yanov, K. L., Grudinin, A. B., Dianov, E. M., Kulevskii, L. A., Prokhorov, A. M. and Khaidarov, D. V. (1987) *Kvantovaya Electron.*, **14**, 2053–5. (*Sov. J. Quantum Electron.*)
Vorob'ev, N. S., Grudinin, A. B., Dianov, E. M., Kozolkin, D. V. and Khaidarov, D. V. (1987) *Pisma Zh. Tech. Fiz.*, **13**, 365–8.
Vysloukh, V. A. and Serkin, V. N. (1983) *Pisma Zh. Eksp. Teor. Fiz.*, **38**, 170–2.
Vysloukh, V. A. and Serkin, V. N. (1984) *Izv. Akad. Nauk SSSR*, **48**, 1777–81.
Vysloukh, V. A., Matveev, A. N. and Petrova, I. Yu. (1989) Decay and interaction of femtosecond multisoliton pulses in Raman-active fibers. *Preprint*, Moscow State University, No. 14.
Wai, P. K. A., Menyuk, C. R., Lee, C. and Chen, H. H. (1986) *Opt. Lett.*, **11**, 464–6.

Wai, P. K. A., Menyuk, C. R., Lee, Y. C. and Chen, H. H. (1987) *Opt. Lett.*, **12**, 628–30.
Weiner, A. M., Heritage, J. P. and Stolen, R. H. (1988) *JOSA B*, **5**, 364–72.
Zakharov, V. E. and Shabat, A. B. (1971) *Zh. Eksp. Teor. Fiz.*, **61**, 118–34. (*Sov. Phys. JETP*, 34 (1972), 62–9.)
Zysset, B., Beaud, P. and Hodel, W. (1987) *Appl. Phys. Lett.*, **50**, 1027–9.

8

Generation and compression of femtosecond solitons in optical fibers

P. V. MAMYSHEV

8.1 Introduction

The traditional methods for soliton generation in optical fibers use laser sources which generate stable transform-limited ultrashort light pulses, the pulse shape and spectrum of which coincide with those of soliton pulses in fibers. For a long time only color-center lasers satisfied these conditions, and these lasers were used in the majority of soliton experiments (see, for example, Mollenauer *et al.*, 1980; Mollenauer and Smith, 1988). The soliton laser (Mollenauer and Stolen, 1984) is also based on a color-center laser. The successes in semiconductor laser technology has made it possible to use laser diodes in recent soliton experiments (see, for example, Iwatsuki *et al.*, 1988).

In this chapter we shall discuss alternative methods for soliton generation in fibers. In these methods the laser radiation coupled into the fiber is not a fundamental soliton at the fiber input, but the fundamental solitons are formed from the radiation due to the nonlinear and dispersive effects in the fiber. Methods for the generation of a single fundamental soliton as well as high-repetition rate (up to THz range) trains of fundamental solitons, which are practically noninteracting with each other, will be described. High-quality adiabatic fundamental soliton compression and the effect of stabilisation of the femtosecond soliton pulse width in fibers with a slowly decreasing second-order dispersion will also be discussed. We shall discuss the problem of adiabatic soliton compression up to a duration of less than 20 fs, so we shall also consider a theoretical approach for the description of ultrashort pulse (USP) propagation through the fiber. Special attention will be paid to the Raman self-scattering effect, because it plays a very important role in femtosecond soliton generation and propagation in optical fibers.

8.2 Raman self-scattering effect: generation of fundamental solitons using the Raman self-scattering of multi-soliton pulses

Effective methods for the soliton generation are methods based on the Raman self-scattering effect (Dianov *et al.*, 1985).

Mollenauer *et al.* (1983) suggested and experimentally realised the 'multi-soliton' pulse compression technique. Theoretical analysis was conducted in the framework of the non-linear Schrödinger (NLS) equation for a complex electric field envelope:

$$\mathrm{i}\,\frac{\partial E}{\partial \xi} - (\tfrac{1}{2})[\operatorname{sign}(k'')_{\omega_0}]\,\frac{\partial^2 E}{\partial \tau^2} + |E|^2 E = 0 \qquad (8.1)$$

where the electric field

$$\Phi(z, t) = \hat{E}(z, t)\exp(-\mathrm{i}\omega_0 t + \mathrm{i}kz)$$

$$E(z, t) = \hat{E}(z, t)/E_0$$

$$\xi = z/z_\mathrm{d}$$

z is the coordinate along the fiber axis

dispersion length $z_\mathrm{d} = t_0^2/|(k'')_{\omega_0}|$

$$k'' = \frac{\partial^2 k}{\partial \omega^2}$$

t = time $\quad \tau = (t - z/v)/t_0 \quad v = (\partial\omega/\partial k)_{\omega_0}$

k is the propagation constant

ω_0 is the mean radiation frequency

E_0^2 is intensity of the fundamental soliton with duration $\tau_0(\mathrm{FWHM}) = 1.763\ t_0$ and equals $2c|(k'')_{\omega_0}|/(\omega_0 N_2 t_0^2)$

N_2 is the non-linear refractive index.

The NLS equation is well known in mathematical physics (see, for example, Zakharov and Shabat, 1971). This equation was first considered for non-linear pulse propagation in fibers by Hasegawa and Tappert in 1973.

For the compression of pulses by the multi-soliton pulse compression technique the length of a fiber is chosen to be equal to the length of the self-compression maximum of these pulses in the fiber. The compressed multi-soliton pulse at the fiber output is a compressed spike against the background of a broad pedestal.

As the NLS equation predicts, the compressed spike is unstable in the fiber due to a non-linear interaction with the pedestal. The spike exists only in the vicinity of the self-compression point of the fiber (the so-called 'focus' point), after which the spike decays into fragments. The evolution of multi-soliton pulses in fibers is somewhat complicated, and the NLS equation predicts restoration of the initial temporal shape and spectrum of pulses in the soliton period fiber length. Note, that according to the NLS equation, the spectra of multi-soliton pulses are always symmetrical in the fibers. Experiments by Stolen *et al.* (1983) on the restoration of 7 ps multi-soliton pulses (number of solitons $N = 2, 3$) in one soliton period showed rather good agreement with the NLS equation-based theory. In these experiments the minimum width of the pulses at the points of maximum self-compression was about 1 ps.

Experiments of Dianov *et al.* (1984, 1985) showed that, in the case of self-compression of multi-soliton pulses with durations of up to approximately 300 fs, pulse evolution after the self-compression point of the fiber differs considerably from the NLS equation-based theory. The parameters of the experiments were the following: a high-repetition rate (up to 1 kHz) parametric light source was used as a pump, which delivered tunable light in the spectral region $\lambda = 1.4$–$1.65\ \mu$m pulses with duration τ_0(FWHM) = 30 ps; fiber length $L = 250$ m, dispersion $D = 15$ ps/nm · km at $\lambda = 1.55\ \mu$m. Before the 'focus' point the spectrum of the multi-soliton pulses was symmetrical, in agreement with the NLS-equation-based theory (see Figures 8.1(*a*) and (*b*)). However, after the 'focus' point of the fiber the spectrum of the multi-soliton pulses becomes asymmetrical and a long-wave Stokes wing appeared in the previously symmetric spectrum (Figure 8.1(*c*) and (*d*)). Measurements of autocorrelation functions in the Stokes wing frequency range showed that stable femtosecond pulses were formed there (Figure 8.2). Further investigations showed that these pulses were fundamental solitons. These experiments contradicted NLS-equation-based theory, which was conventional at that time.

Dianov *et al.* (1985) suggested the following model for the process, which was later confirmed by theoretical calculations (Beaud *et al.*, 1987; Hodel and Weber, 1987; Golovchenko *et al.*, 1989). As the multi-soliton pulse propagates through the fiber, its spectrum begins to broaden as a result of the self-phase modulation effect. Due to the joint action of self-phase modulation and negative group velocity dispersion (GVD) effects the travelling pulse begins to be compressed, forming, in the vicinity of the focus point, an intense narrow (about 100–300 fs) spike against the background of a broad pedestal,

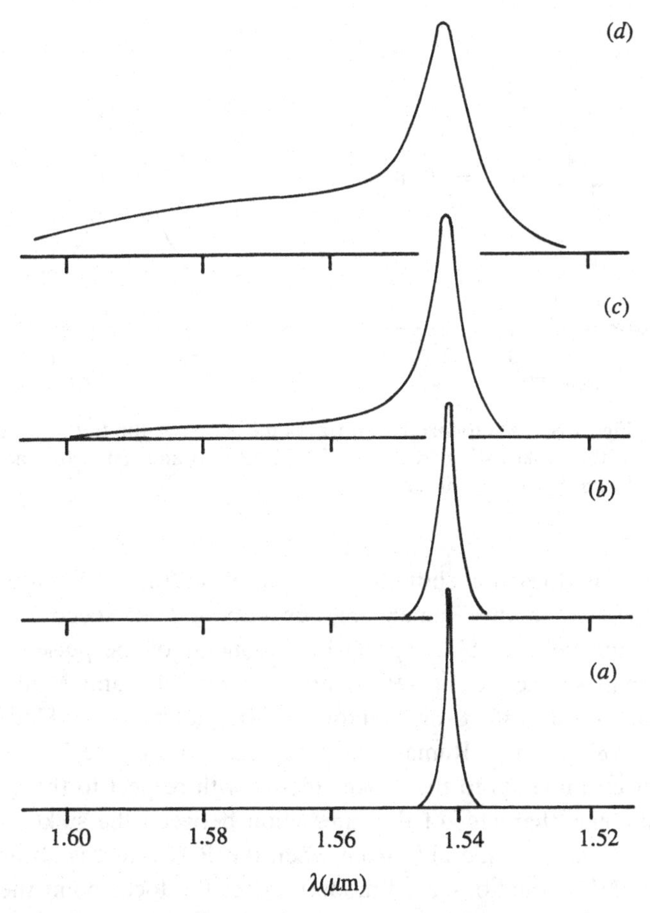

Figure 8.1. Spectra at the output of 250 m fiber for various power of the input 30 ps pulses: (*a*) 50 W, (*b*) 65 W, (*c*) 80 W and (*d*) 900 W. For the cases (*a*) and (*b*) the maximum self-compression fiber length is larger than 250 m; for the cases (*c*) and (*d*) the 'focus' point falls inside the fiber.

as predicted by the NLS equation-based theory. Fused silica has a remarkable property: the continuous spectrum of a stimulated Raman scattering (SRS) gain line stretches essentially from zero up to more than 500 cm^{-1} (see, for example, Stolen, 1979; Stolen *et al.*, 1984). The compressed spike has a broad spectrum and its spectral components are in phase with each other. Under these conditions the Stokes spectral components of the spike are amplified in the field of its anti-Stokes components due to the SRS effect, and the mean frequency of the spike shifts to the Stokes region, while the spectrum of

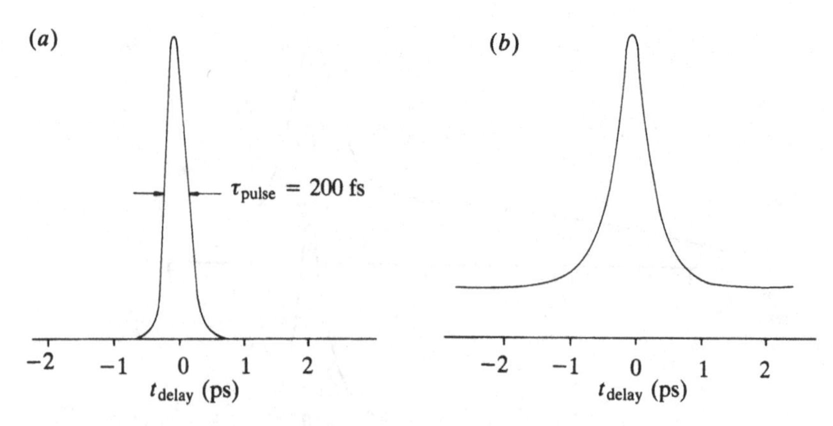

Figure 8.2. Autocorrelation traces measured at the fiber output (*a*) at the Stokes-wing frequency ($\lambda = 1.59\ \mu$m) and (*b*) at the main frequency ($\lambda = 1.54\ \mu$m).

the pedestal does not shift. This effect of shifting the pulse mean frequency due to the Raman amplification of 'red' spectral components in the field of 'blue' spectral components of the pulse is called the Raman self-scattering (RSS) effect. (Mitschke and Mollenauer, 1986 called this effect the soliton self-frequency shift, Hodel and Weber, 1987 called it Raman self-pumping.) Due to the fact that the spike spectrum shifts to the Stokes region with respect to the pedestal spectrum, the character of the interaction between the spike and the pedestal differs considerably from when the RSS effect is absent (i.e. from the NLS equation-based model). After the focus point the spike does not decay, but due to dispersion by the fiber the spike is delayed with respect to the pedestal. The spike shapes into a soliton and after propagation for some time in the fiber separates from the pedestal in the time as well as in the spectral domain.

Using the soliton laser as a source of pulses Mitschke and Mollenauer (1986) carried out quantitative investigations of the soliton mean frequency shift due to the RSS effect. Beaud *et al.* (1987) generated femtosecond solitons by the RSS of 0.83 ps multi-soliton pulses. Using cascade-stimulated Raman scattering and the RSS effect Grudinin *et al.* (1987) generated solitons in the 1.6 μm spectral region under the pumping of a fiber at $\lambda = 1.064\ \mu$m. A number of experiments on the femtosecond soliton generation using the RSS effect have been performed by the Femtosecond Optics Group from Imperial College (see, for example, Gouveia-Neto *et al.*, 1988). Tai *et al.* (1988) observed the decay of multi-soliton pulses influenced by the

RSS effect when the duration of pulses in the fiber focus point lies in the picosecond range. In this case the decay takes place over several soliton periods.

Previous methods for the mathematical description of SRS in fibers were based on a system of two (or more) coupled Schrödinger-type equations for the pump and Stokes waves (see, for example, Lugovoi, 1976; Vysloukh and Serkin, 1983). These methods demanded that the spectra of the pump and Stokes waves be narrow and completely isolated from each other. However in the RSS process the energy exchange takes place inside the continuous spectrum of one pulse, so for the mathematical description of the RSS effect new methods were needed. Gordon (1986) was the first to suggest the correct mathematical model for the RSS effect. In his model Gordon used the formalism of the non-linear response function (Hellwarth *et al.* 1971) and showed that the shift of the fundamental soliton's mean frequency due to the RSS effect is inversely proportional to the fourth order of the soliton pulsewidth, in good agreement with the experiments of Mitschke and Mollenauer (1986). After Gordon's work some theoretical papers on the RSS effect were published: Hodel and Weber, 1987; Beaud *et al.*, 1987; Kodama and Hasegawa, 1987; Haus and Nakazawa, 1987; Blow *et al.*, 1988a,b; Tai *et al.*, 1988, Dianov *et al.*, 1989; Vysloukh *et al.*, 1989. Note that in these works the real dependence of the Raman amplification coefficient on the spectral shift was not taken into account (later we will discuss the theoretical approaches, which were used in these works).

Theoretical consideration of the RSS effect taking into account the spectral dependence on the real Raman gain together with the real part of the non-linear susceptibility was conducted by Golovchenko *et al.* (1989), Tomlinson *et al.* (1989) and Stolen *et al.* (1989). Now we shall consider a more exact equation for the description of non-linear light propagation in fibers.

The molecular vibrations are excited by the field intensity, so the third-order non-linear medium polarisation, which consists of the electron and the Raman contributions, can be written in the form (Hellwarth *et al.*, 1971):

$$P^{(3)}(\tau) \simeq E(\tau) \int_0^\infty F(\theta)|E(\tau - \theta)|^2 \mathrm{d}\theta \tag{8.2}$$

where the response function $F(\theta)$ is determined by the non-linear properties of the medium. In this case the dimensionless non-linear equation for the electric field in the spectral domain has the form

(Mamyshev and Chernikov 1990):

$$\mathrm{i}\,\frac{\partial E(\Delta\omega)}{\partial\xi} - \frac{(\Delta\omega)^2}{2}\,E(\Delta\omega) + \gamma(\Delta\omega)^3 E(\Delta\omega) +$$
$$(1 + \Delta\omega/\Omega_0)\int \mathrm{d}\omega''\chi^{(3)}(\omega'')E(\Delta\omega - \omega'')\int \mathrm{d}\omega' E^*(\omega')E(\omega' + \omega'')$$
$$= 0 \qquad (8.3)$$

where $\Omega_0 = \omega_0 t_0$ is a dimensionless carrier radiation frequency at the fiber input;

$$E(\Delta\omega) = \int \mathrm{d}\tau E(\tau)\exp(\mathrm{i}\Delta\omega\tau)$$

is the spectral component of the electric field envelope; a term $\mathrm{i}\gamma(\partial^3 E/\partial\tau^3)$ describes the third-order dispersion where $\gamma = (\frac{1}{6})(k'''/k''/t_0)$; and the non-linear susceptibility is:

$$\chi^{(3)}(\Delta\omega) = \int_0^\infty \mathrm{d}\tau F(\tau)\exp(\mathrm{i}\Delta\omega\tau) \qquad (8.4)$$

(Note, that in (8.3) and (8.4) the non-linear susceptibility $\chi^{(3)}$ was used in dimensionless form, but in future we shall consider $\chi^{(3)}$ has dimensions.)

The non-linear susceptibility $\chi^{(3)}(\Delta\omega)$ is a complex value, the imaginary part of which is responsible for the RSS effect and the real part for the parametric and self-phase modulation effects.

Unlike previous models (Golovchenko *et al.*, 1989; Tomlinson *et al.*, 1989; and Stolen *et al.*, 1989) equation (8.3) contains a term, which is proportional to $\Delta\omega/\Omega_0$. The term describes the Stokes losses, associated with the material excitations during the RSS process, and dependence of the third-order non-linear effects on the radiation frequency. This term also describes the self-steepening effect (DeMartini *et al.*, 1967; Tzoar and Jain, 1981; Anderson and Lisak, 1983; Golovchenko *et al.*, 1986; Bourkoff *et al.*, 1987). In the time domain equation (8.3) is:

$$\mathrm{i}\,\frac{\partial E}{\partial\xi} - (\tfrac{1}{2})[\mathrm{sign}\,(k'')_{\omega_0}]\,\frac{\partial^2 E}{\partial\tau^2} - \mathrm{i}\gamma\,\frac{\partial^3 E}{\partial\tau^3}$$
$$+ E\int \mathrm{d}\theta F(\theta)|E(\tau - \theta)|^2$$
$$+ \mathrm{i}\sigma\,\frac{\partial}{\partial\tau}\left(E\int \mathrm{d}\theta F(\theta)|E(\tau - \theta)|^2\right) = 0 \qquad (8.5)$$

where $\sigma = 1/\Omega_0 = 1/(\omega_0 t_0)$. One can see that in equation (8.5) the shock term describing the self-steepening effect is half that given in previous papers (see, for example, Tzoar and Jain, 1981; Anderson and Lisak, 1983; Golovchenko *et al.*, 1986; Bourkoff *et al.*, 1987; Dianov *et al.*, 1989). Note that this result coincides with results of Manassah *et al.* (1986).

To solve equation (8.3) (or equation (8.5)) one must obtain the non-linear susceptibility $\chi^{(3)}(\Delta\omega)$ (or the response function $F(\theta)$). These functions can be obtained from the Raman gain spectrum and the value of the non-linear refractive index N_2 in the following way (Hellwarth *et al.*, 1971; Gordon, 1986; Golovchenko *et al.*, 1989; Tomlinson *et al.*, 1989; Stolen *et al.*, 1989).

The third-order non-linear susceptibility can be represented as a sum of two parts: the non-resonant (NR) (the electron non-linear susceptibility) and the resonant (R) (associated with the molecular vibrations – the Raman non-linear susceptibility) parts:

$$\chi^{(3)}(\Delta\omega) = \chi^{(3)}_{\mathrm{NR}} + \chi^{(3)}_{\mathrm{R}}(\Delta\omega) \tag{8.6}$$

it is suggested that $\chi^{(3)}_{\mathrm{NR}}$ is real and independent from $\Delta\omega$; this is a good approximation for the visible and infrared spectral regions, because the electron absorbtion lies in the ultraviolet region. The imaginary part of the resonant susceptibility is anti-symmetric in $\Delta\omega$, and it determines the Raman gain coefficient:

$$\begin{aligned} g(\Delta\omega) &= \mathrm{Im}\,(\chi^{(3)}_{\mathrm{R}}(\Delta\omega))4\pi\omega_0/(c^2 k) \\ &= \mathrm{Im}\,(\chi^{(3)}_{\mathrm{R}}(\Delta\omega))4\pi\omega_0/(cn_0) \end{aligned} \tag{8.7}$$

(c is the velocity of light, n_0 is the linear refractive index). The spectral dependence of the Raman gain coefficient for fused silica $g(\Delta\omega)$ is known from the literature (see, for instance, Stolen (1979)), so using the Kramers–Kronig relationship the real part of the Raman contribution to the non-linear susceptibility can also be obtained (Figure 8.3). From the expresion for the non-linear refractive index

$$N_2 = 2\pi/n_0(\chi^{(3)}_{\mathrm{NR}} + \mathrm{Re}\,(\chi^{(3)}_{\mathrm{R}}(0))) \tag{8.8}$$

and from the known value for fused silica $N_2 = 3.2 \times 10^{-16}\ \mathrm{cm^2/W}$ (Stolen, 1979) one can find $\chi^{(3)}_{\mathrm{NR}} = 4.556\,\mathrm{Re}\,(\chi^{(3)}_{\mathrm{R}}(0))$.

As a result we have calculated the spectral dependence of the third-order non-linear susceptibility and can solve equation (8.3). However if we want to work in the time domain we must calculate the non-linear response function, which is the inverse Fourier transform of the non-linear susceptibility (8.4). After the inverse Fourier transformation of (8.6) we obtain the response function $F(\tau)$:

$$F(\tau) = (1 - \varkappa)\delta(\tau) + \varkappa f(\tau) \tag{8.9}$$

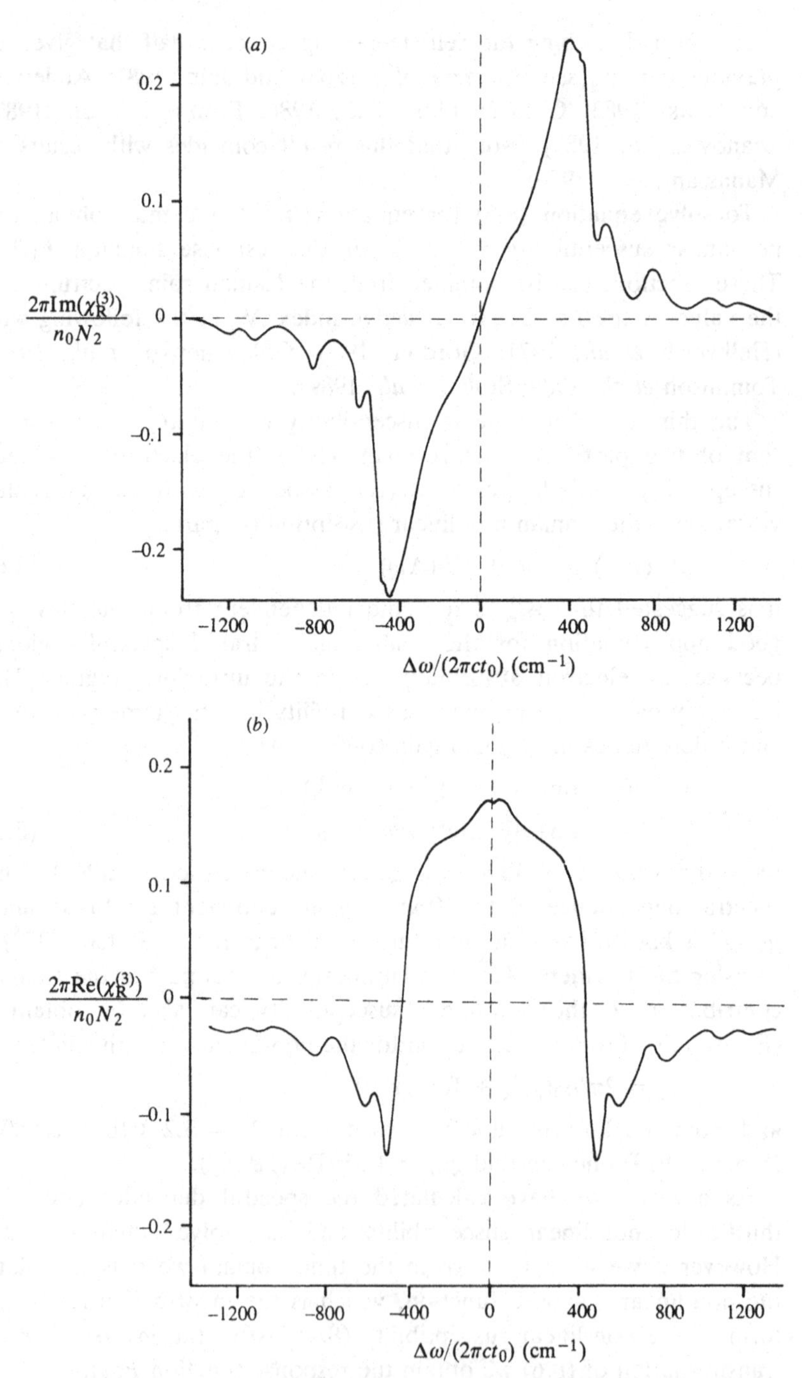

Figure 8.3. The spectral dependence of (a) the imaginary and (b) the real parts of the Raman (nuclear) susceptibility.

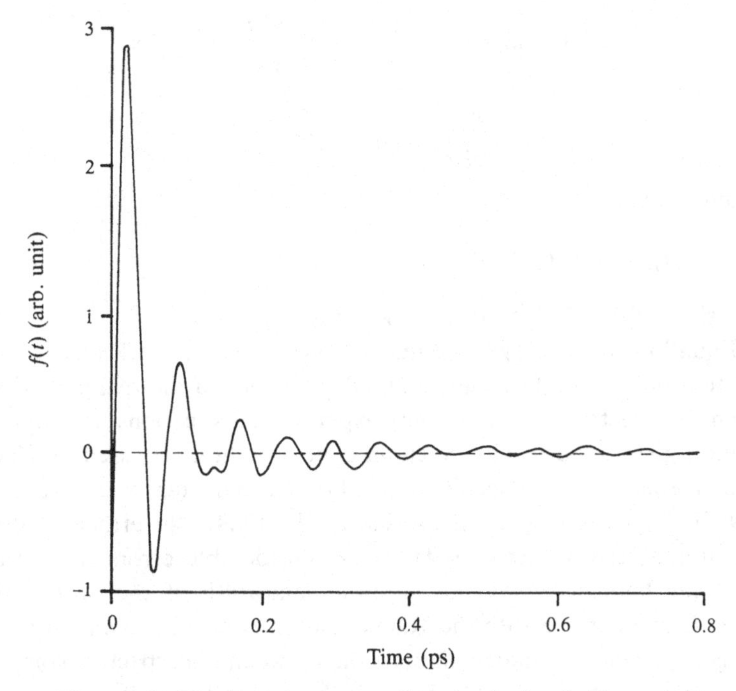

Figure 8.4. The Raman response function in fused silica.

The first term, which is proportional to the delta-function $\delta(\tau)$, is determined by the electron contribution, while the second term is determined by the Raman contribution. The functions $F(\tau)$ and $f(\tau)$ are normalised so that their integrals are a unit. From $\varkappa/(1-\varkappa) = \mathrm{Re}\,(\chi_{\mathrm{R}}^{(3)}(0))/\chi_{\mathrm{NR}}^{(3)}$ we obtain parameter $\varkappa = 0.18$. Figure 8.4 shows the Raman response function of fused silica $f(t)$.

Note that equations (8.3) and (8.5) take into account the real Raman gain spectrum and the associated spectral dependence of the real part of the non-linear susceptibility (Figure 8.3). The spectral dependence of the real part of the non-linear susceptibility plays an important role in the non-linear dynamics of radiation in fibers. For example, it considerably influences the femtosecond soliton propagation (Tomlinson and Stolen, 1989) and the process of the modulational instability (Golovchenko *et al.*, 1990).

Neglecting the last term in equation (8.5) and expanding of $|E(\tau-\theta)|^2$ by Taylor series around time τ and neglecting the second- and higher-order terms we obtain

$$\mathrm{i}\frac{\partial E}{\partial \xi} - (\tfrac{1}{2})[\mathrm{sign}\,(k'')_{\omega_0}]\frac{\partial^2 E}{\partial \tau^2} - \mathrm{i}\gamma\frac{\partial^3 E}{\partial \tau^3} + |E|^2 E - t_{\mathrm{R}} E\frac{\partial |E|^2}{\partial \tau} = 0 \quad (8.10)$$

where

$$t_{\mathrm{R}} = \varkappa \int f(\theta)\theta \mathrm{d}\theta \quad (8.10\mathrm{a})$$

For silica fibers the parameter $T_{\mathrm{R}} = t_{\mathrm{R}} \times t_0 = 3$ fs.

Equation (8.10) suggests a linear dependence of the Raman amplification on the spectral shift, and independence of the real part of $\chi^{(3)}$ from the spectral shift. For many experimental situations this approximation gives good results. Such an approximation was used by Hodel and Weber (1987); Beaud *et al.* (1987); Kodama and Hasegawa (1987); Blow *et al.* (1988a, b); Tai *et al.* (1988). Nevertheless, there are situations when this model gives considerable errors. For example, this happens when the spectral bandwidth of the propagating wave is comparable with the Raman gain bandwidth. Equation (8.10) suggests a strong Raman interaction between the spectral components, when the spectral distance between them is more than 500 cm^{-1}, while in reality this interaction for such spectral shifts is practically absent (see the Raman amplification spectrum of fused silica, Figure 8.3(*a*)).

Haus and Nakazawa (1987) and Dianov *et al.* (1989) used the relaxation-type equation for the non-linear addition to the refractive index to describe the RSS effect. As in equation (8.10), this model gives a considerably overestimated value of the Raman amplification for large spectral shifts, and does not describe the real spectral dependence of the real part of the non-linear susceptibility. Such an approach, which uses the relaxation-type equation for the non-linear addition to the refractive index, is not applicable for the case of silica glass (since the non-linear index consists of electronic and Raman parts) and causes misunderstanding in the interpretation of the phenomenological parameters: Dianov *et al.* (1989) regarded the parameter T_{R} (8.10a) as the relaxation time of the electron non-linearity. But as one can see, T_{R} is not the relaxation time and T_{R} is determined mainly by the Raman non-linearity, namely,

$$T_{\mathrm{R}} = \frac{2\pi}{n_0 N_2}\left[\frac{\mathrm{d}(\mathrm{Im}\,(\chi_{\mathrm{R}}^{(3)}))}{\mathrm{d}(\Delta\omega)}\right]_{\Delta\omega=0} \quad (8.10\mathrm{b})$$

Vysloukh *et al.* (1989) used a system of two coupled equations to describe the RSS effect: the NLS-type equation for the electric field envelope, and a classical oscillator-type equation for the molecular vibrations. Such an approach seems to be more based on comparison with the utilisation of a relaxation-type equation for the non-linear addition to the refractive index. Nevertheless, as the Raman gain line in fused silica consists of a number of inhomogeneously broadened vibrational resonances, this model cannot also describe the real spectral dependence of the Raman gain nor that of the real part of the non-linear susceptibility.

As a result we conclude that the most adequate model for the description of the RSS process in silica fibers is the model based on equation (8.3) (or equation (8.5)), and in our future theoretical considerations we shall use this approach.

With the help of this model we can now consider the formation of femtosecond fundamental solitons by using the decay of multi-soliton pulses caused by the RSS effect. Figures 8.5–8.7 show the results of numerical simulations of the propagation dynamics of 4 ps 10-soliton pulse through the fiber. Up to the focus point the dynamics do not differ practically from those predicted by the conventional NSE-based model: an intense narrow peak of about 100 fs duration is formed against the background of a broad pedestal, the spectrum of the pulse being practically symmetric (Figures 8.5(*a*) and (*b*)). But, as the narrow spike is formed, its mean frequency begins to shift to the Stokes region due to the RSS effect. Under these conditions the spike does not decay (as it would in the case of the NSE), but due to the GVD it is delayed with respect to the pedestal and slides down the pedestal (Figures 8.5(*c*) to (*f*)). Oscillations of the spike intensity and those of its duration during propagation through the fiber are the result of the interaction of the spike with the pedestal (Figures 8.6 and 8.7). As a result the spike becomes temporally and spectrally separated from the pedestal, it is formed into a clean fundamental soliton (Figures 8.5(*e*) and (*f*)). Under further propagation of the soliton its intensity and pulsewidth are constant (in the case of zero third-order dispersion, see Figures 8.6(*a*) and 8.7(*a*)).

Note that we considered the decay of the multi-soliton pulse by the RSS effect from the point of view of the formation of a single femtosecond fundamental soliton. For this reason the main attention was focused on the evolution of the fundamental soliton with maximum intensity, while we did not discuss in detail the evolution and the substructure of the other part of the multi-soliton pulse. This is the part of the multi-soliton pulse we call the 'pedestal'.

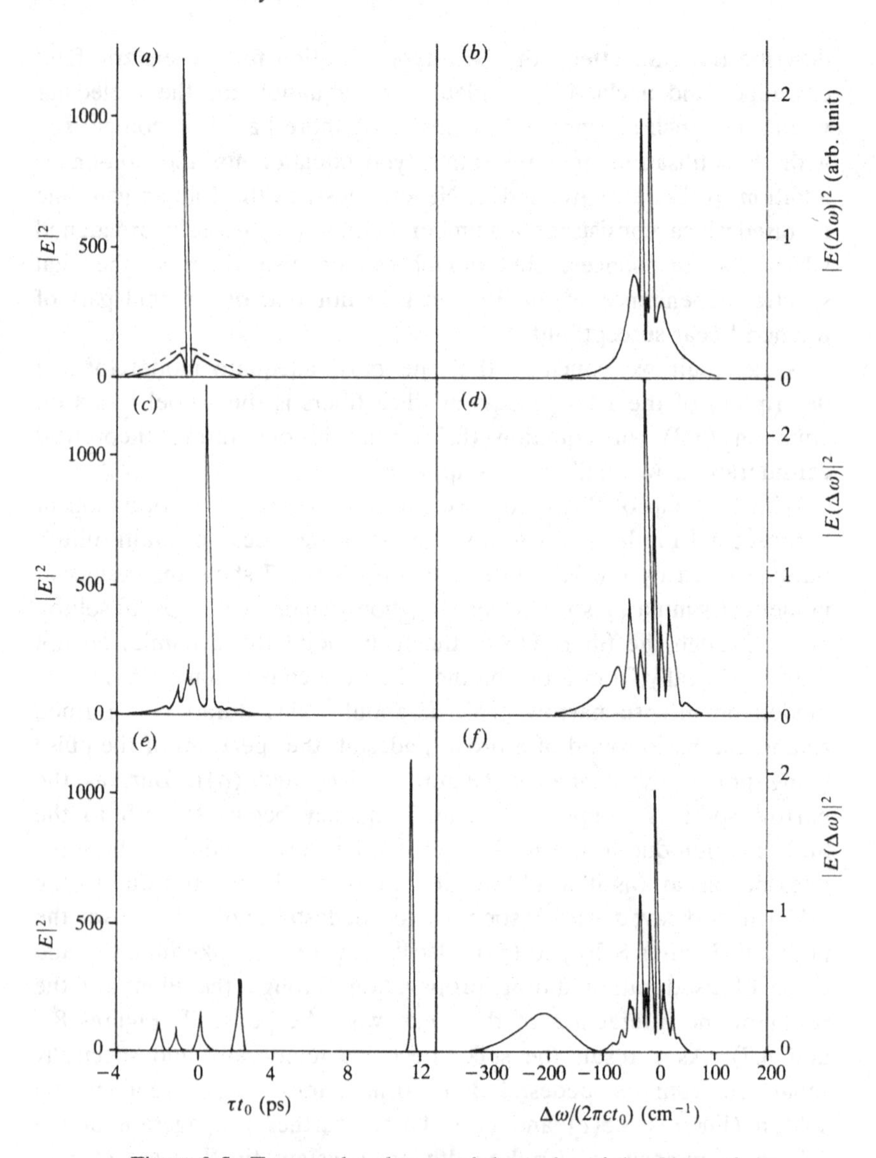

Figure 8.5. Temporal and spectral dynamics of the decay of 10-soliton 4 ps pulse in a fiber caused by the RSS effect ($k''' = 0$). Broken curve, the input pulse; (*a*) and (*b*) the 'focus' point, $\xi = 0.072$; (*c*) and (*d*) $\xi = 0.096$; (*e*) and (*f*) $\xi = 0.183$. One can see that at $\xi = 0.183$ the generated fundamental soliton and the rest of the multi-soliton pulse (the 'pedestal') are completely separated from each other (temporally and spectrally).

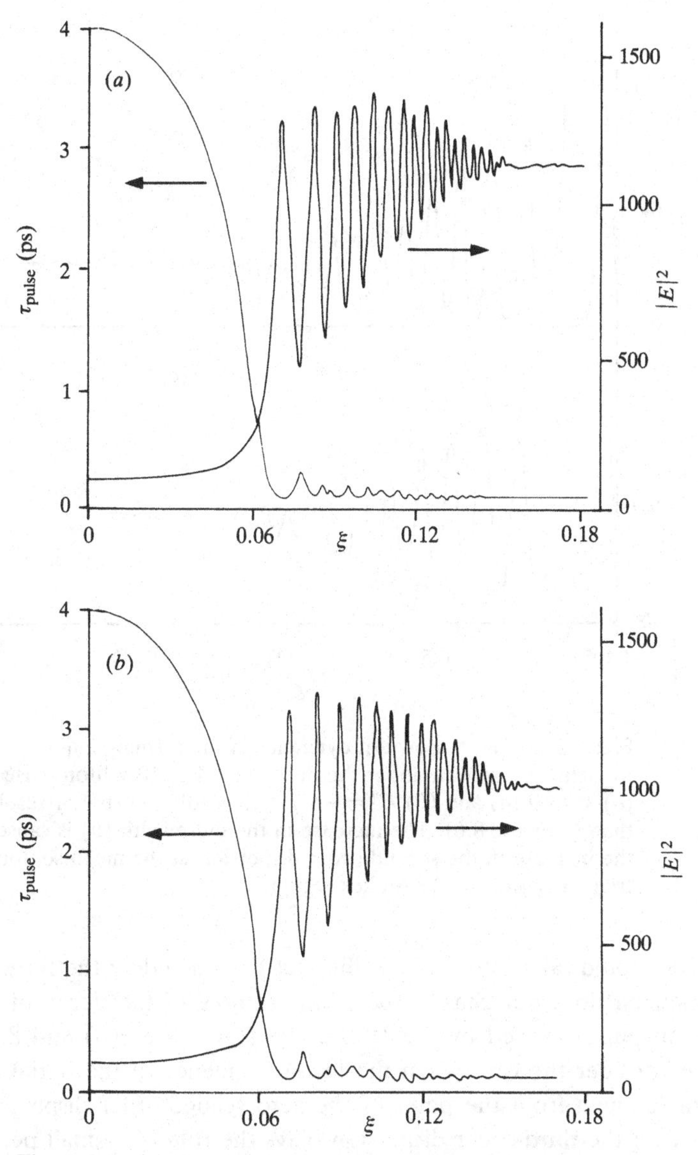

Figure 8.6. (*a*) The dynamics of the pulse peak intensity and the pulsewidth under the decay of 4 ps 10-soliton pulse in a fiber (see also Figure 8.5). Oscillations of the pulsewidth and that of the intensity are the result of the interaction of the forming main fundamental soliton with the rest of 10-soliton pulse ('pedestal'). After $\xi = 0.15$ the soliton is separated from the pedestal and its intensity and pulse width stabilise ($k''' = 0$). (*b*) the same as (*a*), but the third-order dispersion is not zero: $k''' = -k''/(1.93 \times 10^{14}\ \text{s}^{-1})$, it mean that the mean pump frequency is separated from the zero-dispersion wavelength by 1.93×10^{14} rad s^{-1} (or 1025 cm^{-1}).

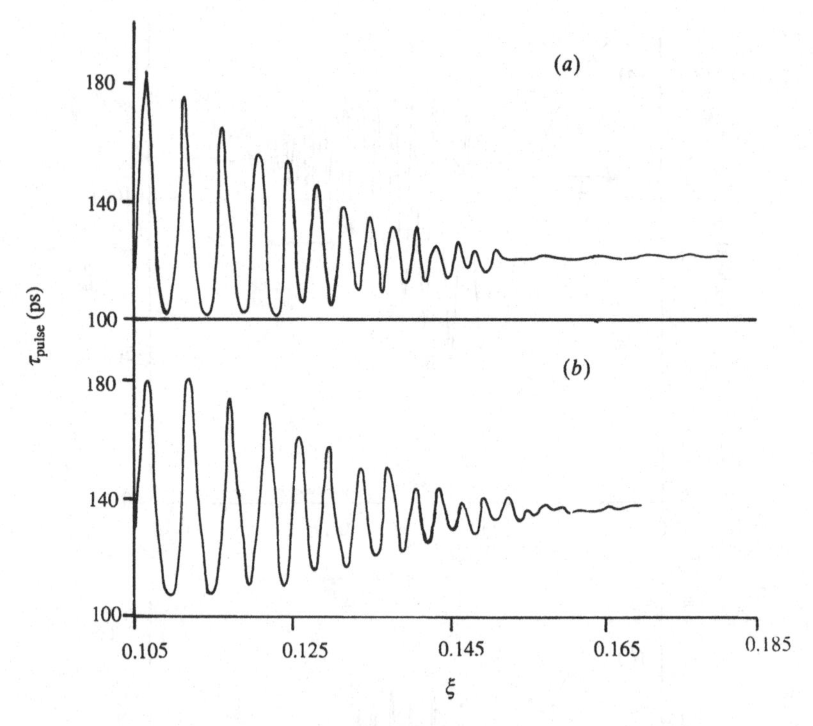

Figure 8.7. The pulsewidth dynamics of the forming main fundamental soliton under the decay of a 4 ps 10-soliton pulse for (*a*) $k''' = 0$ (*a*) and (*b*) $k''' = -k''/(1.93 \times 10^{14}\ \mathrm{s}^{-1})$ (better resolution than in Figure 8.6). The increase in the pulsewidth (*b*) is caused by the increase in the second-order dispersion at the mean soliton frequency shift to the Stokes region.

We should take into account the fact that including the third-order dispersion does not change the main features of the decay of multi-soliton pulses caused by the RSS effect (Figures 8.6(*b*) and 8.7(*b*)). (We consider the case when the mean frequency of the initial pulses is far enough from the point of the zero second-order dispersion. In this case the third-order dispersion plays the role of a small perturbation. Note that soliton propagation in the vicinity of the point of the zero second-order dispersion was investigated, for example, by Vysloukh, 1983; Blow *et al.*, 1983; Agrawal and Potasek, 1986; Wai *et al.*, 1986; Wai *et al.*, 1988; Gouveia-Neto *et al.*, 1988). The third-order dispersion effect results in our case in an increase in the second-order dispersion under the Stokes spectral shifting of the soliton due to the RSS effect. For this reason the solitons formed are broader in comparison with those in the case of the zero third-order dispersion, and they broaden with further propagation through the fiber (Figure 8.7(*b*)).

Note that when the soliton traverses the pedestal it can be Raman-amplified in the field of the pedestal until the down-shift of the soliton mean frequency is less than 500 cm^{-1}. The third-order dispersion broadens solitons and, consequently, retards the RSS effect. For these reasons the cubic dispersion retards the RSS effect and, consequently, prolongs the process of the Raman interaction of the soliton with the pedestal. In some cases this can increase the conversion of pedestal energy into the soliton energy (see also Section 8.5).

When the order of the multi-soliton pulses N is too large modulational instability can develop on the pulses (Tai *et al.*, 1986). Using the method described for the generation of fundamental femtosecond solitons from multi-soliton pulses in the case when modulational instability reaches saturation before the 'focus' point of the multisoliton pulses, at the Stokes frequency one should obtain not a single, but several soliton pulses or even a chaotic sequence of solitons (see the experiments of Vodop'yanov *et al.*, 1987; Islam *et al.*, 1989). For these reasons for the reliable generation of single-soliton pulses by the method described it is better to use pulses with $N < 13$ and a pulsewidth $\tau_p < 10$ ps (see, for example, the experiments of Beaud *et al.*, 1987; Gouveia-Neto *et al.*, 1989).

The key moment in the previously described method for the generation of a stable short soliton pulse is the down-shifting of the compressed pulse's mean frequency due to the RSS effect while another part of the multi-soliton pulse (the pedestal) practically does not experience such a frequency shift. The RSS effect can be regarded as a non-linear filter for short and intense optical pulses. It can be illustrated by the following qualitative example. Imagine that at the input of a Raman-active medium we have two intense pulses (a short one and a broad one) against the background of a pedestal. Only the short intense pulse will experience a considerable Stokes frequency shift since the efficiency of the RSS effect increases with an increase in the pulse intensity and its spectral bandwidth. Neither the pedestal nor the broad pulse, despite of its high intensity, will experience a considerable frequency shift. By filtering the Stokes spectral components at the medium output we can obtain a single short intense pulse without a pedestal. In this sense utilisation of the RSS effect is more effective for the selection of short intense pulses than the non-linear fiber birefringence effect (Stolen *et al.*, 1982) and saturable absorbers.

The sign of the second-order GVD is very important for the efficiency of the RSS process (Golovchenko *et al.* 1989). One can see from equation (8.3), that the RSS process can be regarded as that of scattering the spectral wave components $E(\Delta\omega - \omega'')$ into the $E(\Delta\omega)$ spectral component on the molecular vibrations $Q(\omega'')$:

$$Q(\omega'') \sim \mathrm{Im}\,(\chi^{(3)}(\omega'')) \int \mathrm{d}\omega'' E^*(\omega')E(\omega' + \omega'') \qquad (8.11)$$

In turn, the molecular vibrations $Q(\omega'')$ are excited by every pair of components of the wave spectrum, which are separated from each other by a frequency ω''. That is why the efficiency of the vibration excitation and, consequently, the efficiency of the RSS process at all, will be determined by the mutual phases between the components of the wave spectrum. When the GVD is positive the joint action of the self-phase modulation and dispersion effects makes propagation through the fiber pulse chirp. For this reason the spectral components of the pulse are not in phase, and the efficiency of the RSS effect is low. When the GVD is negative, operation in the due to the soliton regime ensures the spectral components of the pulse are in phase. As a result of this particular feature of solitons, the solitons experience efficiently the RSS effect, and clean fundamental solitons are obtained by the described method.

When the GVD is positive, the influence of the the RSS effect on 'dark' soliton propagation was observed by Tomlinson *et al.* (1989).

8.3 High-quality fundamental soliton compression and soliton pulsewidth stabilisation in fibers with slowly decreasing dispersion

Existing methods enable solitons with a minimum duration of about 50–100 fs to be generated. But there are various applications for which one wants the shortest possible pulses in the near infrared spectral region.

Grudinin *et al.* (1987) and Mitschke and Mollenauer (1987a) obtained pulses of about 20 fs duration by the multi-soliton pulse compression technique. However the pulse compressed by this method has a broad pedestal. This pedestal not only leads to a deterioration in the quality of the pulse and the energy characteristics of the compression, but it also makes the compressed pulse in the fiber unstable due to the non-linear interaction of the pedestal with the compressed pulse. For this reason the compression of fundamental solitons when the compressed pulses remain fundamental solitons is of great interest. Azimov *et al.* (1986) and Blow *et al.* (1987b, 1988a) have shown that such soliton compression can be achieved under the adiabatic (slow) amplification of the solitons in fibers (amplification is small compared with the dispersion length of the soliton). The minimum soliton pulsewidth which can be achieved by this compression is limited by the amplification spectral bandwidth (Blow *et al.* 1988a). For Raman amplification in silica fibers this value is about 100 fs.

Using fibers with a slowly decreasing value of the second-order dispersion along the fiber length (FSDD) enables us to realise effective amplification (Tajima, 1987; Kuehl, 1988; Dianov *et al.*, 1989). This follows from the fact that the NLS equation with slowly varying dispersion

$$\mathrm{i}\frac{\partial E}{\partial \xi} - (\tfrac{1}{2})[\mathrm{sign}\,(k'')_{\omega_0}]\beta(\xi)\frac{\partial^2 E}{\partial \tau^2} + |E|^2 E = 0 \tag{8.12}$$

$$\beta(0) = 1$$

is reduced to a NLS equation with constant dispersion and amplification:

$$\mathrm{i}\frac{\partial V}{\partial \eta} - (\tfrac{1}{2})[\mathrm{sign}\,(k'')_{\omega_0}]\frac{\partial^2 V}{\partial \tau^2} + |V|^2 V + \mathrm{i}\Gamma(\eta)V = 0 \tag{8.13}$$

(where $\Gamma(\eta) = (\frac{1}{2}\beta)\partial\beta/\partial\eta$) by the transformation $E = V(\beta(\xi))^{1/2}$ and $\eta = \int_0^\xi \beta(\xi')\mathrm{d}\xi'$. Note that equation (8.12) suggests that the shape and cross-section of the fiber mode are not changed along the fiber length. Tajima (1987) was the first who suggested the use of FSDD in non-linear fiber optics as they would compensate for soliton broadening in lossy fibers. Kuehl (1988) considered the more general case of soliton propagation in slowly varying axially non-uniform optical fiber. These investigations were based on the NLS model. According to the NSE in the adiabatic approximation (that is the pulse remains as a soliton and its energy equals that of the initial soliton) the soliton compression ratio equals the ratio of the initial value of the second-order dispersion to the final one:

$$\tau_\mathrm{p}(z)/\tau_0(z) = k''(z)/k''(0) \tag{8.14}$$

It should be noted that the spectral bandwidth of effective amplification in the case of soliton compression in FSDD has an infinite value. That is why this method of pulse compression is very promising for the generation of ultrashort pulses.

For our experiments on the compression of femtosecond solitons (Dianov *et al.*, 1989a; Chernikov, Mamyshev and Dianov unpublished), a special 35-m fiber was produced. The dispersion of the first

25 m was constant (13 ps/nm · km at wavelength $\lambda = 1.55\ \mu$m), while the dispersion of the last 10 m of the fiber was close-to-linearly decreasing along the fiber length from the given value to zero. The FSDD was a taped fiber. The outer diameter of the fiber decreased from 160 μm at the input to 110 μm at the fiber end, while the change in the effective mode diameter of the FSDD was less than 6%.

By using the RSS effect on the first 25 m of the fiber, 130 fs solitons were generated at $\lambda = 1.55\ \mu$m (Figure 8.8(*a*)). The NSE predicts an infinite compression ratio for the solitons in the 10-m FSDD (Figure 8.9, curve 1). Nevertheless our experiments showed that only the soliton dynamics on the first 5 m of the FSDD could be described by the NLS model: the soliton compression ratio depended linearly on the fiber length for this piece of fiber, but then the soliton pulsewidth was stabilised at 50 fs (Figures 8.8(*b*) and 8.9), and the mean soliton frequency shifted to the Stokes region. Note the high quality of the compression: the solitons were compressed as a single unit, a pedestal did not appear. The pulses preserved the soliton character of propagation under the compression.

For an explanation of this effect of soliton pulsewidth stabilisation in FSDD we have included the RSS and third-order dispersion effects in the theoretical consideration (Bogatyrev *et al.*, 1989):

$$\mathrm{i}\frac{\partial E}{\partial \xi} - (\tfrac{1}{2})[\operatorname{sign}(k'')_{\omega_0}]\beta(z)\frac{\partial^2 E}{\partial \tau^2} + E\int \mathrm{d}t F(t)|E(\tau - t)|^2 - \mathrm{i}\gamma\frac{\partial^2 E}{\partial \tau^3} = 0 \qquad (8.15)$$

Figure 8.9 (curves 2 and 3) shows the results of numerical calculations of the soliton dynamics in FSDD for the parameters of our experiment. One can see good agreement between theory and experiment. At the first stage of propagation the soliton pulsewidth is somewhat large and the down-shift of the soliton mean frequency due to the RSS effect is negligible (Figure 8.10(*b*), curve 1), the solitons are compressed according to the NLS theory. Compression of solitons results in an increase in the RSS effect, the mean soliton frequency shifts to the Stokes region, where the total dispersion $k''_{\mathrm{tot}}(z)$ increases (in absolute value) due to the cubic dispersion:

$$k''_{\mathrm{tot}}(z) = k''(z) + k'''\Delta\omega_{\mathrm{mean}} \qquad (8.16)$$

($\Delta\omega_{\mathrm{mean}}$ is the soliton mean frequency shift due to the RSS effect). In this case equation (8.14) must be rewritten as follows:

$$\tau_{\mathrm{p}}(z)/\tau_0(z) = k''_{\mathrm{tot}}(z)/k''_{\mathrm{tot}}(0) \qquad (8.17)$$

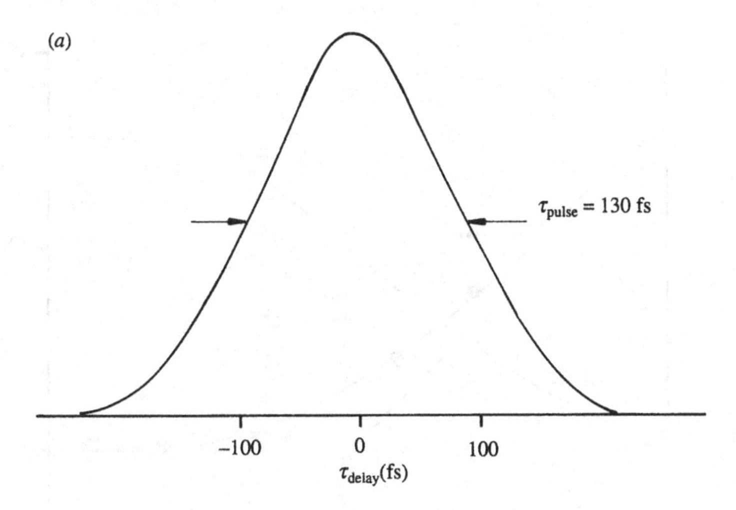

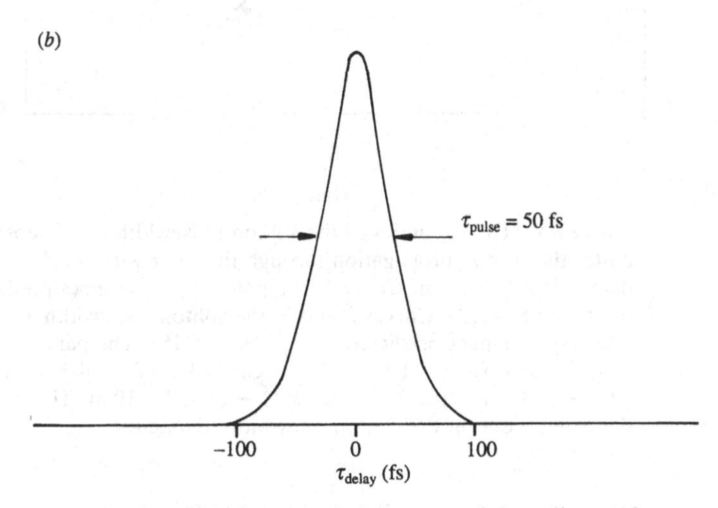

Figure 8.8. Intensity autocorrelation traces of the soliton pulses at (*a*) the input and (*b*) the output of the fiber with slowly decreasing dispersion (experiment).

As a result the soliton compression rate decreases. The soliton pulse-width is stabilised in the FSDD when the changes in second-order dispersion due to the RSS and the cubic dispersion effects fully compensate for the decrease in dispersion in the fiber:

$$\frac{\mathrm{d}k''_{\mathrm{tot}}}{\mathrm{d}z} = 0 \tag{8.18}$$

Using the approximation from Gordon (1986) for the rate of the

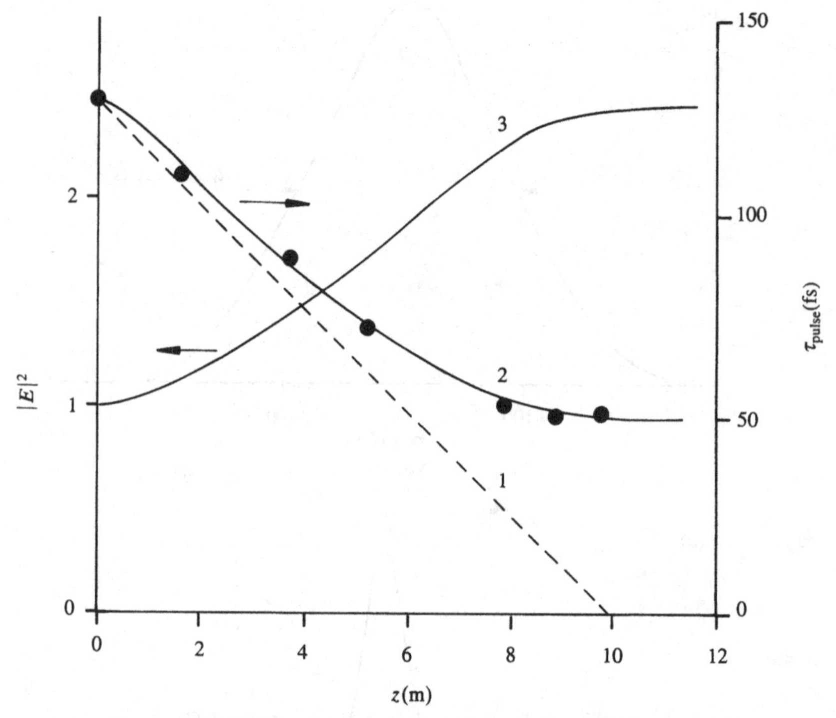

Figure 8.9. The dynamics of the soliton pulsewidth and intensity under the soliton propagation through the fiber with slowly decreasing dispersion. Curve 1, the pulsewidth dynamics predicted by the NLS (8.12); Curves 2 and 3, the soliton pulsewidth and its intensity dynamics predicted by equation (8.15). The parameters are: $\tau_0 = 130$ fs; $k''(0) = -1.69 \times 10^{-28}$ s^2 cm^{-1}; $k''' = 3 \times 10^{-42}$ s^3 cm^{-1}; ($\gamma = -1.73 \times 10^{-2}$); $k''(z) = k''(0)\,(1 - z/l)$, $l = 10$ m. The dots are the experiment on the soliton pulsewidth dynamics.

mean soliton frequency shift due to the RSS effect:

$$\frac{\mathrm{d}\omega_{\text{mean}}}{\mathrm{d}z} = -(\tfrac{8}{15})T_{\mathrm{R}}|k''_{\text{tot}}(z)|/(t_{\mathrm{p}})^4 \tag{8.19}$$

(τ_{pulse}(FWHM) $= 1.763t_{\mathrm{p}}$) and conditions (8.17) and (8.18), one can obtain analytical estimates for the stabilised soliton pulsewidth:

$$\tau_{\text{stab}}(\text{FWHM}) = 1.72(k'''|k''(0)|(\mathrm{d}k''/\mathrm{d}z)^{-1}T_{\mathrm{R}}/\tau_0)^{1/3} \tag{8.20}$$

For the parameters of our experiment this formula gives $\tau_{\text{stab}} = 53$ fs, which is in good agreement with experimental and numerical calculations.

It is seen from (8.20) that τ_{stab} slightly depends on the initial soliton pulsewidth τ_0, that shows high stability of the soliton compression in FSDD. Moreover, using the definite length of FSDD (for the parameters of our FSDD this length is about 6–7 m), the compressed soliton pulsewidths practically equal each other for a wide range of initial soliton pulsewidths (Figure 8.10(*a*)). This fact means that the pulsewidth of the compressed FSDD soliton is determined mainly by the parameters of the FSDD but not by the initial soliton pulsewidth.

In conclusion, it should be emphasised that the effect of the soliton pulsewidth stabilisation in the FSDD is connected with the joint action of the RSS and third-order dispersion effects. On one hand, the soliton pulsewidth stabilisation effect increases the stability of the compression; on the other hand it limits the soliton compression ratio. To obtain solitons with a shorter duration one can utilise FSDD with smaller value of the third-order dispersion (so-called 'dispersion-flattened' fibers). Figure 8.11 shows the results of numerical simulations on the dynamics of 130 fs soliton propagation through the FSDD with the same parameters as in previous examples except for the value of the third-order dispersion. Note the increase in the pulse area at the soliton shortening (Figure 8.11(*b*), curve 4). This fact does not mean a violation of the soliton-like regime of propagation, but it is connected with the decrease in the 'effective' non-linear refractive index at the increase of the pulse spectral bandwidth, as it follows from the spectral dependence of the real part of the Raman contribution to the non-linear susceptibility (Figure 8.3(*b*)) (see also Tomlinson and Stolen, 1989).

Note also that for soliton shortening to less than 50 fs the efficiency of the RSS decreases in comparison with (8.19), because the soliton spectrum becomes comparable with the Raman gain bandwidth (Gordon, 1986). For this reason in the case of the soliton compression to less than 50 fs, expression (8.20) will overestimate the value of τ_{stab} with respect to the one obtained from numerical simulations based on equation (8.15).

Note that we considered the case of $k''' > 0$. But as it follows from equation (8.17), by using special fibers with $k''' < 0$ it is possible to realise adiabatic soliton compression in fibers with $k''(z) = \text{constant}$.

In our previous consideration (equation (8.15)) we did not take into account the Stokes losses associated with the material excitation, that is the difference between the energy of the pump and Stokes photons during the RSS process. Such an approximation is appropriate until the shift of the mean soliton frequency is much smaller than the mean frequency itself. Another effect which must be taken into

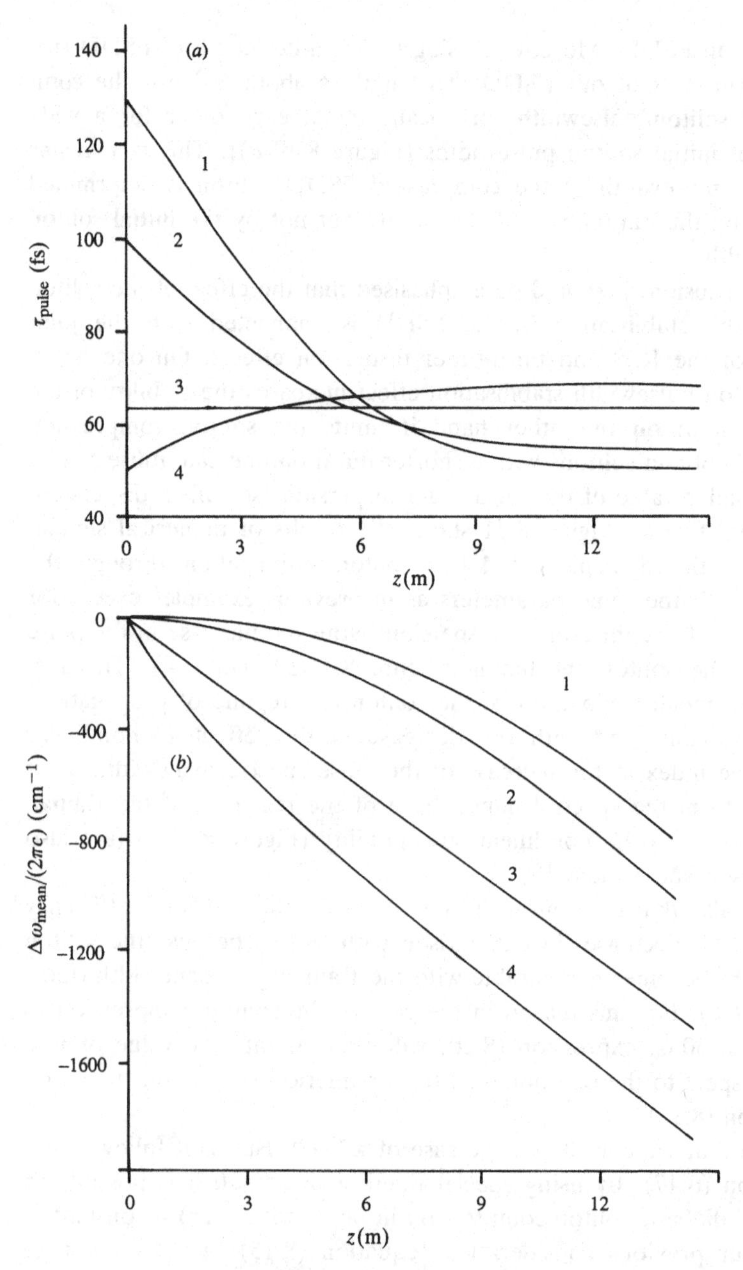

Figure 8.10. Calculated using the adiabatic approximation (equations (8.16), (8.17) and (8.19)). (*a*) Evolution of the soliton pulsewidth and (*b*) its mean spectral shift $\Delta\omega_{\mathrm{MEAN}}$ in a fiber with slowly decreasing dispersion for various input soliton pulsewidths: curves 1, 130 fs; curves 2, 100 fs; curves 3, 62.5 fs; curves 4, 50 fs. The parameters of the FSDD are the same as in Figure 8.9.

account in the case of a considerable mean soliton frequency shift is the linear dependence of non-linear processes connected with the third-order non-linear susceptibility (such as Raman amplification and self-phase modulation) on the frequency of the radiation. To take these effects into account together with the self-steepening effect we shall consider equation (8.5) generalised for FSDD:

$$\mathrm{i}\frac{\partial E}{\partial \xi} + (\tfrac{1}{2})\beta(\xi)\frac{\partial^2 E}{\partial \tau^2} - \mathrm{i}\gamma\frac{\partial^2 E}{\partial \tau^3} + E\int \mathrm{d}\theta F(\theta)|E(\tau-\theta)|^2$$
$$+ \mathrm{i}\sigma\frac{\partial}{\partial \tau}\left(E\int \mathrm{d}\theta F(\theta)|E(\tau-\theta)|^2\right) = 0 \qquad (8.21)$$

For the parameters of Figure 8.11 (curves 3, $\tau_{\mathrm{stab}} = 50$ fs) equation (8.21) predicts practically the same soliton evolution as equation (8.15). Figures 8.11 (curves 2(2)) and 8.12 show the results of numerical calculations of the soliton dynamics in FSDD with the same parameters as in Figure 8.11 (curves 2(1)), but using equation (8.21). A considerable difference in the evolutions for $z > 7$ m is caused by the mean soliton frequency shift due to the RSS effect (at $z = 10$ m the shift is already about $-700\ \mathrm{cm}^{-1}$). Near $z = 10$ m high-quality 18 fs pulses are formed (Figure 8.12). As the adiabatic model of soliton compression in a FSDD predicts, the soliton pulsewidth and intensity must be stabilized for $z > 10$ m for the parameters under consideration. Nevertheless, one can see a decrease in the pulse intensity and an increase in the pulsewidth for $z > 10$ m. This fact is caused by the influence of the k''' dispersion in the neighborhood of the zero point of the k'' dispersion. For soliton propagation in the neighborhood of the zero point of the second-order dispersion some parametric processes take place, and anti-Stokes spectral components are generated in the spectral region of $k'' > 0$ (Wai *et al.*, 1986). In order to decrease the influence of the third-order dispersion one can use an initial soliton with a shorter duration. Figure 8.13 shows a 14 fs pulse (*a*) and its spectrum (*b*) obtained by compression of solitons with $\tau_0 = 65$ fs. Only an experiment can give an answer to whether our model describes the propagation of such short pulses sufficiently well.

Our simulations (Figures 8.11(2), 8.12, 8.13) were conducted for the so-called 'dispersion-flattened' fibers, that is for fibers with a decreased third-order dispersion. Such fibers are especially promising for generating extremely short pulses. Nevertheless, one can obtain high-quality pulses with pulses of about 20 fs duration by soliton compression in a FSDD with a 'normal' value for k''' (see Figure

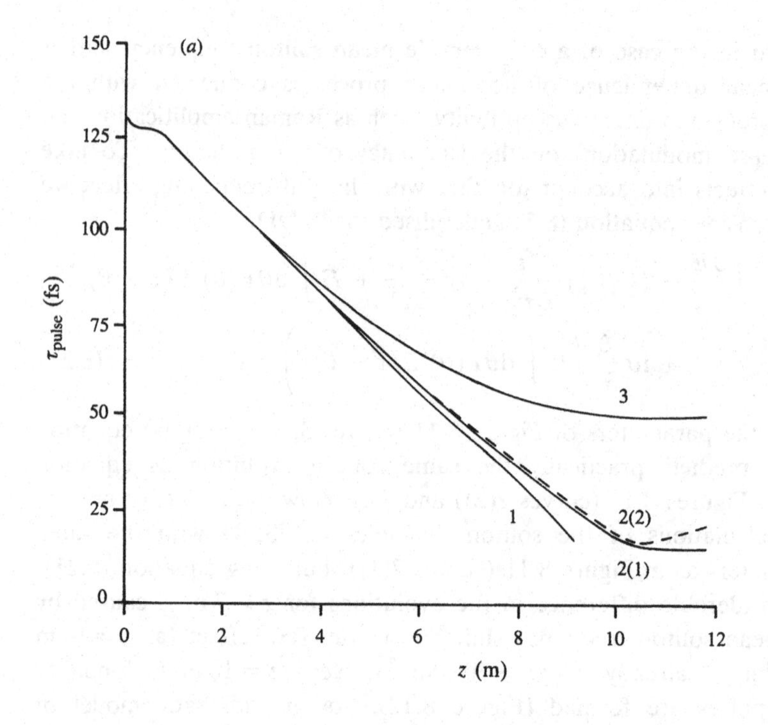
(a)
150
125
100
75
50
25
0
τ_{pulse} (fs)
3
1
2(2)
2(1)
0
2
4
6
8
10
12
z (m)

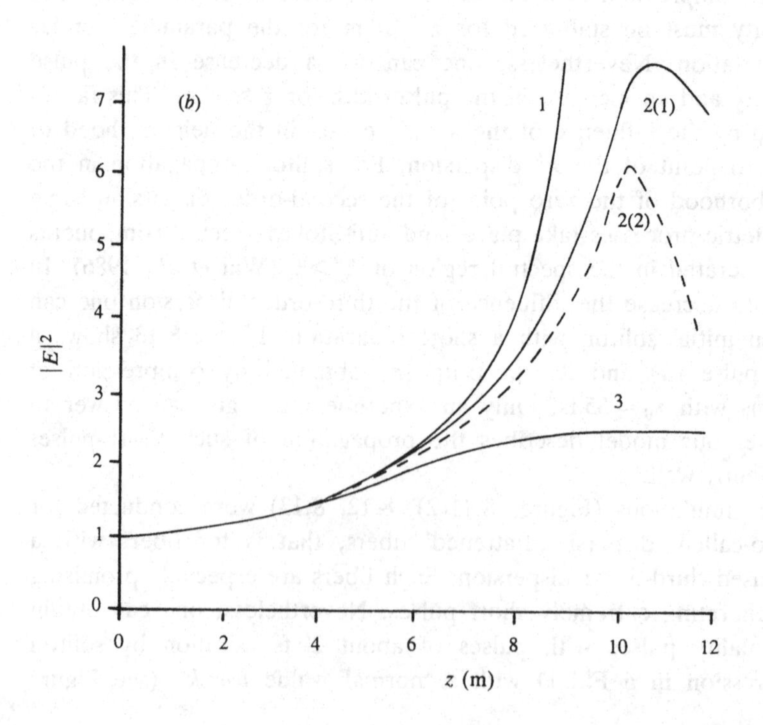
(b)
7
6
5
4
3
2
1
0
$|E|^2$
1
2(1)
2(2)
3
0
2
4
6
8
10
12
z (m)

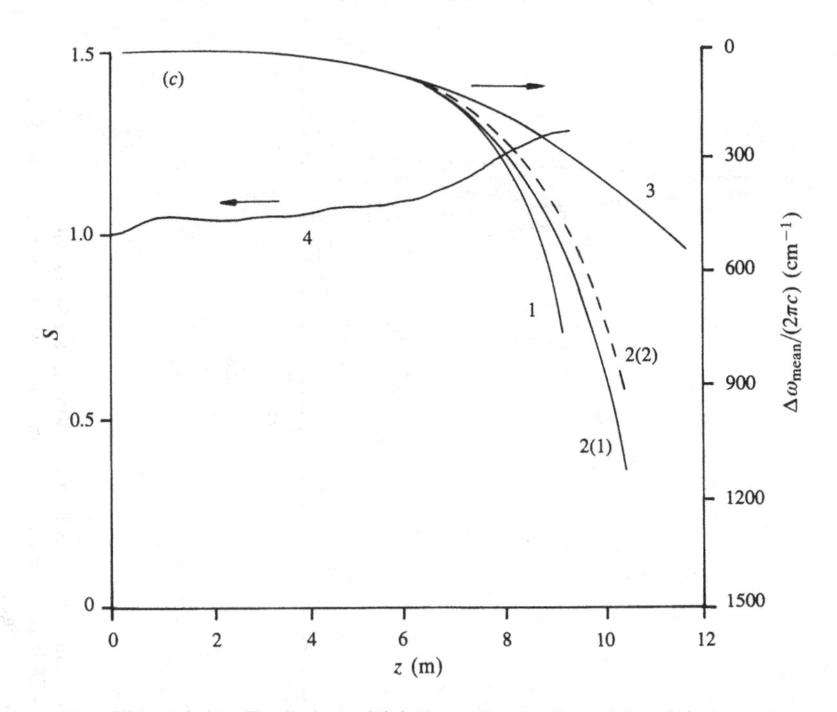

Figure 8.11. Evolution of (*a*) the soliton pulse width, (*b*) the soliton intensity and (*c*) the soliton mean frequency shift in FSDD with parameters of Figure 8.9 except for the values of the third-order dispersion: curves 1, $k''' = 0'$, curves 2(1) and 2(2), $k''' = 1.3 \times 10^{-43}$ s^3/cm ($\gamma = -1.73 \times 10^{-3}$); curves 3, $k''' = 1.3 \times 10^{-42}$ s^3/cm ($\gamma = -1.73 \times 10^{-2}$). In (*c*) (curve 4) the evolution of the product $S = 0.332\, \tau_{pl}^2 |E|^2$ for $k''' = 0$ is also shown. The input soliton pulsewidth for all cases is 130 fs. Curves 1, 2(1) and 3 are calculated in the framework of equation (8.15), and curve 2(2) in the framework of equation (8.21).

8.14). In Figure 8.14 the dimensional parameters are: $k''(0) = 2.18 \times 10^{-28}$ s^2/cm, $k''' = 1.0 \times 10^{-42}$ s^3/cm and the dispersion length $z_d = 3.7$ cm.

8.4 Generation of a high-repetition-rate (up to THz range) train of practically non-interacting solitons

Nowadays methods for generating solitons in fibers have a repetition rate of about 100's MHz. For some applications a much

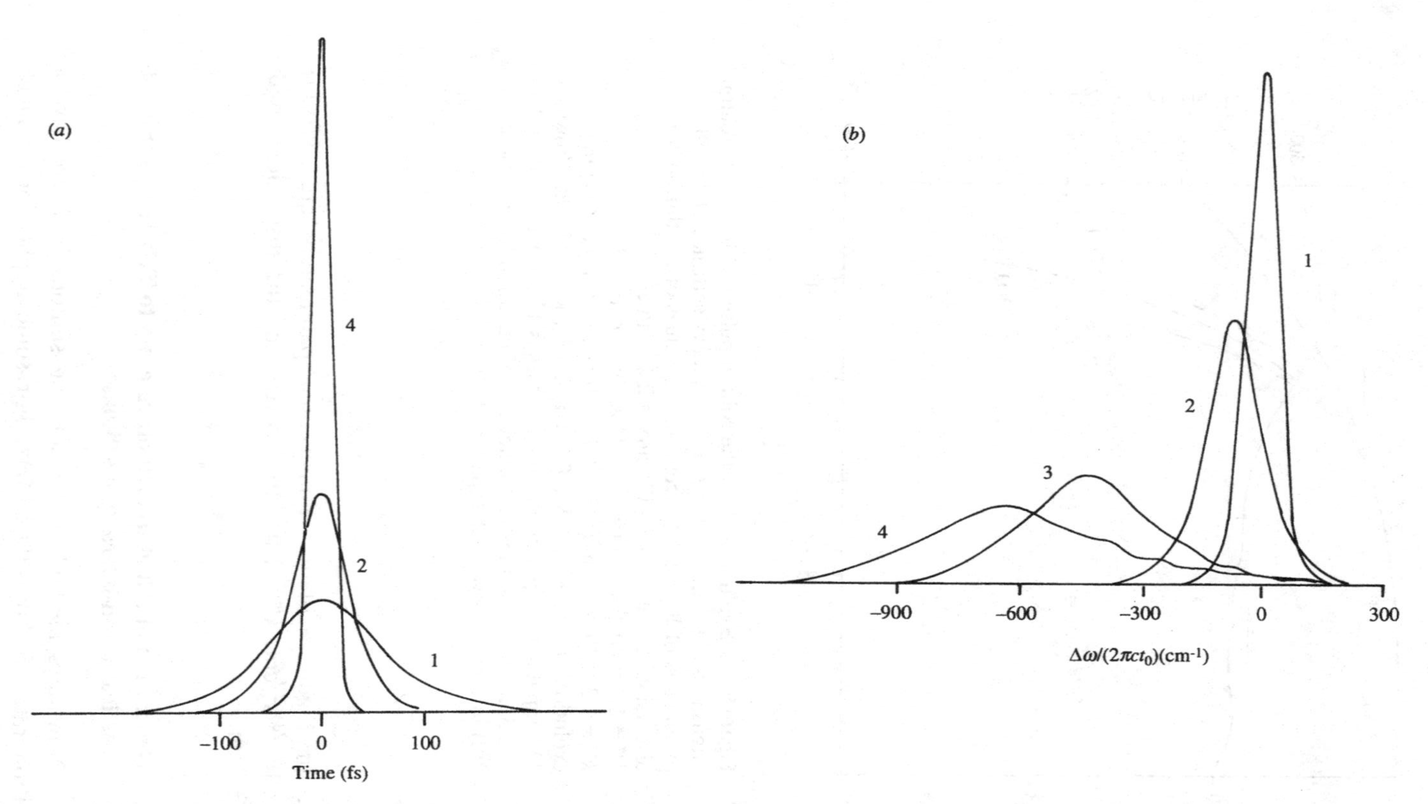

Figure 8.12. (*a*) Temporal and (*b*) spectral evolution of a soliton with an initial pulsewidth of 130 fs in a FSDD for the case 2(2) of Figure 8.11: (1), $z = 0$; (2), $z = 5.4$ m; $\tau_p = 66$ fs; (3), $z = 9$ m, $\tau_p = 28$ fs (only spectrum is shown); (4), $z = 9.8$ m, $\tau_p = 18$ fs. The temporal soliton shifts from their initial positions are not shown in the figure.

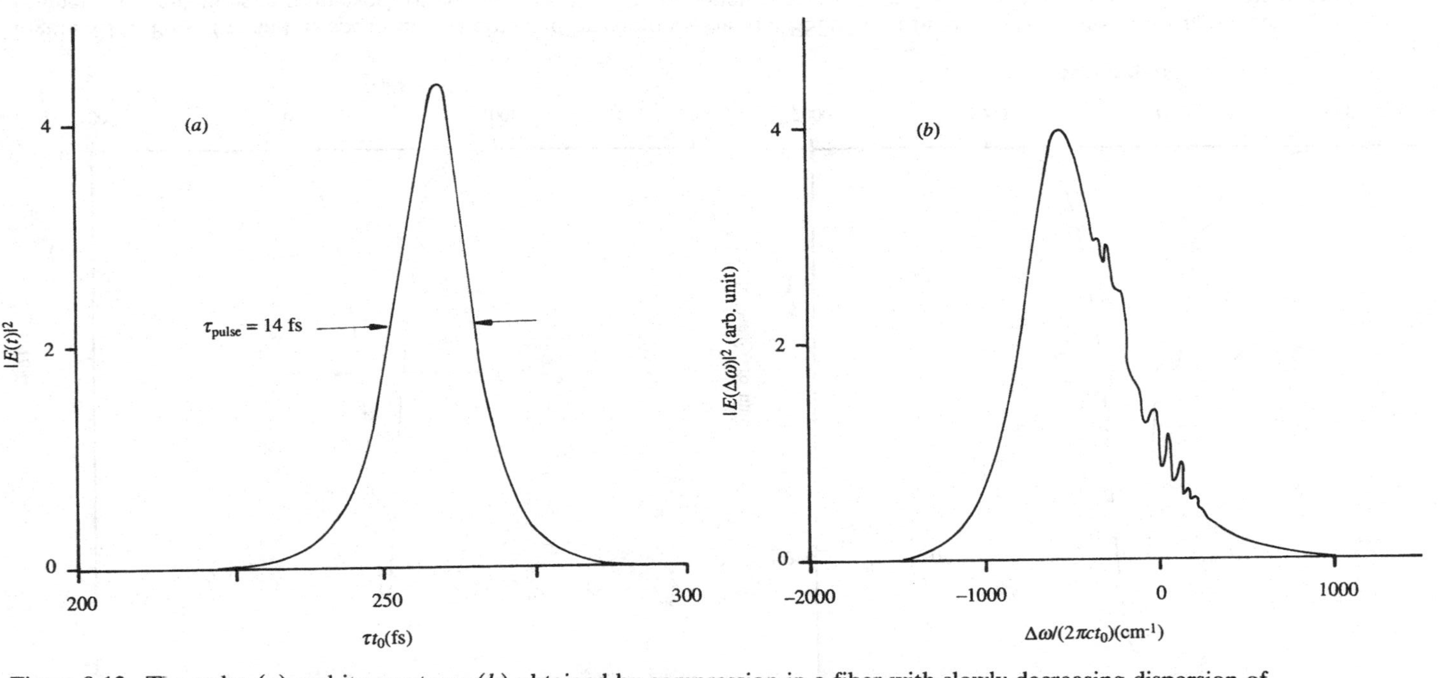

Figure 8.13. The pulse (a) and its spectrum (b) obtained by compression in a fiber with slowly decreasing dispersion of fundamental soliton with $\tau_0 = 65$ fs (numerical simulations in framework of equation (8.21)). The parameters are: $\beta(\xi) = 1 - \xi/20.2$; $\gamma = -3.46 \times 10^{-3}$; the fiber length $\xi = 18.1$.

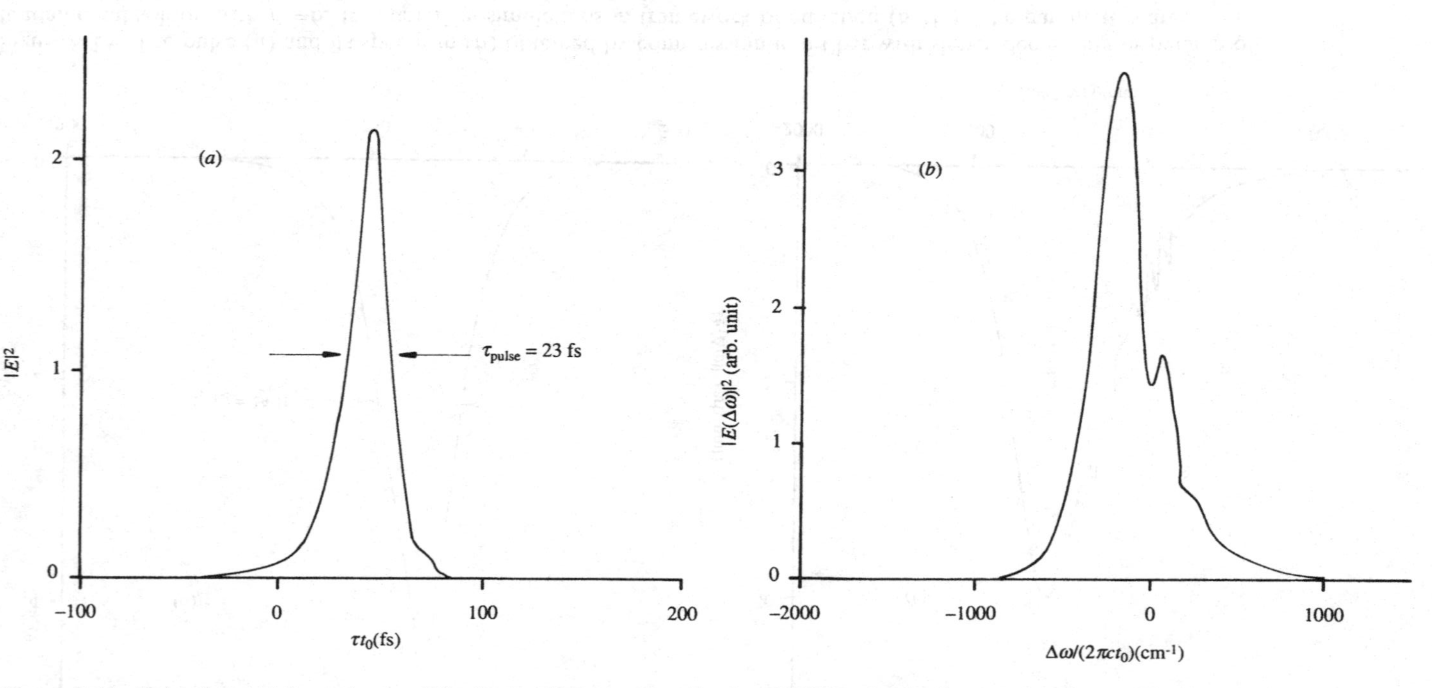

Figure 8.14. Pulse (a) and its spectrum (b) obtained by compression in a FSDD of a fundamental soliton with $\tau_0 = 50$ fs (numerical simulations in framework of equation (8.21). The parameters are: $\beta(\xi) = 1 - \xi/9.3$; $\gamma = -2.7 \times 10^{-2}$; the fiber length $\xi = 6.62$.

higher repetition rate is needed. Using a 2.6 GHz gain-switched 1.3 μm laser diode as a pulse source Iwatsuki *et al.* (1988) observed a soliton-like regime of pulse propagation in optical fibers. Hasegawa (1984) suggested a method for generating a train of ultra-short pulses with a high repetition rate based on the effect of modulational instability of c.w. radiation in optical fibers. Tai *et al.* (1986) were the first to realise this method experimentally and to obtain a train of 0.5 ps pulses with 0.3 THz repetition rate. However, the modulational instability method does not permit the generation of a train of fundamental solitons. The pedestal always exists between the pulses generated by this method. The existence of a pedestal results in the non-linear pulse interaction in the fiber, and the pulse sequence is not stable in the fiber. In principle, the pedestal can be suppressed using non-linear fiber birefringence or other intensity discrimination techniques (see, for example, Stolen *et al.*, 1982; Maier, 1982; Nicolaus *et al.*, 1983; Winful, 1986; Blow *et al.*, 1987a; Gusovskii *et al.*, 1987; Wabnitz and Trillo, 1989; Friberg *et al.*, 1988; Dianov *et al.*, 1989b). In the following section we will show that using the RSS effect for this purpose allows us not only to suppress the pedestal, but also to generate a clean fundamental soliton train. Moreover, using the combined action of the induced modulational instability and the RSS effects one can convert all the pedestal energy into the soliton train.

In the present section we will consider another method for the generation of a high-repetition-rate (up to THz range) stable train of practically non-interacting fundamental solitons. In this method the adiabatic amplification of a periodically modulated c.w. signal in an optical fiber is used and in another such a signal is transmitted through the fibers with slowly decreasing dispersion along the fiber length (FSDD) (Dianov *et al.*, 1989; Bogatyrev *et al.*, 1989). Experiments as well as theoretical models will be described.

We start from the NLS equation with an amplification term:

$$\mathrm{i}\frac{\partial E}{\partial \xi} - (\tfrac{1}{2})[\mathrm{sign}\,(k'')_{\omega_0}]\frac{\partial^2 E}{\partial \tau^2} + |E|^2 E - \mathrm{i}\alpha E = 0 \qquad (8.22)$$

First we shall consider the amplification of a periodic signal in the fiber with $k'' < 0$:

$$E(0, \tau) = a \sin(\pi\tau/T) \qquad (8.23)$$

Note that the initial signal amplitude is smaller than the threshold value for the modulational instability for a period of modulation T. During the signal amplification a train of solitary pulses is generated due to the joint action of self-phase modulation and negative GVD

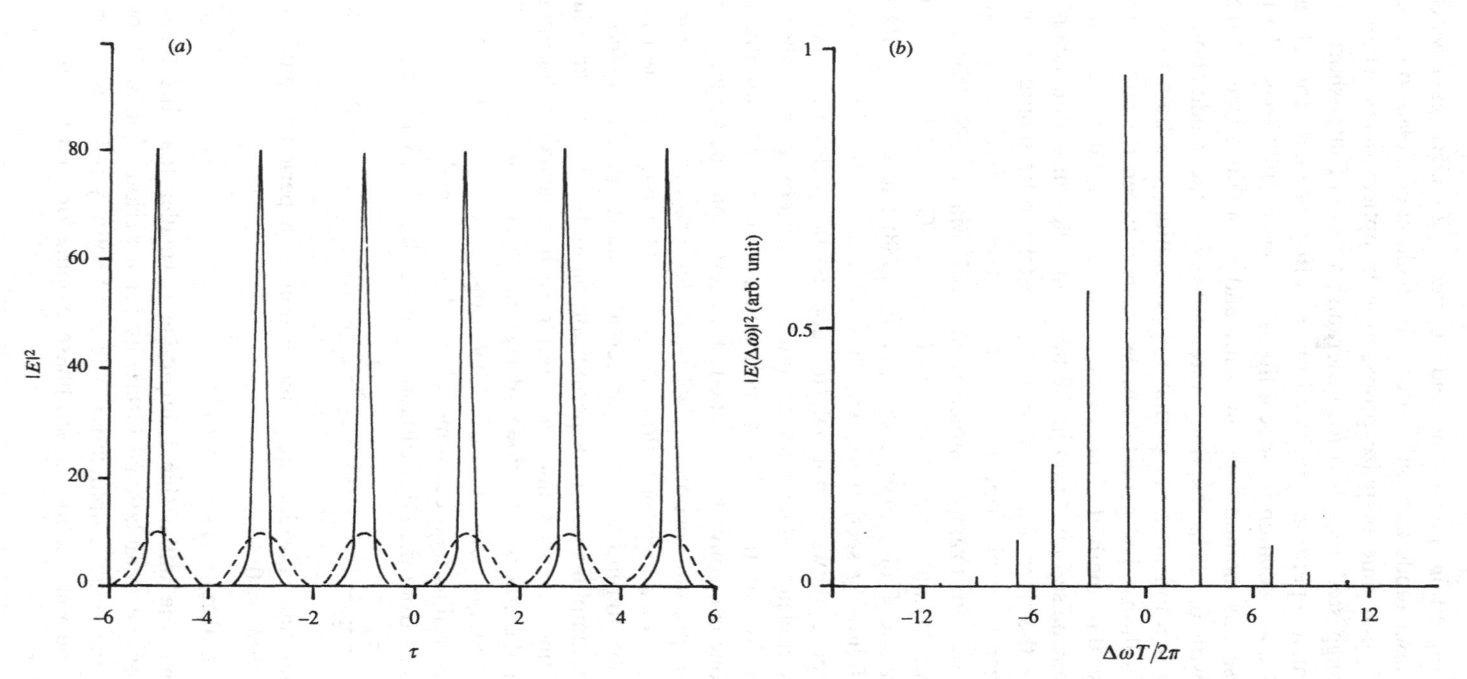

Figure 8.15. (*a*) A train of fundamental solitons formed by amplification of a signal (8.23). The parameters are: $a = 1$, $T = 2$, $\alpha = 0.4$, $\xi = 3.63$. The broken curve is the input signal (in the figure it is magnified by 10). (*b*) The spectrum of the generated soliton train.

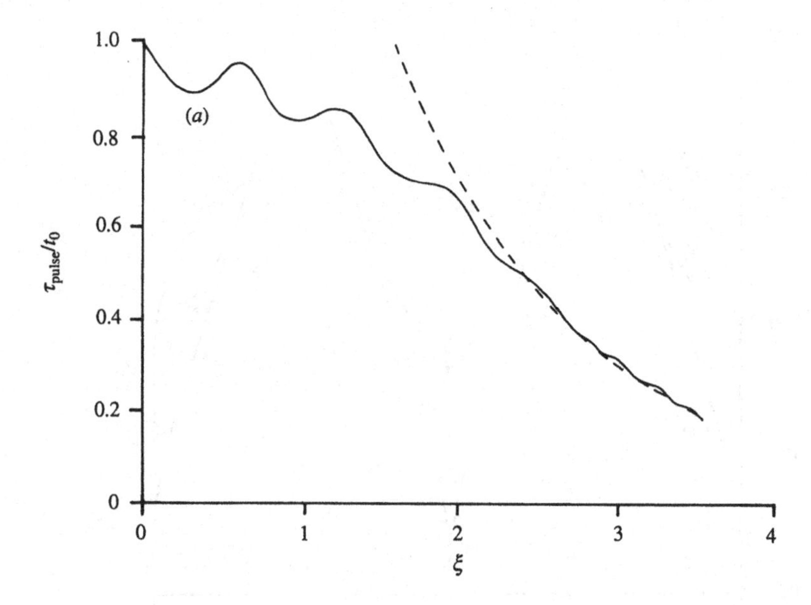

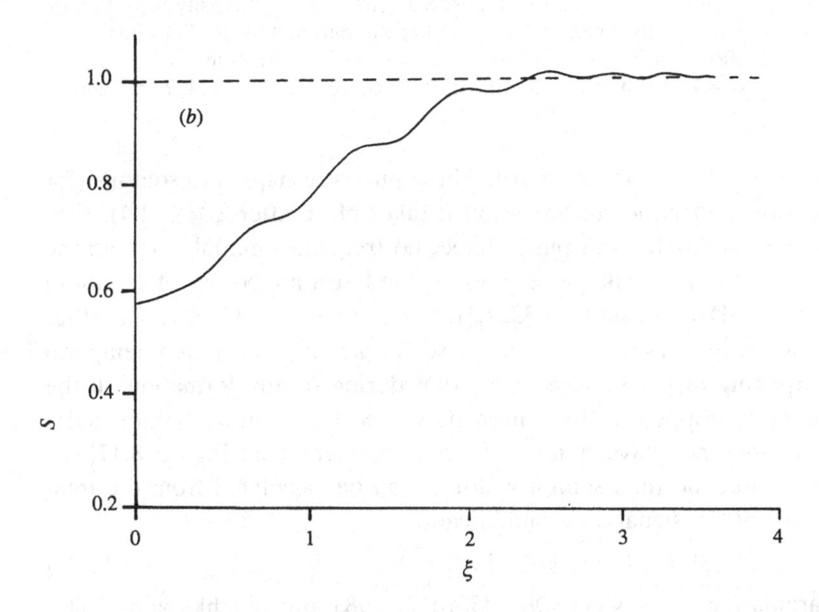

Figure 8.16. Evolution of (*a*) the pulsewidth and (*b*) the parameter $S = 0.322\tau_p^2\,|E|^2$ during the formation of the soliton train full curves, numerical simulations in the framework of equation (8.22), broken curve in (*a*), calculated from expression (8.24).

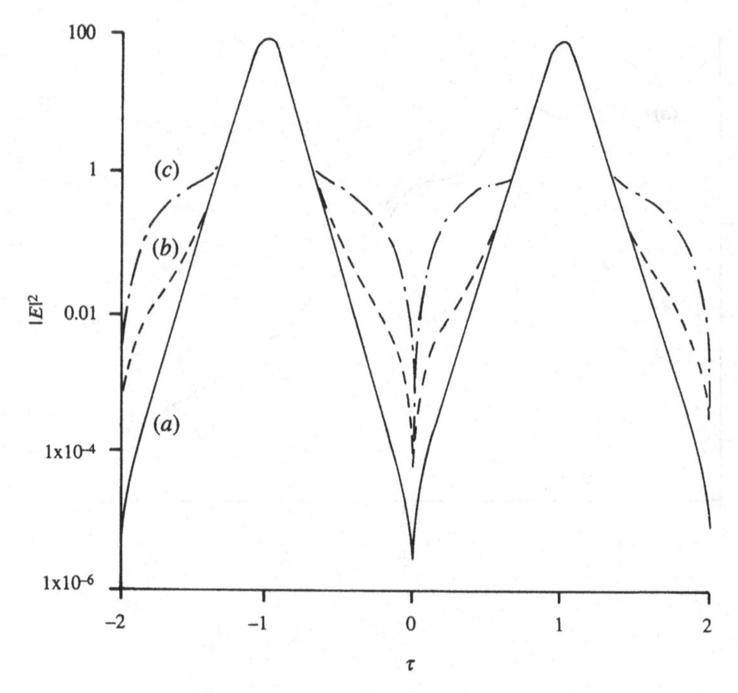

Figure 8.17. Similar to Figure 8.15 (*a*), but the intensity is plotted on a logarithmic scale. Full curve (*a*) the parameters are like those shown in Figure 8.15; broken curve (*b*) the parameters are: $a = 1.414$, $\alpha = 1.5$, $\xi = 0.73$; chain curve (*c*) $a = 1.414$, $\alpha = 3$ and $\xi = 0.36$.

effects (Figures 8.15 and 8.16). These pulses reshape into solitons (for the case under the consideration it takes place after $\xi > 2$–2.4): they have a hyperbolic sech pulse shape, no frequency modulation and the product of the peak pulse intensity and second power of its width τ_p(FWHM) is constant: $0.322\tau_p^2|E|^2 = 1$ (Figures 8.15–8.17). Further amplification results in an increase in the amplitude and temporal sharpening of the solitons. Note that during soliton formation all the energy is trapped in the soliton pulses, so the resulting soliton pulse train does not have a non-soliton component (see Figure 8.17(*a*)). Therefore, the final soliton width τ_p can be calculated from the total energy of the signal after amplification:

$$\tau_p(\xi) = 7.05 \exp(-2\alpha\xi)/(aT^2) \tag{8.24}$$

Karpman and Solov'ev, 1981; Gordon, 1983 and Mitchke and Mollenauer, (1987b) showed that solitons do not practically interact with each other in a fiber if the distance between them is ten (or more) times greater than their pulsewidth. To fulfil this condition a total energy gain in fiber ($\exp(2\alpha\xi) = 9$–18) is sufficient (Figures 8.15–8.17).

It should be noted that when the amplification is non-adiabatic the non-soliton component arises in the train of generated solitons (Figure 8.17). Our computer simulations show that the amplification regime is close to the adiabatic one up to $\alpha T^2 < 6$ (in Figure 8.17(b) $\alpha T^2 = 6$).

After the amplification is 'switched off' the solitons propagate along the fiber practically without interaction with each other, and they do not 'feel' the disappearance of any soliton pulses from the train.

We have chosen as an initial condition the sinusoidally amplitude modulated c.w. signal (8.23). For this signal the amplitude of the electric field has opposite signs for neighboring periods. This condition is very important for the generation of a train of fundamental solitons without a pedestal.

In the case of varying amplification along the fiber $\alpha(\xi)$ expression (8.24) must be rewritten as

$$\tau_{\mathrm{p}}(\xi) = 7.05 \exp\left[-2 \int_0^{\xi} \alpha(\xi')\mathrm{d}\xi'\right] / (aT^2) \tag{8.25}$$

We have estimated the parameters necessary for the realisation of this method. For the cases in Figure 8.17(*a*) and (*b*) respectively, the typical fiber parameters are: for the effective core cross-section area $A_{\mathrm{eff}} = 1.7 \times 10^{-7}\ \mathrm{cm}^2$, dispersion $D = -2\pi ck''/\lambda^2 = 15$ ps/nm.km at the wavelength $\lambda = 1.55\ \mu$m and for the modulation period $Tt_0 = 2$ ps the average initial signal power is 1.25 W (2.5 W) and the fiber length $L = 190$ m (38 m). For $Tt_0 = 20$ ps we have $P = 12.5$ mW (25 mW), $L = 19$ m (3.8 km) and for $Tt_0 = 100$ ps $P = 0.5$ mW (1 mW), $L = 470$ km (96 km). It is necessary to emphasise that, when the dispersion of the fiber decreases, the necessary length of the fiber increases and the initial power of the signal drops. For instance, using a fiber with dispersion $D = 1$ ps/nm.km for $Tt_0 = 2$ ps the length of the fiber increases up to 2.8 km (0.56 km), while the initial power drops to 80 mW (160 mW).

In our previous consideration we did not take into account the higher-order dispersive and RSS effects. As our calculations show (equation (8.6) with the amplification term $-\mathrm{i}\alpha E$), the influence of these effects is negligible for the previous examples with the modulation periods of 20 and 100 ps. For $Tt_0 = 2$ ps, the main features of the formation of the soliton train remain the same as, in the absence of additional effects, the influence of the RSS effect results only in a

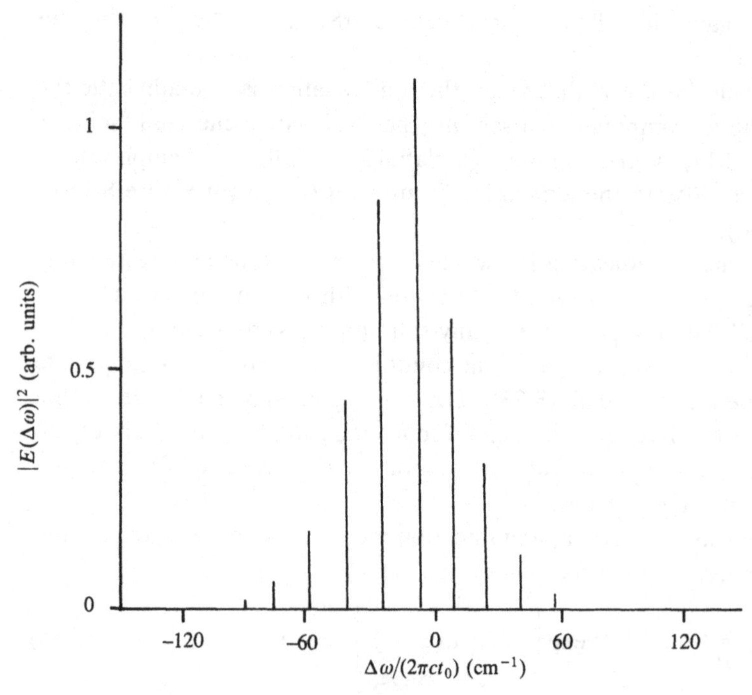

Figure 8.18. The soliton train spectrum shifted to the Stokes region on 18 cm^{-1} due to the RSS effect. The dimensionless parameters are those of Figure 8.15 and $Tt_0 = 2$ ps.

shift of the soliton train mean frequency to the Stokes region on 18 cm^{-1} (Figure 8.18) and 5 cm^{-1} for the cases shown in Figures 8.17(a) and (b) respectively.

Note that at the first stage of the pulse train formation the mean frequency practically does not shift. The main shift of the signal mean frequency takes place after the signal is shaped into a soliton train (for the case in Figures 8.15 and 8.16 it will be for $\xi > 2$). So we can obtain from (8.18) and (8.24) an analytical estimate for the soliton mean frequency shift:

$$\Delta\omega_{\text{mean}} t_0 = t_R(\alpha^2 T/4)^4 \{\exp(8\alpha\xi) - 1\}/(15\alpha) \qquad (8.26)$$

Estimates obtained from (8.26) practically coincide with the results of numerical simulations for the previously described parameters.

For amplification of a signal (8.23) in the spectral region of positive GVD of fibers a train of 'dark' solitons is formed (see Figure 8.19).

For an experimental realisation of these methods active fibers as well as Raman amplification in ordinary fibers can be used. As was

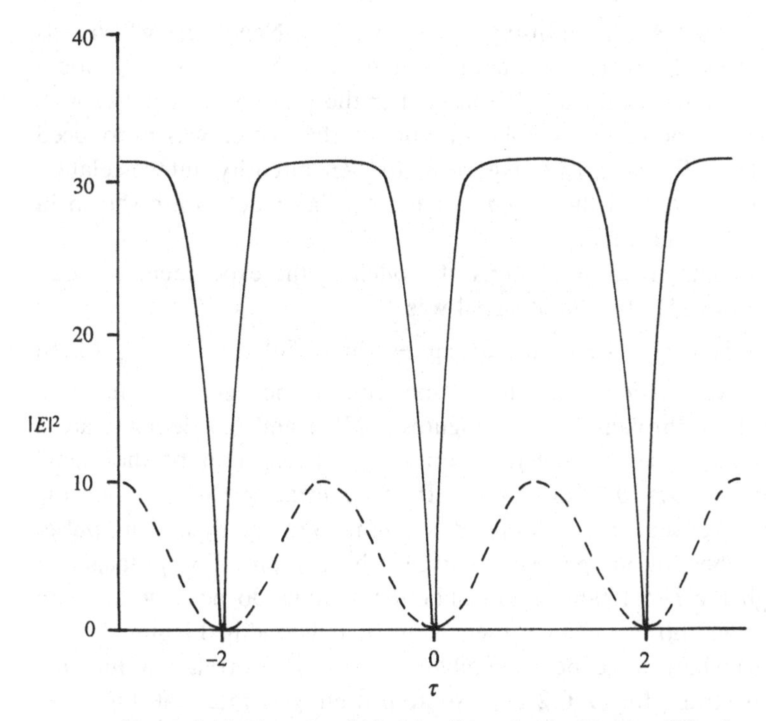

Figure 8.19. A train of 'dark' solitons formed by amplification of a signal (4.2) in the positive fiber GVD region. The parameters are: $a = 1$, $\alpha = 0.4$ and $\xi = 5$. Dotted curve, the input signal (in the figure it is magnified by 10).

shown in the previous section (see equations (8.12) and (8.13)), fibers with slowly decreasing (in absolute value) dispersion along the fiber axis (FSDD) allow us to amplify a signal effectively. So one can use FSDD without amplification for the realisation of the previously described methods. For a constant effective amplification coefficient Γ the dispersion dependence $\beta(\xi)$ must be:

$$\beta(\xi) = 1/(1 + 2\Gamma\xi) \tag{8.27}$$

For our experiments on the formation of a high-repetition-rate soliton train a special 1 km-long fiber was designed and produced (Bogatyrev *et al.*, 1989; Dianov *et al.*, 1990). For the first 0.9 km the dispersion of the fiber slowly decreased from 10 up to 0.83 ps/nm.km as

$$D(z)[\text{ps/nm.km}] = 10/(1 + 2 \cdot 6.12z[\text{km}])$$

The dispersion of the last 0.1 km fiber was constant at 0.83 ps/nm.km. The signal source was a synchronously pumped parametric oscillator synchronously pumped by second harmonic of cw

QS/ML Nd:YAG laser using a crystal of $Ba_2NaNb_5O_{15}$ which was pumped by the second harmonic output of a Q-switched and mode-locked c.w. pumped Nd:YAG laser. For the generation of pulses with sinusoidal modulation a Fabry–Perot interferometer was introduced into the oscillator cavity (Figure 8.20). An intensity autocorrelation function $G_2(t)$ and the spectra of the oscillator pulses are shown in Figure 8.21 (full curves).

In our numerical simulations of modelling the experiment we used equation (8.15). The input signal was

$$E(0, t) = 1.3\cos(\pi t/2)\exp(-t^2 \ln(2)/50) \tag{8.28}$$

The autocorrelation function and spectrum of the signal are shown in Figure 8.21 (broken curves). Figures 8.22(*a*) and (*b*) demonstrate a soliton train and its spectrum after the propagation of the signal through the first 0.9 km of fiber. In the central part of the train the distance between the solitons is ten times greater than their pulsewidth. There is no pedestal between the solitons. For propagation through the last 100 m of the fiber the solitons do not interact with each other, do not change their form or pulsewidth. Figures 8.23(*a*) and (*b*) show an experimentally obtained autocorrelation function and spectrum for a 0.2 THz soliton train generated at the fiber output. Note the good agreement between experiment and theory.

For the experiment described the influence of the Raman self-scattering and third-order dispersion effects is of no significance.

8.5 Generation of a stable high-repetition-rate train of practically non-interacting solitons from a c.w. signal by using the RSS and induced modulational instability effects

The idea of this method is similar to that for the generation of femtosecond solitons under the RSS of multi-soliton pulses, which was considered in Section 8.2.

First we shall consider equation (8.5) without the last term. At the fiber input we have a c.w. signal with a weak sinusoidal modulation (Figure 8.24):

$$E(0, \tau) = a[1 + h\cos(2\pi\tau/T)] \tag{8.29}$$

Due to the modulational instability effect in the vicinity of the focus point of the fiber a train of short pulses against the background of a pedestal is formed (Hasegawa, 1984) (Figures 8.24(*a*) and (*b*)). The further dynamics of the signal propagation differs considerably from the case of modulational instability dynamics in the framework of the

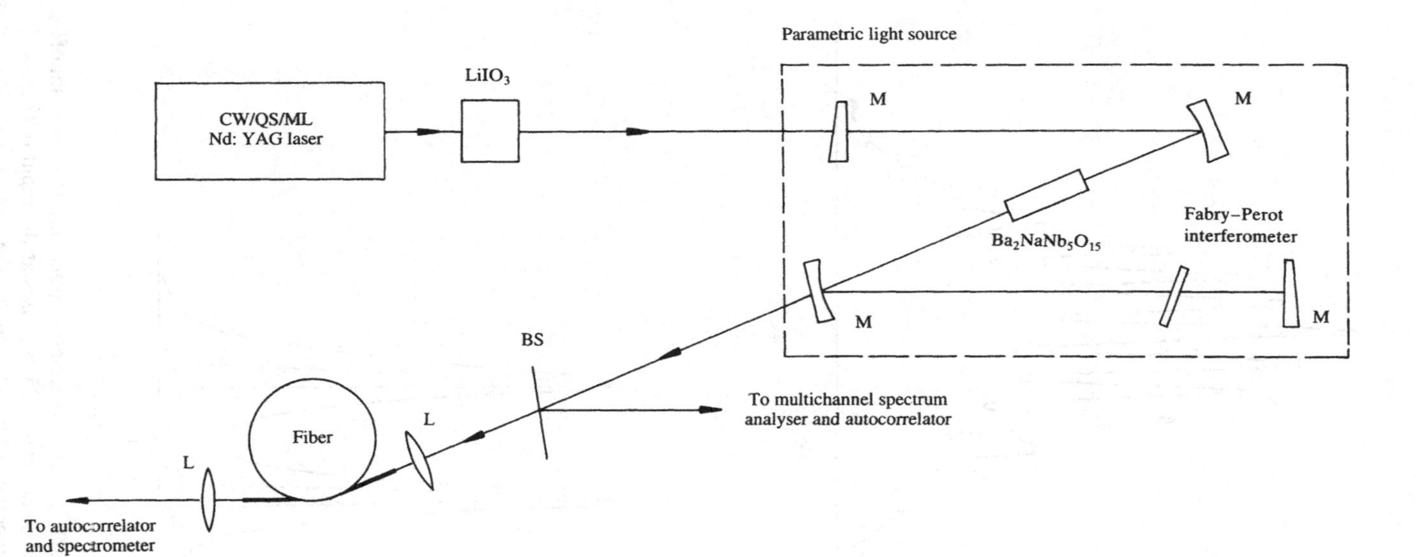

Figure 8.20. Experimental arrangement for the generation of the high-repetition-rate train of solitons. $LiIO_3$, second harmonic crystal, M, mirrors; BS, beam splitter, L, lenses.

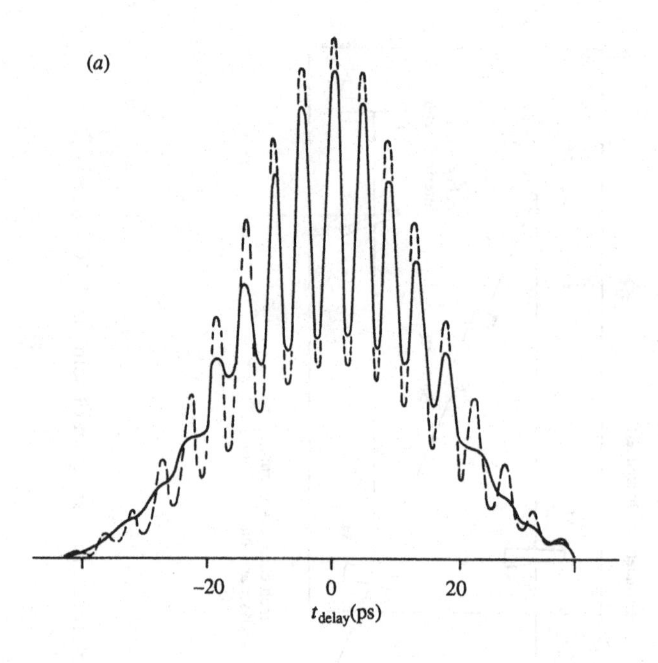

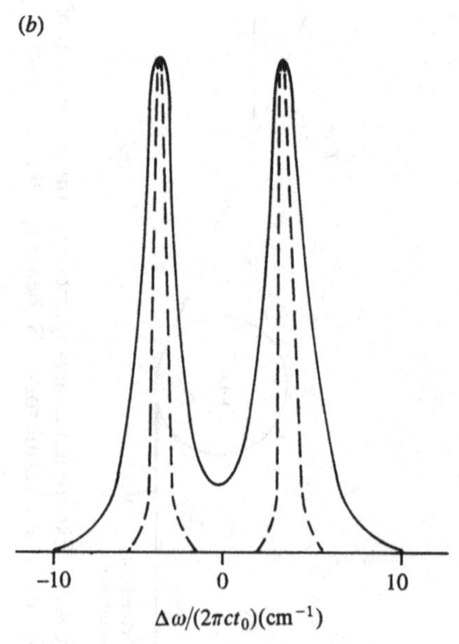

Figure 8.21. (*a*) Autocorrelation trace and (*b*) spectrum of sinusoidally-modulated 25 ps pulses at the input of the FSDD. Full curves, experiment; broken curves, the input signal, which was used in numerical simulations.

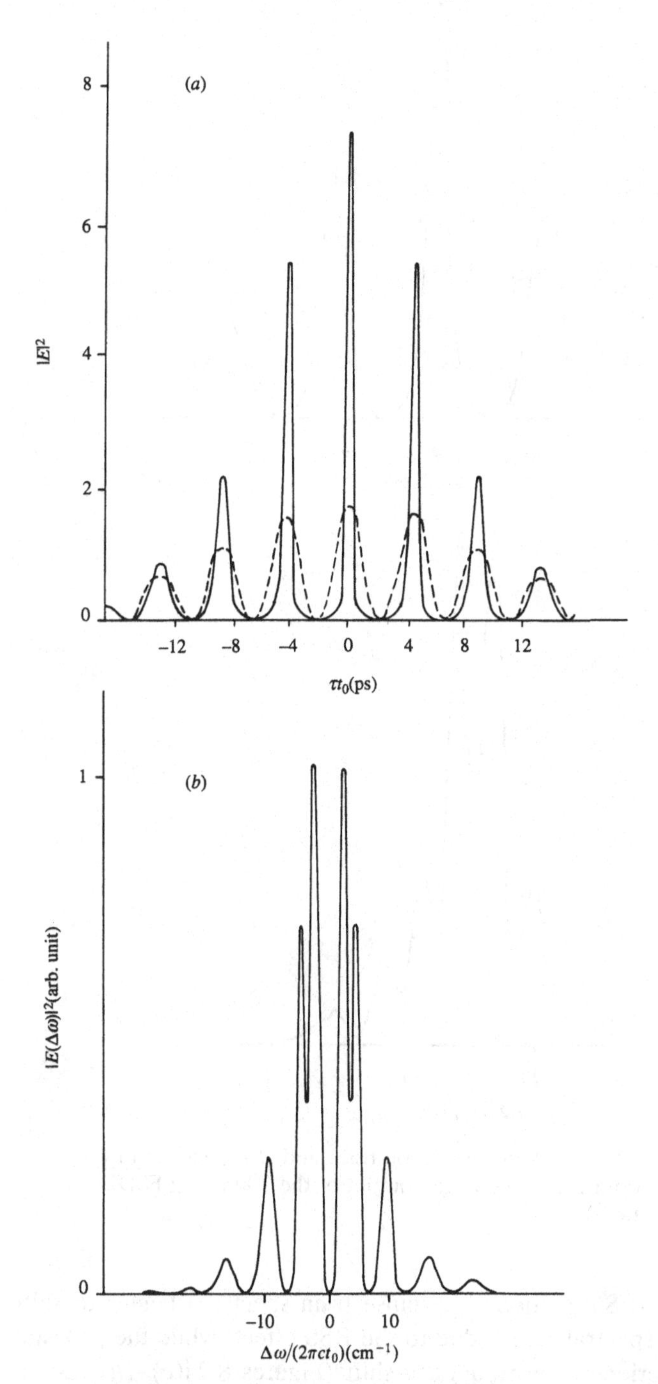

Figure 8.22. (*a*) Soliton train and its (*b*) spectrum generated at the output of the FSDD (theory). Broken curve, the input signal.

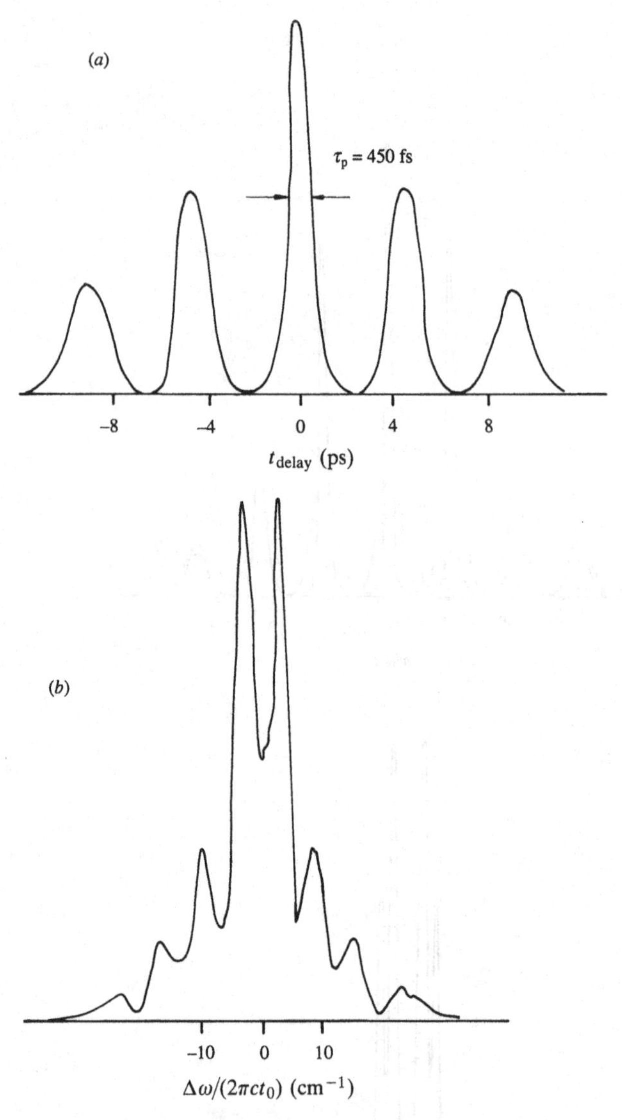

Figure 8.23. (*a*) Autocorrelation trace and (*b*) spectrum of the soliton train generated at the output of the 1-km long FSDD (experiment).

conventional NLS equation. The pulse train spectrum begins to shift to the Stokes spectral region due to the RSS effect, while the pedestal does not experience any frequency shift (Figures 8.24(*c*)–(*h*)). As a result the train of pulses does not decay (as it would be in the absence of the RSS effect (see Hasegawa, 1984)). During further

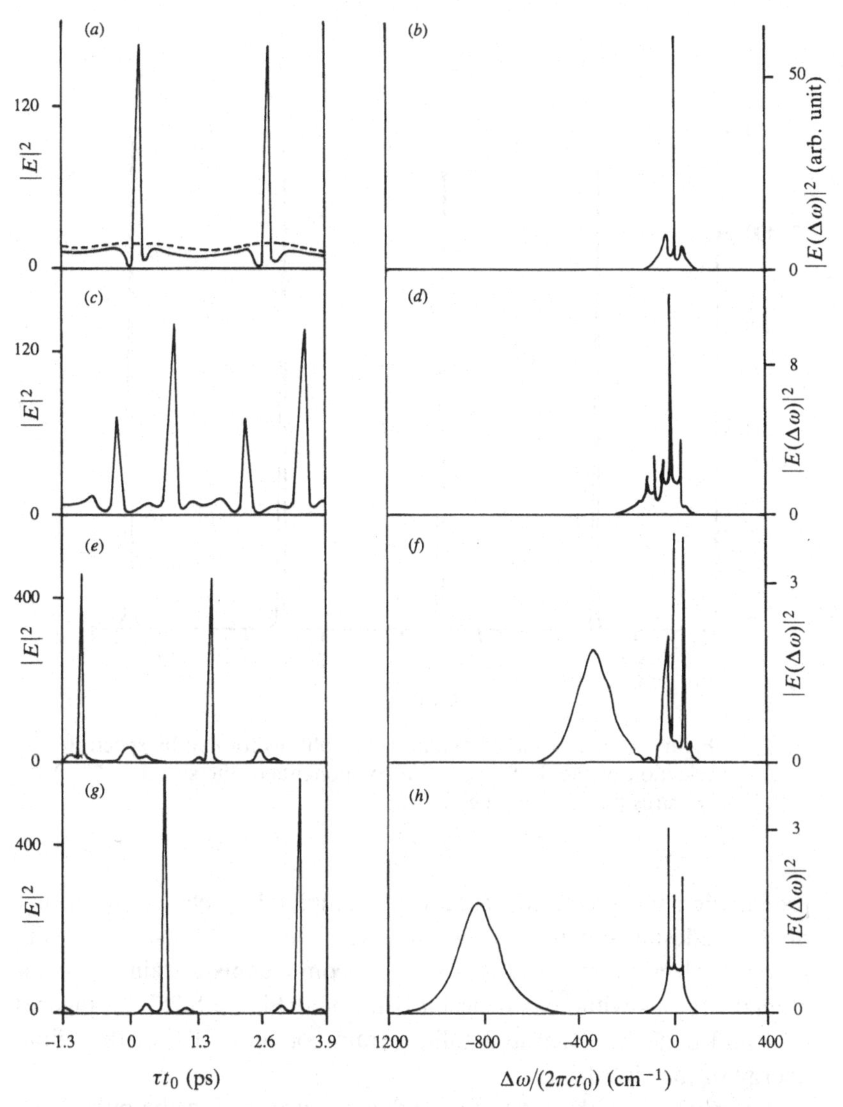

Figure 8.24. Dynamics of the generation of a soliton train from a c.w. signal by the induced modulational instability and the RSS effects. The parameters are: $a = 4$, $h = 0.05$, $T = 4$, $t_0 = 0.65$ ps. The temporal and spectral evolution are shown (a)–(h): broken curve, the input signal, (a) and (b), the 'focus' point, $\xi = 0.41$; (c) and (d), $\xi = 0.58$; (e) and $(f) = \xi = 0.825$; (g) and (h), $\xi = 0.97$. The discrete spectral substructure is not shown. After $\xi = 0.825$ the soliton spectrum is separated from the pedestal spectrum. The soliton train at the Stokes frequency contains 62% of the total energy at $\xi = 0.825$, and 73% at $\xi = 0.97$, the pedestal at pump frequency contains the rest of the signal energy. With further propagation the soliton train energy practically does not increase.

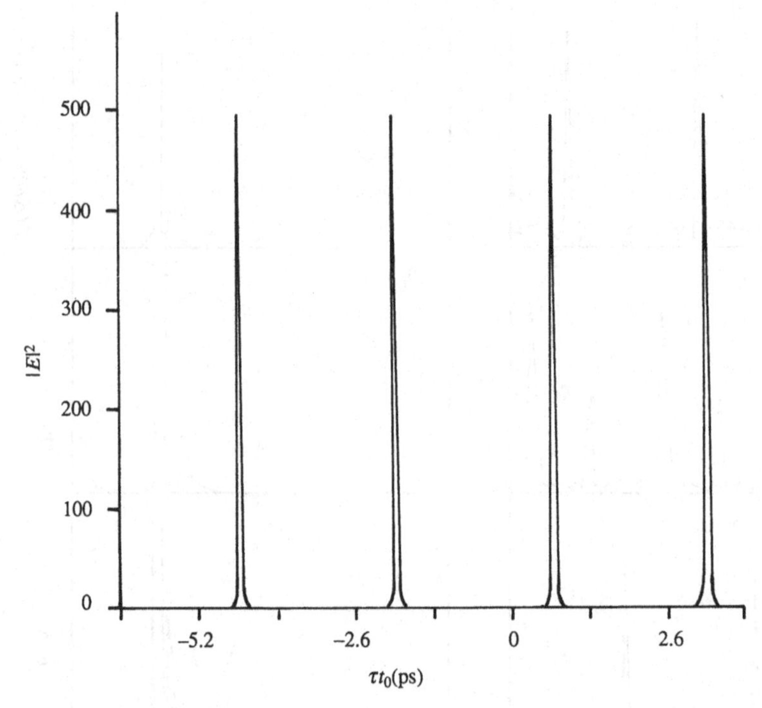

Figure 8.25. A train of fundamental solitons formed by spectral selection of the Stokes spectral components of the signal at $\xi = 0.97$ (Figures 8.24 (*g*) and (*h*)).

propagation the spectrum of the train completely separates from that of the pedestal and the pulses are formed into solitons. Under the spectral selection of the Stokes spectral components a train of fundamental solitons without a pedestal is formed (Figure 8.25). In Figures 8.24 and 8.25 the generated soliton train contains 73% of the initial energy of the signal.

Note that when shifted to the Stokes spectral region the pulse train moves across the pedestal due to dispersion, and the train can be Raman-amplified in the field of the pedestal. By choosing appropriate parameters it is possible to obtain practically 100% energy conversion of the initial signal into the soliton train energy (Figure 8.26).

The Raman amplification of the train in the field of the pedestal takes place only while the soliton spectral shift due to the RSS effect is less than 500 cm^{-1}, that is until the spectral difference between the soliton-train mean frequency and the pedestal mean frequency falls within the Raman gain curve (see Figure 8.3). The third-order disper-

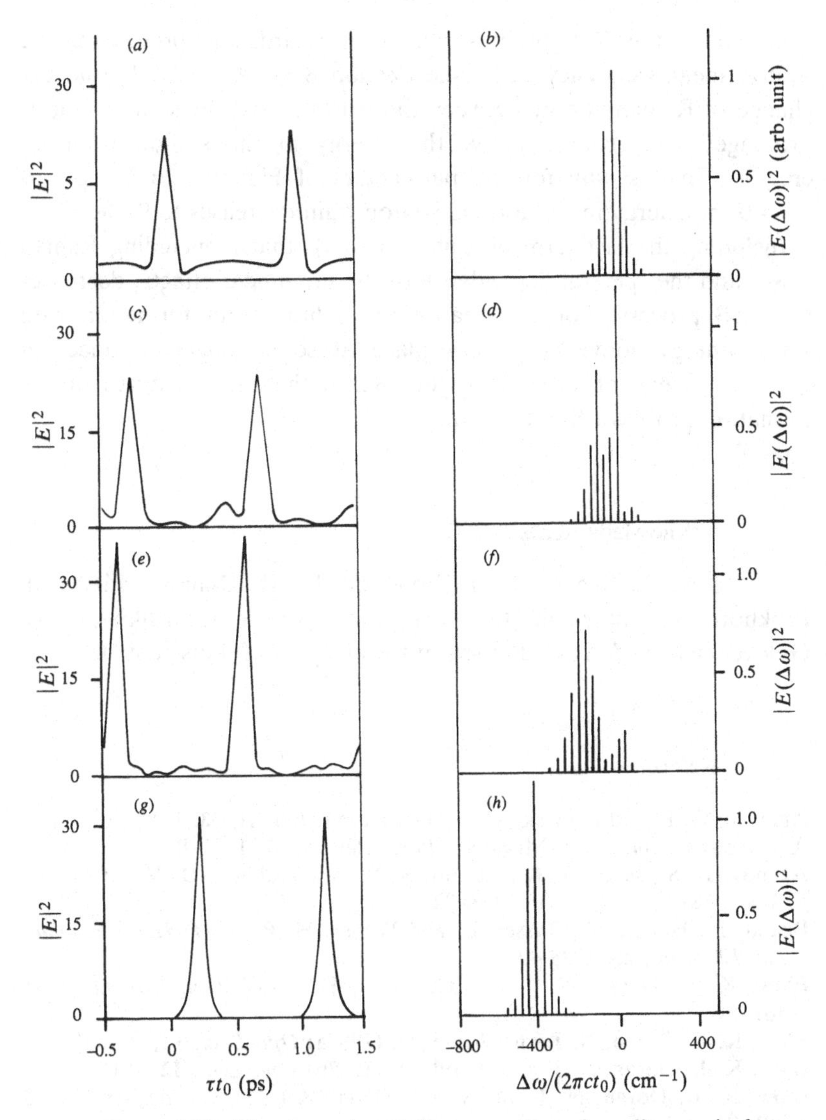

Figure 8.26. The dynamics of the generation of a soliton train from a c.w. signal (8.29) by the induced modulational instability and the RSS effects with 99.9% energy conversion from the input signal energy into the soliton train energy. The parameters are: $a = 2$, $h = 0.05$, $T = 4$, $t_0 = 0.25$ ps, $k''' = -k''/(1 \times 10^{14}\ \mathrm{s}^{-1})$ (the mean pump frequency is separated from the zero-dispersion wavelength by 1×10^{14} rad s^{-1} or 550 cm^{-1}). The temporal and spectral evolution are shown: (*a*) and (*b*), $\xi = 1$; (*c*) and (*d*), $\xi = 2$; (*e*) and (*f*), $\xi = 3$; (*g*) and (*h*), $\xi = 6$.

sion retards the RSS process, that is it retards the process of the soliton mean-frequency shift (see Section 8.3). As a result, the exchange of Raman energy between the pedestal and the soliton train is prolonged and, consequently, the energy in the soliton train increases. That is why for the parameters of Figure 8.26 except for $k''' = 0$ the energy in the formed soliton train decreases to 97%.

Including the last term of equation (8.5), that is including Raman losses and the spectral dependence of the non-linear effects, decreases the RSS process. For this reason the soliton train formation with 100% energy conversion takes place at some longer distance (at $\xi > 7$–8 instead of $\xi > 6$ in Figure 8.26), the main features of the formation process being the same.

Acknowledgments

I would like to thank Professors E. M. Dianov and A. M. Prokhorov for supporting the work, and Drs S. V. Chernikov, E. A. Golovchenko and A. N. Pilipetskii for many fruitful discussions.

References

Agrawal, G. P. and Potasek, M. J. (1986) *Phys. Rev. A,* **33**, 1715–76.

Anderson, D. and Lisak, M. (1983) *Phys. Rev. A,* **27**, 1393–8.

Azimov, B. S., Isaev, S. K., Luzgin, S. N. and Trukhov, D. V. (1986) *Izv. Akad. Nauk, ser. fiz.,* **50**, 2268–72.

Beaud, P., Hodel, W., Zysset, B. and Weber, H. P. (1987) *IEEE J. Quantum Electron.,* **23**, 1938–46.

Blow, K. J., Doran, N. J. and Cummins, E. (1983) *Opt. Commun.,* **48**, 181–5.

Blow, K. J., Doran, N. J. and Wood, D. (1987a) *Opt. Lett.,* **12**, 202–04.

Blow, K. J., Doran, N. J. and Wood, D. (1987b) *Opt. Lett.,* **12**, 1011–13.

Blow, K. J., Doran, N. J. and Wood, D. (1988a) *J. Opt. Soc. Am. B.,* **5**, 381–91.

Blow, K. J., Doran, N. J. and Wood, D. (1988b) *J. Opt. Soc. Am. B.,* **5**, 1301–04.

Bourkoff, E., Zhao, W., Joseph, R. I. and Christodoulides, D. N. (1987) *Opt. Commun.,* **62**, 284–8.

Bogatyrev, V. A., Bubnov, M. M, Dianov, E. M., Kurkov, A. S., Mamyshev, P. V., Miroshnichenko, S. I., Semeonov, S. L., Sysoliatin, A. A. and Chernikov, S. V. (1989) *Topical Meeting on Non-linear Guided-Wave Phenomena: Physics and Applications, 2–4 February 1989, Houston, TX, USA. Technical Digest Series, volume 2* (OSA: Washington, DC), Postdeadline papers, paper PD-9.

DeMartini, F., Towns, C. H., Gustafson, T. K. and Kelley, P. L. (1967) *Phys. Rev.*, **164**, 312–23.

Dianov, E. M., Grudinin, A. B., Khaidarov, D. V., Korobkin, D. V., Prokhorov, A. M. and Serkin, V. N. (1989) *Fiber and Integrated Optics*, **8**, 61–9.

Dianov, E. M., Ivanov, L. M., Mamyshev, P. V. and Prokhorov, A. M. (1989a) *Topical Meeting on Non-linear Guided-Wave Phenomena: Physics and Applications, 2–4 February 1989, Houston, TX, USA. Technical Digest Series, volume 2*, paper FA-5, pp. 157–60. (OSA: Washington, DC).

Dianov, E. M., Ivanov, L. M., Mamyshev, P. V. and Prokhorov, A. M. (1989b) *IEEE J. Quantum Electron.*, **25**, 828–35.

Dianov, E. M., Karasik, A. Ya., Mamyshev, P. V., Onischukov, G. I., Prokhorov, A. M., Stel'makh, M. F. and Fomichev, A. A. (1984) *Pis'ma Zh. Eksp. Theor. Fiz.*, **40**, 148–50. (*JETP. Lett.*, **40**, 903–06.)

Dianov, E. M., Karasik, A. Ya., Mamyshev, P. V., Prokhorov, A. M., Serkin, V. N., Stel'makh, M. F. and Fomichev, A. A. (1985) *Pis'ma Zh. Eksp. Teor. Fiz.*, **41**, 242–4. (*JETP. Lett.*, **41**, 294–7.)

Technical Digest of OFC-86, Atlanta, USA, 1986, paper WG7, pp. 106–07 (OSA/IEEE: New York).

Dianov, E. M., Kurkov, A. S., Mamyshev, P. V., Prokhorov, A. M., Sysoliatin, A. A., Gurianov, A. N., Devyatykh, G. G. and Miroshnichenko, S. I. (1990) Single-mode fiber with chromatic dispersion varying along the length. Accepted for the Optical Fiber Communication conference OFC-90, San Francisco, Paper WM-7.

Dianov, E. M., Mamyshev, P. V., Prokhorov, A. M. and Chernikov, S. V. (1989) *Opt. Lett.*, **14**, 1008–10.

Friberg, S. R., Weiner, A. M., Silberberg, Y., Sfez, B. G. and Smith, P. W. (1988) *Opt. Lett.*, **13**, 904–06.

Golovchenko, E. A., Dianov, E. M., Prokhorov, A. M. and Serkin, V. N. (1986) *Dokl. Akad. Nauk,* **288**, 851–6.

Golovchenko, E. A., Dianov, E. M., Karasik, A. Ya., Mamyshev, P. V., Pilipetskii, A. N. and Prokorov, A. M. (1989) *Kvantovaya Elektronika (Moscow)*, **16**, 592–4 (*Sov. J. Quantum Electron.*, **19** (3.)

Golovchenko, E. A., Mamyshev, P. V., Pilipetskii, A. N. and Dianov, E. M. (1990) *IEEE J. Quant. Electr.* **26**, 1815–20.

Gordon, J. P. (1983) *Opt. Lett.*, **8**, 596–8.

Gordon, J. P. (1986) *Opt. Lett.*, **11**, 662–4.

Gouveia-Neto, A. S., Faldon, M. E. and J. R. Taylor (1988) *Opt. Lett.*, **19**, 770–2.

Gouveia-Neto, A. S., Gomes, A. S. L. and J. R. Taylor (1989) *IEEE J. Quantum Electron.*, **24**, 332–40.

Gouveia-Neto, A. S., Gomes, A. S. L. and J. R. Taylor (1989) *Opt. Lett.*, **14**, 514–16.

Grudinin, A. B., Dianov, E. M., Korobkin, D. V., Prokhorov, A. M., Serkin, V. N. and Khaidarov, D. V. (1987) *Pis'ma Zh. Eksp. Teor. Fiz.*, **45**, 211. (*JETP. Lett.*, **50**, 260.)

Gusovskii, D. D., Dianov, E. M., Maier, A. A., Neustreuv, V. B., Osiko, V. V., Prokhorov, A. M., Sitarskii, K. Yu. and Scherbakov, I. A. (1987) *Kvantovaya Elektronika (Moscow)*, **14**, 1144–7. (*Sov. J. Quantum Electron.*, **17**, 724.)

Iwatsuki, K., Takada, A. and Saruwari, M. (1988) *XVI Int. Conf. on Quantum Electronics, IQEC-88, Tokyo, 1988*, Postdeadline papers, paper PD-14, pp. 34–5.
Islam, M. N., Sucha, G., Bar-Joseph, I., Wegener, M., Gordon, J. P. and Chemla, D. S. (1989) *Topical Meeting on Non-linear Guided-Wave Phenomena: Physics and Applications, 2–4 February 1989, Houston, TX, USA. Technical Digest Series, volume 2*, paper FA-3, pp. 149–152 (OSA: Washington, DC).
Hasegawa, A. (1984) *Opt. Lett.*, **9** 288–91.
Hasegawa, A. and Tappert, F. (1973) *Appl. Phys. Lett.*, **23**, 142–4.
Haus, H. A. and Nakazawa, M. (1987) *J. Opt. Soc. Am. B*, **4**, 652–60.
Hellwarth, R. W., Owyoung, A. and George, N. (1971) *Phys. Rev. A*, **4**, 2342.
Hodel, W. and Weber, H. P. (1987) *Opt. Lett.*, **12**, 924–6.
Karpman, V. I. and Solov'ev, V. V. (1981) *Physica D*, **3**, 487.
Kodama, Y. and Hasegawa, A. (1987) *IEEE J. Quantum Electron.*, **23**, 510–24.
Kuehl, H. H. (1988) *J. Opt. Soc. Am. B*, **5**, 709–13.
Lugovoi, V. N. (1976) *Zh. Eksp. Teor. Fiz.*, **71**, 1307–19. (*Sov. Phys. JETP*, **44**, 683.)
Maier, A. A. (1982) *Kvantovaya Elektronika (Moscow)*, **9**, 2296–302. (*Sov. J. Quantum Electron.*, **12**, 190.)
Mamyshev, P. V. and Chernikov, S. V. (1990) *Opt. Lett.*, **15**, 1076–8.
Manassah, J., Mustafa, M., Alfano, R. R. and Ho, P. (1986) *IEEE J. Quantum Electron.*, **22**, 197–204.
Mitchke, F. M. and Mollenauer, L. F. (1986) *Opt. Lett.*, **11**, 659–61.
Mitchke, F. M. and Mollenauer, L. F. (1987a) *Opt. Lett.*, **12**, 355–7.
Mitchke, F. M. and Mollenauer, L. F. (1987b) *Opt. Lett.*, **12**, 407–09.
Mollenauer, L. F., Gordon, J. P. and Islam, M. N. (1986) *IEEE J. Quantum Electron.*, **22**, 157–73.
Mollenauer, L. F. and Smith, K. (1988) *Opt. Lett.*, **13**, 675–7.
Mollenauer, L. F. and Stolen, R. H. (1984) *Opt. Lett.*, **9**, 13–15.
Mollenauer, L. F., Stolen, R. H. and Gordon, J. P. (1980) *Phys. Rev. Lett.*, **45**, 1095–98.
Mollenauer, L. F., Stolen, R. H., Gordon, J. P. and Tomlinson, W. J. (1983) *Opt. Lett.*, **8**, 289–91.
Nicolaus, B., Grischkowsky, D. and Balant, A. C. (1983) *Opt. Lett.*, **8**, 189–91.
Stolen, R. H. (1979) In S. E. Miller and A. G. Chynoweth (eds.) *Optical Fibre Telecommunications*, chap. 5, pp. 125–50 (Academic: New York).
Stolen, R. H., Botineau, J. and Ashkin, A. (1982) *Opt. Lett.*, **7**, 512–14.
Stolen, R. H., Gordon, J. P., Tomlinson, W. J. and Haus, H. A. (1989) *J. Opt. Soc. Am. B*, **6**, 1159–66.
Stolen, R. H., Lee, C. and Jain, R. K. (1984) *J. Opt. Soc. Am. B*, **1**, 652.
Stolen, R. H., Mollenauer, L. F. and Tomlinson, W. J. (1983) *Opt. Lett.*, **8**, 186–8.
Tai, K., Hasegawa, A. and Bekki, N. (1988) *Opt. Lett.*, **13**, 392–4.
Tai, K., Hasegawa, A. and Tomita, A. (1986) *Phys. Rev. Lett.*, **56**, 135–8.
Tai, K., Tomita, A., Jewell, J. L. and Hasegawa, A. (1986) *Appl. Phys. Lett.*, **49**, 236–8.

Tajima, K. (1987) *Opt. Lett.*, **12**, 54–6.
Tomlinson, W. J. and Stolen, R. H. (1989) *Seventh Int. Conf. on Integrated Optics and Optical Fiber Communication, 18–21 July 1989, Kobe, Japan. Technical digest, volume 4*, paper 21C4-3, pp. 72–3.
Tomlinson, W. J., Stolen, R. H., Hawkins, R. J. and Weiner, A. M. (1989) *Topical Meeting on Non-linear Guided-Wave Phenomena: Physics and Applications, 2–4 February 1989, Houston, TX, USA. Technical Digest Series, volume 2*, paper FA-5, pp. 132–5 (OSA: Washington, DC).
Tzoar, N. and Jain, M. (1981) *Phys. Rev. A*, **23**, 1266–70.
Vodop'yanov, K. L., Grudinin, A. B., Dianov, E. M., Kulevskii, L. A., Prokhorov, A. M. and Khaidarov, D. V. (1987) *Kvantovaya Elektronika (Moscow)*, **14**, 2053–5. (*Sov. J. Quantum Electron.*, **17**, 1311.)
Vysloukh, V. A. (1983) *Kvantovaya Elektronika (Moscow)*, **10**, 1688–90. (*Sov. J. Quantum Electron.*, **13**, 1113.)
Vysloukh, V. A., Matveev, A. N. and Petrova, I. Yu. (1989) Decay and interaction of femtosecond multisoliton pulses in Raman-active fibers, *Preprint*, Moscow State University, No. 14.
Vysloukh, V. A. and Serkin, V. N. (1983) *Pis'ma Zh. Eksp. Teor. Fiz.*, **38**, 170–2. (*JETP. Lett.*, **38**, 199.)
Wabnitz, S. and Trillo, S. (1989) *Topical Meeting on Non-linear Guided-Wave Phenomena: Physics and Applications, 2–4 February 1989, Houston, TX, USA. Technical Digest Series, volume 2*, paper FD-1, pp. 198–201 (OSA: Washington, DC).
Wai, P. K. A., Menyuk, C. R., Lee, Y. C. and Chen, H. H. (1986) *Opt. Lett.*, **11**, 464–6.
Wai, P. K. A., Menyuk, C. R., Chen, H. H. and Lee, Y. C. (1988) *IEEE J. Quantum Electron.*, **24**, 373–81.
Winful, H. G. (1986) *Opt. Lett.*, **11**, 33–5.
Zakharov, V. A. and Shabat, A. B. (1971) *Zh. Eksp. Teor. Fiz.*, **61** 118–34. (*Sov. Phys. JETP*, **34**, 62–9.)

9

Optical fiber solitons in the presence of higher-order dispersion and birefringence

CURTIS R. MENYUK AND PING-KONG A. WAI

9.1 Introduction

Since the original work of Mollenauer *et al.* (1980) announcing the observation of the optical fiber solitons earlier predicted by Hasegawa and Tappert (1973), these solitons have been the continued focus of experimental and theoretical attention. They have possible application to both long-haul communication (Hasegawa and Kodama, 1981; Mollenauer *et al.*, 1986) and switches (Doran and Wood, 1988; Trillo *et al.*, 1988). A salient feature of these solitons is their extraordinary robustness. From a mathematical standpoint, solitons are special solutions of equations which can be solved using nonlinear spectral transform methods (Ablowitz and Segur, 1981). This class of equations is itself quite special. While it includes the nonlinear Schrödinger (NLS) equation

$$\mathrm{i}\frac{\partial u}{\partial \xi} + \frac{1}{2}\frac{\partial^2 u}{\partial s^2} + |u|^2 u = 0 \tag{9.1}$$

which is often used to model fibers, it does not include the equations which result when higher order dispersion, birefringence, finite radial effects, the Raman effect, the Brillouin effect and linear attenuation are taken into account. All these effects play an important role in some situations; nonetheless, solitons often persist when these effects are important.

A clear distinction exists between effects like attenuation and the Raman effect on one hand and effects like higher-order dispersion and birefringence on the other. The former effects lead to a steady, continual change in the fiber parameters as the soliton propagates in the fiber. Attenuation leads to a decrease in the amplitude and an increase in the width, while the Raman effect lead to a decrease in the central frequency. By contrast, higher-order dispersion and birefringence lead to a small, finite change in the soliton parameters which quickly saturates as the soliton propagates. As a consequence, a deviation from the NLS equation must be quite large before it can be observed.

What is the source of this distinction? Higher-order dispersion and birefringence are autonomous, Hamiltonian deformations while attenuation and the Raman effect are not (Menyuk *et al.*, 1987). It appears to be universally the case that autonomous, Hamiltonian deformations of integrable field equations have little impact on solitons (Menyuk, 1986). From a physical standpoint, the soliton can be viewed as a non-linear eigenmode; autonomous, Hamiltonian perturbations alter but do not destroy the eigenmode, much as is the case when linear systems, such as solid state waveguides, are perturbed non-linearly (Burns and Milton, 1988).

This robustness has important practical consequences. In a soliton communication system, it is advantageous to operate as close to the zero dispersion point as possible. Solitons are created by a balance between non-linearity and dispersion. Lowering the dispersion reduces the required non-linearity and thus the power that an individual soliton must contain. In present-day experiments solitons are generated by use of color-center or Nd:YAG lasers. In a real system, however, these would have to be replaced by laser diodes, and lowering the power requirement is helpful. We (Wai *et al.*, 1986, 1987) have found that when light is injected precisely at the zero dispersion point, a large portion of the injected power ($\simeq$ 60%) down-shifts in frequency and forms a soliton; the rest up-shifts and disperses. Hence, solitons exist remarkably close to the zero dispersion point. These considerations determine the minimum power requirement for a soliton of a given pulse duration.

Real fibers always have some residual birefringence. For some applications, it is useful to make this birefringence large (Kaminow, 1981). The robustness of solitons suggests that they should still form in the presence of substantial birefringence, and we have shown that indeed that is the case (Menyuk, 1987a, 1988). Our results indicate that when a pulse consisting of both polarisations is injected into an optical fiber, each polarisation shifts its central frequency in response to the other polarisation just enough so that the two polarisations self-trap and move down the fiber at the same average velocity. From a physical standpoint, this self-trapping occurs when the non-linear scale length is less than the scale length for intermodal dispersion between the two birefringent modes.

The predicted soliton formation near the zero dispersion point has been confirmed experimentally by Gouveia-Neto *et al.* (1988), and the predicted soliton formation in birefringent fibers has been confirmed experimentally by Islam *et al.* (1989). Recent work, carried out at AT&T Bell Laboratories, indicates that the basic results which have been obtained for birefringent fibers continue to hold when the birefringence is randomly varying; one must however determine an effective birefringence, averaging the rapidly varying birefringence over the non-linear scale length.

Recently, the use of optical fibers in switches has been considered by a number of authors. Two particularly attractive configurations are the directional coupler studied by Trillo *et al.* (1988) and the interferometer studied by Shirasaki *et al.* (1988). In the latter case, a switching pulse is used to shift the phase of the signal pulse in one of the interferometer arms through the use of self-phase modulation. The signal pulse is in one polarisation, and the switching pulse is in the other. The advantage of using solitons is that they do not change their shape when interacting with other pulses. Thus, if the signal pulse is a soliton, it will, in principle, undergo no change in shape, merely changing its position and phase. A difficulty with this picture is that light propagating in linearly birefringent fibers satisfies equations which are not integrable. As a consequence, a soliton of one polarisation when interacting with a switching pulse of the other polarisation develops a shadow; a portion of the other polarisation becomes slaved to the soliton and travels with it even after interaction with the switching pulse. The shadow substantially changes the central frequency and thus the phase variation of the soliton. To avoid this variation, we have proposed using elliptically birefringent fibers whose birefringence angle is 35°. At this angle, the light pulses satisfy Manakov's equation which is integrable (Manakov, 1974), and the solitons will not form shadows.

In Section 9.2, we derive the basic equations which describe higher-order dispersion and birefringence. In Section 9.3, we discuss the origins of soliton robustness. In Section 9.4, we discuss higher order dispersion. In Section 9.5, we discuss birefringence, and in Section 9.6, we discuss the effect of random variations. Finally, in Section 9.7, we discuss switches.

9.2 Coupled non-linear Schrödinger equation

For simplicity, we will present here a derivation of the basic equations which assumes plane wave propagation. The method for taking into account the detailed geometry is described by Kodama (1985). Geometric effects do not change the basic structure of the equations, but merely alter their coefficients somewhat.

The starting point is Maxwell's wave equation which may be written for plane waves in the form

$$\frac{\partial^2 \boldsymbol{E}}{\partial z^2} - \frac{1}{c^2}\frac{\partial^2 \boldsymbol{D}}{\partial t^2} = 0 \tag{9.2}$$

where $\boldsymbol{E}$ is the electric field, $\boldsymbol{D}$ is the dielectric response, c is the speed of light, and z and t are propagation distance and time. Both $\boldsymbol{E}$ and $\boldsymbol{D}$ are the observed fields which are real, not complex.

Our first task is to relate $\boldsymbol{D}$ to $\boldsymbol{E}$. We write, as usual,

$$\boldsymbol{D} = \boldsymbol{E} + 4\pi\boldsymbol{P} \tag{9.3}$$

where $\boldsymbol{P}$ is the polarisability. We shall assume that the linear response of the medium is anisotropic, so that the medium is birefringent along the z-direction. Hence, considering only the linear response, $\boldsymbol{P}$ and $\boldsymbol{E}$ are related through a tensor $\mathbf{X}$, such that

$$\boldsymbol{P}(z, t) = \int_{-\infty}^{t} \mathbf{X}(t - t')\cdot\boldsymbol{E}(z, t')\mathrm{d}t' \tag{9.4}$$

The non-linear response will be treated separately. It is a consequence of causality that $\boldsymbol{P}$ at time t can only depend on $\boldsymbol{E}$ at earlier times.

We now consider the Fourier transforms of $\boldsymbol{E}$, $\boldsymbol{P}$ and $\mathbf{X}$. In general, given a quantity $X(z, t)$, we shall define the transform $\widetilde{X}(z, \omega)$ such that

$$\widetilde{X}(z, \omega) = \int_{-\infty}^{\infty} X(z, t)\exp(\mathrm{i}\omega t)\mathrm{d}t \tag{9.5}$$

from which it follows that

$$X(z, t) = \frac{1}{2\pi}\int_{-\infty}^{\infty} \widetilde{X}(z, \omega)\exp(-\mathrm{i}\omega t)\mathrm{d}\omega \tag{9.6}$$

We also define the quantities

$$\widetilde{X}^{+}(z, \omega) = \begin{cases} \widetilde{X}(z, \omega) & \omega > 0 \\ 0 & \omega < 0 \end{cases} \tag{9.7}$$

and $\widetilde{X}^{-}(z, \omega) = \widetilde{X}(z, \omega) - \widetilde{X}^{+}(z, \omega)$. The corresponding quantities $X^{+}(z, t)$ and $X^{-}(z, t)$ are then defined by equation (9.6). Although $X(z, t) = X^{+}(z, t) + X^{-}(z, t)$ is real X^{+} and X^{-} are individually complex and conjugate to each other.

In this paper, we will assume that the non-zero contribution to $\boldsymbol{E}$ and $\boldsymbol{D}$ comes from a small region in ω-space surrounding some carrier frequency ω_0 and another small region surrounding its opposite $-\omega_0$.

The reality of $E(z, t)$ implies $\tilde{E}(z, -\omega) = \tilde{E}^*(z, \omega)$. Hence, if $E(z, \omega)$ is non-zero near $\omega = \omega_0$, it must be non-zero near $\omega = -\omega_0$. This assumption is equivalent to the slowly varying envelope approximation.

Since equation (9.4) is in the form of an autocorrelation, it may be re-written

$$\tilde{\boldsymbol{P}}^+(z, \omega) = \tilde{\mathbf{X}}^+(\omega) \cdot \tilde{\boldsymbol{E}}^+(z, \omega) \tag{9.8}$$

At any frequency ω, $\tilde{\mathbf{X}}^+$ will have two orthonormal eigenvectors $\hat{\boldsymbol{e}}_1$ and $\hat{\boldsymbol{e}}_2$ which satisfy the relations

$$\hat{\boldsymbol{e}}_1 \cdot \hat{\boldsymbol{e}}_1^* = \hat{\boldsymbol{e}}_2 \cdot \hat{\boldsymbol{e}}_2^* = 1 \qquad \hat{\boldsymbol{e}}_1 \cdot \hat{\boldsymbol{e}}_2^* = 0 \tag{9.9}$$

Analogous quantities indicating the eigenvectors of $\tilde{\mathbf{X}}^-$ may be defined. Writing now,

$$\begin{aligned} \tilde{\boldsymbol{E}}^+ &= \tilde{E}_1^+ \hat{\boldsymbol{e}}_1 + \tilde{E}_2^+ \hat{\boldsymbol{e}}_2 \\ \tilde{\boldsymbol{P}}^+ &= \tilde{P}_1^+ \hat{\boldsymbol{e}}_1 + \tilde{P}_2^+ \hat{\boldsymbol{e}}_2 \end{aligned} \tag{9.10}$$

we obtain

$$\tilde{P}_1^+ = \chi_1 \tilde{E}_1^+ \qquad \tilde{P}_2 = \chi_2 \tilde{E}_2^+ \tag{9.11}$$

where χ_1 and χ_2 are the eigenvalues corresponding to $\hat{\boldsymbol{e}}_1$ and $\hat{\boldsymbol{e}}_2$. Specifying $\chi_1(\omega)$, $\chi_2(\omega)$, $\hat{\boldsymbol{e}}_1(\omega)$, and $\hat{\boldsymbol{e}}_2(\omega)$ is equivalent to specifying $\tilde{\mathbf{X}}(\omega)$. The linear dispersion relations corresponding to the eigenmodes are given by

$$\begin{aligned} k(\omega) &= \frac{\omega}{c}[1 + 4\pi\tilde{\chi}_1(\omega)]^{1/2} \\ l(\omega) &= \frac{\omega}{c}[1 + 4\pi\tilde{\chi}_2(\omega)]^{1/2} \end{aligned} \tag{9.12}$$

Equations (9.8) and (9.12) are general and do not depend on the assumption that $\boldsymbol{E}^+(\omega)$ is zero outside a small range surrounding $\omega = \omega_0$ frequency.

We now use this assumption, and we also suppose that within this frequency range we may set $\hat{\boldsymbol{e}}_1(\omega) = \hat{\boldsymbol{e}}_1(\omega_0)$ and $\hat{\boldsymbol{e}}_2(\omega) = \hat{\boldsymbol{e}}_2(\omega_0)$ which is equivalent to ignoring linear mode coupling. It follows from this latter assumption that

$$P_1^+(z, t) = \int_{-\infty}^{t} \chi_1(t - t') E_1^+(z, t') dt' \tag{9.13a}$$

$$P_2^+(z, t) = \int_{-\infty}^{t} \chi_2(t - t') E_2^+(z, t') dt' \tag{9.13b}$$

We now write

$$P_1^+(z, t) = \rho(z, t)\exp(ik_0z - i\omega_0t)$$
$$E_1^+(z, t) = U(z, t)\exp(ik_0z - i\omega_0t) \qquad (9.14)$$

where $k_0 = k(\omega_0)$ is determined from the dispersion relation, equation (9.12). Because the spectra of P_1^+ and E_1^+ are concentrated near $\omega = \omega_0$, the spectra of ρ and U are concentrated near $\omega = 0$. In effect, ρ and U are the envelopes of P_1^+ and E_1^+; the rapid variation at frequency ω_0 has been removed. One then finds

$$\rho(z, t) = \frac{1}{2\pi}\int_{-\infty}^{\infty}\tilde{\chi}_1(\omega + \omega_0)\tilde{U}(z, \omega)\exp(-i\omega t)d\omega \qquad (9.15)$$

Since $\tilde{U}$ is zero outside a small region surrounding $\omega = 0$, we may approximate $\tilde{\chi}_1$ by its Taylor expansion

$$\tilde{\chi}_1(\omega + \omega_0) \simeq \tilde{\chi}_1(\omega_0) + \chi_1'(\omega_0)\omega + \tfrac{1}{2}\chi_1''(\omega_0)\omega^2 + \tfrac{1}{6}\chi_1'''(\omega_0)\omega^3 \qquad (9.16)$$

where $\chi_1'(\omega_0) = d\chi_1/d\omega$, $\chi''(\omega_0) = d^2\chi_1/d\omega^2$, and $\chi'''(\omega_0) = d^3\chi_1/d\omega^3$, all evaluated at $\omega = \omega_0$. Substituting equation (9.16) into equation (9.15) and evaluating the Fourier transform, we obtain

$$\rho(z, t) = \tilde{\chi}_1U(z, t) + i\tilde{\chi}_1'\frac{\partial U(z, t)}{\partial t} - \frac{1}{2}\tilde{\chi}_1''\frac{\partial^2U(z, t)}{\partial t^2} - \frac{i}{6}\tilde{\chi}_1'''\frac{\partial^3U(z, t)}{\partial t^3} \qquad (9.17)$$

where $\tilde{\chi}_1$, $\tilde{\chi}_1'$, $\tilde{\chi}_1''$ and $\tilde{\chi}_1'''$ are all evaluated at $\omega = \omega_0$. If we let

$$D_1^+(z, t) = \Delta(z, t)\exp(ik_0z - i\omega_0t) \qquad (9.18)$$

we then find

$$\Delta(z, t) = \tilde{\varepsilon}_1'U(z, t) + i\tilde{\varepsilon}_1'\frac{\partial U(z, t)}{\partial t} - \frac{1}{2}\tilde{\varepsilon}_1''\frac{\partial^2U(z, t)}{\partial t^2} - \frac{i}{6}\tilde{\varepsilon}_1'''\frac{\partial^3U(z, t)}{\partial t^3} \qquad (9.19)$$

where

$$\tilde{\varepsilon}_1(\omega) = 1 + 4\pi\tilde{\chi}_1(\omega) \qquad (9.20)$$

and its derivatives are evaluated at $\omega = \omega_0$. We have thus determined D_1^+ in terms of E_1^+. We may similarly determine D_2^+ in terms of E_2^+. Noting that $D_1^- = D_1^{+*}$ and $D_2^- = D_2^{+*}$, we see that the linear portion of $\boldsymbol{D}$ is completely determined in terms of $\boldsymbol{E}$.

In equation (9.12), we have written the dispersion relation which corresponds to forward propagating waves. Since Maxwell's wave equation, equation (9.2), is second order in z, it will also have backward propagating solutions which correspond to choosing a negative sign in equation (9.12). It is, however, a consequence of our assumption that the frequency spectrum of E_1^+ is concentrated near $\omega = \omega_0$ that the wavenumber spectrum is concentrated near $k = k_0$ or $k = -k_0$. In other words, except for a brief transient, an optical pulse consists entirely of forward-going or backward-going waves. We shall assume that the optical pulse consists of forward-going waves with no loss in generality.

It is now possible to reduce Maxwell's wave equation so that it is only first order in z. We substitute E_1^+ and D_1^+ into Maxwell's equation and eliminate the term $\partial^2 U/\partial z^2$ in favor of terms containing only time derivatives by expanding Maxwell's equation in order of the number of derivatives. This expansion is permissible when the slowly varying assumption is valid (Kodama, 1985). We omit a detailed description of the lengthy algebra involved in this procedure; it yields

$$\mathrm{i}\frac{\partial U}{\partial z} + \mathrm{i}k'\frac{\partial U}{\partial t} - \frac{1}{2}k''\frac{\partial^2 U}{\partial t^2} - \frac{\mathrm{i}}{6}k'''\frac{\partial^3 U}{\partial t^3} = 0 \tag{9.21a}$$

$$\mathrm{i}\frac{\partial V}{\partial z} + \mathrm{i}l'\frac{\partial V}{\partial t} - \frac{1}{2}l''\frac{\partial^2 V}{\partial t^2} - \frac{\mathrm{i}}{6}l'''\frac{\partial^3 V}{\partial t^3} = 0 \tag{9.21b}$$

where k, l and their derivatives are given by equation (9.12) and, by analogy to equation (9.14),

$$V(z, t) = E_2^+(z, t)\exp(\mathrm{i}l_0 - \mathrm{i}\omega_0 t)$$

It is no accident that the coefficients of the time derivatives just involve derivatives of the dispersion relation. This result can be made apparent by using a Green function or Fourier–Laplace transform approach. The Fourier transform of Maxwell's wave equation for $\tilde{E}_1^+$ yields

$$\frac{\partial^2 \tilde{E}_1^+}{\partial z^2} + k^2(\omega)\tilde{E}_1^+ = 0 \tag{9.22}$$

If we write the Laplace transform in the form

$$\bar{E}_1^+(k, \omega) = \int_0^\infty \tilde{E}_1(z, \omega)\exp(-\mathrm{i}kz)\mathrm{d}z \tag{9.23}$$

where $\mathrm{Im}(k) < 0$, we find that equation (9.22) becomes

$$[k^2(\omega) - k^2]\bar{E}_1^+ = \tilde{E}_0' + \mathrm{i}k\tilde{E}_0 \tag{9.24}$$

where $\tilde{E}_0' = \partial\tilde{E}(z, \omega)/\partial z$ and $\tilde{E}_0 = E(z, \omega)$ are evaluated at $z = 0$. Demanding that our light pulse consist of only forward-going waves is equivalent to demanding that $E_0' = \mathrm{i}k(\omega)E_0$. In this case, equation (9.24) becomes

$$i[k - k(\omega)]\bar{E}_1^+ = \tilde{E}_0 \tag{9.25}$$

Expanding $k(\omega)$ in a Taylor series about the frequency $\omega = \omega_0$ yields

$$i[(k - k_0) - k'(\omega - \omega_0) - \tfrac{1}{2}k''(\omega - \omega_0)^2 \\ - \tfrac{1}{6}k'''(\omega - \omega_0)^3]\bar{E}_1^+ \simeq \tilde{E}_0 \tag{9.26}$$

Using the definition of $U(z, t)$, one may verify that equation (9.26) is just the Fourier–Laplace transform of equation (9.21a). This approach yields the linear wave equation more easily than the approach previously described where one directly eliminates $\partial^2 U/\partial z^2$ in the z-domain; however, this approach does not generalise in any simple way to non-linear problems, while the previous approach does.

We turn now to consideration of the non-linear contribution to the polarisabililty $\boldsymbol{P}$. We shall suppose that no second-order non-linearity appears, so that the lowest order non-linearity is third order. We shall also suppose that the medium is only weakly anisotropic, so that the non-linear response can be considered isotropic. Leaving aside second harmonic generation, which only appears in special circumstances, both these assumptions apply to optical fibers. The non-linear polarisability must have the form

$$\boldsymbol{P}(z, t) = \int_{-\infty}^{t} dt_1 \int_{-\infty}^{t} dt_2 \int_{-\infty}^{t} dt_3 \chi(t - t_1, t - t_2; t - t_3) \\ \cdot [\boldsymbol{E}(z, t_1)\cdot\boldsymbol{E}(z, t_2)]\boldsymbol{E}(z, t_3) \tag{9.27}$$

This combination of $\boldsymbol{E}$-vectors is the only combination which is invariant under rotations and mirror reflections. From the form of equation (9.27), it follows that $\chi(\tau_1, \tau_2; \tau_3)$ is invariant under the interchange $\tau_1 \leftrightarrow \tau_2$ but not necessarily under the interchanges $\tau_1 \leftrightarrow \tau_3$ and $\tau_2 \leftrightarrow \tau_3$. Since the spectrum of $\boldsymbol{E}$ is concentrated primarily in small spectral regions surrounding $\omega = \omega_0$ and $\omega = -\omega_0$, it follows that $\boldsymbol{P}$ will be concentrated primarily in spectral regions surrounding $\omega = -3\omega_0$, $-\omega_0$, ω_0 and $3\omega_0$. Waves in optical fibers which propagate at $\omega = \omega_0$ cannot propagate at $\omega = 3\omega_0$; hence, we can ignore the contributions of $\boldsymbol{P}$ at $\pm 3\omega_0$ to the electric field. Designating $\boldsymbol{P}^+$ as the contribution to $\boldsymbol{P}$ concentrated near $\omega = \omega_0$, we find that it consists of all combinations of the $\boldsymbol{E}$-field containing two positive and one negative contribution. It follows that

$$\boldsymbol{P}^+(z, t) = \int_{-\infty}^{t} dt_1 \int_{-\infty}^{t} dt_2 \int_{-\infty}^{t} dt_3 \chi(t - t_1, t - t_2; t - t_3) \\ \cdot \{2[\boldsymbol{E}^+(z, t_1)\cdot\boldsymbol{E}^-(z, t_2)]\boldsymbol{E}^+(z, t_3) + [\boldsymbol{E}^+(z, t_1)\cdot\boldsymbol{E}^+(z, t_2)]\boldsymbol{E}^-(z, t_3)\} \tag{9.28}$$

In the Fourier domain equation (9.28) becomes

$$\begin{aligned}\tilde{\boldsymbol{P}}^{+}(z,\omega) = \frac{1}{(2\pi)^2}\int_{-\infty}^{\infty}\mathrm{d}\omega_1\int_{-\infty}^{\infty}\mathrm{d}\omega_2\tilde{\chi}(\omega_1,\omega_2;\omega_3)\\ \cdot\{2[\tilde{\boldsymbol{E}}^{+}(z,\omega_1)\cdot\tilde{\boldsymbol{E}}^{-}(z,\omega_2)]\tilde{\boldsymbol{E}}^{+}(z,\omega_3)\\ +[\tilde{\boldsymbol{E}}^{+}(z,\omega_1)\cdot\tilde{\boldsymbol{E}}^{+}(z,\omega_2)]\tilde{\boldsymbol{E}}^{-}(z,\omega_3)\}\end{aligned}\tag{9.29}$$

where $\omega_3 = \omega - \omega_1 - \omega_2$. The first term in equation (9.29) is concentrated in the spectral region $\omega_1 = -\omega_2 = \omega_3 = \omega_0$. The second term in equation (9.29) is concentrated in the spectral region $\omega_1 = \omega_2 = -\omega_3 = \omega_0$. If we Taylor expand $\tilde{\chi}(\omega_1, \omega_2; \omega_3)$ just as when deriving the linear response, but retain only the lowest order contribution, we obtain

$$\begin{aligned}\tilde{\boldsymbol{P}}^{+}(z,\omega) = \frac{1}{(2\pi)^2}\int_{-\infty}^{\infty}\mathrm{d}\omega_1\int_{-\infty}^{\infty}\mathrm{d}\omega_2\{2a[\tilde{\boldsymbol{E}}^{+}(z,\omega_1)\\ \cdot\tilde{\boldsymbol{E}}^{-}(z,\omega_2)]\tilde{\boldsymbol{E}}^{+}(z,\omega_3)\\ +b[\tilde{\boldsymbol{E}}^{+}(z,\omega_1)\cdot\tilde{\boldsymbol{E}}^{+}(z,\omega_2)]\tilde{\boldsymbol{E}}^{-}(z,\omega_3)\}\end{aligned}\tag{9.30}$$

where $a = \tilde{\chi}(\omega_0, -\omega_0; \omega_0)$ and $b = \tilde{\chi}(\omega_0, \omega_0; -\omega_0)$. Neglecting higher-order contributions is equivalent to neglecting the contribution to the non-linear polarisability of terms which include the factors $\partial U/\partial t$, $\partial V/\partial t$, and higher-order envelope derivatives. This assumption is valid in optical fibers as long as the optical pulses are longer than several hundred femtoseconds. Returning to the time domain yields

$$\begin{aligned}\boldsymbol{P}^{+}(z,t) = 2a[\boldsymbol{E}^{+}(z,t)\cdot\boldsymbol{E}^{-}(z,t)]\boldsymbol{E}^{+}(z,t)\\ +b[\boldsymbol{E}^{+}(z,t)\cdot\boldsymbol{E}^{+}(z,t)]\boldsymbol{E}^{-}(z,t)\end{aligned}\tag{9.31}$$

It should be emphasised that in deriving equation (9.31), we have assumed that the field *envelopes* vary slowly compared to the dielectric response times, not the $\boldsymbol{E}$-field itself. When the dielectric response times are so fast that they may be regarded as instantaneous, i.e., they are much greater than ω_0^{-1}, then

$$\tilde{\chi}(\omega_0, -\omega_0; \omega_0) = \tilde{\chi}(\omega_0, \omega_0; -\omega_0) = \tilde{\chi}(0, 0; 0)\tag{9.32}$$

so that $a = b$ and the number of independent Kerr coefficients is reduced from two to one. To make this point explicit, we return to equation (9.28) and note that if $\chi(\tau_1, \tau_2; \tau_3) \to 0$ so rapidly that the variation of $\boldsymbol{E}$ can be neglected, then

$$\begin{aligned}\boldsymbol{P}^{+}(z,t) = \int_{-\infty}^{t}\mathrm{d}t_1\int_{-\infty}^{t}\mathrm{d}t_2\int_{-\infty}^{t}\mathrm{d}t_3\chi(t-t_1, t-t_2, t-t_3)\\ \{2[\boldsymbol{E}^{+}(z,t)\cdot\boldsymbol{E}^{-}(z,t)]\boldsymbol{E}^{+}(z,t)+[\boldsymbol{E}^{+}(z,t)\cdot\boldsymbol{E}^{+}(z,t)]\boldsymbol{E}^{-}(z,t)\}\end{aligned}\tag{9.33}$$

Noting that

$$\int_{-\infty}^{t} dt_1 \int_{-\infty}^{t} dt_2 \int_{-\infty}^{t} dt_3 \chi(t - t_1, t - t_2; t - t_3) = \chi(0, 0; 0) \tag{9.34}$$

and comparing equation (9.33) with equation (9.31), we arrive at equation (9.32). In optical fibers, the non-linear dielectric response can be viewed as instantaneous and one does find that $a = b$. The reduction in the number of independent Kerr coefficients from two to one when the non-linear dielectric response becomes instantaneous is implicit in the previous results of Maker and Terhune (1965).

Recalling that the unit vectors $\hat{e}_1 \equiv \hat{e}_1(\omega_0)$ and $\hat{e}_2 \equiv \hat{e}_2(\omega_0)$ define the eigenmodes, and using the orthogonality relations, equation (9.9), we find

$$\begin{aligned} P_1^+ &= 2a[E_1^+ E_1^- + E_2^+ E_2^-]E_1^+ + b[E_1^+ E_1^+(\hat{e}_1 \cdot \hat{e}_1) \\ &\quad + 2E_1^+ E_2^+(\hat{e}_1 \cdot \hat{e}_2) + E_2^+ E_2^+(\hat{e}_2 \cdot \hat{e}_2)][E_1^-(\hat{e}_1^* \cdot \hat{e}_1^*) \\ &\quad + E_2^-(\hat{e}_1^* \cdot \hat{e}_2^*)] \end{aligned} \tag{9.35}$$

The term in which a appears does not depend on the eigenmode structure, but the term in which b appears does. Hence, the strength of the non-linear mode coupling will depend on the eigenmode structure. In general, a Kerr medium may be elliptically birefringent. Choosing $\hat{e}_x$ along the major axis of the birefringence ellipse and $\hat{e}_y$ along the minor axis of the birefringence ellipse, with no loss of generality, we find that we may write

$$\hat{e}_1 = \frac{\hat{e}_x + ir\hat{e}_y}{(1 + r^2)^{1/2}} \qquad \hat{e}_2 = \frac{r\hat{e}_x - i\hat{e}_y}{(1 + r^2)^{1/2}} \tag{9.36}$$

where, letting $r = \tan(\theta/2)$, we find that $\hat{e}_1 \cdot \hat{e}_1 = \hat{e}_2 \cdot \hat{e}_2 = \cos\theta$ and $\hat{e}_1 \cdot \hat{e}_2 = \sin\theta$. A linearly birefringent fiber corresponds to $r = 0$ and $\theta = 0$; while a circularly birefringent fiber corresponds to $r = 1$ and $\theta = \pi/2$. In the former case one finds $\hat{e}_1 = \hat{e}_x$ and $\hat{e}_2 = -i\hat{e}_y$. This choice of eigenvectors differs from the usual choice, $\hat{e}_1 = \hat{e}_x$ and $\hat{e}_2 = \hat{e}_y$, but this difference leads to no change in the evolution equations. In the latter case, one finds $\hat{e}_1 = (\hat{e}_x + i\hat{e}_y)/\sqrt{2}$ and $\hat{e}_2 = (\hat{e}_x - i\hat{e}_y)/\sqrt{2}$ which is standard. Noting that $E_1^- = E_1^{+*}$ and $E_2^- = E_2^{+*}$, equation (9.35) now becomes

$$\begin{aligned} P_1^+ &= (2a + b\cos^2\theta)|E_1^+|^2 E_1^+ + (2a + 2b\sin^2\theta)|E_2^+|^2 E_1^+ \\ &\quad + b\cos\theta\sin\theta(E_1^+)^2 E_2^- + 2b\cos\theta\sin\theta|E_1^+|^2 E_2^+ \\ &\quad + b\cos^2\theta(E_2^+)^2 E_1^- + b\cos\theta\sin\theta|E_2^+|^2 E_2^+. \end{aligned} \tag{9.37}$$

Using the definitions for the wave envelopes, we conclude

$$\begin{aligned}\rho = {} & (2a + b\cos^2\theta)|U|^2U + (2a + 2b\sin^2\theta)|V|^2U \\ & + b\cos^2\theta V^2U^*\exp[-2i(k_0 - l_0)z] \\ & + b\cos\theta\sin\theta\{U^2V^*\exp[i(k_0 - l_0)z] \\ & + (2|U|^2 + |V|^2)V\exp[-i(k_0 - l_0)z]\}.\end{aligned} \tag{9.38}$$

In high birefringence fibers all but the first two terms in equation (9.38) are rapidly oscillating and will be dropped henceforth (Menyuk, 1987b).

We now combine equation (9.38) which gives the non-linear polarisability with equation (9.17) which gives the linear polarisability and substitute the result into Maxwell's wave equation. We assume that the non-linear contribution is of the same order as the dispersive contribution because solitons are obtained when these two contributions balance. Substituting the total polarisability into Maxwell's wave equation and reducing the equation so that it is first order in z, just as in the strictly linear case, we obtain

$$\begin{aligned} & i\frac{\partial U}{\partial z} + i\Gamma U + ik'\frac{\partial U}{\partial t} - \frac{1}{2}k''\frac{\partial^2 U}{\partial t^2} - \frac{i}{6}k'''\frac{\partial^3 U}{\partial t^3} \\ & \quad + (2a' + b'\cos^2\theta)|U|^2U \\ & \quad + (2a' + 2b'\sin^2\theta)|V|^2U = 0\end{aligned} \tag{9.39}$$

where $a' = a/2k_0$ and $b' = b/2k_0$. In a similar fashion it can be shown that

$$\begin{aligned} & i\frac{\partial V}{\partial z} + i\Gamma V + il'\frac{\partial V}{\partial t} - \frac{1}{2}l''\frac{\partial^2 V}{\partial t^2} - \frac{i}{6}l'''\frac{\partial^3 V}{\partial t^3} \\ & \quad + (2a' + 2b'\sin^2\theta)|U|^2V \\ & \quad + (2a' + b'\cos^2\theta)|V|^2V = 0\end{aligned} \tag{9.40}$$

where we assume that $a/2k_0 \simeq a/2l_0$ and $b/2k_0 \simeq b/2l_0$.

We now reduce equations (9.39) and (9.40) to normalized form. To do so, we assume light is propagating in the anomalous dispersion regime where $k'' < 0$ and $l'' < 0$. We also assume, as is appropriate for optical fibers, that the small difference between k'' and l'' may be neglected and that $k' - l' \simeq (k_0 - l_0)/\omega_0$. Letting

$$k'' = l'' = -\frac{\lambda_0}{2\pi c^2}D(\lambda_0) \tag{9.41}$$

we define

$$\xi = \frac{\pi z}{2z_0} \qquad z_0 = \frac{\pi^2c^2t_0^2}{D(\lambda_0)\lambda_0} \qquad t_0 = 0.568\tau$$

$$s = \frac{1}{t_0}\left(t - \frac{z}{v_g}\right) \qquad \bar{v}_g = \frac{2}{k' + l'}$$

$$u = (2a' + b'\cos^2\theta)^{1/2}U \qquad v = (2a' + b'\cos^2\theta)^{1/2}V$$

$$\delta = \frac{k' - l'}{2|k''|}t_0 \qquad \beta = \frac{\pi c \Delta n}{D(\lambda_0)\lambda_0}t_0$$

$$\beta = \frac{k'''}{6|k''|t_0} \qquad \gamma = \frac{2}{\pi}\Gamma z_0$$

$$R = \frac{8\pi c}{\lambda_0}t_0 \qquad B = \frac{2a' + 2b'\sin^2\theta}{2a' + b'\cos^2\theta}$$

where τ is the FWHM pulse intensity and $\Delta n = (k_0 - l_0)c/\omega_0$ is the difference between the indices of refraction. With these definitions equations (9.39) and (9.40) become

$$\mathrm{i}\frac{\partial u}{\partial \xi} + \mathrm{i}\gamma u + \mathrm{i}\delta\frac{\partial u}{\partial s} + \frac{1}{2}\frac{\partial^2 u}{\partial s^2} - \mathrm{i}\beta\frac{\partial^3 u}{\partial s^3} + (|u|^2 + B(|v|^2)u = 0 \qquad (9.42a)$$

$$\mathrm{i}\frac{\partial v}{\partial \xi} + \mathrm{i}\gamma v - \mathrm{i}\delta\frac{\partial v}{\partial s} + \frac{1}{2}\frac{\partial^2 v}{\partial s^2} - \mathrm{i}\beta\frac{\partial^3 v}{\partial s^3} + (B|u|^2 + |v|^2)v = 0 \qquad (9.42b)$$

Two special cases are of interest to us. The first is when pulses propagate near the zero dispersion point but birefringence can be ignored. In this case, it is sufficient to keep one envelope variable u, and equation (9.42) becomes

$$\mathrm{i}\frac{\partial u}{\partial \xi} + \mathrm{i}\gamma u + \frac{1}{2}\frac{\partial^2 u}{\partial s^2} - \mathrm{i}\beta\frac{\partial^3 u}{\partial s^3} + |u|^2 u = 0 \qquad (9.43)$$

This case will be considered in Section 9.4. The second case is when birefringence plays an important role, but higher order dispersion does not. In this case, equation (9.42) becomes

$$\begin{aligned} &\mathrm{i}\frac{\partial u}{\partial \xi} + \mathrm{i}\gamma u + \mathrm{i}\delta\frac{\partial u}{\partial s} + \frac{1}{2}\frac{\partial^2 u}{\partial s^2} + (|u|^2 + B|v|^2)u = 0 \\ &\mathrm{i}\frac{\partial v}{\partial \xi} + \mathrm{i}\gamma v - \mathrm{i}\delta\frac{\partial v}{\partial s} + \frac{1}{2}\frac{\partial^2 v}{\partial s^2} + (B|u|^2 + |v|^2)v = 0 \end{aligned} \qquad (9.44)$$

From the definition of B, we find that when $\theta = 0$, corresponding to linear birefringence, $B = \frac{2}{3}$. This case will be considered in Section 9.5. When $\cos^2\theta = 2\sin^2\theta$, corresponding to $\theta \simeq 35°$, we find that $B = 1$. This case will be considered in Section 9.7.

9.3 Robustness of solitons

Many, if not most, physical systems exhibit chaotic behavior in at least some parameter regimes. Such systems are appropriately modeled by equations like the Navier–Stokes equation which has turbulent solutions at high Reynolds numbers and is used to study fluids. In many important cases, however – and these cases certainly include optical fibers – physical systems exhibit nice, coherent behavior over a wide range of parameters. That is particularly the case in devices like optical fibers which are useful for something, as opposed to systems which are handed to us by nature, since one usually wants the device to behave in a nice predictable manner.

The NLS equation

$$\mathrm{i}\frac{\partial u}{\partial \xi} + \frac{1}{2}\frac{\partial^2 u}{\partial s^2} + |u|^2 u = 0 \tag{9.45}$$

which is often used to model optical fibers always exhibits coherent behavior. The reason is that it falls into that special class of systems which is integrable using spectral transform methods (Ablowitz and Segur, 1981) as was first shown by Zahkarov and Shabat (1972).

Spectral transform methods are well understood when linear equations are to be solved. In this case the spectral transform just amounts to the Fourier transform, i.e. the decomposition of the wave into its separate frequency components. As an example, we may consider equations (9.45) without the non-linear contribution,

$$\mathrm{i}\frac{\partial u}{\partial \xi} + \frac{1}{2}\frac{\partial^2 u}{\partial s^2} = 0 \tag{9.46}$$

This equation describes linear dispersion in an optical fiber. Its direct solution, along the broken line shown in Figure 9.1 would require a numerical approach. Instead, we use an eigenmode decomposition of equation (9.46) which can be obtained through use of the Fourier transform. Letting

$$\tilde{u}(\xi, \omega) = \int_{-\infty}^{\infty} \frac{\mathrm{d}s}{2\pi} \exp(\mathrm{i}\omega s) u(\xi, s) \tag{9.47}$$

we find

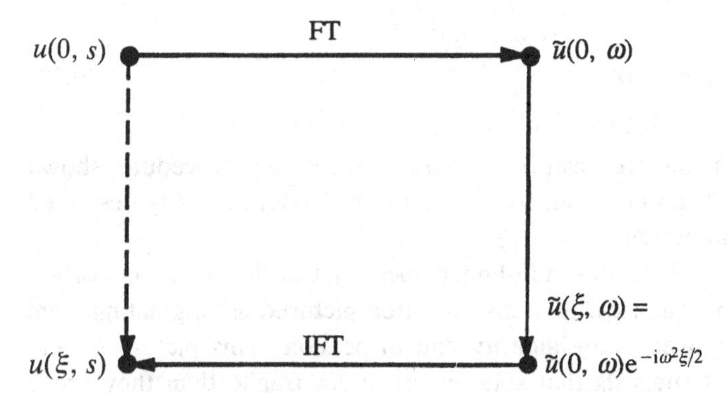

Figure 9.1. Schematic illustration of the way in which the frequency transform and its inverse can be used to solve the dispersive wave equation which models linear dispersion in optical fibers.

$$\mathrm{i}\frac{\partial \tilde{u}}{\partial \xi} - \frac{\omega^2}{2}\tilde{u} = 0 \qquad (9.48)$$

which can be immediately integrated to yield

$$\tilde{u}(\xi, \omega) = \tilde{u}(0, \omega)\exp(-\mathrm{i}\omega^2\xi/2) \qquad (9.49)$$

Thus, to solve equation (9.46), we use the three-step approach shown schematically in Figure 9.1. We carry out the frequency transform using equation (9.47) at $\xi = 0$, we determine how the frequency transform evolves from equation (9.49), and we then inverse transform to determine the temporal behavior at a given ξ. Physically, each eigenmode or frequency travels at a different velocity as the pulse travels along the fiber. Hence, the pulse has a tendency to spread, leading to dispersion.

An analogous, albeit far more complicated spectral transform exists for the NLS equation. There is a quantity $r(\xi, \omega)$ which is roughly analogous to the Fourier transform, but is not identical unless $\int_{-\infty}^{\infty}|u|\mathrm{d}s \ll 1$. In particular, ω is not frequency. Additionally, there are a number of discrete pairs $[\zeta_i(\xi), C_i(\xi)]$ of complex numbers. There can be any number, including zero, of these pairs, depending on the initial conditions. Schematically, as $|u|$ increases from zero, we find that $|r(\xi, \omega)|$ increases and at some point a pair $[\zeta_1(\xi), C_1(\xi)]$ pops into existence. As $|u|$ continues to increase, $|r(\xi, \omega)|$ oscillates rather than increasing steadily, and at intervals new pairs pop into existence. At some point, the $[\zeta_i(\xi), C_i(\xi)]$ pairs dominate the behavior of the system. Physically, $r(\xi, \omega)$ corresponds to dispersive waves while each $[\zeta_i(\xi), C_i(\xi)]$ corresponds to a soliton. These spectral variables evolve simply,

$$\begin{aligned} r(\xi, \omega) &= r(0, \omega) \exp(-\mathrm{i}\omega^2 \xi/2) \\ \zeta_i(\xi) &= \zeta_i(0) \\ C_i(\xi) &= C_i(0) \exp(2\mathrm{i}\zeta_i^2 \xi) \end{aligned} \tag{9.50}$$

implying that the NLS can be solved in a three-step procedure, shown in Figure 9.2, precisely analogous to the procedure already described for the linear system.

The existence of this non-linear spectral transform has important physical consequences. Solitons are often pictured as originating from a balance between non-linearity and dispersion. This picture is very useful, but it suggests that solitons are more fragile than they prove to be in practice. In the picture just presented, solitons appear as true eigenmodes of the optical fiber, along with the dispersive wave components. Changing the initial conditions slightly changes the spectral decomposition slightly, but the basic existence of solitons is not affected. This non-linear spectral decomposition appears extremely unintuitive at first. We are only used to decomposing the time domain into frequency. By contrast, we are used to using different spectral decomposition of space, depending on geometry – Bessel functions for cylindrical geometry, spherical harmonics for spherical geometry, and so on. Of particular interest is the spectral decomposition for planar waveguides which has both bounded modes analogous to solitons and radiation modes analogous to the dispersive waves. The point is that non-linear propagation in optical fibers does not respect frequency; frequencies can shift both up and down. The propagation does respect these new spectral variables.

These considerations explain why solitons are robust when the initial conditions are altered, but we need now to understand why solitons are robust when the equations are altered. To attack this problem, we first note that the NLS equation is a Hamiltonian system. Its Hamiltonian may be written

$$H = -\frac{\mathrm{i}}{2} \int_{-\infty}^{\infty} \mathrm{d}s \left(\frac{\partial u}{\partial s} \frac{\partial u^*}{\partial s} - |u|^4 \right) \tag{9.51}$$

Letting $q \equiv u$ and $p \equiv u^*$, we find

$$\frac{\partial q}{\partial \xi} = \frac{\delta H}{\delta p} \qquad \frac{\partial p}{\partial \xi} = -\frac{\delta H}{\delta q} \tag{9.52}$$

as must be the case for a Hamiltonian system. The derivatives $\delta x/\delta y$ are functional derivatives. The NLS equation is often referred to as an infinite-dimensional Hamiltonian system because each point in s can be considered a separate degree-of-freedom. The NLS equation is

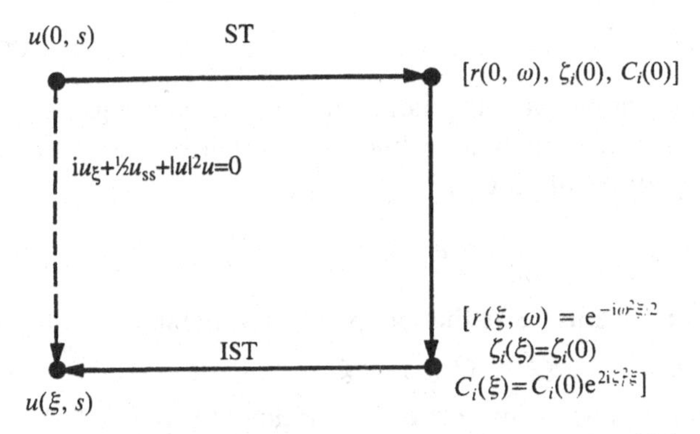

Figure 9.2. Schematic illustration of the way in which a nonlinear spectral transform and its inverse can be used to solve the NLS.

also integrable. When one states that a Hamiltonian system is integrable, one generally means that a canonical transformation exists which yields a Hamiltonian independent of the new coordinates, depending only on the new momenta. This point of view seems different from that of the previous section where we said that the NLS equation could be solved by making a spectral transformation; in fact, these two points of view are equivalent. The spectral transformation turns out to be a canonical transformation which yields a Hamiltonian only depending on the momenta.

Before demonstrating this point explicitly, it is useful to turn to a simpler example to explain how these canonical transformations work. They are important because when integrable field equations with autonomous, Hamiltonian perturbations are considered, it is possible to find an infinite series of canonical transformations which eliminates order-by-order the dependence on the coordinates as long as the initial conditions contain at most one soliton. Autonomous perturbations are perturbations which have no explicit dependence on space or time. This result explains qualitatively why integrable behavior is rugged under the influence of autonomous, Hamiltonian deformations. (At least when the deformations are small!)

The example we will consider is a simple, finite-dimensional system

$$H = \sum_i \frac{\omega_i}{2} (p_i^2 + q_i^2) \tag{9.53}$$

The canonical transformation $(p_i, q_i) \to (P_i, Q_i)$, where

$$p_i = (2P_i)^{1/2} \cos Q_i, \qquad q_i = (2P_i)^{1/2} \sin Q_i \tag{9.54}$$

reduces the Hamiltonian to the desired form

$$H = \sum_i \omega_i P_i \tag{9.55}$$

which depends only on the momenta. As a consequence, the momenta are constant in time, while the coordinates vary linearly. Writing the equations of motion,

$$\frac{\partial P_i}{\partial \xi} = 0 \qquad \frac{\partial Q_i}{\partial \xi} = \omega_i \tag{9.56}$$

where, as before, we use ξ as the independent variable, we obtain,

$$P_i = P_{i,0} \qquad Q_i = Q_{i,0} + \omega_i \xi \tag{9.57}$$

where $P_{i,0}$ and $Q_{i,0}$ are constants of integration. In similar fashion, if we make the transformation $u \to [P(\omega), Q(\omega), P_j, Q_j]$, where, in terms of the spectral data,

$$P(\xi) = \frac{\mathrm{i}}{2\pi} \ln [1 + |r(\omega)|^2] \qquad Q(\omega) = \arg r(\omega) \tag{9.58}$$

$$P_i = 2\mathrm{i}\zeta_i \qquad Q_i = -\ln C_i$$

we find that the transformed Hamiltonian becomes

$$H = \int_{-\infty}^{\infty} \mathrm{d}\omega \left[\frac{\omega^2}{2} P(\omega)\right] + \frac{\mathrm{i}}{6} \sum_i P_i^3 \tag{9.59}$$

which only depends on the momenta. Hence, just as in the previous case, the momenta are constant in ξ while the coordinates vary linearly.

Suppose now that we perturb the finite-dimensional system by adding cubic terms to the Hamiltonian,

$$H = \sum_i \frac{\omega_i}{2} (p_i^2 + q_i^2) + a p_i^3 + b p_1^2 q_1 + \ldots \tag{9.60}$$

In the limit where p_i and q_i are small, this perturbation only makes a small contribution to the Hamiltonian. As long as all the ω_i are incommensurable, it is possible to find a canonical transformation, using the Lie transform method or the Poincaré–von Zeipel method, which eliminates the cubic terms at the expense of introducing fourth and higher order terms (Lichtenberg and Lieberman, 1983), i.e. there exists a transformation $[p_i, q_i] \to [\tilde{p}_i, \tilde{q}_i]$, such that our Hamiltonian becomes

$$H = \sum_i \frac{\omega_i}{2} (\tilde{p}_i^2 + \tilde{q}_i^2) + \tilde{a} \tilde{p}_1^4 + \ldots \tag{9.61}$$

The fourth-order terms can then be eliminated by making another, analogous transformation, and we may continue in this fashion order-by-order. Physically, this series of transformations is possible because

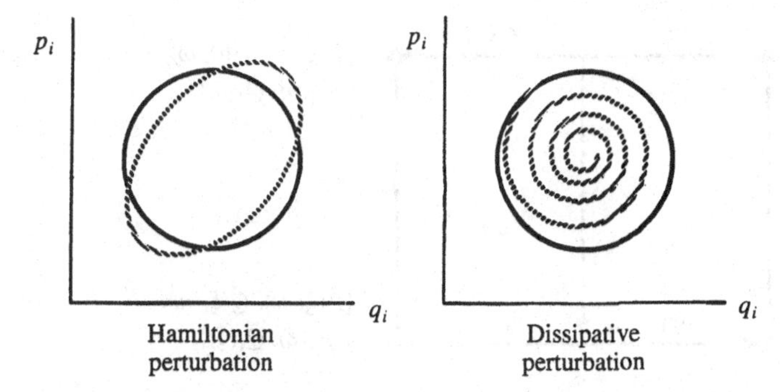

Figure 9.3. The effect of Hamiltonian and dissipative perturbations. A Hamiltonian perturbation slightly deforms the trajectory, but it remains neutrally stable. While the trajectory does not generally close on itself in the original coordinates, it does in appropriately perturbed coordinates. A dissipative perturbation, no matter how small, leads to a spiral trajectory which ultimately falls into the origin.

when q_i and p_i are sufficiently small, the effect of the cubic perturbations, for the vast majority of initial conditions, is to deform the orbit of the pair (p_i, q_i) without destroying its neutral stability. By contrast, a dissipative perturbation, no matter how small, will lead to a fundamental change in orbit topology as shown qualitatively in Figure 9.3.

A similar series of transformations exists for autonomous, Hamiltonian perturbations of integrable field equations when the initial conditions only include a single soliton. Consider the original transformation $u \to [P(\omega), Q(\omega), P_1, Q_1]$, where we have written P_1 and Q_1 to indicate that we are only considering initial conditions which contain at most one soliton. This transformation yields quantities which evolve linearly in ξ when u_ξ is given by the NLS equation. That is no longer the case once the equations are perturbed. However, at any given order, the canonical transformations yield a new set of quantities $[\tilde{P}(\omega), \tilde{Q}(\omega), \tilde{P}_1, \tilde{Q}_1]$ which evolve linearly in ξ *through the order to which we are working*, i.e.

$$\begin{aligned} \tilde{P}(\omega) &= \tilde{P}_0(\omega) \qquad & \tilde{Q}(\omega) &= \tilde{Q}_0(\omega) + \Omega(\omega)t \\ \tilde{P}_1 &= \tilde{P}_{1,0} \qquad & \tilde{Q}_1 &= \tilde{Q}_{1,0} + \Omega_1 t \end{aligned} \tag{9.62}$$

where $\tilde{P}_0(\omega)$, $\tilde{Q}_0(\omega)$, $\tilde{P}_{1,0}$, $\tilde{Q}_{1,0}$, $\Omega(\omega)$, and Ω_1 are all constant in ξ. Hence, just as in the integrable case, it is possible to integrate the equation in a fixed number of steps, as shown schematically in Figure 9.4, independent of the length ξ over which one wishes to determine

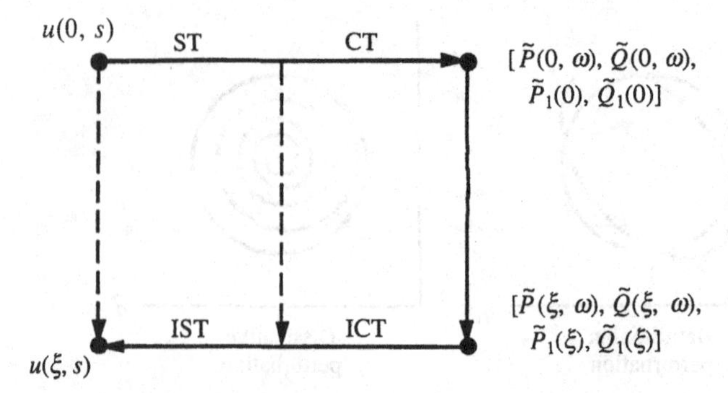

Figure 9.4. Schematic illustration of the integration procedure for the perturbed NLS when the perturbations are autonomous and Hamiltonian. One first makes a spectral transformation followed by a series of canonical transformations to arrive at variables which evolve linearly in ξ. Having calculated the new variables at ξ, one reverses the original sequence of transformations to determine u.

the solution. Why then are these perturbed equations not also considered integrable? The reason is that in general this series of transformations is only asymptotic; so, it does not converge, and only a finite number of transformations can be usefully made. Nonetheless, a soliton exists at every order which implies that to the extent that this asymptotic transformation is useful, solitons will exist. That is expected to usually be the case, implying that solitons are usually robust.

This approach is analogous in important respects to the way in which coupled mode theory is used to determine the effect of nonlinearities on waveguide junctions (Burns and Milton, 1988). The eigenmodes are coupled together by the perturbations and new eigenmodes, decoupled to whatever order we are working – which in practice almost always means lowest non-trivial order, must be found.

9.4 Higher-order dispersion

In Section 9.1, we discuss the possibility of utilising solitons in optical communication systems. Since non-linearity is used to counter the effect of anomalous dispersion in a soliton system, it is desirable to operate near the zero dispersion point (Hasegawa and Kodama, 1981). The power required for creating solitons there is significantly lower. Additionally, propagation through two fibers, one just above and the other just below the zero dispersion point, was used by Blow *et al.* (1985) to achieve significantly cleaner pulse compression than is possible by using a single fiber in the anomalous-dispersion regime (Mollenauer *et al.*, 1983). Thus operation near the zero dispersion point allows one to achieve low-power, clean pulse compression. Also, it has been shown that third-order dispersion can be used to reduce the mutual interaction between solitons (Chu and Desem, 1985). Near the zero dispersion point, the third-order dispersion plays a crucial role, and its effects on pulse evolution cannot be neglected. In the following, we first study the effect of third-order dispersion on the fundamental soliton, the two-soliton and the three-soliton solutions of the NLS equation. We find that a resonance occurs at the frequency where the combined effect of second- and third-order dispersion is zero. The soliton pulse radiates energy at this frequency. While this resonant radiation has only a minor effect on the fundamental soliton, it is detrimental to the N-solitons. We show that both the two-solitons and three-solitons break up into their constituent soliton components after certain threshold values of β $(= k'''/6|k''|t_0)$ are exceeded. Seeing the robustness of the fundamental solitons in the presence of higher order dispersion, we look for soliton propagation at the zero dispersion point. We find numerically that a pulse launched at the zero dispersion point up-shifts part of its energy into the normal dispersion regime, and down-shifts the rest into the anomalous dispersion regime. The latter portion develops into a soliton-like pulse. Using a numerical shooting method, we obtain a criterion for the existence of solitary pulses near the zero dispersion point. The threshold power for such solitons is ten times smaller than that in the anomalous dispersion regime.

The equation governing the non-linear pulse propagation near the zero dispersion point

$$\mathrm{i}\frac{\partial u}{\partial \xi} + \frac{1}{2}\frac{\partial^2 u}{\partial s^2} + |u|^2 q = -\mathrm{i}\gamma u + \mathrm{i}\beta \frac{\partial^3 u}{\partial s^3} \tag{9.63}$$

has been derived in Section 9.2. The left-hand side of equation (9.63) is the NLS equation. Fundamental soliton or N-soliton $(N > 1)$ solutions can be obtained using the initial conditions $q(0, s) = N\,\mathrm{sech}\,(s)$, where N is an integer. The fundamental soliton propagates without change in pulse shape, while the N-solitons oscillate with a period of $\xi = \pi/2$, corresponding to a distance of,

$$z_0 = \pi t_0^2/2|k''| \tag{9.64}$$

The N-soliton is a bound state of N fundamental soliton. However, there is no binding energy between the solitons in the N-soliton, i.e. the total energy of the N-soliton is equal to the sum of the energies of its constituents solitons. The N one-solitons simply move with the same speed. The power required to excite an N-soliton is N^2 times that of the fundamental soliton. The critical power for launching the fundamental soliton is

$$P_0 = \tfrac{1}{2} v_g |U_0|^2 \varepsilon_0 S n_0^2 \tag{9.65}$$

$$= \frac{\varepsilon_0 n_0 c |k''| \lambda S}{2\pi n_2 t_0^2}$$

where S is the cross-sectional area of the fiber, n_0 is the refractive index, U_0 is the peak electric field, and $\varepsilon_0 = 8.85 \times 10^{-12}$ F/m. From equation (9.65), it follows that the peak power decreases as one approaches the zero dispersion point. The normalised third-order dispersion coefficient, on the other hand, would increase. As an example, we consider the single-mode fiber parameters discussed by Marcuse (1980). Taking a linear fit to the k'' and k''' curves near the zero dispersion point $\lambda_0 = 1.27\ \mu$m, k'' and β can be expressed as

$$k'' = 87.5 \times [1.27 - \lambda(\mu\text{m})]\text{ps}^2/\text{km} \tag{9.66}$$

$$\beta = 3.5 \times 10^{-4} \frac{[\lambda(\mu\text{m}) - 0.86]}{t_0(\text{ps})[1.27 - \lambda(\mu\text{m})]}$$

Letting $S = 20\ \mu\text{m}^2$ and considering a 1-ps pulse (full width at half-maximum intensity, FWHM) corresponding to $t_0 = 0.57$ ps, we find that $P_0 = 1.1$ W when $\lambda - \lambda_0 = 0.03\ \mu$m and $P_0 = 0.4$ W when $\lambda - \lambda_0 = 0.01\ \mu$m. At the same time, β on the right-hand side of equation (9.63) increases from 0.009 to 0.026.

In equation (9.63), we include the effect of higher order dispersion, but do not consider higher-order non-linear correction such as the shock term,

$$\frac{2in_2}{c} \frac{\partial |U|^2}{\partial t} U$$

In normalised units, the shock term is given by $i(2/\omega_0 t_0) u \partial |u|^2 / \partial s$. In the example above, the coefficient $2/\omega_0 t_0$ is about 10^{-3}. Therefore, we would not expect it to become important until the pulsewidth is less than a picosecond.

We integrate equation (9.63) numerically, using initial conditions $q(0, s) = N \operatorname{sech}(s)$, with $N = 1, 2, 3$ at various values of β. The numerical method employed here is the propagating-beam or split-step Fourier method (Press *et al.*, 1986). Artificial damping are added at both ends of the integration domain far away from the region of interest in order to simulate the infinite boundary conditions. It

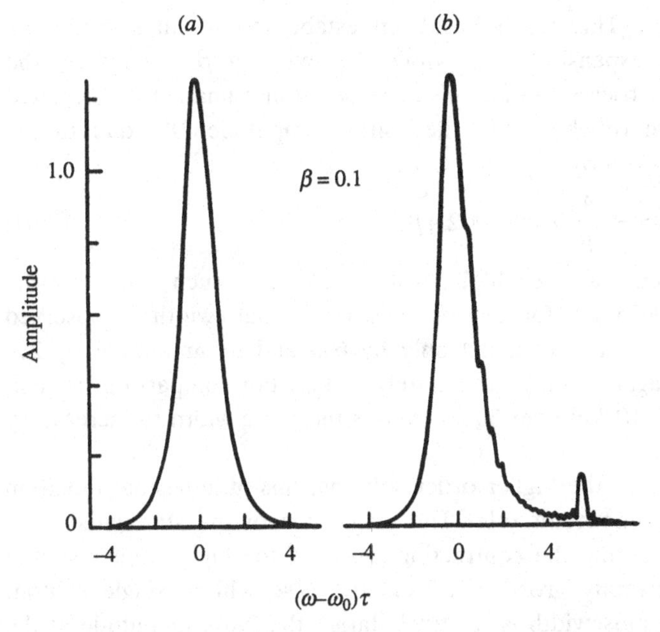

Figure 9.5. The amplitude in the frequency domain of the one-soliton with $\beta = 0.1$ at (*a*) $\xi = 0.0$, (*b*) $\xi = 5\pi$. A resonance peak is seen at $(\omega - \omega_0)t_0 \approx 1/2\beta$. Reprinted by permission of the Optical Society of America.

damps away any radiation which arrives there. We double the number of grid points and halve the step size in selected cases to ensure accuracy. We now consider in detail the effect of the third-order dispersion. The perturbation is observed to have little effect on the fundamental soliton. Shown in Figure 9.5 is the Fourier transform of $u(s)$ before and after propagation over a distance $\xi = 5\pi$ for the one-soliton with $\beta = 0.1$. It is observed that radiation is stimulated resonantly at a frequency $(\omega - \omega_0)t_0 \approx \frac{1}{2}\beta$, at which the dispersive terms in equation (9.63) cancel. Similar results are obtained when $N = 2$, 3 and at arbitrary values of β, although the amplitude of the radiation rapidly diminishes as β decreases. The effect of this radiation on the fundamental soliton and the higher-order solitons is quite different. In the case of the fundamental soliton, when linear attenuation is neglected, the magnitude of the excited radiation increased steadily. The loss in energy through this radiation results in changes in the soliton amplitude. However, the decrease is very small. In fact, it is exponentially small. It can be shown (Wai *et al.*, 1989) that the third-order dispersion only shifts the soliton parameters, but does not destroy them. This result has been established to all orders of a perturbation expansion in polynomial powers of β. Therefore, the effect of the observed radiations must be infinitesimal. It is calculated that the rate of change in the soliton amplitude (A) due to the radiation is given by

$$\frac{\partial A}{\partial \xi} = -\frac{44}{\beta^3}\exp(-\pi/2A\beta) \tag{9.67}$$

Hence, even at a high value of β, such as $\beta = 0.1$ ($\lambda - \lambda_0 = 0.0025\ \mu$m for a 1-ps pulse), the pulsewidth is observed numerically to have changed only by 6% and its amplitude by 2% after a propagation of approximately 30 km. For comparison, a dissipation of 0.2 dB/km over 30 km causes the pulse width to increase by a factor of 4.

In the case of the higher order solitons, this infinitesimal radiation play a more important role. The magnitude of the stimulated radiation jumps during the contracting phase of the higher order soliton instead of steadily growing as was the case with a single soliton. Initially, the pulsewidth is relatively large, the pulse amplitude at the resonant point $(\omega - \omega_0)t_0 = \frac{1}{2}\beta$ is exponentially small, hence the amount of radiation stimulated is minute. These jumps occur when the pulsewidth contracts or equivalently pulse bandwidth in frequency domain expands, leading to a larger amplitude at the resonant frequency. The increase in radiation leads to sharp changes in the parameters of the constituent solitons of the N-soliton. In particular, their group delays are shifted differently. Since there is no binding energy, higher order solitons are unstable to such group delays difference among its constituents. We observed a threshold value of β above which the higher order solitons are broken into their constituent solitons after one to three soliton periods. These solitons then separate asymptotically.

Figure 9.6 shows the breakup of a two-soliton as a function of the distance travelled. After the first contraction, radiation is excited at and around the resonant frequency. The radiation then leaves the main peak in the positive s-direction. The group delay can be calculated using the linear dispersion relation since the amplitude of the pulse is small there, i.e.

$$k = -\tfrac{1}{2}\omega^2 + \beta\omega^3$$

$$\frac{\partial k}{\partial \omega}(\tfrac{1}{2}\beta) = \tfrac{1}{4}\beta$$

The measured group delay agrees with these estimates. Shortly there-

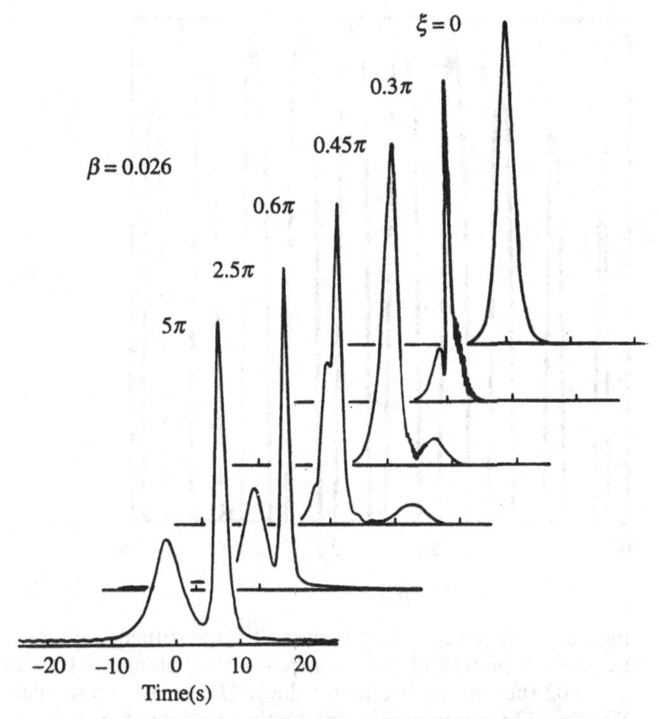

Figure 9.6. Break-up of a two-soliton at $\beta = 0.026$. The pulse at different values of z/z_0 is normalised to the same height for comparison. Reprinted by permission of the Optical Society of America.

after, the pulse separates into its constituent solitons. In the case of the two-solitons, they are one-solitons with amplitudes one and three respectively. Figures 9.7 and 9.8 plot the time separations of two peaks of the two-soliton as a function of distance travelled along the fiber. Below the threshold value β_0, shown in Figure 9.7 the maximum separation of the two peaks oscillate with distance, but the two-soliton is stable. When the integration length is doubled to 10π, there is still no sign of break-up. Above the threshold, shown in Figure 9.8, the two peaks break up into two solitons after one oscillation, which then separate at constant speed. As β increases further, the speed of separation increases but is not linear in β. We have made similar studies for the three-soliton, and the results are qualitatively the same. Above the threshold, which is smaller than that of the two-soliton, the three-soliton breaks up into a fundamental soliton and a two-soliton. The threshold values of β and the corresponding physical parameters are given in Table 9.1. Note the large

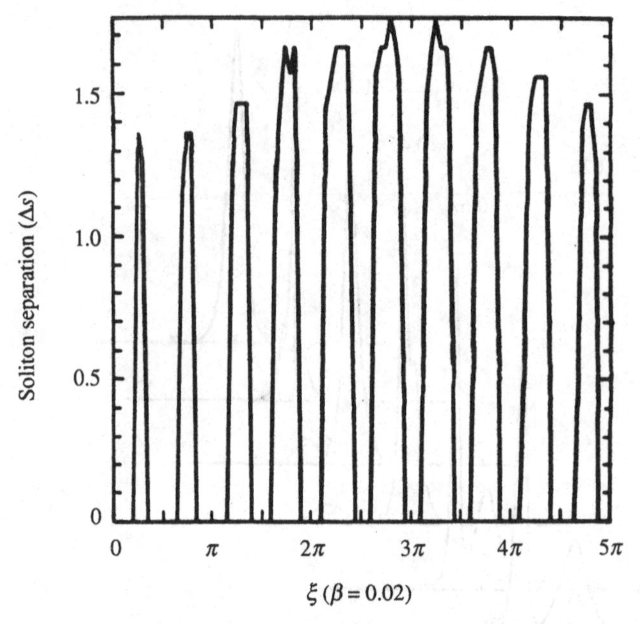

Figure 9.7. The separation between the constituent solitons of a two-soliton plotted *versus* distance travelled along the fiber at $\beta = 0.02$ (below the threshold value). The separation oscillates. Reprinted by permission of the Optical Society of America.

decrease in β_0 as N increases from two to three. It is because in the pulse width contracting stage, the three-soliton spread out more in frequency domain than does the two-soliton. Since the wings of the soliton's spectrum decreases exponentially with $(\omega - \omega_0)t_0$, the three-soliton overlaps the resonant frequency significantly more than the two-soliton for the same β. This trend is expected to continue with increasing N. When the loss, which is 0.4 dB/km in a standard fiber and as low as 0.2 dB/km in a fiber whose zero dispersion point has been shifted, is included, it is found to have a negligible effect on the threshold values.

In summary, we have found that the NLS equation models pulse propagation in optical fibers quite well, even for pulses whose wavelengths are close to the zero dispersion point. For fundamental solitons, their shapes undergo no significant distortion until $\lambda - \lambda_0 < 0.001\ \mu\text{m}$. Two- and three-solitons are significantly affected only by higher order dispersion, assuming that $\lambda - \lambda_0 \approx 0.01\ \mu\text{m}$ when the pulse length is less than or equal to 1 ps, indicating that low-order solitons are suitable for communication applications when the injec-

Table 9.1. *The threshold values of β_0 for the two-soliton and three-soliton. Reprinted by permission of the Optical Society of America.*

Type of soliton	β_0	FWHM (ps)	$\lambda - \lambda_0(\mu m)$
Two-soliton	0.022	1.0	0.01
Three-soliton	0.006	1.0	0.03
Three-soliton	0.006	3.0	0.01

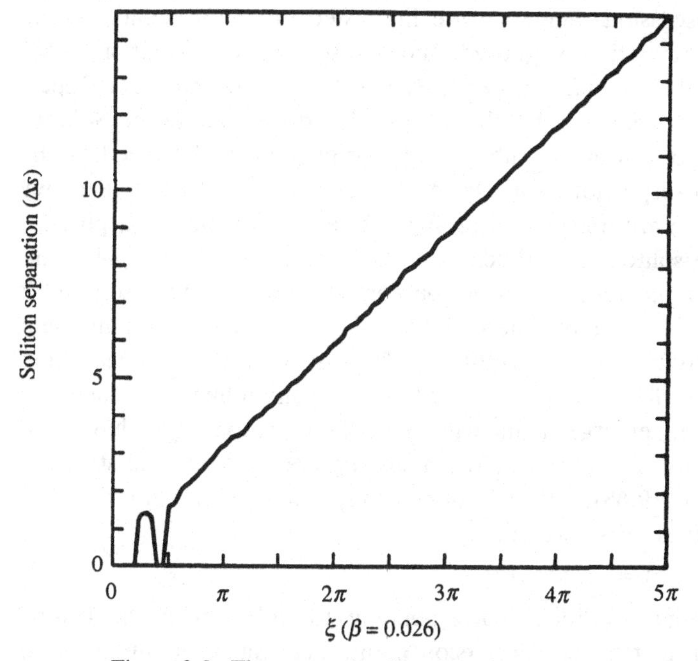

Figure 9.8. The separation between the constituent solitons of a two-soliton plotted *versus* distance travelled along the fiber at $\beta = 0.026$ (above the threshold value). The separation increases linearly with z/z_0 after the first oscillation. Reprinted by permission of the Optical Society of America.

tion pulse length is greater than or equal to 1 ps and $\lambda - \lambda_0 \gtrsim 0.01\ \mu m$. They are suitable in any case for use in pulse-compression applications since the break-up occurs in all cases after the first contraction.

After seeing how robust the fundamental soliton is, we proceed to investigate whether solitons can propagate at the zero dispersion point. If feasible, this scheme combines the low power requirement of propagation at the zero dispersion point with the robustness of the soliton approach. If a pulse is launched at or near the zero dispersion point, the NLS equation which has been used to model phenomena in the anomalous dispersion regime is no longer adequate. Using the slowly varying envelope approximation at zero dispersion point, where $k''(\omega_0) = 0$, one can show that

$$\mathrm{i}\frac{\partial u}{\partial \xi} - \mathrm{i}\frac{1}{6}\frac{\partial^3 u}{\partial s^3} + |u|^2 u = \mathrm{i}\gamma u \tag{9.68}$$

where $\gamma = \Gamma t_0^3/k'''$. Equation (9.68) will be the focus of the subsequent discussion. Since we are interested in soliton solutions, the effect of attenuation is ignored, and γ is set to zero. Equation (9.68) is very similar to equation (9.63) except that the second-order dispersion term is replaced by the third-order dispersion term. Solitons would be possible if the cubic non-linearity could be balanced by the third-order dispersion. An important property of solitons is that they emerge from arbitrary initial profiles. As a consequence, it is possible to observe solitons experimentally when neither the initial pulse amplitude nor the initial pulse shape corresponds to a pure soliton. It also provides a simple method to determine whether a non-linear equation contains soliton solutions. The equation is numerically integrated with different initial conditions. If soliton-like non-dispersive wave packets emerge, it indicates that the equation might have soliton solutions. Therefore, before analysing the mathematical structure of equation (9.68), we will numerically simulate it with different initial profiles. First, we take

$$u(0, s) = A_0 \operatorname{sech}(s)$$

as our initial condition, corresponding to pulses with the central wavelength at the zero dispersion point. The initial amplitude A_0 is varied and the subsequent pulse evolution is followed up to $\xi = 5\pi$.

In general, dispersive waves appear as multiple peaks at the leading edge of the pulse (a consequence of $k''' > 0$; for the opposite sign, the dispersive waves would appear at the trailing edge) and eventually separate from the main peak. The normalised distance required for the separation decreases with increasing amplitude. The peak that is left behind is a soliton-like pulse; it propagates without change in shape. Both the soliton and the dispersive waves move in the forward direction. Figure 9.9 shows the pulse amplitude for a hyperbolic secant initial profile with $A_0 = 2$ at $\xi = 3\pi$. The dispersive wave appears as a pedestal to the central soliton peak. It is much more informative to study the evolution into soliton and dispersive waves in the frequency domain. The initial spectrum splits into two peaks at

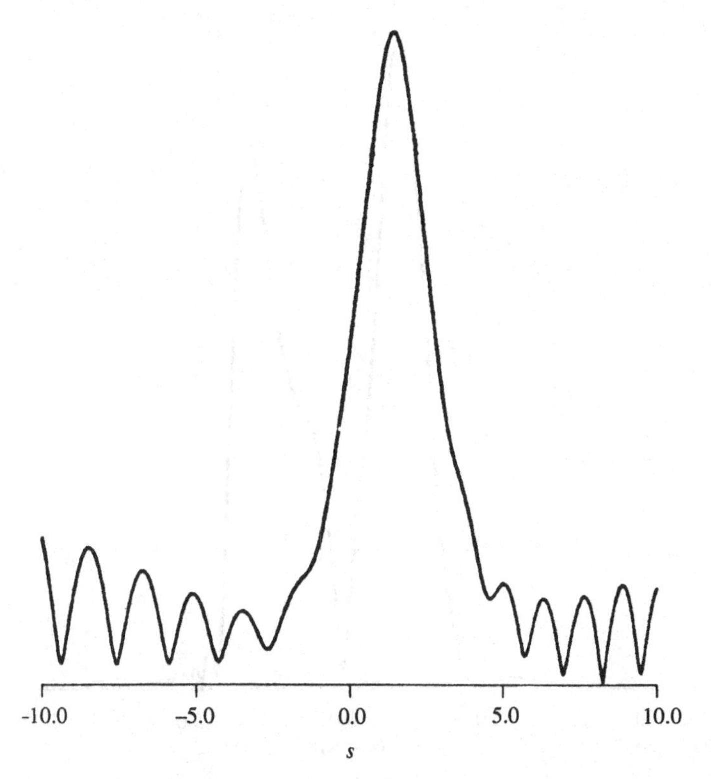

Figure 9.9. The pulse amplitude for a hyperolic secant initial profile with $A_0 = 2$, at $\xi = 10$. The soliton peak is surrounded by a pedestal consisting of dispersive waves. Reprinted by permission of the Optical Society of America.

either side of the zero dispersion frequency ω_0. This splitting occurs at $\xi \approx 10/A_0^2$ when $A_0 > 1.5$ and at somewhat smaller lengths at lower values of A_0. The peak in the anomalous regime corresponds to the soliton part while the peak in the normal regime corresponds to the dispersive wave component. There is a small interaction between the two components which shifts part of the pulse from the normal to the anomalous regime. The existence of the soliton, however, does not depend on the dispersive wave component. When we numerically filter out the normal spectrum immediately after the formation of the two-peak structure in the spectrum, we find that the remaining peak in the anomalous region propagates with very little radiation. The soliton contains approximately 60% of input power for all the cases considered. Figure 9.10 shows the two-peak structure of the spectrum for $A_0 = 2$ at $\xi = 2$. The shift in frequency of the soliton and dispersive waves from ω_0, is different, with the latter

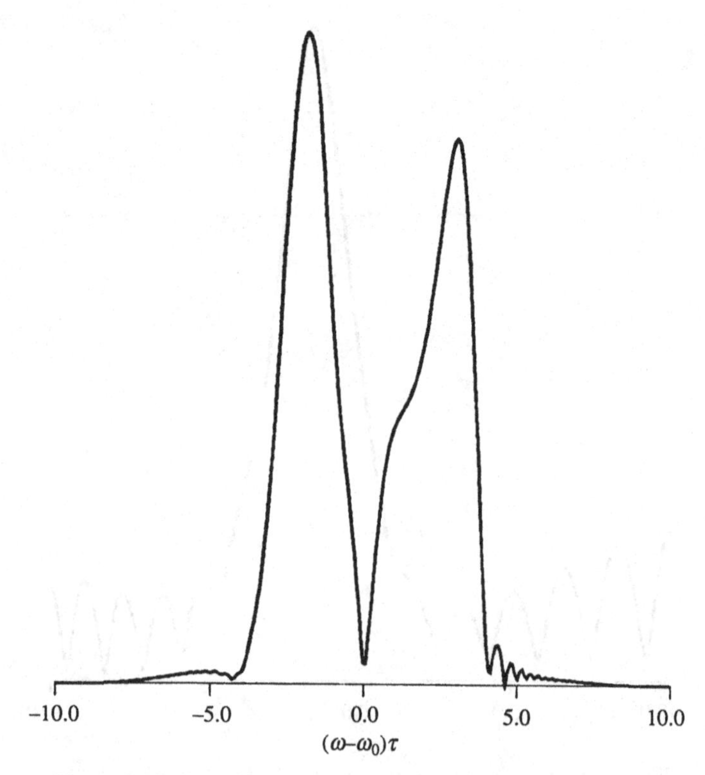

Figure 9.10. The frequency spectrum at $\xi = 2$ for $A_0 = 2$. The peak in the anomalous regime corresponds to the soliton while that in the normal regime corresponds to dispersive waves. Reprinted by permission of the Optical Society of America.

about 1.7 times larger than the former. Since the group delay is given by

$$\partial k/\partial\omega = \omega^2/2$$

the dispersive wave component travels faster and breaks away from the soliton. Figure 9.11 plots the frequency shift of the solitons which emerge at different values of A_0. This figure shows that the shift varies almost linearly with A_0. Hence, the dispersive wave component separates from the soliton at shorter distances for larger A_0.

We have also investigated Gaussian pulses in a similar fashion. The evolution is qualitatively similar in both cases, although solitons emerge over a length about a factor of two larger. We have also included the effect of attenuation. Since for hyperbolic secant initial conditions, the soliton forms over a length $\xi \approx 10/A_0^2$ when $A_0 > 1.5$, attenuation will not affect the soliton's emergence as long as $\gamma < A_0^2/10$.

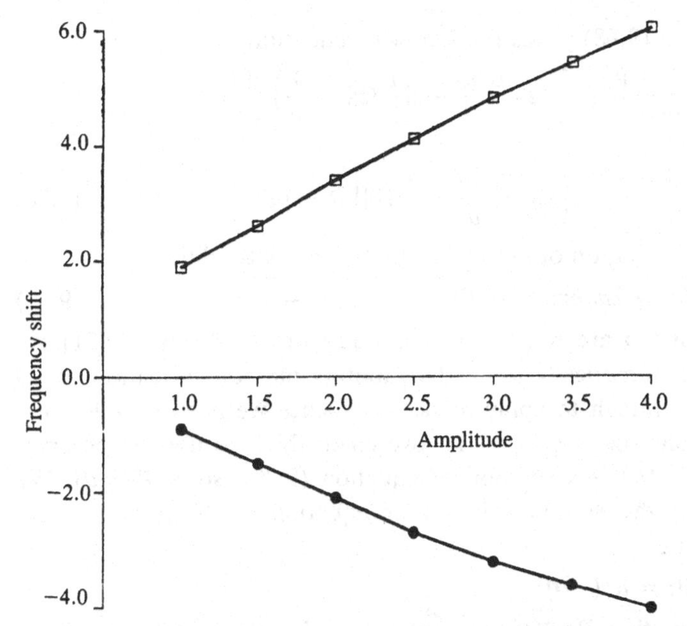

Figure 9.11. The frequency shifts of the solitons (full circles) and the dispersive waves (open squares) plotted *versus* initial amplitudes of hyperbolic secant profiles for the ideal case. Reprinted by permission of the Optical Society of America.

Next, we try to find the explicit form of the soliton solution of equation (9.68). We apply both Painlevé analysis (Ablowitz *et al.*, 1978, 1980) and the Chen's, Lee, and Liu's (1979) procedure to equation (9.68). Both indicate non-integrability. Since it would be difficult to obtain an analytic solution, we resort to numerical means to obtain the waveform of the soliton. First, we reduce equation (9.68) from a partial differential equation to an ordinary differential equation by making the travelling wave assumption,

$$u(\xi, s) = \tilde{u}(\theta)\exp(\mathrm{i}k_0\xi - \mathrm{i}\Omega_0\theta) \tag{9.69}$$

where $\theta = s - \xi/v$, v is the speed of the pulse, and k_0 and Ω_0 are the shift in normalised wavenumber and frequency of the soliton respectively. The quantity Ω_0 takes into account the variation in the central frequency near the zero dispersion point. When Ω_0 is large and negative, the soliton is deep inside the anomalous dispersion regime, and the second-order dispersion dominates. The quantity $\tilde{u}(\theta)$ can then be determined from the NLS equation with the third-order dispersion treated as a perturbation. Substitution of equation (9.69) into equation (9.68) yields the following equation,

$$-\mathrm{i}\frac{1}{6}\frac{\mathrm{d}^3\tilde{u}}{\mathrm{d}\theta^3} - \frac{1}{2}\Omega_0\frac{\mathrm{d}^2\tilde{u}}{\mathrm{d}\theta^2} + \mathrm{i}\left(\frac{1}{2}\Omega_0^2 - \frac{1}{v}\right)\frac{\mathrm{d}\tilde{u}}{\mathrm{d}\theta}$$

$$-\left(k_0 - \frac{\Omega_0}{v} - \frac{1}{6}\Omega_0^3\right)\tilde{u} + |\tilde{u}|^2\tilde{u} = 0 \tag{9.70}$$

We seek the solution of $\tilde{u}(\theta)$ with the boundary condition,

$$\tilde{u}(\theta) \to 2a_0 \exp(-\sigma|\theta|) \qquad \text{as } |\theta| \to \infty \tag{9.71}$$

where a_0 and σ are real constants. Equations (9.70) and (9.71) are then solved numerically using the shooting method. Equation (9.71) however, is diffcult to apply numerically since we need to follow the recessive solution as $\theta \to \infty$. To overcome this, we use the observation that if $\tilde{u}(\theta)$ is a solution of equation (9.70), so is $\tilde{u}^*(-\theta)$. We impose that the soliton solution of equation (9.70) possesses this symmetry, i.e.

$$\tilde{u}(\theta) = \tilde{u}^*(-\theta) \tag{9.72}$$

which means that $\mathrm{Re}\,\tilde{u}(\theta)$ is a symmetric function, while $\mathrm{Im}\,\tilde{u}(\theta)$ is an antisymmetric function of θ. Then, equation (9.71) can be reformulated as follows

$$\tilde{u}(\theta) \to 2a_0 \exp(\sigma\theta) \qquad \text{as } \theta \to -\infty \tag{9.73}$$

$$\frac{\partial}{\partial\theta}[\mathrm{Re}\,\tilde{u}](0) = 0 \tag{9.74}$$

$$\frac{\partial^2}{\partial\theta^2}[\mathrm{Im}\,\tilde{u}](0) = 0 \tag{9.75}$$

Numerically, equations (9.74) and (9.75) are written as

$$\frac{\partial}{\partial\theta}[\ln(\mathrm{Re}\,\tilde{u})](0) < \delta \tag{9.76}$$

$$\frac{\partial}{\partial\theta}\left[\ln\left(\frac{\partial\,\mathrm{Im}\,\tilde{u}}{\partial\theta}\right)\right](0) < \delta \tag{9.77}$$

It is found that solutions are possible if a_0, σ and Ω_0 are chosen such that

$$a_0^2 \approx \left[1 - \frac{5}{6}\left(\frac{\sigma}{\Omega_0}\right)^2\right]\sigma^2|\Omega_0| \tag{9.78}$$

$$0 > \frac{\sigma}{\Omega_0} > -0.24 \tag{9.79}$$

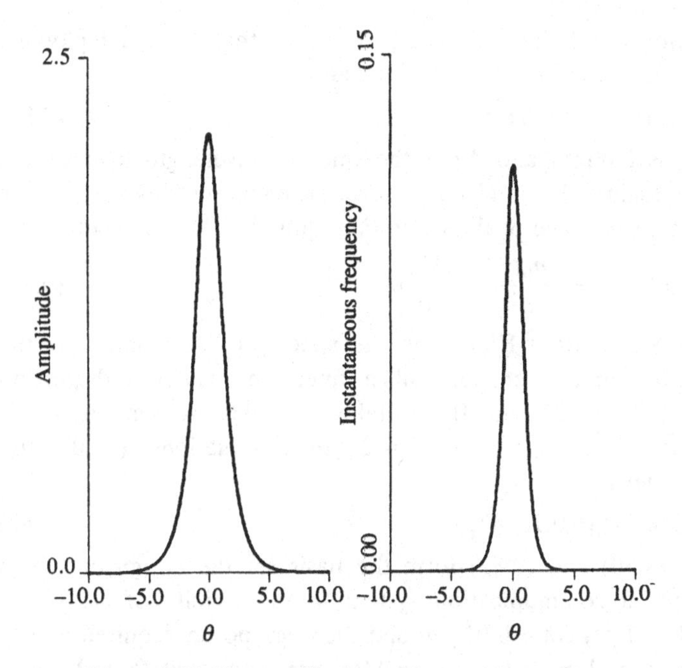

Figure 9.12 The amplitude and instantaneous frequency of a soliton of the ideal equation near the zero dispersion point. The bandwidth $\sigma = 1$ and frequency shift $\Omega_0 = 4.5$. Reprinted by permission of the Optical Society of America.

The parameter δ is varied from 10^{-2} to 10^{-4}, and there is little change in the relations (9.78) and (9.79). Equation (9.79) agrees with previous results, i.e. the shift in carrier frequency Ω_0 is non-zero and always into the anomalous dispersion region. Figure 9.12 plots the solutions for the amplitude $|\tilde{u}(\theta)|$ and the normalised instantaneous frequency $-\mathrm{Im}\{\mathrm{d}\ln[\tilde{u}(\theta)]/\mathrm{d}\theta\}$ for $\sigma = 1$ and $\Omega_0 = 4.5$. Notice that the instantaneous frequency varies with θ, in contrast to the soliton solution of the NLS equation which is a constant. We then verify that this solution is a soliton by using it as the initial profile and integrating equation (9.68) directly using the split-step Fourier method. The pulse propagates without observably changing its shape. Hence, the effect of the initial frequency splitting is to shift part of the pulse into the anomalous dispersion regime. Subsequent evolution of the down-shifted pulse is described by the NLS equation with higher order dispersion as a perturbation. Recall that the third-order dispersion excites an infinitesimal radiation. Equation (9.79) in effect limits the amount of radiation the soliton pulse can have for a given frequency Ω_0 away from the zero dispersion point.

The bandwith of the soliton equals σ, so that in actual physical quantities, equation (9.79) can be written as

$$(2\tau_0)\Delta\lambda > 6\ \text{nm ps} \tag{9.80}$$

where $2\tau_0$ is FWHM and $\Delta\lambda$ is the shift in wavelength towards the anomalous regime. For a given pulsewidth, equation (9.80) imposes a lower limit on $\Delta\lambda$. The peak power P_0 required to launch a soliton is

$$P_0(2\tau_0)^3 \approx \frac{\varepsilon n_0 c\lambda k'''}{\pi n_2}\left|\frac{\Omega_0}{\sigma}\right| S \tag{9.81}$$

where $\varepsilon = 8.85 \times 10^{-12}$ F/m, c is the speed of light, and S is the effective core area. For pure silica fiber, the material dispersion vanishes at $\lambda_0 = 1.27\ \mu$m. If we take $k''' = 0.08\ \text{ps}^3/\text{km}$, $n_0 = 1.5$, $n_2 = 1.22 \times 10^{-22}(\text{m/V})^2$ and $S = 20\ \mu\text{m}^2$, equations (9.80) and (9.81) together give

$$P_0(2\tau_0)^3 \geqslant 0.16\ \text{W}(\text{ps})^3 \tag{9.82}$$

Equations (9.80) and (9.82) form the basis for the design of a high bit-rate optical communication system in the vicinity of λ_0. If the pulsewidth is 1 ps, $\Delta\lambda \approx 0.01\ \mu$m and the peak power required is only 0.16 W. For a 2 ps pulse, $\Delta\lambda \approx 0.005\ \mu$m and only 20 mW peak power is required. The wavelength shift is automatically produced by the fiber as long as the initial central frequency of the pulse is at or slightly below the zero dispersion point and should be relatively insensitive to variations of the zero dispersion point along the fiber length. For a 2 ps pulse, the dispersion and the non-linearity become effective over 100 km, and the use of solitons is only of interest over greater propagation lengths. At these long lengths, this approach must be used in conjunction with Raman amplification.

9.5 Linear birefringence

Birefringence can lead to pulse splitting which would be bad in communication applications. Typical fibers have values $\Delta n \simeq 10^{-6}$–10^{-5}, where Δn is the change in the index of refraction between the two polarisations (Kaminov, 1981). Physically, for a 5 ps FWHM pulse, that corresponds to a walk-off over 20 km of between 13 and 130 times the FWHM pulse width. Two solutions to this problem suggest themselves. First, one might use polarisation-preserving fiber; however, the doping process which produces this fiber adds substantially to both the attenuation and the cost. Second, one might carefully control the injected pulse so that only one polarisation is present. As long as the birefringent beat length is short compared to the twist length of the axes of polarisation, a single polarisation will be preserved. However, these requirements appear to be difficult to meet over 20 km of propagation length and to maintain stably over time.

In this section, we show that the Kerr effect which stabilises solitons against spreading due to dispersion also stabilises them against splitting due to birefringence. In effect, the partial pulses in each of the two polarisations trap each other and move together as one unit. Ignoring an oscillatory component, one finds in the frequency domain that the central frequency of each partial pulse shifts just enough so that the group velocity of each becomes the same as the other.

We have shown in Section 9.2 that the equations which describe pulse propagation in a linearly birefringent optical fiber are

$$\mathrm{i}\frac{\partial u}{\partial \xi} + \mathrm{i}\gamma u + \mathrm{i}\delta\frac{\partial u}{\partial s} + \frac{1}{2}\frac{\partial^2 u}{\partial \xi^2} + \left(|u|^2 + \frac{2}{3}|v|^2\right)u = 0 \quad (9.83)$$

$$\mathrm{i}\frac{\partial v}{\partial \xi} + \mathrm{i}\gamma v - \mathrm{i}\delta\frac{\partial v}{\partial s} + \frac{1}{2}\frac{\partial^2 v}{\partial \xi^2} + \left(\frac{2}{3}|u|^2 + |v|^2\right)v = 0$$

Recalling that typical fibers have values of Δn concentrated in the range 10^{-6}–10^{-5}, we find $\delta = 0.3$–3.0.

In order to determine the amplitude which at a given value of δ is sufficient to stabilise injected pulses against splitting, we solve equation (9.83) numerically. We use a semi-spectral approach with edge-damping to eliminate any continuous wave component which reaches the ends of the simulation region. We have checked our results in selected instances by doubling the number of node points, doubling our simulation region, altering the edge damping, and verifying that our results do not change. In all our numerical studies, we chose as our initial conditions

$$\begin{aligned} u(\xi = 0) &= A\cos\alpha\,\mathrm{sech}\, s \\ v(\xi = 0) &= A\sin\alpha\,\mathrm{sech}\, s \end{aligned} \quad (9.84)$$

The angle α thus determines the relative strengths of the partial pulses in each of the two polarisations.

We begin by considering the case $\alpha = 45°$ which corresponds to equal initial amplitudes in both polarisations. In Figures 9.13 and 9.14, we show $s^u_{\max}(\xi)$, the values of s at which u has a maximum, as a function of distance along the fiber. When $\alpha = 45°$, it follows from equation (9.83) that $v(\xi, s) = u(-\xi, s)$ and, hence, $s^v_{\max}(\xi) = -s^u_{\max}(\xi)$ where $s^v_{\max}(\xi)$ is the value of s at which v has a maximum. Hence, if $s^u_{\max}(\xi)$ increases without bound, then the two partial pulses

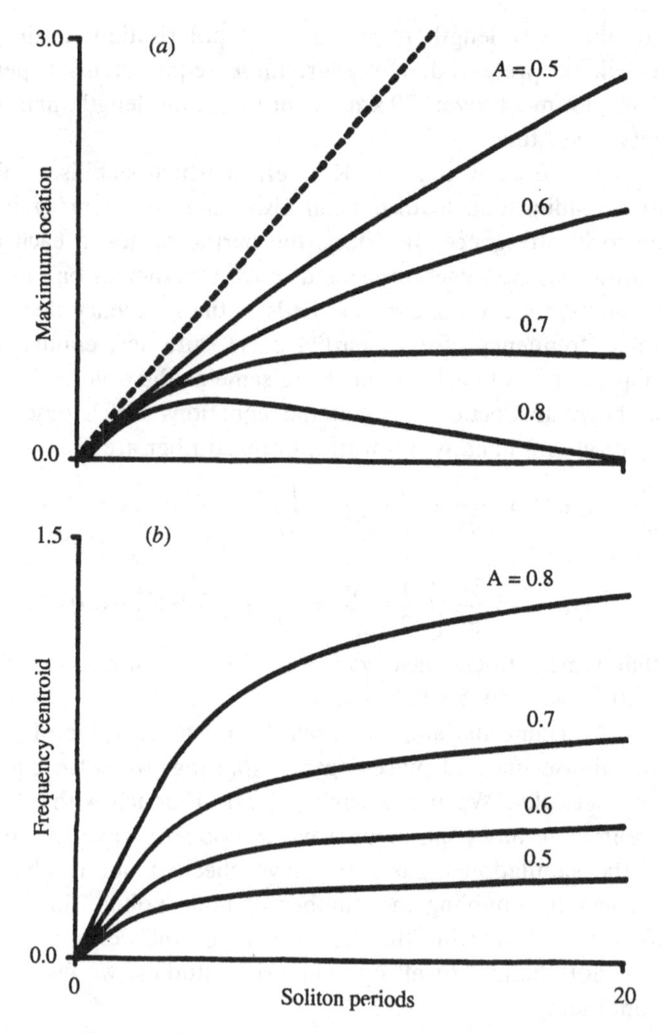

Figure 9.13. Variation of (*a*) the maximum location $s^u_{\max}$ and (*b*) the frequency centroid ω^u_{cent} with distance along the fiber measured in soliton periods ($\delta = 0.15$, $\gamma = 0.0$). Reprinted by permission of the Optical Society of America.

separate. Conversely, if $s^u_{\max}(\xi)$ is bounded, then the two partial pulses stay together. In addition, we define the Fourier transformed quantities

$$\tilde{u}(\xi, \omega) = \frac{1}{2\pi} \int_{-\infty}^{\infty} ds \exp(i\omega s) u(\xi, s) \tag{9.85}$$

$$\tilde{v}(\xi, \omega) = \frac{1}{2\pi} \int_{-\infty}^{\infty} ds \exp(i\omega s) v(\xi, s)$$

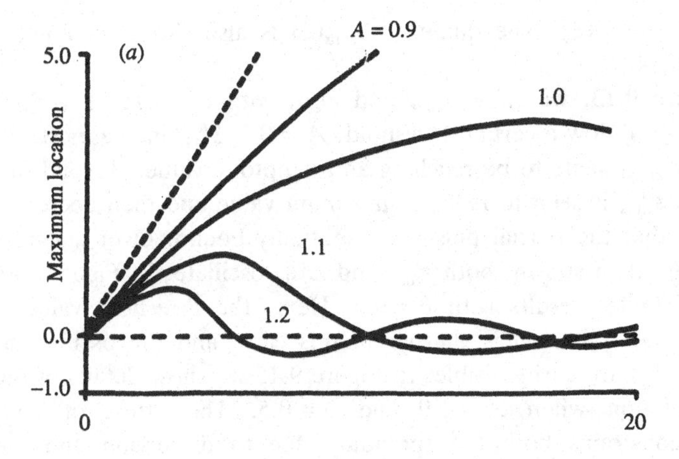

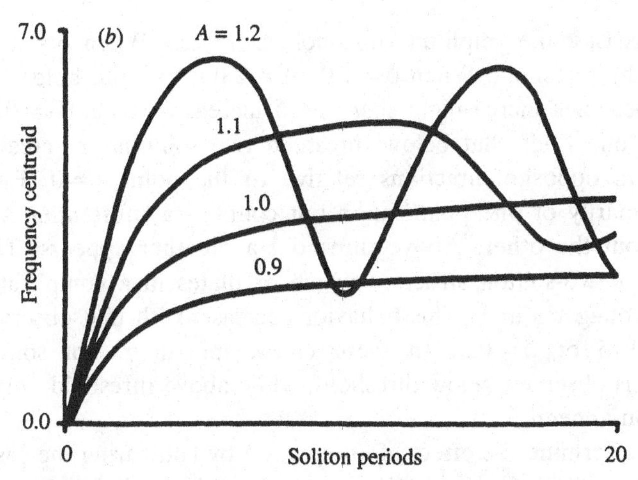

Figure 9.14. Variation of (*a*) the maximum location s^u_{max} and (*b*) the frequency centroid ω^u_{cent} with distance along the fiber ($\delta = 0.5$, $\gamma = 0.0$). Reprinted by permission of the Optical Society of America.

from which we may define in turn the frequency centroids,

$$\omega^u_{\text{cent}}(\xi) = \frac{\int_{-\infty}^{\infty} \mathrm{d}\omega\omega|\tilde{u}(\xi, \omega)|^2}{\int_{-\infty}^{\infty} \mathrm{d}\omega|\tilde{u}(\xi, \omega)|^2} \tag{9.86}$$

$$\omega^v_{\text{cent}}(\xi) = \frac{\int_{-\infty}^{\infty} \mathrm{d}\omega\omega|\tilde{v}(\xi, \omega)|^2}{\int_{-\infty}^{\infty} \mathrm{d}\omega|\tilde{v}(\xi, \omega)|^2}$$

Noting that $v(\xi, s) = u(-\xi, s)$, it follows that $\tilde{v}(\xi, \omega) = \tilde{u}(-\xi, \omega)$ and

$\omega^v_{\mathrm{cent}}(\xi) = -\omega^u_{\mathrm{cent}}(\xi)$. The quantity $\omega^u_{\mathrm{cent}}(\xi)$ is also shown in Figures 9.13 and 9.14.

In Figure 9.13, we show $s^u_{\max}$ and ω^u_{cent} with $\delta = 0.15$ at various values of A. Below a certain threshold, $A = 0.7$, $s^u_{\max}$ increases steadily and ω^u_{cent} appears to be reaching an asymptotic value. Beyond this threshold, $s^u_{\max}$ is seen to reach a maximum value and then decrease, indicating that the partial pulses are mutually bound. Over a longer length scale than shown, both $s^u_{\max}$ and ω^u_{cent} oscillate. In Figure 9.14 we show similar results with $\delta = 0.5$. Here, the threshold value is $A = 1.0$. The transition is sharper in this case, and the oscillations when $A > 1.0$ are easily visible. In Figure 9.15 we show details of the pulse evolution when $A = 1.0$ and $\delta = 0.5$. The effect of nonlinearity constrains both the spreading due to dispersion and the splitting due to birefringence. The accompanying frequency shift is also visible.

As δ increases, the amplitude threshold increases. When $\delta = 0.75$, $A = 1.5$ is the threshold. When $\delta = 1.0$, $A = 2.0$ is the threshold. The threshold becomes increasingly sharp as δ increases. When $\delta = 0.75$ or $\delta = 1.0$, one finds that below threshold two solitons are created and move in opposite directions relative to the point $s = 0$. Each consists primarily of one polarisation but contains a substantial contribution from the other. Above threshold a breather appears. This breather is a two-soliton structure which oscillates in a complicated way as it propagates in ξ. This behavior contrasts with that observed when $\delta = 0.15$ or $\delta = 0.5$. In these cases, spreading, not soliton generation, is observed below threshold, while above threshold only a single soliton appears.

We now determine the effect of attenuation by comparing the cases $\gamma = 0.0$ and $\gamma = 0.0105$; the latter case correspond to 2 db/km for 5 ps pulses. In general, we find that the evolution is essentially unchanged by attenuation during the first ten-soliton periods but differs substantially at distances greater than twenty-soliton periods. In Figure 9.16 we display this comparison with $A = 1.0$ and $\delta = 0.5$. Only the quantity $u(\xi, s)$ is exhibited, but the shape of $v(\xi, s)$ can be immediately inferred from the relation $v(\xi, s) = u(\xi, s)$. One of the effects of increasing the attenuation is to increase slightly the amplitude threshold beyond which the two partial pulses are mutually bound. When $\gamma = 0.0$, the two partial pulses initially move apart as they spread due to dispersion, but, beyond some point, they begin to move together again as if they were tied by a spring. The nonlinearity acts to sharpen the total pulse as well. When it has travelled

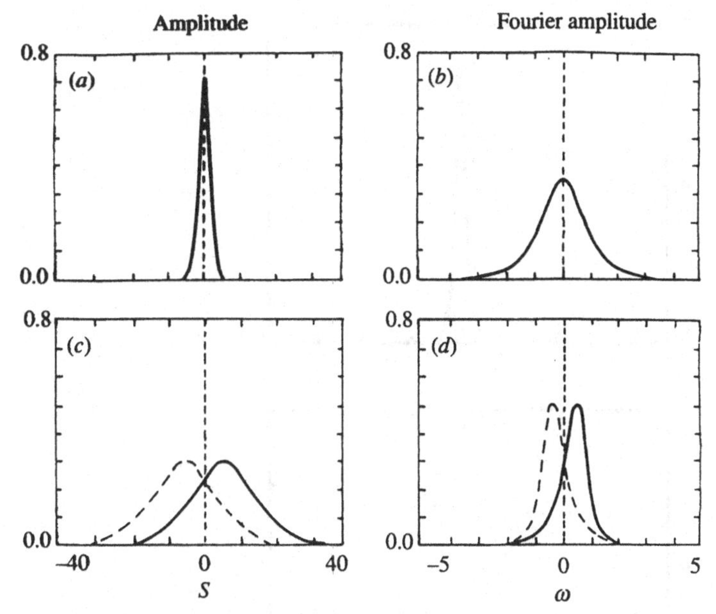

Figure 9.15. Details of the pulse evolution ($A = 1.0$, $\delta = 0.5$, $\gamma = 0.0$). Full lines indicate u and $\tilde{u}$; broken lines indicate v and $\tilde{v}$. (a) $u(s)$ and $v(s)$, $\xi = 0$; (b) $\tilde{u}(\omega)$ and $\tilde{v}(\omega)$, $\xi = 0$; (c) $u(s)$ and $v(s)$, $\xi = 5\pi$ (ten-soliton periods); (d) $\tilde{u}(\omega)$ and $\tilde{v}(\omega)$, $\xi = 5\pi$. Reprinted by permission of the Optical Society of America.

fifty-soliton periods, one finds that the maxima of the partial pulses are almost on top of each other and that there is a central portion, almost as sharp as the original component, sitting on a broad background of dispersive waves. By contrast, the attenuated pulse decreases too rapidly for the non-linearity to have a chance to effect the partial pulses, and they simply separate and spread. This difference is quite visible in Figure 9.17 where $s^u_{\rm max}$ is plotted in the two cases.

As A is increased, we have already seen that the oscillation periods of $s^u_{\rm max}$ and $\omega^u_{\rm cent}$ rapidly decrease. As a consequence, when $A = 1.1$, the non-linearity is already large enough for the two partial pulses to be bound together. When $\delta = 0.75$ or $\delta = 1.0$, the threshold difference introduced by attenuation is almost undetectable. The threshold results in all the cases we considered are summarised in Table 9.2.

In order to study what happens in the general case when $\alpha \neq 45°$, it is useful to define the partial pulsewidths w^u and w^v. Writing

$$s^u_{\rm ave} = \frac{\int_{-\infty}^{\infty} {\rm d}s\, s|u(s, \xi)|^2}{\int_{-\infty}^{\infty} {\rm d}s|u(s, \xi)|^2} \qquad (s^2)^u_{\rm ave} = \frac{\int_{-\infty}^{\infty} {\rm d}s\, s^2|u(s, \xi)|^2}{\int_{-\infty}^{\infty} {\rm d}s|u(s, \xi)|^2} \tag{9.87}$$

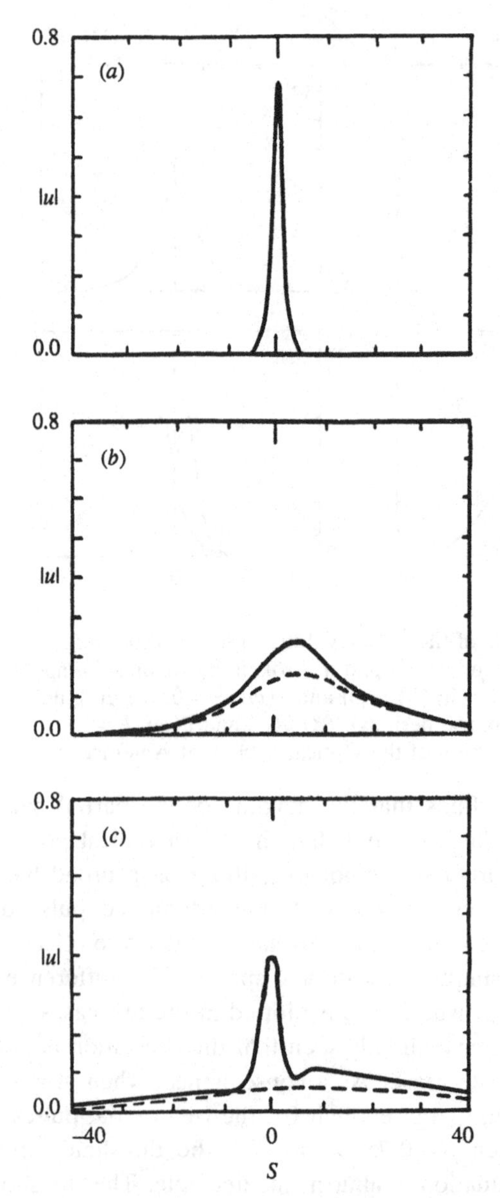

Figure 9.16. Pulse evolution ($A = 1.0$, $\delta = 0.5$). Full lines indicate $\gamma = 0.0$; broken lines indicate $\gamma = 0.0105$. (*a*) $\xi = 0$, (*b*) $\xi = 10\pi$ (twenty-soliton periods), (*c*) $\xi = 25\pi$. Reprinted by permission of the Optical Society of America.

Table 9.2. *The threshold values of A at which the Kerr non-linearity is sufficient to compensate for linear birefringence at $\alpha = 45°$. Reprinted by permission of the Optical Society of America.*

δ	γ	A
0.15	0.0	0.7
0.15	0.0105	0.8
0.5	0.0	1.0
0.5	0.0105	1.1
0.75	0.0	1.5
0.75	0.0105	1.5
1.0	0.0	2.0
1.0	0.0105	2.0

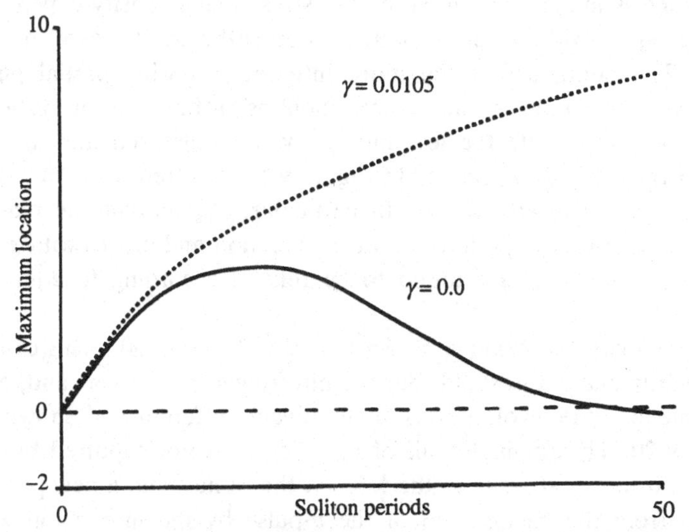

Figure 9.17. Variation of the maximum location $s^u_{\max}$ ($A = 1.0$, $\delta = 0.5$). Reprinted by permission of the Optical Society of America.

$$s^v_{\text{ave}} = \frac{\int_{-\infty}^{\infty} \mathrm{d}s\, s|v(s, \xi)|^2}{\int_{-\infty}^{\infty} \mathrm{d}s|v(s, \xi)|^2} \qquad (s^2)^v_{\text{ave}} = \frac{\int_{-\infty}^{\infty} \mathrm{d}s\, s^2|v(s, \xi)|^2}{\int_{-\infty}^{\infty} \mathrm{d}s|v(s, \xi)|^2}$$

we let

$$w^u = [(s^2)^u_{\text{ave}} - (s^u_{\text{ave}})^2]^{1/2} \tag{9.88}$$

$$w^v = [(s^2)^v_{\text{ave}} - (s^v_{\text{ave}})^2]^{1/2}$$

No special symmetry relations were found which relate $s^v_{\max}$ to $s^u_{\max}$ or w^v to w^u at arbitrary values of α, but there is a special relation between ω_{cent}^{v} and ω_{cent}^{u} when attenuation can be neglected,

$$\omega_{\text{cent}}^{v}(\xi) = -\cot^2 \alpha \omega_{\text{cent}}^{u}(\xi) \tag{9.89}$$

The first case that we consider in detail is $\delta = 0.15$, $A = 0.8$, $\alpha = 30°$. The spatial evolution is shown in Figure 9.18. Two points should be apparent. First, the peak of the v-pulse travels to the right along with the peak of the u-pulse; in effect, the smaller partial pulse has been captured by the larger. Second, while the width of the u-pulse has increased, the width of the v-pulse has increased even more. Indeed, there was a strong tendency throughout all our computations for the largest amplitude structure, whether a soliton or a breather, to end up primarily in the polarisation which is initially more intense while the continuum ends up primarily in the polarisation which is initially less intense. We stress that intensity-dependent birefringence, without dispersion, is not sufficient to explain this result. The point is that the more intense, less wide partial pulse remains *consistently* in one polarisation as a function of distance along the fiber, while the less intense, wider pulse remains *consistently* in the other polarisation. In Figure 9.19, the trends we have just pointed out are clearly visible. In particular, s_{max}^{v} initially decreases, but, after two-soliton periods, reverses direction and moves with s_{max}^{u}. Also ω_{cent}^{u} and ω_{cent}^{v} are found to be related according to equation (9.89).

The next case we examine is $\delta = 0.5$, $A = 1.1$, $\alpha = 30°$. This case is also slightly above threshold, but the birefringence is larger, and, as a consequence, the evolution is qualitatively different, as shown in Figure 9.20. The original peak of the v-pulse is not captured by the u-pulse; it moves steadily to the left. At the same time, a new peak is created from the background of the v-pulse by the interaction with the peak of the u-pulse. This new peak moves to the left with the u-pulse. The v-pulse thus consists of two parts – a dispersive wave component which moves to the left, spreading and diminishing in amplitude as it moves, and a portion which contributes to the soliton. The effect on the widths is clearly visible in Figure 9.21. The width of the v-pulse grows steadily while the width of the u-pulse is almost constant. Once again, we find that ω_{cent}^{u} and ω_{cent}^{v} are related by equation (9.89).

Similar results are found when $\alpha = 15°$ (Menyuk, 1988).

The designation of an exact threshold is somewhat arbitrary in these cases as no sharp transition in the pulse behavior is observed. As the amplitude is increased at 30° with $\delta = 0.5$, one finds that when

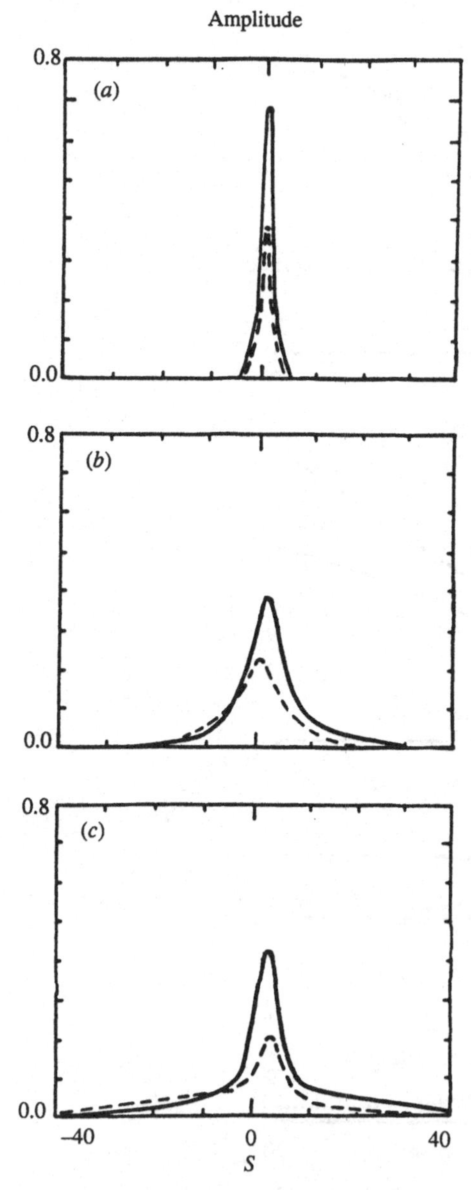

Figure 9.18. Pulse evolution ($A = 0.8$, $\delta = 0.15$, $\gamma = 0.0$, $\alpha = 30°$). Full lines indicate the u-polarisation; broken lines indicate the v-polarisation. (a) $\xi = 0$, (b) $\xi = 5\pi$ (ten-soliton periods) and (c) $\xi = 10\pi$. Reprinted by permission of the Optical Society of America.

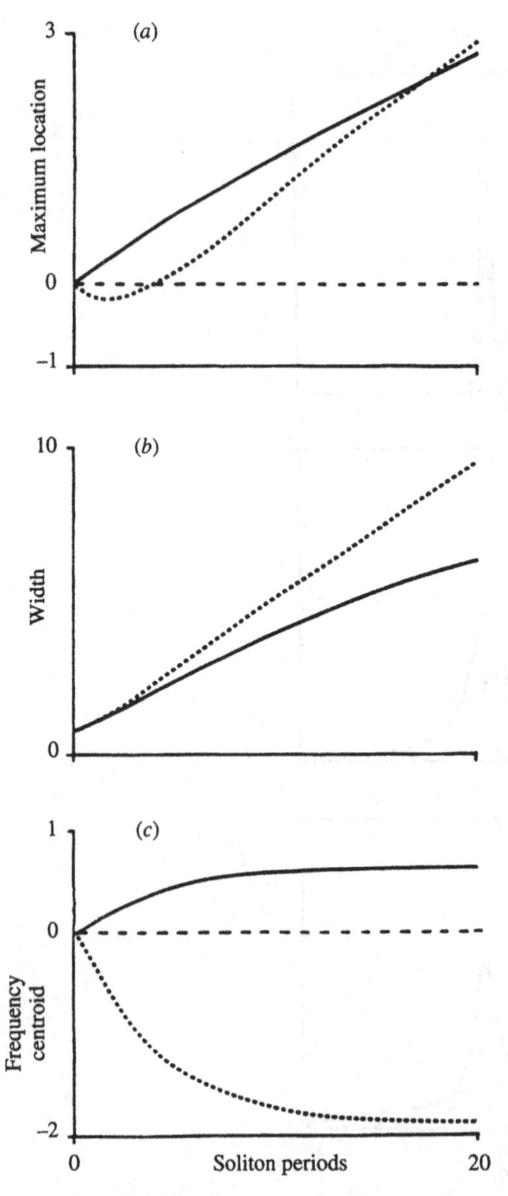

Figure 9.19. Parameter variation with distance along the fiber ($A = 0.8$, $\delta = 0.15$, $\gamma = 0.0$, $\alpha = 30°$). (*a*) The maximum locations s^u_{max} and s^v_{max}, (*b*) the pulsewidths w^u and w^v and (*c*) the frequency centroids ω^u_{cent} and ω^v_{cent}. Full lines indicate the *u*-polarisation; dotted lines indicate the *v*-polarisation. Reprinted by permission of the Optical Society of America.

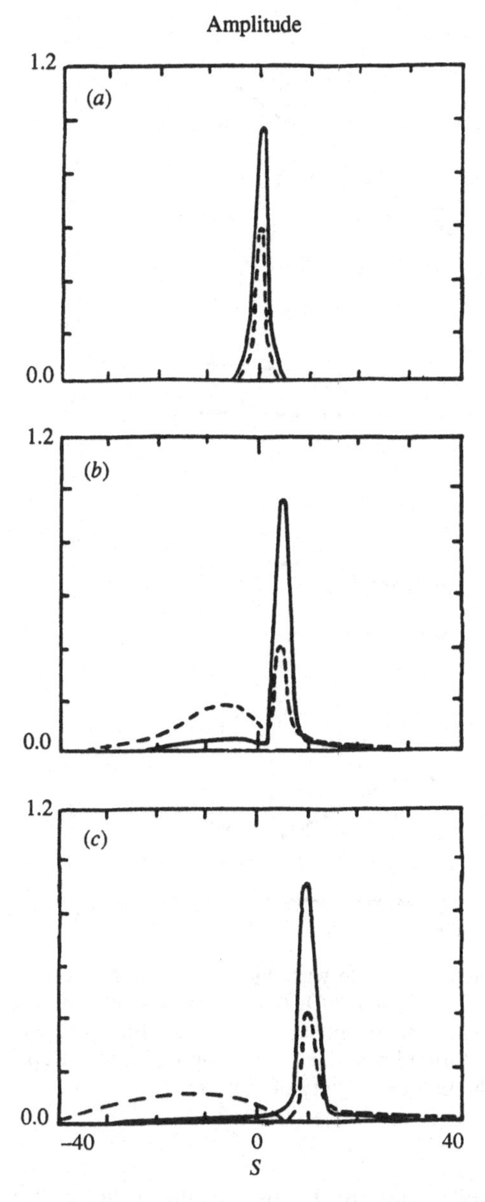

Figure 9.20. Pulse evolution ($A = 1.1$, $\delta = 0.5$, $\gamma = 0.0$, $\alpha = 30°$). Full lines indicate the u-polarisation; broken lines indicate the v-polarisation. (a) $\xi = 0$, (b) $\xi = 5\pi$ (ten-soliton periods) and (c) $\xi = 10\pi$. Reprinted by permission of the Optical Society of America.

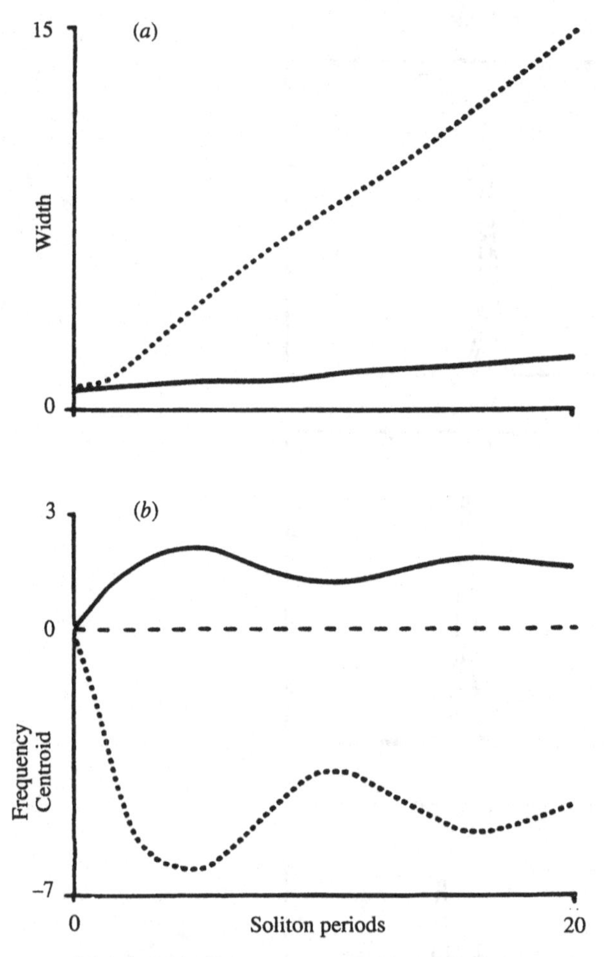

Figure 9.21. Parameter variation with distance along the fiber ($A = 1.1$, $\delta = 0.5$, $\gamma = 0.0$, $\alpha = 30°$). (*a*) The pulsewidths w^u and w^v, and (*b*) the frequency centroids ω^u_{cent} and ω^v_{max}. Full lines indicate the u-polarisation; dotted lines indicate the v-polarisation. Reprinted by permission of the Optical Society of America.

$A = 0.8$, a double-peaked structure forms on the v-pulse but no soliton is created. When $A = 0.9$, the fraction of the v-pulse which ends up travelling with the u-pulse is larger than before, and a soliton is created. The fraction of the v-pulse which travels with the u-pulse continues to increase as A increases. If we choose as our threshold the amplitude required to generate a soliton, we find $A = 0.9$. We can fix the amplitude and vary α instead to determine the limit beyond which solitons can no longer form. Setting $A = 0.9$, we find that when $\alpha \geqslant 40°$, only a continuum is produced. In general, as we decrease α, the amplitude at which solitons first form decreases, reducing to $A = 0.5$ at $\alpha = 0°$ where the NLS equation becomes valid.

9.6 Axial inhomogeneity

In this section, we discuss the effect of variation of fiber parameters, particularly chromatic dispersion, along the fiber axis. Chromatic dispersion, or more precisely the coefficient of second-order dispersion (k''), is composed of material dispersion and waveguide dispersion. The former depends only on the refractive index and the operating wavelength while the latter is also a function of fiber parameters. The wavelength at which the material and waveguide dispersion balance each other is called the zero dispersion point. For an ideal fiber, its parameters are assumed to be axially uniform, hence the zero dispersion wavelength is constant. In practice, waveguide parameters, such as the core radius and the index difference between the core and the cladding in a step-index fiber, fluctuate along the fiber. Consequently, at a fixed carrier wavelength, cancellation between the two types of dispersion over the whole length of the fiber is not possible. In other words, the zero dispersion point fluctuates due to the axial variations. A pulse launched at the zero dispersion wavelength at one point is subjected to both normal and anomalous dispersion when it travels along the fiber.

Equation (9.68) is, therefore, no longer sufficient. It has to be modified to include the effect of axial variations. We assume that the inhomogeneity along the axial direction is small so that $|\lambda \partial \ln(\varepsilon)/\partial z| \ll 1$, where λ is the carrier wavelength, and ε is the dielectric constant. In this case, equations (9.39) and (9.40) hold except that the coefficients k', k'' and k''' are now dependent on the axial coordinate due to the fluctuation. We note that k'' cannot be set to zero as in the case of an ideal fiber. However, since the variation of k''' is small compared to its average value, we can safely set k''' to a constant. Equations (9.39) and (9.40) are normalised as in Section 9.2, except for the time variable, the transformation is modified to

$$s = \left[t - \int^{z} k'(\tilde{z})\mathrm{d}\tilde{z} \right] / t_0 \tag{9.90}$$

to account for the axial dependence of the group delay $k'(z)$. If the group delay is independent of z, equation (9.90) reduces to the original equation. The normalised equation at the zero dispersion point, including the effect of axial inhomogeneity, is given by

$$\mathrm{i}\frac{\partial u}{\partial \xi} - \frac{1}{2} f(\xi) \frac{\partial^2 u}{\partial s^2} - \mathrm{i}\frac{1}{6}\frac{\partial^3 u}{\partial s^3} + |u|^2 u = 0 \tag{9.91}$$

where

$$f(\xi) = t_0 k''/k'''$$

Apart from the modification of the group delay, the effect of axial variation is to reintroduce the second-order dispersion term. Its coefficient takes on both positive and negative values. Equation (9.91) forms the basis of the following analysis. Although equation (9.91) is obtained using heuristic arguments, it can be derived rigorously starting from Maxwell's equations and using the reductive perturbation method developed by Taniuti (1974) and Kodama (1985).

From relation (9.79), the soliton has a frequency shift given by $\Omega_0 \approx 4\sigma$. The normalised second-order dispersion coefficient (f) introduced by this frequency shift is $\Omega_0/\sigma \approx 4$. Hence, fluctuation of $f(\xi)$ on the order of unity is expected to affect the propagation of solitons in equation (9.91). We proceed to estimate the magnitude of the variation of the second-order coefficient. An explicit expression for the coefficient can be obtained from the exact derivation. For a step-index fiber, assuming a transverse electric field, the variation of the second-order coefficient at the zero dispersion point has been estimated to be (Jeunhomme, 1983),

$$\delta k'' = -4\frac{\delta\Delta n}{\Delta n} - 12\frac{\delta a}{a}\ (\mathrm{ps})^2/\mathrm{km} \tag{9.92}$$

where a is the core radius, Δn is the index difference, and $\delta\Delta n$ and δa are the corresponding fluctuations. For example, a variation of the core radius by 1% (Jaeger, 1979) and a negligible variation of the index difference would give a variation of $\pm 0.12\ (\mathrm{ps})^2/\mathrm{km}$ in chromatic dispersion. The magnitude of the variations of the normalised coefficient is given by $|f(\xi)| = 1.5 t_0(\mathrm{ps})$. For a 1 ps pulse, $|f(\xi)| = 1.5$. Therefore, the contribution from the second-order dispersion term in equation (9.91) is not negligible.

Another important parameter governing the axial variation is the correlation length of the fluctuations. The outer diameter of an optical fiber can be controlled carefully in the drawing process through a feedback controlled system using non-destructive and non-contacting measurement methods such as the scanning beam technique, or the forward scattering method (Jaeger, 1979; Marcuse, 1981). On the other hand, the core diameter cannot be monitored directly in the pulling process. If one assumes that the ratio of the core to the cladding diameter is strictly maintained, the variations in the core diameter are then proportional to those in the outer diameter. The diameter variations can originate from vibration in the drawing machine, thermal instabilities such as turbulent cooling of the hot gases, or long-term temperature drift. They can also result from mechanical problems such as variations in a lead screw that is feeding the preform, or from eccentric drawing drums, or from some cogging effect in a gear mechanism (Jaeger, 1979). They give rise to correlation lengths on the order of metres or less. In addition, long periodic variations can result from composition variations in the preform caused by the fabrication process. Variations in the index difference also arise from the fabrication process and possible diffusion of dopant materials. The correlation length in normalised coordinates is given by, $L_c = k'''l_c/t_0^3 = 0.1 \times l_c(\text{km})/t_0^3(\text{ps})$, where L_c is the normalised correlation length and l_c is the correlation length in physical units. For a 1 ps pulse and $l_c \simeq 1$ m, $L_c \simeq 10^{-4}$. The normalised correlation length can, therefore, be very short. We are going to demonstrate that if the correlation length of axial variations is sufficiently short, soliton propagation is possible even when the strength of the fluctuation, $|f(\xi)|$, is large. The axial variation has almost no effect on the evolution characteristics.

First, we consider the different length scales which play a role in the fiber. Two of these are the carrier wavelength and the core radius. They are of the same order, and are much smaller than any other length scales in the fiber. We now introduce the concept of characteristic dispersion length, which is defined as the distance at which the effect of dispersion becomes important. There are two different characteristic dispersion lengths, corresponding to the second- and the third-order dispersion. The second-order dispersion length is defined as $l_s = t_0^2/|k''|$, and the third-order dispersion length is given by $l_t = t_0^3/|k'''|$. In the case of interest in this chapter, the second-order dispersion length l_s fluctuates between a minimum value $l_{sm} = t_0^2/|k''|_{max}$ corresponding to the maximum value of k'', and infinity, corresponding to $k'' = 0$. Equation (9.91) is normalised to the third-order dispersion length. Notice that the normalised second-order coefficient of equation (9.91) is the ratio between the third- and second-order dispersion length, $f(\xi) = l_t/l_s$.

We now look at the different parameter regimes of equation (9.91).

In the regime where the third-order dispersion length is much shorter than the minimum second-order dispersion length, i.e. $l_t \ll l_{sm}$, or equivalently, $f(\xi) \ll 1$, the effect of axial inhomogeneity can be treated as a small perturbation. The evolution characteristics of the fiber are not changed significantly in this case. Soliton propagation near the fluctuating zero dispersion point is not affected. However, this regime corresponds to very stringent control of the fiber parameters such as the core radius. For example, if one requires that $|f(\xi)| = 0.1$ for a 1 ps pulse, the tolerance on the core radius would be 0.1%, an unrealistic demand. This regime can also be achieved by using very short pulses ($t_0 \ll 1$), but then the higher-order non-linear effects which are ignored in equation (9.91) would become important. Equation (9.91) would no longer be sufficient. Another parameter regime of interest is when the third-order dispersion length is much shorter than the correlation length, $l_t \ll l_c$. The variation of the second-order dispersion with the axial coordinate is very slow and can be assumed to be constant over the dispersion length. Therefore, if the carrier wavelength is chosen initially at $k'' \approx 0$, the chromatic dispersion would stay small for a long time. The solitons would adjust adiabatically to the slowly changing dispersion. However, for a 1 ps pulse the third-order dispersion length, l_t, is on the order of kilometres implying a correlation length of the order of hundreds of kilometres, an unrealistic number for most fibers.

The most interesting, and also physically important, regime of parameters is when the correlation length is much shorter than both the minimum second-order dispersion length and the third-order dispersion length, i.e. $l_c \ll l_{sm}$ and $l_c \ll l_t$. Soliton propagation in this regime is possible, and its evolution characteristics are similar to that of the ideal case. This result might seem a bit surprising because the minimum second-order dispersion length can be very small so that the magnitude of the axial variations ($|f(\xi)|$) is very large. However, since the correlation length is assumed to be much shorter than the minimum second-order dispersion length, the coefficient $f(\xi)$ changes sign many times in a distance of length l_{sm}. In other words, the zero dispersion point varies about the zero value so fast that its effect is averaged out, and the pulse is governed by the weaker third-order dispersion.

To illustrate these arguments, let us assume the normalised second-order coefficient is of the form

$$f(\xi) = \frac{1}{\delta} g\left(\frac{\xi}{\Delta}\right) \tag{9.93}$$

where $\Delta = l_c/l_t$, $\delta = l_{sm}/l_t$, and $g(\xi)$ is a function with a maximum amplitude of one. Let us assume that the correlation length is much shorter than the minimum second-order dispersion length. For example, let $\Delta = \delta^2$. Since $\Delta \ll \delta \ll 1$, the criteria $l_c \ll l_{sm}$, and $l_c \ll l_t$ is satisfied. The variation of the second-order dispersion in this case is both large and rapid. We now apply the method of multiple length scale expansions (Nayfeh, 1974) to equations (9.92) and (9.93). We first assume that the envelope $u(\xi)$ depends on both the fast (η_0) and the slow (η_1) variables, defined as

$$\eta_0 = \xi/\delta^2 \qquad (9.94)$$
$$\eta_1 = \xi.$$

Substitution of equations (9.94) into (9.91) and (9.93) gives

$$\frac{i}{\delta^2}\frac{\partial u}{\partial \eta_0} + i\frac{\partial u}{\partial \eta_1} - \frac{1}{2\delta}\,g(\eta_0)\,\frac{\partial^2 u}{\partial s^2} - i\frac{1}{6}\frac{\partial^3 u}{\partial s^3} + |u|^2 u = 0 \qquad (9.95)$$

We then write $u(\eta_0, \eta_1, s)$ as a series in the small parameter δ,

$$u(\eta_0, \eta_1, s) = \sum_{n=0}^{\infty} \delta^n u^{(n)}(\eta_0, \eta_1, s) \qquad (9.96)$$

We substitute equation (9.96) into equation (9.95) and collect terms at the same order of δ. At order δ^{-2}, we have

$$i\,\frac{\partial u^{(0)}}{\partial \eta_0} = 0 \qquad (9.97)$$

Therefore, $u^{(0)}$ is independent of the fast variable. Its slow variable dependence is determined at higher order. At order δ^{-1}, we have

$$i\,\frac{\partial u^{(1)}}{\partial \eta_0} - \frac{1}{2}\,g(\eta_0)\,\frac{\partial^2 u^{(0)}}{\partial s^2} = 0 \qquad (9.98)$$

from which we obtain

$$u^{(1)} = -i\bar{g}(\eta_0)\,\frac{\partial^2 u^{(0)}}{\partial s^2} + \bar{u}^{(1)}(\eta_1, s) \qquad (9.99)$$

where $\bar{g}(\eta_0) = \frac{1}{2}\int^{\eta_0} g(\eta)\mathrm{d}\eta$, and $\bar{u}^{(1)}(\eta_1, s)$ is a function of the slow variable only. At order 1, we have

$$i\,\frac{\partial u^{(2)}}{\partial \eta_0} + i\frac{\partial u^{(0)}}{\partial \eta_1} - \frac{1}{2}\,g(\eta_0)\,\frac{\partial^2 u^{(1)}}{\partial s^2} - i\frac{1}{6}\frac{\partial^3 u^{(0)}}{\partial s^3} + |u^{(0)}|^2 u^{(0)} = 0 \qquad (9.100)$$

from which we obtain the fast time dependence of $u^{(2)}$.

$$u^{(2)}(\eta_0, \eta_1, s) = \mathrm{i}\left\{\mathrm{i}\,\frac{\partial u^{(0)}}{\partial \eta_1} - \mathrm{i}\,\frac{1}{6}\,\frac{\partial^3 u^{(0)}}{\partial s^3} + |u^{(0)}|^2 u^{(0)}\right\}\eta_0$$

$$-\frac{1}{2}\bar{g}^2\,\frac{\partial^4 u^{(0)}}{\partial s^4} - \mathrm{i}\bar{g}\,\frac{\partial^2 \bar{u}^{(1)}}{\partial s^2} + \mathrm{i}\bar{u}^{(2)} \qquad (9.101)$$

where $\bar{u}^{(2)}$ is only a function of the slow variable. From equation (9.101), we see that the function $u^{(2)}(\eta_0, \eta_1, s)$ grows linearly with the fast variable η_0. At a distance of $\eta_0 \sim 1/\delta^2$ or $\xi = 1$, i.e. the third-order dispersion length, $u^{(0)}$ and $u^{(1)}$ have the same order of magnitude. The expansion in equation (9.96) therefore becomes invalid. In order to extend the region of validity of the series expansion, we eliminate the secularity by demanding that its coefficient be zero. Hence,

$$\mathrm{i}\,\frac{\partial u^{(0)}}{\partial \eta_1} - \mathrm{i}\,\frac{1}{6}\,\frac{\partial^3 u^{(0)}}{\partial s^3} + |u^{(0)}|^2 u^{(0)} = 0 \qquad (9.102)$$

which is exactly the same as equation (9.68) (with $\gamma = 0$). Therefore, at leading order, the pulse evolves as if there is no axial variation. Similarly, at the next order of δ, we find that $\bar{u}^{(1)}(\eta_1, s)$ is governed by the linearised form of equation (9.102). To first order in δ, the pulse envelope is given by

$$u(\eta_0, \eta_1, s) = u^{(0)}(\eta_1, s) + \delta\left\{-\mathrm{i}\bar{g}(\eta_0)\,\frac{\partial^2 u^{(0)}}{\partial s^2} + \bar{u}^{(1)}(\eta_1, s)\right\} \qquad (9.103)$$

The modification in the pulse envelope due to the axial variation of the fiber only appears at the first order in the small parameter δ, even though the amplitude of the fluctuation of the normalised chromatic dispersion is very large ($\sim 1/\delta$). Further calculations show that the secularities which appear at higher order can be eliminated in a similar fashion. Equation (9.103) has a range of validity which is at least of order $\xi \sim \delta^{-2}$. In this argument, the assumption $\Delta = \delta^2$ is not essential. The derivation remains valid as long as $\Delta \ll \delta$. In fact, numerical simulations show that soliton propagation is still possible even when $l_c \sim l_{sm}$ ($\Delta \sim \delta$). Our estimates are conservative because we overestimate the effect of the second-order dispersion length by using the maximum second-order dispersion.

Next we integrate equation (9.91) numerically using different functions for the normalised second-order coefficient $f(\xi)$. In practice, the fiber variations are very complex. If the fluctuation in the drawing process is random, or partially random, the waveform of the axial variation can be represented as a superposition of random Fourier components. We first study the case where $f(\xi)$ consists of a single Fourier component, i.e. $f(\xi) = f_0 \sin(\kappa\xi + \chi)$, where f_0, κ and χ are constant parameters. This model allows us to study the effect of different amplitudes and fluctuation lengths in detail. The initial pulse is chosen to be $2\,\mathrm{sech}\,(s)$ in all cases that we report here. This choice is somewhat arbitrary; however, if the initial pulse amplitude is too weak ($\lesssim 1$), the pulse will not evolve into a soliton, and if the initial amplitude is too large ($\gtrsim 4$), the pulse may break up into multiple solitons. Studies that we have carried out on Gaussian waveforms yield qualitatively similar results.

The amplitude and the wavenumber of the fluctuations are varied between 10^{-2} and 10^2. The equation is integrated up to $\xi = 15$. While this represents 10^3 cycles if $\kappa = 500$, it is only one-tenth of a cycle for the case $\kappa = 0.05$. We do not integrate past $\xi = 15$ because with a pulsewidth on the order of 1 ps, $\xi = 15$ already represents 200 km, and other effects such as dissipation have to be taken into account. The results obtained agree with previous analysis. When f_0 is small ($\lesssim 0.1$), the evolution of the initial pulse is essentially the same at all wavenumbers κ. If the fluctuation length is very long ($\kappa \lesssim 0.1$), the evolution is sensitive to the initial phase χ. For example, if $\chi = 0$, the pulse broadens for large f_0 ($\gtrsim 10$) and no soliton emerges. In this case, the zero dispersion point moves towards longer wavelengths and the pulse is in the normal dispersion regime. In this regime, the non-linearity and the dispersion both work to broaden the pulse. If $\chi = \pi$, the zero dispersion point moves towards shorter wavelength leaving the pulse in the anomalous dispersion regime. The pulse first contracts due to the interplay between the anomalous dispersion and the non-linearity. A soliton is formed. As the strength of the dispersion increases, the pulse begins to broaden and finally disperses away. Figure 9.22(a) shows the pulse shape for $f_0 = 10$, $\kappa = 0.4$, $\chi = 0$ after $\xi = 1$, and Figure 9.22(b) gives the corresponding frequency spectrum. The pulse has broadened considerably, and there is no splitting in the frequency spectrum. The initial pulse $2\,\mathrm{sech}\,(s)$ is shown in the broken lines for comparison. When $\chi = \pi$, as shown in Figures 9.23(a) and (b), the soliton peak is in the negative frequency part of the spectrum. When ξ increases, the soliton starts to broaden due to increase in chromatic dispersion. When the variations are very short ($\kappa \gtrsim 10$), the initial phase is not essential, and the propagation characteristics are almost indistinguishable from those of the ideal case. We find that solitons propagate even in a regime where the

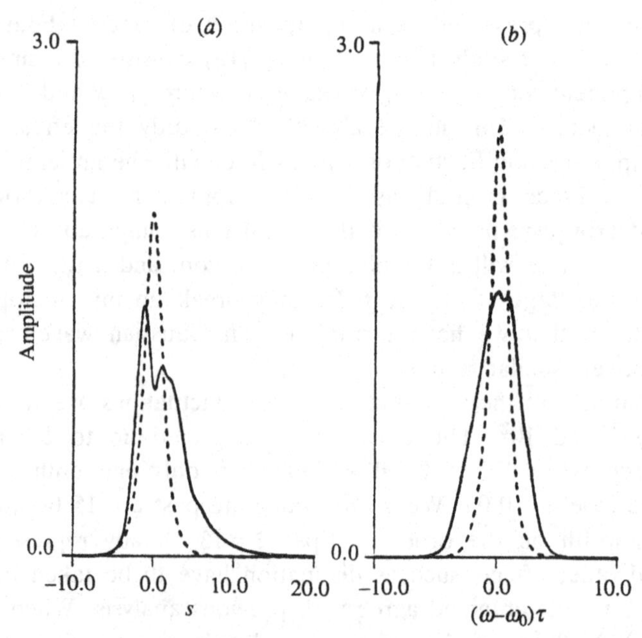

Figure 9.22. (*a*) The pulse shape of an initial profile 2 sech(s) at $\xi = 1$, the axial inhomgeneity is given by $f_0(\xi) = 10\sin(0.4\xi)$. (*b*) The corresponding frequency spectrum. The pulse is broadened and distorted. There is no splitting in the frequency spectrum. The initial pulse is shown in the broken line for comparison. Reprinted by permission of the IEEE.

analytic arguments are not applicable, i.e. when the fluctuations amplitude is large but the correlation length is not short enough. For example, in the case of $f_0 = 100$, and $\kappa = 40$, the pulse evolves differently; it takes $\xi = 6$ for the two-peak structure to develop in the frequency spectrum while it takes only $\xi = 1.5$ for the ideal case. However, a soliton with a somewhat different amplitude and pulse-width finally emerges.

We now simulate the axial fluctuations by using the following series (Rice, 1954),

$$f(\xi) = f_0 \sum_{n=0}^{N} \exp(-\alpha r_n^2) \sin(\kappa_n \xi + \chi_n) \tag{9.104}$$

where $\kappa_n = \kappa_0 - r_n \Delta\kappa$, r_n is a random number $-1 \leqslant r_n \leqslant 1$, and χ_n is the random phase. Equation (9.104) represents a finite Fourier series with N wavenumbers randomly distributed between two cut-offs at $\kappa_0 \pm \Delta\kappa$. The cut-offs are introduced because physically there are neither very long nor very short correlations in a fiber. The mag-

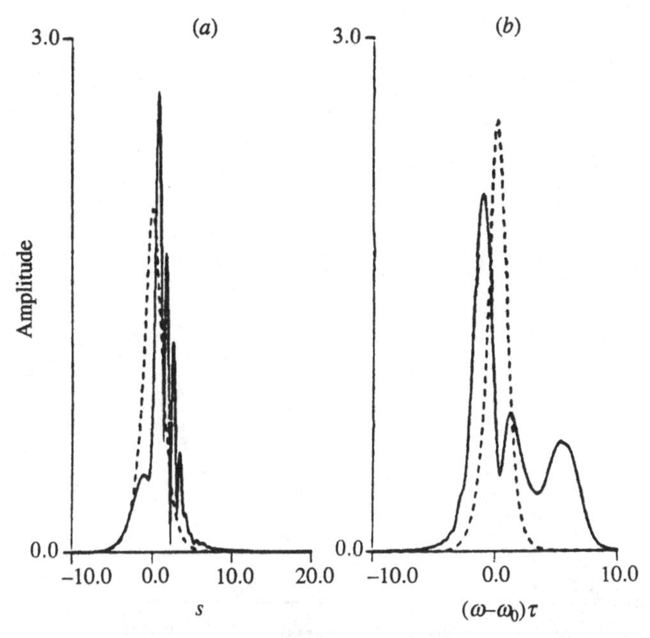

Figure 9.23. (*a*) The pulse shape of an initial profile 2 sech(s) at $\xi = 1$, the axial inhomgeneity is given by $f_0(\xi) = -10\sin(0.4\xi)$. (*b*) The corresponding frequency spectrum. The soliton can be identified in the negative part of the frequency spectrum. The initial pulse is shown in the broken line for comparison. Reprinted by permission of the IEEE.

nitude of the Fourier components are Gaussian modulated with the maximum value f_0 at κ_0. The half-width of the modulation is controlled by the parameter α. Figure 9.24 shows one realisation of the Fourier component distribution and the corresponding amplitudes for the case $\alpha = 10$, $N = 40$, $f_0 = 10$, $\kappa_0 = 400$ and $\Delta\kappa = 400$. Figure 9.25 gives the corresponding variations of $f(\xi)$ as a function of ξ.

The parameters $\Delta\kappa$ is chosen to be κ_0 for all the cases reported here. It cuts off short wavelength fluctuation but does not affect the long wavelength variations. When f_0 is small ($\lesssim 0.1$), the results again resemble those of the ideal case for all center wavenumbers κ_0, modulation bandwidths α, and numbers of Fourier components N; solitons emerge from the initial pulse. If f_0 is not small ($\gtrsim 10$), the results divide roughly into two categories – those of large and small α. If the spectrum of the axial fluctuation is sharp, i.e. α is large, the evolution of the initial pulse is similar to the results where there is only a single Fourier component with amplitude f_0 and wavenumber

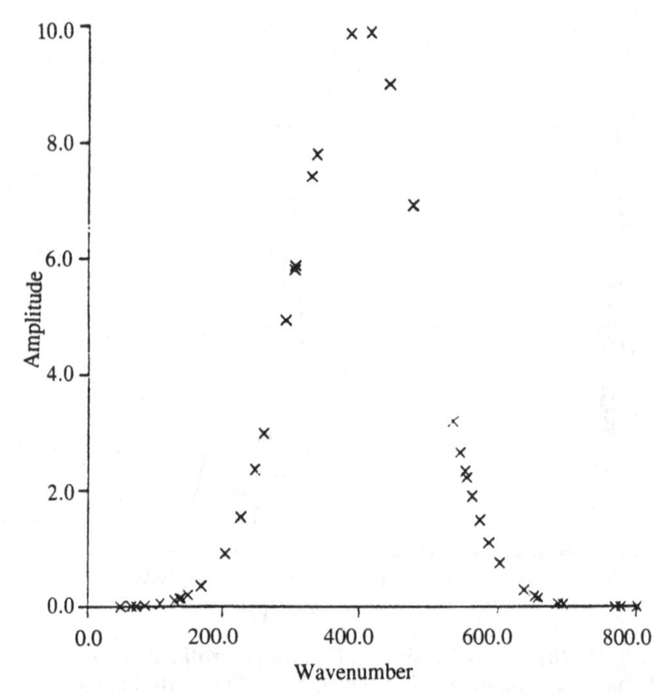

Figure 9.24. One realisation of the spectrum of the axial inhomogeneity with 40 Fourier components, modulation bandwidth $\alpha = 10$, center wavenumber $\kappa_0 = 400$, $f_0 = 10$, and cut-off wavenumbers amplitude at 0 and 800. Reprinted by permission of the IEEE.

κ_0. Solitons emerge if $\kappa_0 \gg f_0$. For example, Figure 9.26 gives the pulse amplitude at $\xi = 10$ for the axial variation shown in Figures 9.24 and 9.25, i.e. the case with $\alpha = 10$, $N = 40$, $f_0 = 10$, $\kappa_0 = 400$ and $\Delta\kappa = 400$. It is the same as the result shown in Figure 9.9 for the ideal case. The variation of the evolution characteristics among different realisations of the Gaussian-modulated random wavenumbers are small. We have used both 40 and 100 Fourier modes; there is no significant change in behavior. When the axial variation has a broad spectrum (α small), there are relatively large contributions from the long wavelength components. In this case, the results differ from the case where only a single component at the central wavenumber κ_0 is kept. The evolution of the pulse is dominated by the long wavelength components and their initial phases. The initial pulse broadens and no soliton emerges.

We have shown in Section 9.4 that for an axially uniform fiber, it is possible to launch solitons near the zero dispersion point of single-

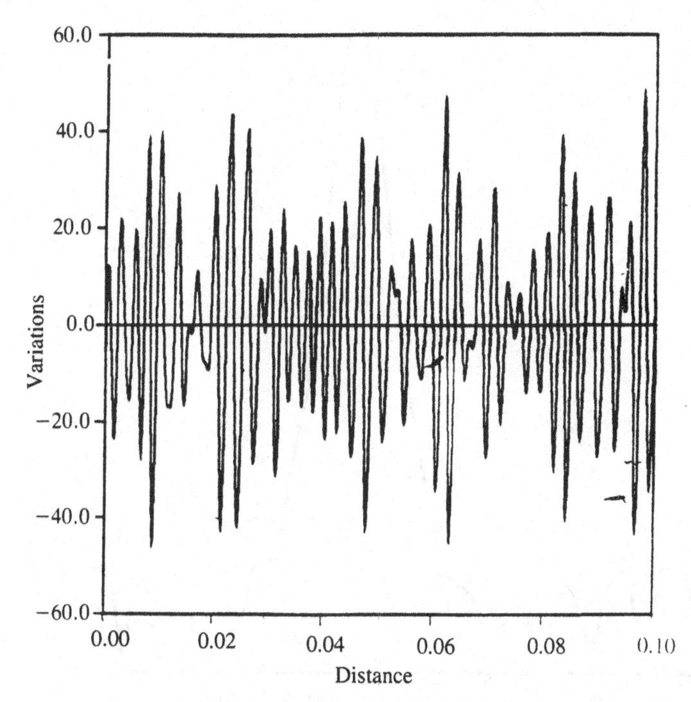

Figure 9.25. The axial variation $f_0(\xi)$ as a function of ξ for the spectrum shown in Figure 9.24. Reprinted by permission of the IEEE.

mode optical fibers. If the effect of axial inhomogeneity is included, we note that the variations generally have large amplitudes. A physically important regime of parameters has been found analytically, and extended by results from numerical simulations, in which soliton propagation is nonetheless possible. If the correlation length of the axial inhomogeneity is short compared with the second- and the third-order dispersion lengths, the effect of the fluctuations is averaged out. The evolution characteristics of the pulse are then described by the ideal equation.

9.7 Soliton switches

Recent interests in optical fibers switching devices have produced a number of different designs such as the directional coupler (Trillo *et al.,* 1988), the optical-fiber Kerr gate (Kitayama *et al.,* 1985), and the interferometer (LaGasse *et al.,* 1988). In general, intensity induced switching generates undesirable side effects like

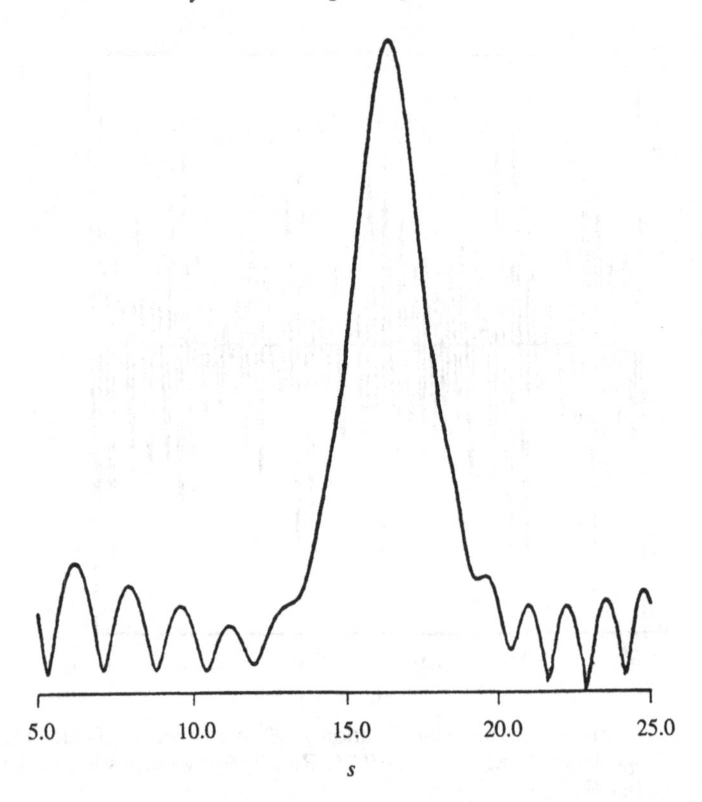

Figure 9.26. The pulse amplitude of an initial profile 2 sech(s) at $\xi = 10$. The axial variation is given by that shown in Figures 9.24 and 9.25 for $N = 40$, $\alpha = 10$, $\kappa_0 = 400$, $f_0 = 10$, and $\Delta\kappa = 400$. The soliton pulse is identical to that shown in Figure 9.10 for the ideal case. Reprinted by permission of the IEEE.

pulse reshaping and breaking which degrade the performance of the device. For example, in a dual-core non-linear coupler, this limits the achievable energy transfer at the switching power to about 50%. In this section, we study a fiber optical switch in an interferometer configuration. The switch is illustrated schematically in Figure 9.27. A signal pulse in the form of a soliton is introduced at one polarisation. It is divided into two and each portion goes down one arm of a Mach–Zehnder interferometer. In one of these arms, a switching pulse is introduced at the other polarisation. The switching pulse would shift the phase of the signal pulse in that arm such that when it is recombined with the signal pulse from the other arm, they may interfere destructively rather than constructively. The advantage of using solitons is that they do not change their shapes when interacting

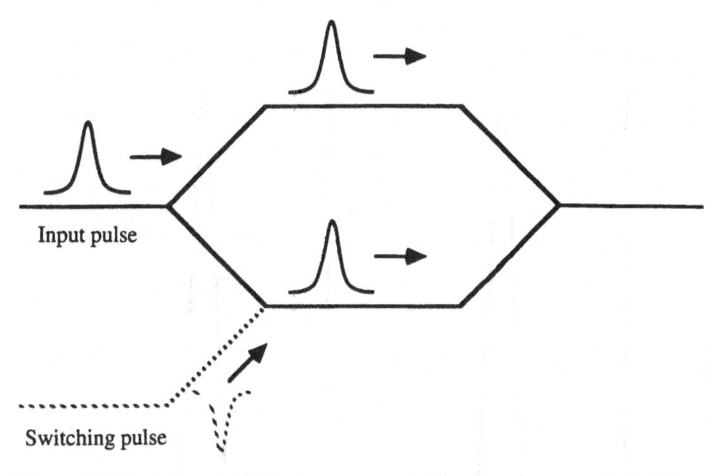

Figure 9.27. Schematic of the proposed interferometer switch. The polarisation of the switching pulse (broken curve) is orthogonal to that of the signal pulse (full curve).

with other pulses. They merely undergo a shift in position and phase. The equations governing pulse evolution in a birefringence fiber have been derived in Section 9.2, they are

$$\mathrm{i}\frac{\partial u}{\partial \xi} + \mathrm{i}\gamma u + \mathrm{i}\delta\frac{\partial u}{\partial s} + \frac{1}{2}\frac{\partial^2 u}{\partial s^2} + (|u|^2 + B|v|^2) = 0 \quad (9.105)$$

$$\mathrm{i}\frac{\partial v}{\partial \xi} + \mathrm{i}\gamma v - \mathrm{i}\delta\frac{\partial v}{\partial s} + \frac{1}{2}\frac{\partial^2 v}{\partial s^2} + (B|u|^2 + |v|^2)v = 0$$

Recall from the definition that $B = \frac{2}{3}$ for a linearly birefringent fiber ($\theta = 0$), and $B = 2$ for a circularly birefringent fiber ($\theta = 90°$). Since the length of the interferometer is much shorter than the typical dissipation length scale, we set $\gamma = 0$ in the following discussion. Notice that if both the signal pulse and the switching pulse are of the same polarisation, i.e. either $u = 0$, or $v = 0$, equations (9.105) are reduced to the familiar NLS. However, this case is of little interest in switch application. For an elliptically birefringent fiber and pulses of different polarisation, equations (9.105) in general are not integrable. As a result, after interacting with the switching pulse, the soliton develops a shadow; a portion of the other polarisation becomes slaved to the soliton and travels with it even after the interaction. For example, the pulse envelope of the signal pulse travelling in a linearly birefringent fiber ($B = \frac{2}{3}$) before and after the interaction is plotted in Figure 9.28. Note that part of the switching pulse at the other polarisation (broken curve) is attached to the soliton after switching. More

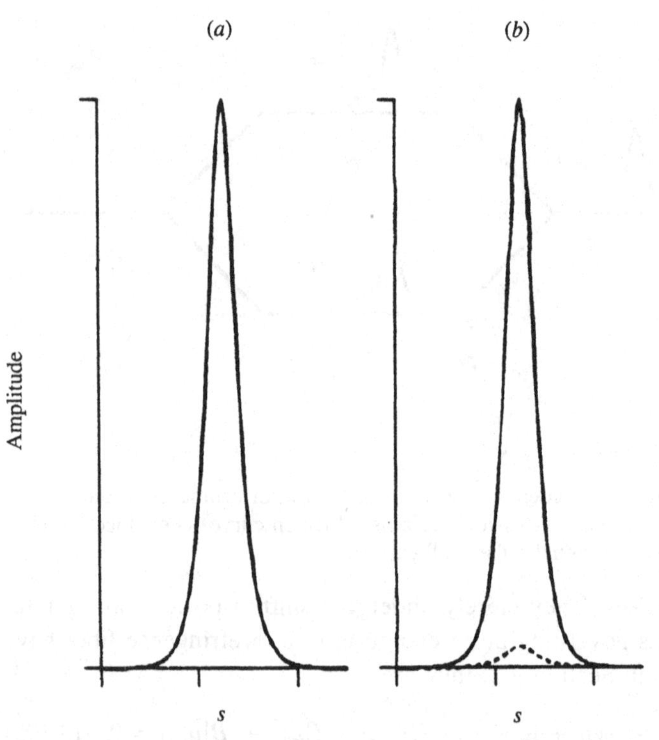

Figure 9.28. The pulse amplitude of a soliton travelling in a linearly birefringent fiber ($\theta = 0$) (a) before and (b) after interaction with a switching pulse is plotted. A shadow is observed after the interaction. The initial profile for the signal pulse is 2 sech(s), and that of the switching pulse is sech(s). The parameter δ is 1.

important, apart from the expected phase shift and position shift, the time delay and the central frequency of the soliton are also changed. In Figure 9.29, we plot the time and the phase shifts of the soliton *versus* the distance travelled along the fiber. After the initial jump due to the interaction with the switching pulse, the phase and time shifts continue to change linearly with distance, indicating that the frequency and time delay of the signal pulse are also affected. These changes are devastating to the performance of the interferometer switch since its output would depend on where the pulses from the two arms are recombined.

The problems of shadows can be avoided if one uses elliptically birefringent fibers with birefringence angle $\approx 35°$ ($B = 1$). At this angle, equation (9.105) is known as the Manakov equation which is integrable (Manakov, 1974). No shadows are formed and only the

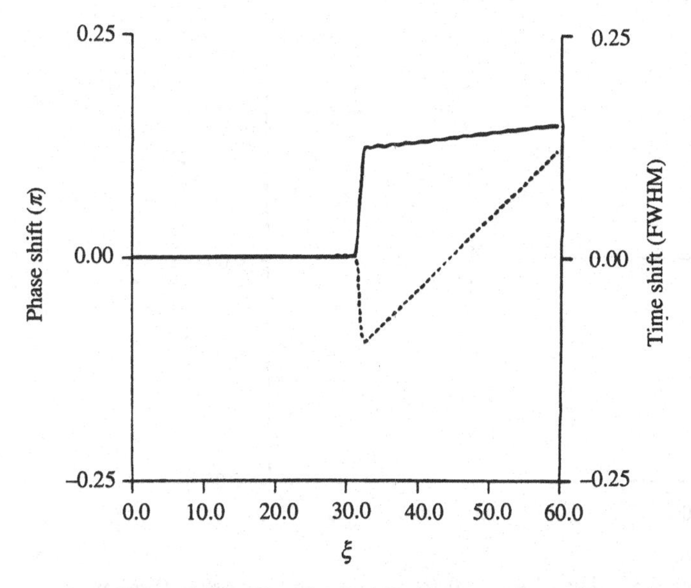

Figure 9.29. The phase shift (full curve) and time shift (broken curve) of the soliton in Figure 9.28 are plotted *versus* the distance along the fiber. The phase shift and time shift continue to change linearly after the interaction with the switching pulse.

time shift and phase shift of the soliton are changed by the interaction. Figure 9.30 shows the phase and time shift of the soliton along the fiber. The curves stay constant after the jumps. The shift in position and phase can, in principle, be calculated from the spectral transform method since the equations are now integrable. In particular, if both the signal pulse and the switching pulse are solitons, it can be shown that the phase shift is given by

$$\tan \psi_1 = \frac{4\delta A_2}{4\delta^2 + A_1^2 - A_2^2} \tag{9.106}$$

and the time shift is

$$\Delta s = \frac{1}{A_1} \ln \left[\frac{4\delta^2 + (A_1 + A_2)^2}{4\delta^2 + (A_1 - A_2)^2} \right] \tag{9.107}$$

where A_1 and A_2 are the amplitude of the signal pulse and the switching pulse respectively. Equations (9.106) and (9.107) are plotted in Figure 9.31 for unit signal pulse strength and different values of δ. Both the phase and time shifts decrease as the relative speed δ between the solitons increases. As the strength of the switching pulse increases, the phase shift increases and approaches π for large amplitudes. By contrast the time shift first increases with A_2 but

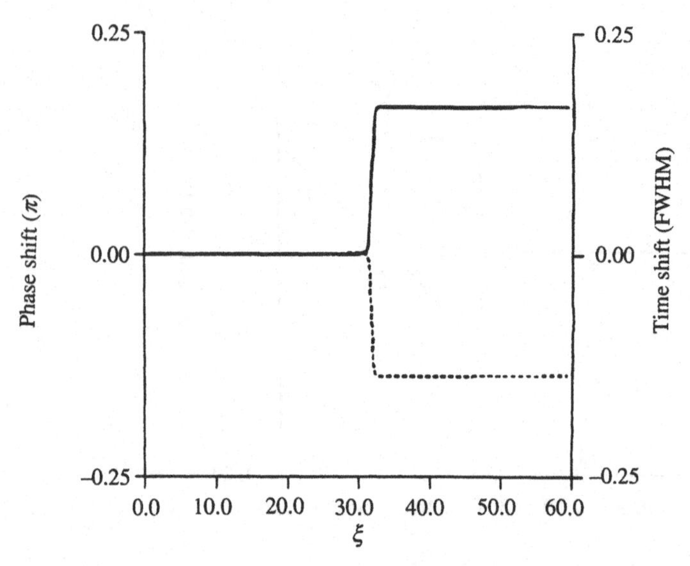

Figure 9.30. The phase shift (full curve) and time shift (broken curve) of a soliton travelling in a fiber with birefringence angle ($\theta \approx 35°$) are plotted. After the jump induced by the interaction with the switching pulse, the time shift and phase shift remain constant. The initial pulse profiles and δ are the same as those in Figure 9.28.

eventually approaches zero. Though not necessary, it is also advantageous to use solitons for the switching pulse since their pulse shape do not change. Other types of switching profiles, a Gaussian for example, would evolve or even split after injection into the interferometer, thus complicating the performance of the switch. The difficulty of achieving π phase shift by one switching pulse can be easily overcome by using two, each shifting the phase of the signal pulse by $\pi/2$. However, even with a π phase shift, total destructive interference would not occur when the signals from both arms of the interferometer recombine due to the time offset introduced by the switching action. In Figure 9.32, we plot the amplitude of the recombined signal pulse for different values of time offset. The original signal pulse is also plotted for comparison. Hence, for a signal-to-noise ratio of 1 to 10, the time offset introduced by switching should be less than $\frac{1}{4}$ FWHM. In general, equations (9.106) and (9.107) can be used to determine the parameters of the switching pulses for a given contrast ratio.

In summary, we have shown that the proposed interferometer switch utilising solitons is feasible if the birefringence angle is chosen to be 35°. At other birefrigence angles, shadows attach to the signal

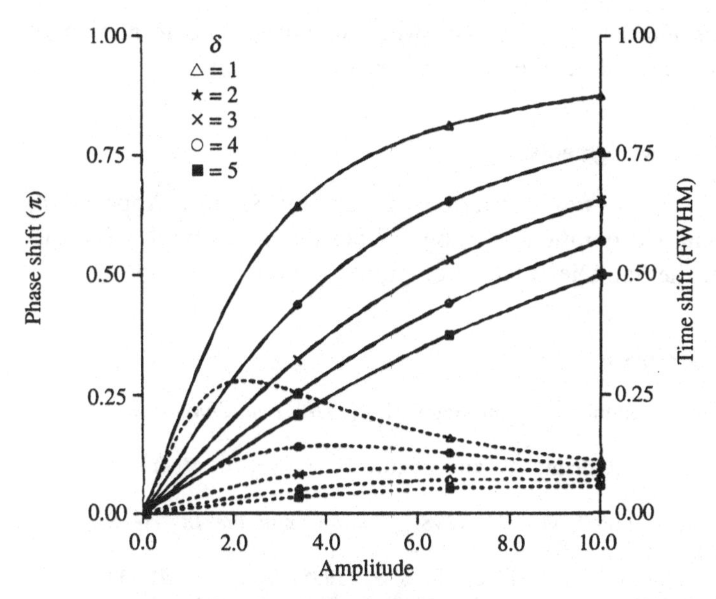

Figure 9.31. The phase shift (full curve) and time shift (broken curve) of a soliton resulting from interaction with another soliton are plotted for different values of relative speed δ. The birefringence angle is 35°. The amplitude of the signal pulse is chosen to be unity.

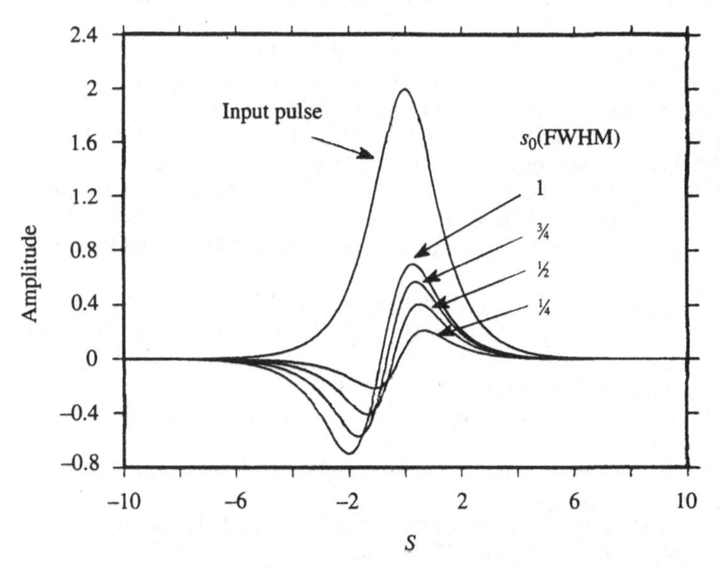

Figure 9.32. The resultant amplitude of two solitons with a π-phase difference are plotted with different time offset (s_0). The sum of the two solitons is graphed for comparison.
$f(s) = \text{sech}(s) - \text{sech}(s - s_0)$.

pulse after interacting with the switching pulse. The formation of shadows severely affect the performance of the switch.

Acknowledgments

Our work was supported in part by Science Applications International Corporation. Portions of the numerical work were carried out at the San Diego Supercomputing Center.

References

Ablowitz, M. J., Ramani, A. and Segur, H. (1978) *Lett. Nuovo Cimento,* **23**, 333–8.
Ablowitz, M. J., Ramani, A. and Segur, H. (1980) *J. Math. Phys.,* **21**, 715–21; 1006–15.
Ablowitz, M. J. and Segur, H. (1981) *Solitons and the Inverse Scattering Transform* (SIAM: Philadelphia, PA).
Blow, K. J., Doran, N. J. and Nelson, B. P. (1985) *Opt. Lett.,* **10**, 393–5.
Burns, W. K. and Milton, A. F. (1988) In T. Tamir (ed.), *Guided-Wave Optoelectronics,* pp. 89–144 (Springer: Berlin).
Chen, H. H., Lee, Y. C. and Liu, C. S. (1979) *Physica Scripta*, **20**, 490–2.
Chu, P. L. and Desem, C. (1985) *Electron. Lett.,* **21**, 228–9.
Doran, N. J. and Wood, D. (1988) *Opt. Lett.,* **13**, 56–8.
Gouveia-Neto, A. S., Faldon, M. E. and Taylor, J. R. (1988) *Opt. Lett.,* **13**, 770–2.
Hasegawa, A. and Kodama, Y. (1981) *Proc. IEEE,* **69**, 1145–50.
Hasegawa, A. and Tappert, F. (1973) *Appl. Phys. Lett.,* **23**, 142–4.
Islam, M. N., Gordon, J. P. and Poole, C. D. (1989) *Technical Digest of the Conference on Lasers and Electro-Optics 1989*, postdeadline paper PD-19. (Optical Society of America: Washington, DC).
Jaeger, R. E. (1979) In B. Bendow and S. S. Mitra (eds.), *Fiber Optics, Advances in Research and Development*, pp. 33–53 (Plenum: New York).
Jeunhomme, Luc B. (1983) *Single-Mode Fiber Optics* (Marcel Dekker, New York).
Kaminow, I. P. (1981) *IEEE J. Quantum Electron.,* **17**, 15–22.
Kitayama, K. I., Kimura, Y. and Seikai, S. (1985) *Appl. Phys. Lett.,* **46**, 317–19.
Kodama, Y. (1985) *J. Stat. Phys.,* **39**, 597–614.
LaGasse, M. J., Liu-Wong, D., Fujimoto, J. C. and Haus, H. A. (1989) *Opt. Lett.,* **14**, 311–13.
Lichtenberg, A. J. and Lieberman, M. A. (1983) *Regular and Stochastic Motion,* Chap. 2 (Springer: New York).
Maker, P. D. and Terhune, R. W. (1965) *Phys. Rev. A,* **137**, 801–18.
Manakov, S. V. (1974) *Sov. Phys. JETP*, **38**, 248–53.
Marcuse, D. (1981) *Principles of Optical Fiber Measurement* (Academic Press: New York).
Menyuk, C. R. (1986) *Phys. Rev. A,* **33**, 4367–74.
Menyuk, C. R. (1987a) *Opt. Lett.,* **12**, 614–16.

Menyuk, C. R. (1987b) *IEEE J. Quantum Electron.*, **23**, 174–6.
Menyuk, C. R. (1988) *J. Opt. Soc. Am. B*, **5**, 392–402.
Menyuk, C. R., Wai, P. K. A., Chen, H. H. and Lee, Y. C. (1987) In F. Dressel (ed.), *Published Proc. Fourth Conf. on Applied Mathematics and Computing: US Army Report 87-1*, pp. 373–86 (US Government Printing Office: Washington DC).
Mollenauer, L. F., Stolen, R. H. and Gordon, J. P. (1980) *Phys. Rev. Lett.*, **45**, 1095–8.
Mollenauer, L. F., Stolen, R. H., Gordon, J. P. and Tomlinson, W. J. (1983) *Opt. Lett.*, **8**, 289–90.
Mollenauer, L. F., Gordon, J. P. and Islam, M. N. (1986) *IEEE J. Quantum Electron.*, **22**, 157–73.
Nayfeh, A. (1973) *Perturbation Methods* (Wiley: New York).
Press, W. H., Flannery, B. P., Teukolsky, S. A. and Vetterling, W. T. (1986) *Numerical Recipe* (Cambridge University Press: Cambridge).
Rice, S. O. (1954) In N. Wax (ed.), *Selected Papers on Noise and Stochastic Processes*, pp. 133–294 (Dover: New York).
Shirasaki, M., Haus, H. A. and Liu Wong, D. (1987) *Lasers and Electro-optics, Technical Digest Series* 14, pp. 284–6.
Taniuti, T. (1974) *Prog. Theor. Phys. Suppl.*, **55**, 1–35.
Trillo, S., Wabnitz, S., Wright, E. M. and Stegeman, G. I. (1988) *Opt. Lett.*, **13**, 672–4.
Wai, P. K. A., Menyuk, C. R., Chen, H. H. and Lee, Y. C. (1986) In R. Fleming and A. E. Siegman (eds.), *Ultrafast Phenomena V*, pp. 65–7 (Springer: New York).
Wai, P. K. A., Menyuk, C. R., Chen, H. H. and Lee, Y. C. (1987) *Opt. Lett.*, **12**, 628–30.
Wai, P. K. A., Chen, H. H. and Lee, Y. C. (1989) *Phys. Rev. A*, **41**, 426–39.
Zakharov, V. E. and Shabat, A. B. (1972) *Sov. Phys. JETP*, **34**, 62–9.

10

Dark optical solitons

A. M. WEINER

10.1 Introduction

Although soliton phenomena arise in many distinct areas of physics, the single-mode optical fiber has been found to be an especially convenient medium for their study. As described in the previous chapters of this book, soliton propagation of bright optical pulses has been verified in a number of elegant experiments performed in the negative group velocity dispersion (GVD) region of the spectrum; most recently, transmission of 55-ps optical pulses through 4000 km of fiber was achieved, by use of a combination of non-linear soliton propagation to avoid pulse spreading and Raman amplification to avoid losses (Mollenauer and Smith, 1988). For positive dispersion ($\lambda < 1.3\ \mu$m in standard single-mode fibers), bright pulses cannot propagate as solitons and the interaction of the non-linear index with GVD leads to spectral and temporal broadening of the propagating pulses. These effects form the basis for the fiber-and-grating pulse compressor (Nakatsuka *et al.*, 1981; Grischkowsky and Balant, 1982; Tomlinson *et al.*, 1984), which was utilised to produce the shortest optical pulses (6 fs) ever reported (Fork *et al.*, 1987). For both signs of GVD, the experimental results are in quantitative agreement with the predictions of the non-linear Schrödinger equation (NLS).

Although bright solitons are allowed only for negative GVD, the NLS admits other soliton solutions for positive GVD (Hasegawa and Tappert, 1973; Zakharov and Shabat, 1973). These solutions are 'dark pulse solitons', consisting of a rapid dip in the intensity of a broad pulse of a c.w. background. The fundamental dark soliton is an anti-symmetric function of time, with an abrupt π phase shift and zero intensity at its center. Other dark solitons with a reduced contrast and a lesser, more gradual phase modulation also exist. Although the existence of dark pulse solitons in optical fibers was first predicted in the early 1970s (Hasegawa and Tappert, 1973), due to difficulty in generating the required input dark pulses, experimental verification was reported only very recently. Nevertheless, adequate experimental techniques for shaping ultrashort light pulses have now been developed, and these techniques have been applied to study the basic properties of dark solitons in fibers (Emplit *et al.*, 1987; Krokel *et al.*, 1988; Weiner *et al.*, 1988; 1989). The results of these studies are the subject of this chapter.

This chapter is structured as follows. In Section 10.2 we introduce the basic properties of dark solitons and summarise the predicted differences between bright and dark solitons. In Section 10.3 we discuss experimental verification of dark soliton propagation; included in this section is a description of the pulse-shaping techniques which made these experiments possible. In Section 10.3 we also describe propagation of dark pulses at power levels higher than those required for soliton propagation; in particular, we discuss the formation of multiple black and gray solitons from single dark input pulses and report on temporal and spectral self-shifts of dark solitons due to stimulated Raman scattering. Finally, in Section 10.4 we summarise.

10.2 Basic properties of dark optical solitons

10.2.1 *The non-linear Schrödinger equation*

As we have seen in previous chapters, pulse propagation in single-mode fibers in the anomalous dispersion regime is well described by the NLS equation. Dark pulse propagation in fibers in the normal dispersion regime is also described by this equation. In the following we will introduce the variables, the normalisations, and the specific form of the NLS equation we will use for our treatment of dark optical solitons.

It was shown in Chapter 1 that the pulse envelope function $a(z, \tau)$ satisfies the non-linear equation

$$\mathrm{i}\left(\frac{\partial a}{\partial z} + k_1 \frac{\partial a}{\partial \tau}\right) = \frac{-k_2}{2} \frac{\partial^2 a}{\partial \tau^2} + \kappa |a|^2 a \tag{10.1}$$

where z is the distance along the fiber, τ is the time, and $a(z, \tau)$ is related to the electric field (with carrier) by

$$E(z, \tau) = \mathrm{Re}\{a(z, \tau) \exp(\mathrm{i}(\omega_0 \tau - k_0 z))\} \tag{10.2}$$

In this expression ω_0 and k_0 are the angular frequency and the propagation constant of the carrier, respectively. The other parameters, defined by $k_1 = \partial k/\partial\omega$, $k_2 = \partial^2 k/\partial\omega^2$, and $\kappa = k_0 n_2/2n_0$, are related respectively to the group velocity, dispersion and the non-linearity. n_2 is the non-linear index, defined by $n = n_0 + \frac{1}{2}n_2|E|^2$, and n_0 is the low intensity refractive index of the fiber core. Loss is not included in equation (10.1); this is an approximation which is valid for several recent demonstrations of dark soliton propagation (Krokel *et al.*, 1988; Weiner *et al.*, 1988; 1989).

Equation (10.1) is transformed into a dimensionless NLS equation

$$\frac{\partial u}{\partial(z/z_0)} = i\frac{\pi}{4}\left[\pm\frac{\partial^2 u}{\partial(t/t_0)^2} - 2|u|^2 u\right] \tag{10.3}$$

by introducing the normalised variables z/z_0, t/t_0 and $u = a/E_0$, where

$$t = \tau - k_1 z \tag{10.4a}$$

$$z_0 = \frac{\pi t_0^2}{2|k_2|} = \frac{\pi^2 c^2 t_0^2}{\lambda|D(\lambda)|} \tag{10.4b}$$

and

$$E_0 = \left(\frac{\pi}{2\kappa z_0}\right)^{1/2} = \left(\frac{\lambda}{2n_2 z_0}\right)^{1/2} \tag{10.4c}$$

In these equations t is the retarded time measured in a reference frame moving along the fiber at the group velocity, and t_0 is a time normalisation which can be chosen arbitrarily (although it usually will be most convenient to select a value of t_0 which is characteristic of the input pulse to the fiber). $D(\lambda) = \lambda^2 d^2n/d\lambda^2$ is the dimensionless dispersion parameter; λ is the vacuum wavelength and c is the speed of light. The sign associated with the $\partial^2 u/\partial(t/t_0)^2$ term is positive for positive or normal dispersion ($D(\lambda) > 0$); this is a necessary condition for the existence of dark solitons and will be assumed throughout the remainder of this chapter. For $D(\lambda) < 0$, the negative sign is used; this is the regime of bright solitons.

These normalisations have been used extensively both in the study of soliton propagation in fibers in the anomalous dispersion regime as well as in the study of compression of visible pulses (Tomlinson *et al.*, 1984). Physically, z_0 is the characteristic length in which a low intensity pulse of duration t_0 will broaden by approximately a factor of two due to dispersion acting alone. The power P_0 associated with the characteristic field E_0 is given by

$$P_0 = \frac{cn_0|E|^2 A_{\text{eff}}}{8\pi} = \frac{\lambda c n_0 A_{\text{eff}}}{16\pi n_2 z_0} \tag{10.5}$$

where A_{eff} is the effective area of the fiber core. For the power P_0, corresponding to $|u|^2 = 1$, a pulse propagating a distance z_0 (with dispersion turned off) would experience a $\pi/2$ non-linear phase shift and a spectral broadening of roughly a factor of two.

10.2.2 *Basic dark soliton solutions to the NLS equation*

Single dark soliton solutions to the NLS equation consist of a rapid dip in the intensity of a constant background (Hasegawa and Tappert, 1973; Zakharov and Shabat, 1973). Representative dark soliton solutions, as well as the fundamental bright soliton, are sketched in Figure 10.1. The fundamental dark soliton, which we will refer to as a black soliton (Tomlinson *et al.*, 1989), is written as follows.

$$u(z, t) = A \tanh\left(\frac{|A|t}{t_0}\right) \exp\left(\frac{-\mathrm{i}\pi|A|^2 z}{2z_0}\right) \tag{10.6}$$

The black soliton is an anti-symmetric function of time, with an abrupt π phase shift at $t = 0$. In contrast, the bright soliton is an even function of time, with a constant phase across the entire pulse.

In addition to the black soliton, there also exists a more general dark soliton solution which describes a continuous range of lower contrast dark solitons (Hasegawa and Tappert, 1973; Zakharov and Shabat, 1973). Here we write this solution in a form which is somewhat different than (but equivalent to) that which has appeared in the literature, namely:

$$u(z, t) = A\left\{v - \mathrm{i}\tanh\left(\frac{|A|t}{t_0} - \frac{\pi|A|^2 vz}{2z_0}\right)\right\} \cdot \exp\left[\frac{-\mathrm{i}\pi|A|^2(1 + v^2)z}{2z_0}\right] \tag{10.7}$$

Here v is taken to be a real number. This more general solution consists of a tangent hyperbolic dark pulse in quadrature with a constant background level. Equation (10.7) can be rewritten in a form which emphasises the intensity and the phase profile of the dark soliton, as follows (Tomlinson *et al.*, 1989)

$$u(z, t') = \frac{A}{|B|}\left[1 - B^2 \operatorname{sech}^2\left(\frac{|A|t'}{t_0}\right)\right]^{1/2} \cdot \exp\left\{\mathrm{i}\phi\left(\frac{|A|t'}{t_0}\right) - \mathrm{i}\,\frac{\pi|A|^2 z}{2B^2 z_0}\right\} \tag{10.8a}$$

where

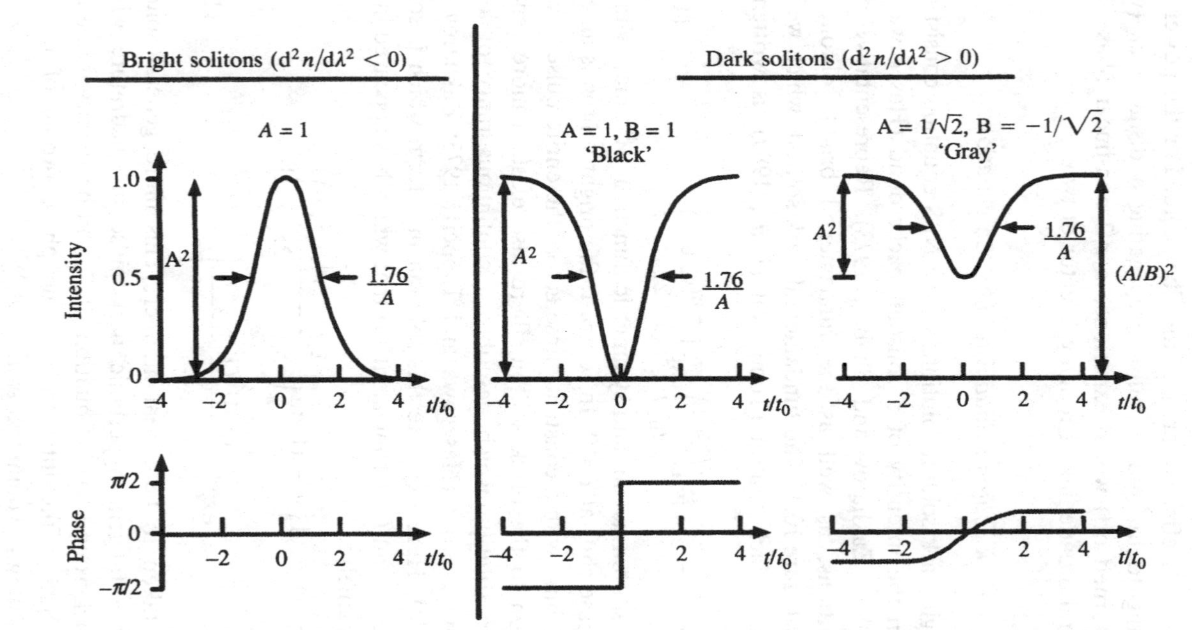

Figure 10.1. Intensity and phase, as functions of normalised time, for bright and dark solitons (from Tomlinson *et al.*, 1989).

$$\phi(\xi) \equiv \sin^{-1}\left[\frac{-B\tanh(\xi)}{(1 - B^2\,\mathrm{sech}^2(\xi))^{1/2}}\right]† \tag{10.8b}$$

$$B \equiv \frac{\pm 1}{(1 + \nu^2)^{1/2}} \tag{10.8c}$$

and

$$\frac{t'}{t_0} \equiv \frac{t}{t_0} - \frac{\pi}{2}|A|\frac{(1 - B^2)^{1/2}}{B}\frac{z}{z_0} \tag{10.8d}$$

In equation (10.8c) we take $B > 0$ for $\nu > 0$ and $B < 0$ for $\nu < 0$. The intensity of the solitons defined by equations (10.7) and (10.8) does not generally dip all the way to zero; the contrast ratio is determined by the parameter B (note the right-hand side of Figure 10.1). We will refer to pulses with $|B| < 1$ as gray solitons, with $|B|$ as the blackness parameter (Tomlinson *et al*, 1989). In the special case $|B| = 1$, equation (10.8) reduces to equation (10.6), and we regain the black soliton. For both black and gray solitons, the normalised peak intensity of the soliton (actually the depth) is given by $I_p = |A|^2$. The full-width at half-peak (or depth) of the intensity is $\tau_p = 1.76/|A|$ (in units of t/t_0), so that the product $I_p\tau_p^2 = 1.76^2$ is a constant for all solitons (Tomlinson *et al.*, 1989).

From equation (10.8) we see that gray solitons, like black solitons, have a non-trivial phase profile. Gray solitons have a phase profile which is an anti-symmetric function of time, but with a smaller and more gradual phase shift than black solitons (see Figure 10.1). The total phase shift across a gray soliton is $2\sin^{-1}|B|$. For Gray solitons the time-dependent phase shift described by equation (10.8b) represents an effective frequency shift, and therefore dark pulses of the form given by equation (10.8) propagate at velocities slightly different from the group velocity of the background. The rate at which gray solitons walk off from the background depends on B (or ν) and increases with decreasing blackness. For $B < 0$ a gray soliton will advance with respect to the background; conversely, for $B > 0$, the gray soliton will lag behind the background. The phase function of dark solitons is a major reason why these solitons are difficult to study experimentally; only recently has it been possible to create individual dark pulses with the required phase variation.

10.2.3 *Dark solitons with finite-duration background pulses*

Although the theoretical dark soliton solution to the NLS contains a background of infinite extent, in all experimental investiga-

†Equation (10.8b) has a negative sign which does not appear in Weiner *et al.* (1989) and Thurston and Weiner (1991). Equation 5(b) in the former and (9) in the latter contain sign errors which are corrected by the above.

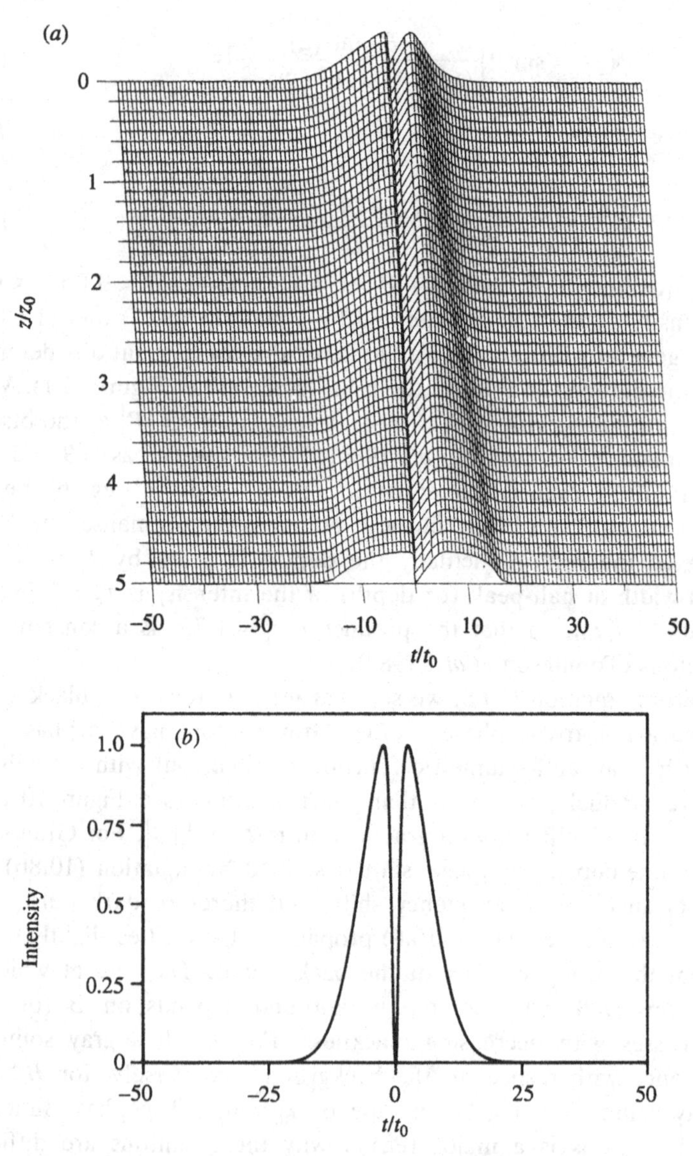

Figure 10.2. (*a*) Perspective plot showing pulse intensity *versus* time, as a function of distance along the fiber. This plot is for an odd-symmetry input dark pulse, given by equation (10.9), with amplitude $A = 1$ (from Weiner *et al*. (1988b)). (*b*) Intensity profile of the input pulse $z/z_0 = 0$ (*c*) Intensity profile for the pulse at $z/z_0 = 5$.

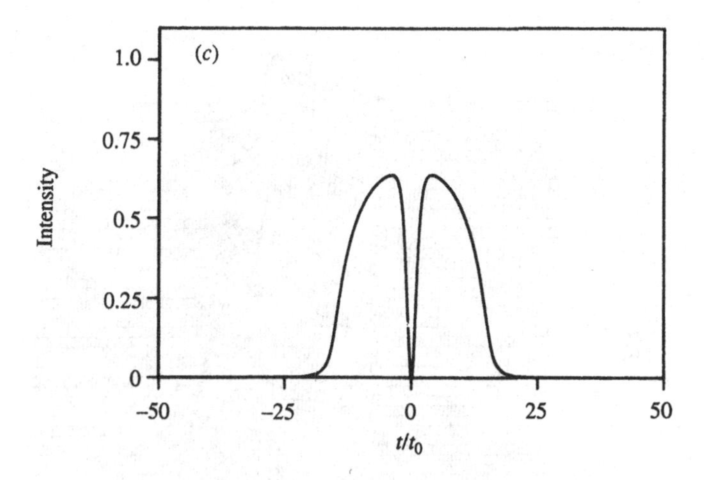

tions performed to date, finite duration background pulses have been used. For the experiments which we will discuss in the most detail in this chapter (Weiner *et al.*, 1988; 1989), it was most convenient to generate 100–200 fs duration dark pulses on background pulses 1–4 ps in duration. Because the background pulses chirp as they propagate, it is necessary to test whether dark pulses can exhibit stable soliton propagation with finite-extent background pulses. In this section we describe the results of numerical simulations which indicate that very stable soliton behavior can be expected even with a rapidly evolving background pulse (Tomlinson *et al.*, 1989). These simulations will also serve to provide a graphic illustration of the propagation behavior of various types of dark solitons.

The simulation results described later were obtained by solving the NLS equation (10.3), by the beam propagation method (Fleck *et al.*, 1978) as described in Tomlinson *et al.* (1989). The input pulse for the simulations in a dark pulse with blackness $B = 1$, superimposed on a Gaussian background, as follows

$$u(t, z = 0) = A \tanh(t/t_0) \exp(-(t/15t_0)^2) \qquad (10.9)$$

The duration of the background pulse is approximately ten times that of the input dark pulse. For $A \simeq 1$, the input pulse should generate the fundamental dark soliton, except for possible perturbations due to the finite extent background pulse. The propagation results for $A = 1$ are shown in Figure 10.2 over the range $0 \leqslant z/z_0 \leqslant 5$ (Tomlinson *et al.*, 1989; Weiner *et al.*, 1988b). The background pulse, which is not a soliton, broadens and loses intensity as it propagates; and the phase plot of the output pulse (not shown) indicates a substantial chirping of the background. Nevertheless, although the dark pulse does broaden somewhat, otherwise it propagates almost without distortion.

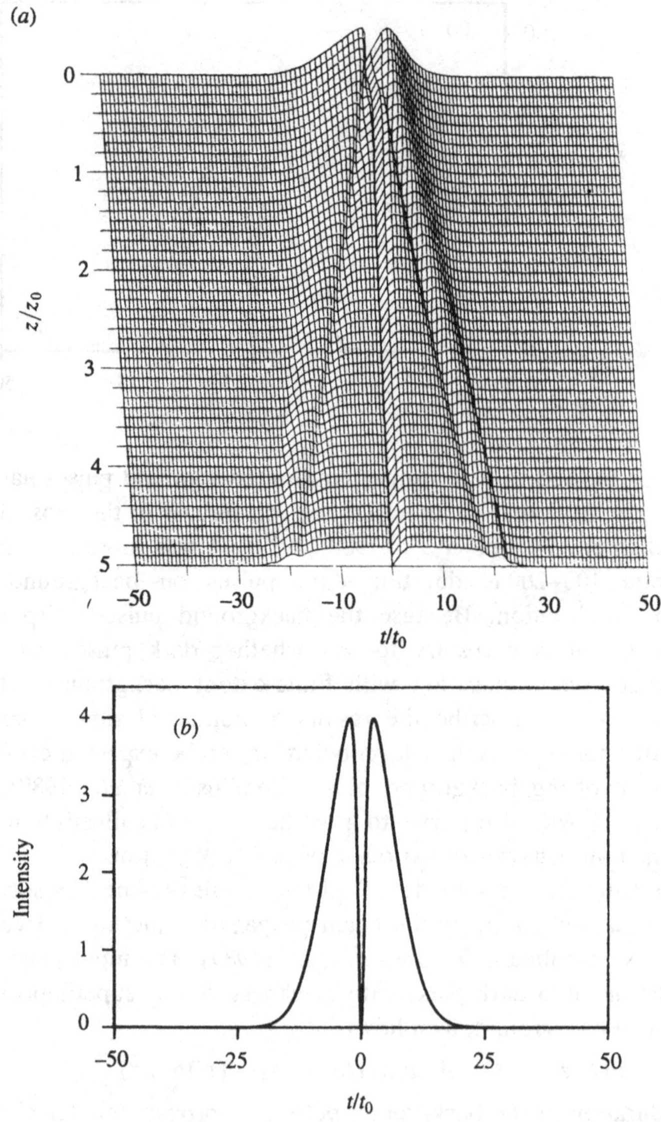

Figure 10.3. (*a*) Perspective plot for an odd-symmetry input dark pulse, with amplitude $A = 2$. (b) and (c) Intensity profiles for the input pulse ($z/z_0 = 0$) and for $z/z_0 = 5$, respectively.

The output dark pulse still dips to zero intensity and still has an abrupt π shift (not shown) at its center. Further simulations out to a distance $z/z_0 = 15$ show that the dark pulse continues to propagate

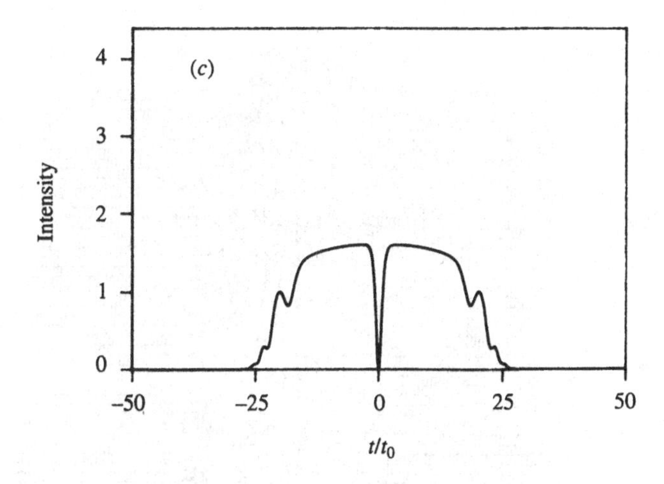

undistorted as a black soliton. As the background spreads and continues to decrease in intensity, the dark pulse broadens adiabatically, so that the product $I_p \tau_p^2 = 1.76^2$ is maintained. The constancy of the $I_p \tau_p^2$ product indicates that soliton propagaton is maintained despite the chirp and the decreasing intensity of the background pulse (Tomlinson *et al.*, 1989). Thus, propagation of dark pulses with finite-extent background pulse is analogous to propagation of bright solitons in the presence of loss (Doran and Blow 1983; 1985a); in both cases the pulse evolves adiabatically to maintain its soliton character.

Figure 10.3 shows the results of a similar simulation performed for $A = 2$, corresponding to an intensity four times that required for fundamental dark soliton propagation. In this situation, the dark pulse initially narrows, since the increased intensity corresponds to a soliton which is narrower than the input dark pulse. The excess (negative) energy of the input is shed in the form of two gray solitons which separate symmetrically from the central dark pulse as they propagate. At even higher input intensities, the input dark pulse spawns multiple pairs of gray solitons, with different blackness parameters, in addition to central black soliton. This is quite different from the case of bright solitons, for which multiple soliton bound states are formed at higher input intensities. This difference results from the fact that for positive GVD, the non-linear term in equation (10.3) corresponds to a repulsive potential between solitons (Blow and Doran, 1985b). The evolution of the initial dark pulse into multiple black and gray solitons has been analysed analytically in Zhao and Bourkoff (1989) and will be described in more detail later in this chapter.

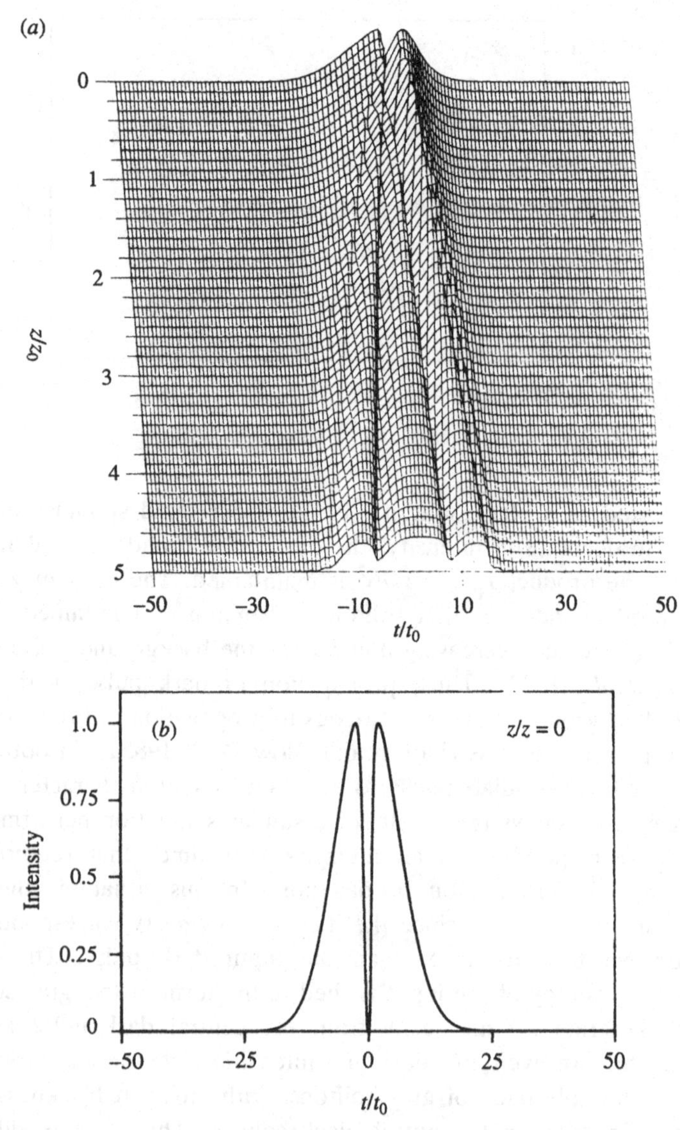

Figure 10.4. (a) Perspective plot for an even-symmetry input dark pulse, with amplitude $A = 1$ (from Weiner *et al*. (1988b). (b) and (c) Intensity profiles for the input pulse ($z/z_0 = 0$) and for $z/z_0 = 5$, respectively.

Simulation results are shown in Figure 10.4 for the case of an even symmetry input pulse on a finite duration background (Tomlinson *et al*., 1989; Weiner *et al*., 1988b). The input pulse is as given by

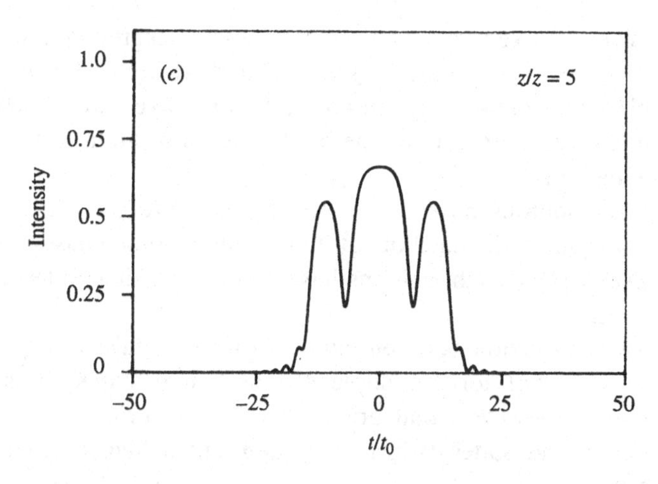

equation (10.9) with $A = 1$, except that $\tanh(t/t_0)$ is replaced by $|\tanh(t/t_0)|$. An even symmetry input pulse is predicted to split into a complementary pair of gray soliton (Blow and Doran, 1985b) and this is clearly indicated in Figure 10.4. A phase plot of the output pulse (not shown) confirms that the generated pulse pair has the proper gray soliton phase profile. The simulation results are in qualitative agreement with experimental measurements described later in this chapter which also demonstrate splitting of an even input pulse into a pair of gray solitons (Krokel *et al.*, 1988; Weiner *et al.*, 1988a). These observations emphasise the fact that an input dark pulse without the required soliton phase profile is not a soliton.

10.2.4 *Differences between dark and bright optical solitons*

Our discussion has revealed several fundamental differences between dark and bright optical solitons, and these differences should be kept in mind when considering experimental investigations of dark soliton propagation. The distinct properties of dark solitons (compared to bright solitons) are now summarised.

(1) Dark solitons occur only for positive (or normal) GVD which occurs for wavelengths shorter than 1.3 μm in standard telecommunications fiber. Bright solitons occur only for negative (or anomalous) dispersion.
(2) Dark solitons consist of a momentary decrease in the intensity of a constant background. Bright solitons, on the other hand, consist of localised packet of light with zero background.

(3) Given a fixed optical frequency and background intensity, there exists a continuous range of dark solitons with different blackness parameters. In contrast, for a fixed frequency and intensity, there is only one bright soliton solution to the NLS equation.
(4) Dark solitons have a non-trivial phase profile, which is an anti-symmetric function of time with a total phase shift of $2\sin^{-1}|B|$. Bright solitons have a constant, time-independent phase.
(5) The interaction between dark solitons is repulsive; dark solitons do not form mulitple soliton bound states. It is well known, however, that bright solitons can form multiple soliton bound states (which are often called higher order solitons).

10.3 Experimental verification of dark soliton propagation

Experimental methods for controlling the shape of ultrashort optical pulses have recently been developed; and by utilising these methods to generate dark pulses, researchers have been able to investigate propagation of dark solitons in optical fibers. To date measurements have been performed at visible wavelengths where fiber loss is quite high compared to that experienced in bright soliton experiments performed at 1.3 or 1.55 μm. This imposes an additional experimental requirement: the dark pulses should have subpicosecond durations so that the characteristic length for soliton propagation z_0 is much shorter than the attenuation length in the fiber. The first measurements of non-linear dark pulse propagation, reported in 1987, utilised dark pulses 5 ps long; unfortunately, because z_0 exceeded the attenuation length, clear indications of soliton propagation were not observed (Emplit *et al.*, 1987). Subsequent experiments utilising subpicosecond duration, dark input pulses have resulted in successful demonstrations of dark soliton propagation (Krokel *et al.*, 1988; Weiner *et al.*, 1988a; 1989); and these are described in the remainder of this chapter.

10.3.1 *Non-linear propagation of even-symmetry dark pulses*

We first discuss experiments, performed by Krokel *et al.* (1988),. which started with even-symmetry dark input pulses. The experiments were made possible by an 'ultrafast light-controlled optical fiber modulator' (Halas *et al.*, 1987) based on the optical Kerr effect. The modulator consisted of a 15-mm length of low-birefringence single-mode optical fiber with a polariser at the output. The inputs to the modulator are low intensity, 100-ps signal pulses from a mode-locked and frequency-doubled Nd:YAG laser (532 nm wavelength) and high-intensity, subpicosecond gating pulses from a compressed and amplified subpicosecond dye laser system (600 nm wavelength). The dye laser gating pulses cause a momentary non-linear polarisation rotation of the longer duration signal pulses, and this polarisation is translated into an intensity modulation by the polariser. By suitable adjustment of the polarisations, 0.3-ps dark pulses (or holes), with a contrast of approximately 50%, were produced on the signal pulses. Because the optical fiber Kerr gate does not provide phase control, the phase of the resultant dark pulses is approximately constant with time. Thus, the input pulses used in (Krokel *et al.*, 1988) do not exhibit the phase profile required for individual dark solitons. Instead, at appropriate input powers, these even-symmetry input pulses are expected to break up, forming a complementary pair of gray solitons (Blow and Doran, 1985b).

Non-linear pulse propagation measurements were performed by coupling the linearly polarised beam of 532-nm dark pulses from the modulator into a 10-m length of single-mode, polarisation-preserving optical fiber. The intensity profile of the pulses emerging from the fiber was measured by cross-correlation with 0.3-ps probing pulses split off from the 600-nm gating beam driving the modulator. Figure 10.5 shows cross-correlation traces of the output pulses obtained for various coupled input powers (Krokel *et al.*, 1988). At low power where non-linear effects are negligible (Figure 10.5(*a*)), the output dark pulses have broadened by an order of magnitude, and a pronounced ringing structure has developed on the background. These effects are explained by considering the input dark pulse to consist of a short pulse π radians out of phase with a c.w. background at the carrier frequency. Due to linear dispersion, the short pulse broadens and acquires a chirp. The intensity modulation on the output pulse is then understood as the interference between the broadened, frequency-swept pulse and the carrier. Increasing the input power causes the central hole to split into two clearly resolved dips. At 20 W input power (Figure 10.5(*d*)), two clearly resolved dark pulses with a duration of 0.6 ps and a measured contrast of approximately 20% are evident.

In order to help interpret the experimental results (Figure 10.5),

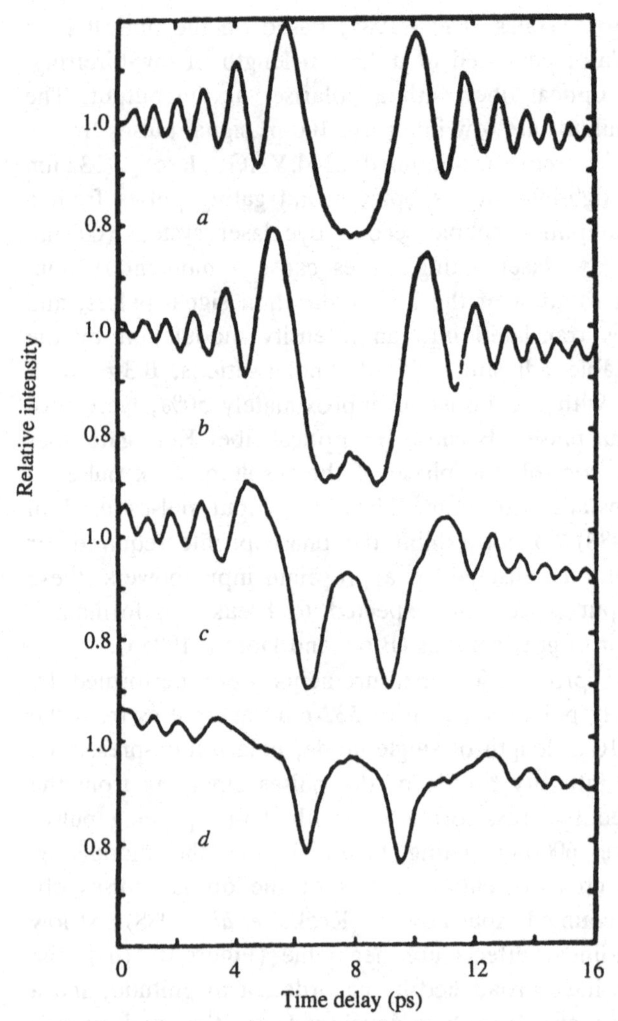

Figure 10.5. Experimental cross-correlation measurements of pulses emerging from 10 m optical fiber, for even-symmetry input pulses (from Krokel *et al*. (1988)). The input powers are: (*a*) 0.2 W, (*b*) 2 W, (*c*) 9 W and (*d*) 20 W.

numerical simulations of the plane-wave NLS equation were perfomed for the case of a 10-m fiber and an even-symmetry input pulse (Krokel *et al*., 1988). Computed results are shown in Figure 10.6. At low power the input dark pulse broadens substantially, and a strong ringing is obtained. As the power is increased, the broadened central dark pulse splits into a well separate pair of subpicosecond dark pulses, and the amplitude of the ringing is reduced. The simulations

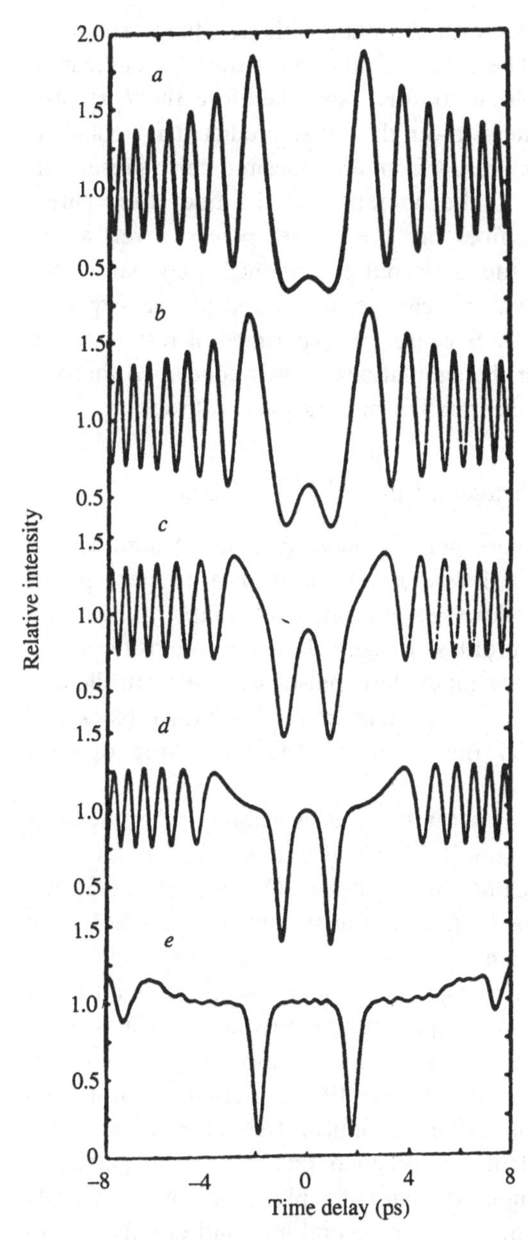

Figure 10.6. (*a*)–(*d*) Computed pulse shapes emerging from a 10 m fiber for even-symmetry input pulses. The input powers are: (*a*) 0.2 W, (*b*) 4 W, (*c*) 18 W and (*d*) 40 W. (*e*) Computed pulse shape emerging from a 20 m fiber for an input power of 40 W. (From Krokel *et al*. (1988).)

are in qualitative agreement with the data, although there are some quantitative differences. (The observed soliton contrast is significantly less than the predicted soliton contrast, and therefore the $I_p\tau_p^2$ product evident from the data is lower than that predicted for solitons. Some of this difference may result from an undesired chirp present on the input dark pulse due to the operation of the fiber Kerr gate.) Simulations for a 20-m fiber at the same power level as in Figure 10.6(d) show that the dark pulses continue to separate for longer fiber lengths, but with no change in pulse shape, as expected for solitons (Figure 10.6(e)). Because the experimental results are in qualitative agreement with the simulations, it was concluded that the data are consistent with the formation of dark-pulse solitons.

10.3.2 *Observation of individual black and gray solitons*

We next discuss more detailed investigations of dark soliton propagation (Weiner *et al.*, 1988a; 1989) based on a different pulse-shaping technique which allows generation of ultrashort pulses with nearly arbitrary temporal profiles. Because both the intensity profile and the phase profile of the input dark pulses can be controlled, it becomes possible to demonstrate propagation of individual black and gray solitons and to verify the predicted phase structure of dark optical solitons.

The experimental apparatus for these dark soliton experiments is shown schematically in Figure 10.7 (Weiner *et al.*, 1988a). A colliding-pulse mode-locked (CPM) ring dye laser (Valdmanis, 1985) and a copper vapor laser pumped dye amplifier system (Knox, 1984) serve as a source of 75-fs, 620-nm pulses at a repetition rate of 8.6 kHz. Pulse-shaping is achieved by spatial masking within a temporally non-dispersive lens and grating apparatus, consisting of pair of gratings situated at the focal planes of a unit magnification, confocal lens pair (Weiner *et al.*, 1988c; d; Froehly, 1983). The first grating and lens spatially separate the different optical frequency components which make up the incident femtosecond pulse, and the patterned mask then adjusts the amplitudes and the phases of these spatially dispersed frequency components. The second lens and grating recombine the masked frequency components to produce a pulse whose temporal profile is the Fourier transform of the pattern transferred by the mask onto the spectrum. Thus, the problem of generating an input pulse with a desired temporal profile is reduced to the problem of fabricating a mask patterned according to the Fourier transform of

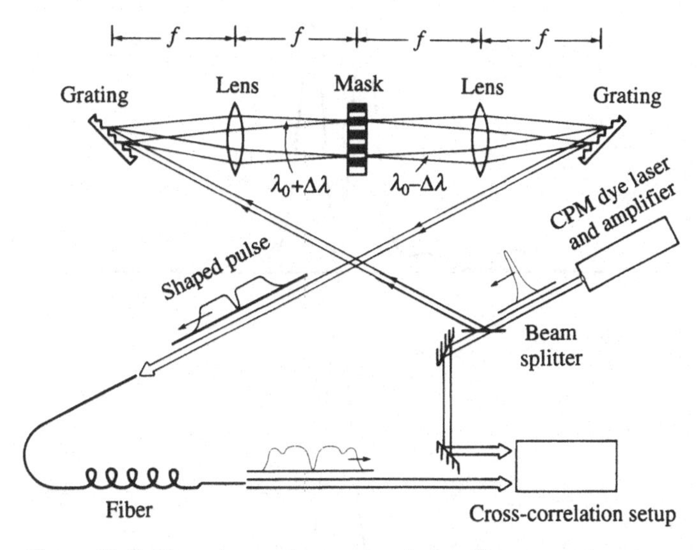

Figure 10.7. Experimental setup for dark soliton experiments utilising Fourier transform pulse-shaping apparatus (from Weiner *et al.* (1988a)).

the desired pulse shape. Using this technique, input dark pulses closely resembling the expected shape of black and gray solitons, as well as even-symmetry input dark pulses, have been generated. Experiments were performed by launching appropriately shaped dark pulses into a 1.4-m length of single-mode, polarisation-preserving fiber. The intensity profile of the input pulses and the pulses emerging from the fiber were measured by cross-correlation, with the use of 75-fs pulses directly from the amplifier as the probe.

In order to investigate propagation of the fundamental (black) dark soliton, hyperbolic-tangent dark pulses were synthesised on a broader Gaussian background pulse with a duration ten times that of the dark pulse (see equation (10.9)). The spectrum corresponding to this odd-symmetry input pulse, plotted as the full line in Figure 10.8(a), is a doubly peaked, anti-symmetric function of frequency. In order to generate the required spectrum, two separate masks were used: the first an amplitude mask consisting of a variable transmission metal film on a fused silica substrate, and the second a phase mask etched to introduce a relative phase shift of π onto half of the spectrum. Both masks were patterned by using standard microlithographic fabrication techniques (Weiner *et al.*, 1988c, d). The power spectrum of the resultant shaped pulse, measured with a spectral resolution of approximately 0.17 nm, is plotted as the dotted line in Figure 10.8(a).

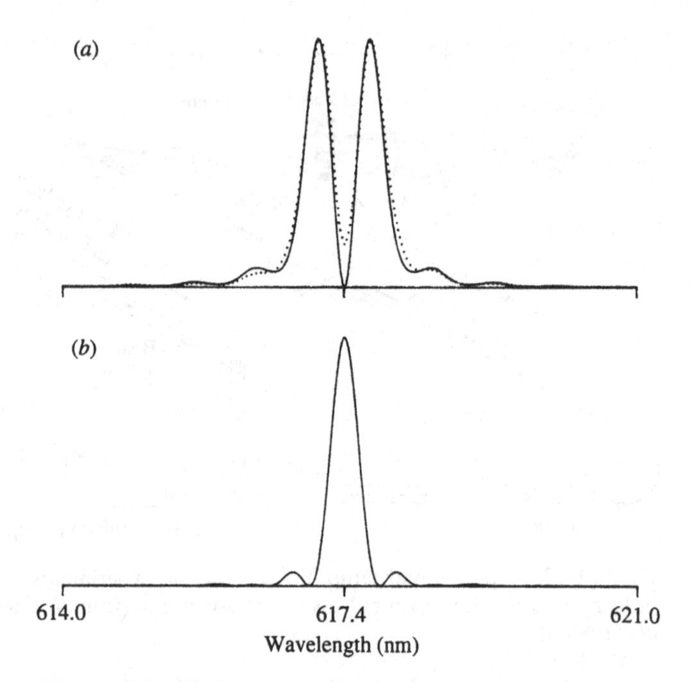

Figure 10.8. (*a*) Calculated (full line) and measured (dotted line) power spectra of odd-symmetry dark input pulse. (*b*) Calculated power spectrum of even-symmetry dark input pulse (from Weiner *et al*. (1988a)).

The excellent agreement between the actual and targeted spectra gives an indication of the precision available with this technique.

Other dark input pulses, such as even-symmetry input pulses and input pulses suitable for launching individual gray solitons, were generated in a similar manner. The calculated power spectrum for an even-symmetry dark pulse is shown in Figure 10.8(*b*). The spectrum (and the corresponding amplitude and phase masks) are each symmetric functions of frequency. The spectrum for a gray soliton input pulse is intermediate between that for an odd-symmetry black soliton and for an even-symmetry pulse. (More precisely, the spectrum is split into two main peaks of different amplitudes, with the ratio of the peak amplitudes depending on the blackness parameter B). The various types of dark input pulses used in these experiments are easily distinguished due to the clear differences in their spectra.

Figure 10.9(*a*) shows an intensity cross-correlation measurement of an odd-symmetry input pulse used for investigating the fundamental dark soliton. The duration of the central hole is 185 fs full width at half maximum (FWHM) of the intensity, and the background dura-

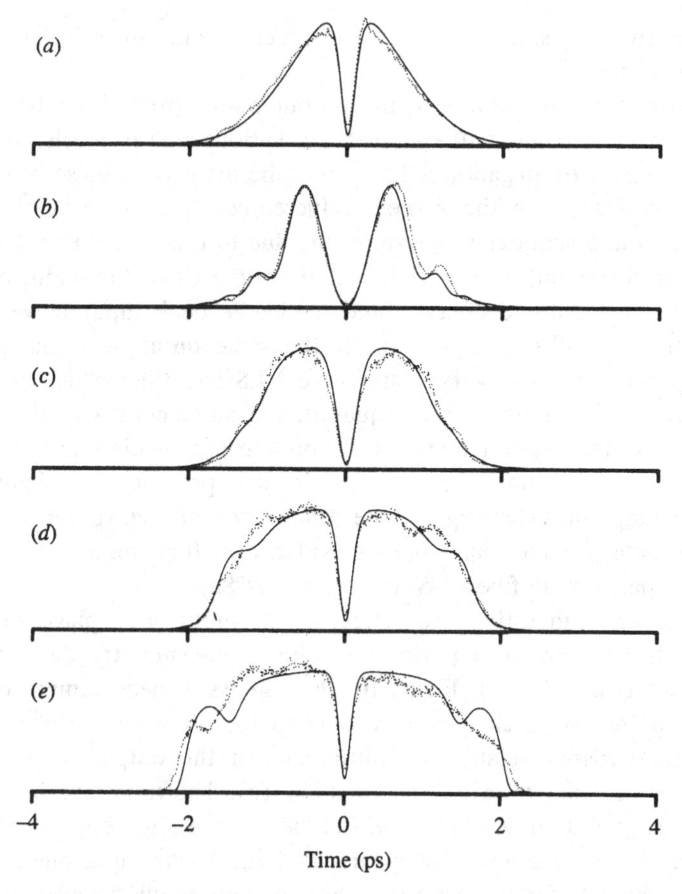

Figure 10.9. Measured (dotted lines) and calculated (full lines) cross-correlation data for odd-symmetry dark input pulse: (*a*) input dark pulse; (*b*)–(*e*) pulses emerging from 1.4 m fiber for peak input power of (*b*) 1.5 W, (*c*) 52.5 W, (*d*) 150 W and (*e*) 300 W (from Weiner *et al*. (1988a)).

tion is 1.76 ps FWHM. A comparison with the calculated input pulse shape (full line in Figure 10.9(*a*)) shows that the actual pulse is an excellent replica of the target input pulse. For a 185 fs input dark pulse, the corresponding t_0 in equation (10.9) is 105 fs, and the characteristic dispersion length z_0 is calculated to be approximately 32 cm, using $D(\lambda) = 0.0493$ (corresponding to 265 ps/nm · km). Thus, the ratio of the fiber length to z_0 is approximately 4.4, which is sufficient for observation of soliton propagation. The soliton power is expected to be $P_0 \approx 190$ W, computed using the known non-linear coefficient for silica, $n_2 = 1.1 \times 10^{-13}$ esu, and an effective area $A_{\text{eff}} \approx 12.6\ \mu\text{m}^2$.

Cross-correlation traces of the output pulses from the 1.4-m fiber are shown in Figures 10.9(*b*)–(*e*) for various power levels. At the lowest power propagation is linear and the input dark pulse broadens to over 600 fs. As the power is increased, the background pulse broadens and acquires a square profile due to the combined effects of the non-linear index and GVD. At the same time, the width of the output dark pulse decreases; and at 300 W peak input power, the output dark pulse is of essentially the same duration as the input. Computer solutions to the plane-wave NLS equation, which are also plotted in Figure 10.9, are in quantitative agreement with the data. Thus, the dark pulse undergoes soliton-like propagation and emerges from the fiber unchanged, even in the presence of significant broadening and chirping of the finite-duration background pulse. These data provide unambiguous evidence of fundamental dark soliton propagation in fibers (Weiner *et al.*, 1988a).

In order to investigate the effect of the input pulse phase profile, experiments were also performed with even-symmetry dark pulses (Weiner *et al.*, 1988a). Figure 10.10(*a*) shows a measurement of the input pulse: an even-symmetry dark pulse on a square-like background. Cross-correlation measurements of the output pulses from the fiber are shown in Figures 10.10(*b*)–(*e*). The trend are similar to those reported in Krokel *et al.* (1988) – see Figure 10.5. At low power the central hole disappears, and the background pulse is reshaped by interference with the chirped, temporally broadened dark pulse. As the power is increased, two low contrast dark pulses are formed, with a pulse separation of 2.3 ps at an input power of 285 W. Again, the data agree closely with numerical solutions to the NLS equation. These experiments demonstrate the crucial importance of the dark-pulse phase profile: an odd dark pulse propagates undistorted as a soliton, as predicted by theory (Hasegawa and Tappert, 1973), while an even dark pulse, which is not a soliton solution to the NLS, splits into a pair of shallow gray solitons.

The data in Figures 10.9 and 10.10 demonstrate the propagation of an individual black soliton and the evolution of an even-symmetry input pulse into a complementary pair of gray solitons, respectively. In addition, experiments have also been performed which show generation and propagation of individual gray solitons (Thurston and Weiner, 1991). Some of the data are plotted in Figure 10.11. The top traces are cross-correlation measurements of two different input

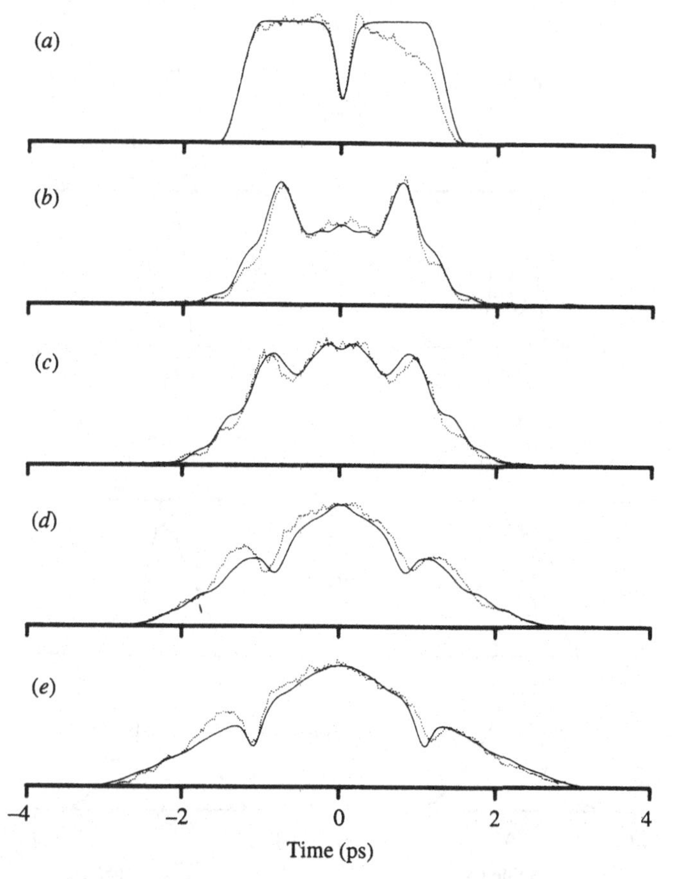

Figure 10.10. Measured (dotted lines) and calculated (full lines) cross-correlation data for even-symmetry dark input pulse: (*a*) input dark pulse; (*b*)–(*e*) pulses emerging from 1.4 m fiber for peak input power of (*b*) 2.5 W, (*c*) 50 W, (*d*) 150 W, (*e*) 285 W (from Thurston and Weiner (1991)).

pulses to the 1.4-m length fiber. The intensity profiles of the two inputs are nearly identical: a 130-fs duration dark pulse with approximately 50% contrast ($B^2 = 0.5$) superimposed on a square-like background pulse 1.76 ps in duration. The difference between the two inputs consists in the sign of the B parameter in equation (10.8). The data in the left-hand column correspond to an input pulse with a positive-going phase modulation, i.e. $B < 0$, while the data in the right-hand column correspond to an input pulse with a negative-going phase modulation ($B > 0$). From equation (10.8) we expect that gray solitons with $B < 0$ will have a velocity which is faster than that of

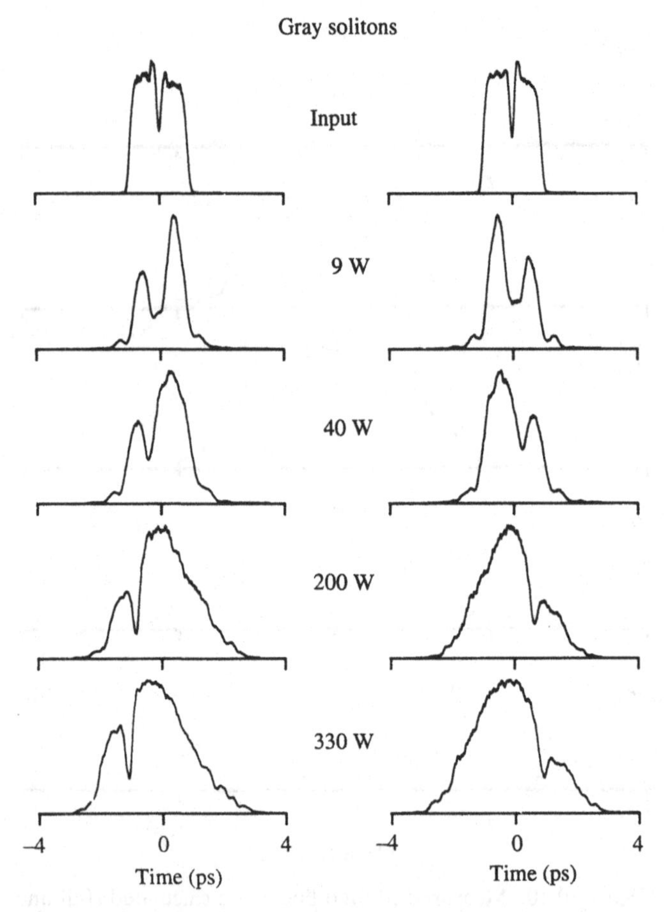

Figure 10.11. Measured cross-correlation data for gray solitons, both for input pulses and for pulses emerging from 1.4 m fiber. The input power levels are as marked. Left side: data for input pulses with $B < 0$. Right side: data for input powers with $B > 0$ (from Thurston and Weiner (1991)).

the background pulse, whereas solitons with $B > 0$ will have velocity slower than that of the background. This prediction is confirmed by the data. At low power the dark pulses emerging from the fiber are substantially broadened; but as the power is increased, the duration of the output dark pulses continues to decrease, as expected for solitons. For the pulse corresponding to $B < 0$, the dark pulse emerges from the fiber earlier than does the bulk of the background pulse, while for the pulse with $B > 0$, the converse is true. Thus, by controlling the phase profile of the input dark pulse, it is indeed possible to control the velocity of the resultant gray soliton.

10.3.3 *Soliton propagation at high powers*

At this point we consider the behavior of dark pulses with background powers higher than those required for soliton propagation. In Figure 10.3 we showed the expected behavior of an odd-symmetry input pulse with an intensity four times that required for soliton propagation ($A = 2$ in equation (10.9)). As the pulse propagates down the fiber, it evolves into a complementary pair of gray solitons which separate symmetrically from a narrow black soliton left behind at the center. Such behavior is already evident in the experimental data shown in Figure 10.9(*e*) for 300 W input power. High power propagation of tangent-hyperbolic dark input pulses of the form given by equation (10.9) was studied numerically by Hawkins *et al*. (1988) and Zhao *et al*. (1989) found analytical expressions for the contrast, duration and temporal location of the various solitons which are formed. For a background field with an amplitude A, the input field, $A \tanh(t/t_0)$, evolves into a central black soliton of the form $A \tanh(At/t_0)$, together with $N - 1$ pairs of complementary gray solitons with blackness parameters given by

$$B = \pm \frac{|A| - n}{|A|} \qquad \text{for } n = 1, 2, \ldots, N - 1 \qquad (10.10)$$

where N is the smallest integer $\geqslant |A|$.

Experiments have been performed to test dark soliton propagation at high powers (Weiner *et al*., 1989). The input pulses and the experimental arrangement were identical to those discussed in the previous section in connection with measurements performed with odd-symmetry input pulses (see Figure 10.9). Cross-correlation traces of the pulses emerging from the 1.4-m length of fiber are shown in Figure 10.12 for input powers ranging between 7.5 and 975 W. For powers below approximately 300 W, the data are in quantitative agreement with the predictions of the standard NLS equation, and the fundamental dark soliton is clearly observed (Figure 10.12(*b*)). For higher powers, however, the data deviate significantly from the predicted behavior. As the power is increased, the central soliton progressively loses contrast, shifts to later times, and finally broadens. Furthermore, only single gray soliton, which appears near the rising edge of the background pulses and which grows narrower and deeper as the central soliton broadens and loses contrast, becomes evident. The unpredicted features in the data – the shift, broadening and loss of contrast of the central black soliton and the appearance of a single

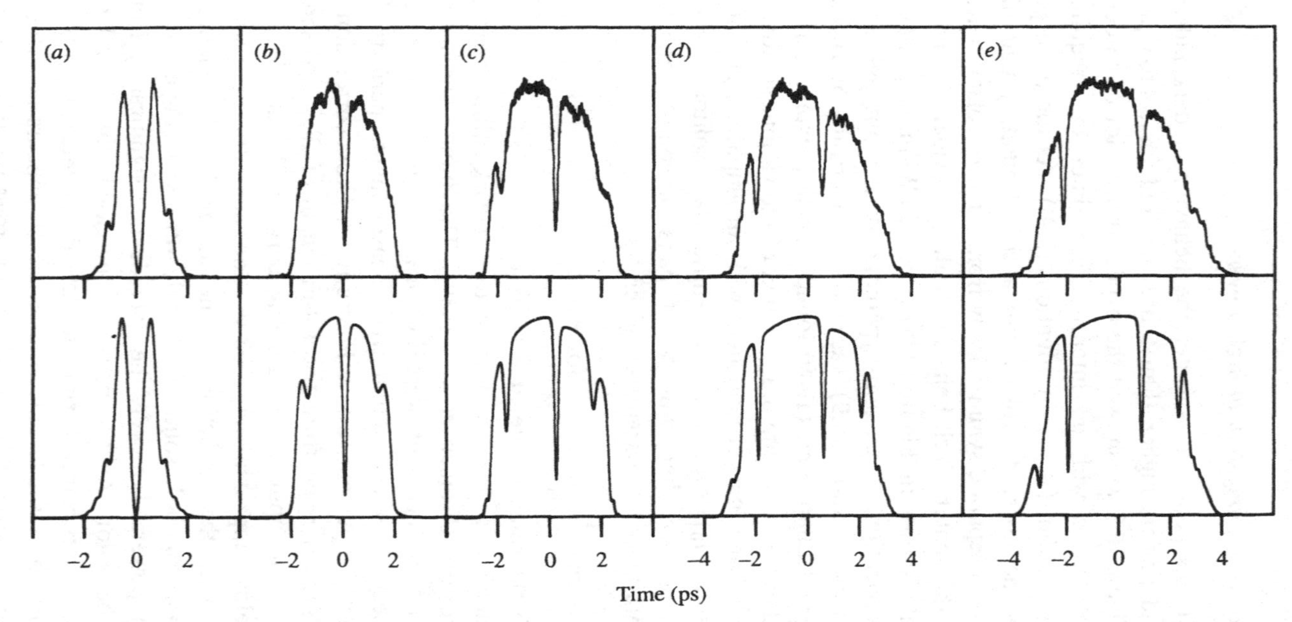

Figure 10.12. Top traces: cross-correlation measurements of dark pulses emerging from 1.4 m fiber. Bottom traces: theoretical cross-correlation traces, calculated by using a modified NLS that includes the Raman contribution to the non-linear index. The input peak powers are: (*a*) 7.5 W, (*b*) 225 W, (*c*) 450 W, (*d*) 750 W and (*e*) 975 W. The same horizontal scale is used for all the plots (from Weiner *et al*. (1989)).

gray soliton instead of a gray soliton pair – signify a breakdown of the standard NLS equation under the relevant experimental conditions.

In addition to the temporal data, power spectra of the fiber output pulses were also measured, and these are plotted as the upper traces in Figure 10.13. At low power the spectrum exhibits two peaks symmetrically located about a central hole (Figure 10.13(*a*)). According to the standard NLS equation, the spectrum should broaden symmetrically as the power is increased. Instead, the data show that, for increasing power, the spectra broaden preferentially toward the red, with the central hole shifting toward the blue. At the highest powers a second hole, corresponding to the pronounced gray soliton in the time-domain data, appears on the red of the spectrum. The blue shift of the central hole and the excess spectral broadening toward the red are caused by Raman amplification of longer wavelength spectral components by the shorter wavelength components. Analogous Raman self-pumping causes spectral red shifts (Mitschke and Mollenauer, 1986; Dianov *et al.*, 1985; Gordon 1986) and fissioning (Tai *et al.*, 1988; Hodel and Weber, 1987) of subpicosecond duration fundamental and higher-order bright solitons, respectively; these effects are described in detail in previous chapters in this book.

The data can be modeled by including a time-dependent Raman response function as part of the non-linear refractive index in the NLS equation. This time-domain formulation is fully equivalent to the usual approach in which Raman scattering is treated in the frequency domain. The treatment of the non-linear polarisation as a response function was introduced in (Hellwarth *et al.*, 1971); derivation of the Raman response function for silica-core fibers and a procedure for its inclusion in a modified NLS equation are given in Stolen *et al.* (1989). Briefly, the response function is the Fourier sine transform of the Raman gain spectrum and comprises a series of rapidly damped oscillations, with an initial period of 75 fs, corresponding to the inverse of the gain peak at 440 cm^{-1}. Temporal profiles and power spectra of the pulses emerging from the fiber, calculated using this modified NLS equation, are plotted as the lower traces in Figures 10.12 and 10.13, respectively. The calculated spectra quantitatively reproduce the blue shift of the central hole, the detailed shape of the spectra, and the position and strength of the second hole that appears at longer wavelengths. The computed temporal profiles are also in striking agreement with the data. Both the temporal shift

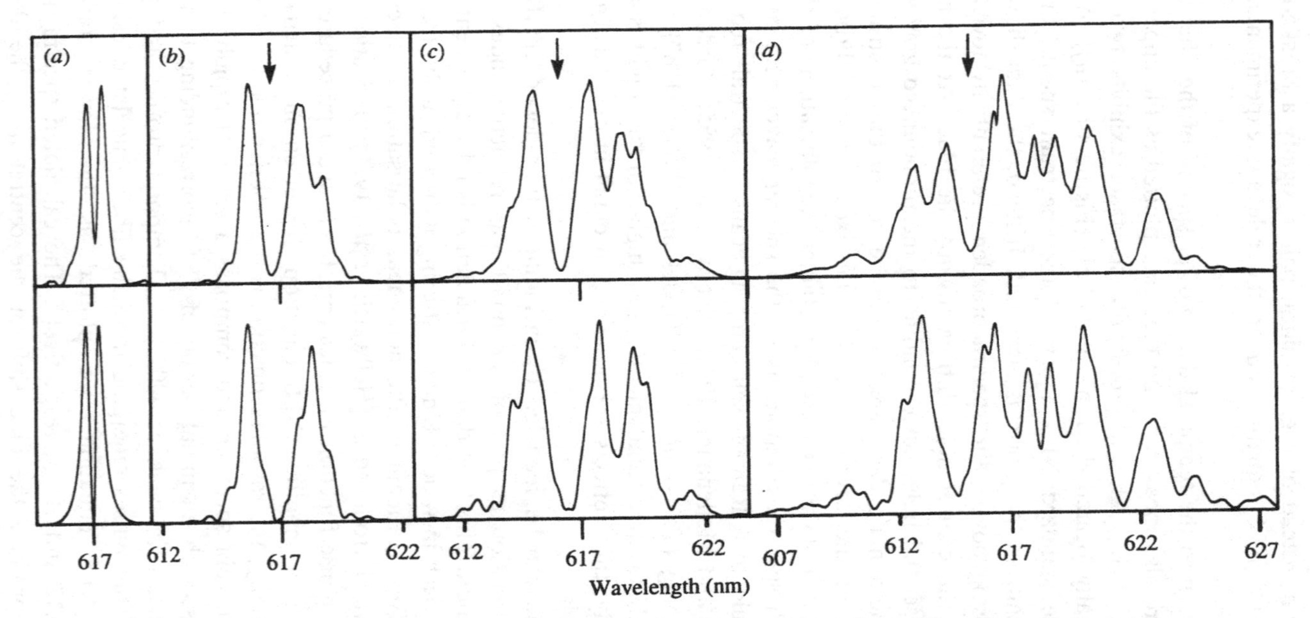

Figure 10.13. Top traces: measured power spectra of dark pulses emerging from 1.4 m fiber. Bottom traces: theoretical power spectra, calculated by using the modified NLS equation. The input peak powers are: (*a*) 7.5 W, (*b*) 225 W, (*c*) 450 W and (*d*) 975 W. In (*b*)–(*d*) the arrow marks the location of the central hole, which shifts to the blue as the power is increased. The same horizontal scale is used for all the plots (from Weiner *et al*. (1989)).

of the central soliton and the enhancement of the negative-time gray soliton at the expense of its positive-time counterpart are clearly evident. Furthermore, the computations accurately predict the temporal positions of the various solitons in the data for all experimental power levels. The excellent agreement evident in Figures 10.12 and 10.13 between theory and experiment is obtained without the use of adjustable parameters.

It is of interest to consider how these spectral and temporal self-shifts of dark optical solitons scale with the physical parameters. Gordon (1986) has used a perturbational approach to analyse the bright soliton self-frequency shift, and a similar approach may be applied to dark solitons. As a result, it is found that self-shifts of dark solitons scale with fiber length, pulse duration, etc., in the same way that these shifts scale for bright solitons. For example, spectral and temporal self-shifts for dark solitons (as well as for bright solitons (Gordon, 1986)) are expected to scale linearly and quadratically, respectively, with the fiber length. This point was investigated by comparing the data in Figure 10.9 measured for a 1.4-m fiber with similar data measured for a longer, 4.1-m fiber (Weiner *et al.*, 1989). As expected, temporal shifts considerably larger than those in Figure 10.9 were observed. At approximately 300 W input power, for example, the shift of the central soliton was approximately 500 fs, compared with approximately 50 fs for the 1.4-m fiber, in rough agreement with a quadratic length dependence. For a fiber of fixed length (in *real* units), both bright (Gordon, 1986) and dark soliton self-shifts are expected to scale as the inverse fourth power of the soliton duration (i.e. as τ_p^{-4}). Thus, for dark solitons with durations on a picosecond time scale, temporal and spectral self-shifts caused by the Raman contribution to the non-linear index should be much less significant than they are in Figures 10.12 and 10.13 for 200-fs dark solitons.

10.4 Summary and outlook

In summary, we have discussed dark soliton solutions to the NLS equation which describes non-linear pulse propagation in optical fibers in the positive GVD regime, and we have described experimental verifications of predicted dark soliton behavior. Our discussion has revealed that dark solitons have many properties which are quite different from the corresponding properties of bright solitons. These include some simple points, such as the fact that dark solitons possess a non-zero background and occur for positive rather than negative GVD, as well as some less obvious points – for example, the phase profile of dark solitons, the fact that dark solitons come in various shades (gray and black), and the fact that dark solitons do not form multiple soliton bound states.

Despite these differences, the fact that bright solitons show great promise for applications in high bit-rate optical communication systems suggests that dark solitons should also be considered for this application. One can suppose that the phase shifts built into dark solitons may make dark solitons useful as the basis for high bit-rate coherent optical communications system. Dark solitons could bring the same advantage to a coherent communication system which bright solitons bring to direct detection systems – namely, the elimination of performance degradation caused by dispersive pulse spreading. Of course, the only dark soliton experiments to date have been performed in the visible where high loss and high dispersion have resulted in fiber lengths measured in metres and soliton powers measured in hundreds of watts. Fortunately this is not a fundamental problem. Dark soliton propagation should be possible at any wavelength where GVD is positive; this includes wavelengths just short of 1.3 μm (the zero dispersion wavelength for standard telecommunications fibers), at which fiber loss and fiber dispersion should both be very low. This would allow dark soliton transmission over kilometres or tens of kilometres of fiber at power levels in the milliwatt range. Furthermore, by using suitable dispersion-shifted fibers, it should be possible to achieve positive GVD and dark soliton transmission even at 1.55 μm, the wavelength at which silica fibers exhibit their minimum loss.

The ultimate bit rate in proposed bright soliton communication systems is limited in part by soliton–soliton interactions and by timing jitter incurred as a result of Raman amplification, and this would also be true of a dark soliton communication system. However, theoretical studies of dark soliton interactions (Blow and Doran, 1985b) and of the stability of dark solitons (Zhao and Bourkoff, 1989a; Gredeskul and Kivshar, 1989a,b) are to date quite limited, and no experimental investigations of dark soliton collisions have yet been undertaken. Similarly, although one numerical study of Raman amplification of dark solitons to overcome fiber loss was recently published (Zhao and Bourkoff, 1989b), the phase noise and the timing jitter which would be introduced as a result of the amplification process have not been addressed; neither have any experimental investigations been attempted. Clearly, many interesting and relevant questions about dark optical solitons remain to be settled.

Acknowledgement

It is my pleasure to thank Dr Philippe Emplit for careful proof reading of the equations.

References

Blow, K. J. and Doran N. J. (1985a) *Opt. Commun.*, **52**, 367.
Blow, K. J., and Doran N. J. (1985b) *Phys. Lett.,* **107A**, 55.
Dianov, E. M., Karasik, A. Ya., Mamyshev, P. V., Prokhorov, A. M., Serkin, V. N., Stel'makh, M. F. and Fomichev, A. A. (1985) *JETP Lett.*, **41**, 294.
Doran, N. J. and Blow, K. J. (1983) *IEEE J. Quantum Electron.*, **19**, 1883.
Emplit, P., Hamaide, J. P., Reynaud, F., Froehly, C. and Barthelemy, A. (1987) *Opt. Commun.*, **62**, 374.
Fleck, J. A., Morris, J. R. and Bliss E. S., (1978). *IEEE J. Quantum Electron.*, **14**, 353.
Fork, R. L., Brito Cruz, C. H., Becker, P. C. and Shank, C. V. (1987) *Opt. Lett.*, **12**, 483.
Froehly, C., Colombeau, B. and Vampouille, M. (1983) In E. Wolf (ed.), *Progress in Optics, volume 10*, pp. 115–21 (North-Holland: Amsterdam).
Gordon, J. P. (1986) *Opt. Lett.*, **11**, 662.
Gredeskul, S. A. and Kivshar, Y. S. (1989a) *Phys. Rev. Lett.*, **62**, 977.
Gredeskul, S. A. and Kivshar, Y. S. (1989b) *Opt. Lett.*, **14**, 1281.
Grischkowsky, D. and Balant, A. C. (1982) *Appl. Phys. Lett.*, **41**, 1.
Halas, N. J., Krokel, D. and Grischkowsky, D. (1987) *Appl. Phys. Lett.*, **50**, 886.
Hasegawa, A. and Tappert, F. (1973) *Appl. Phys. Lett.*, **23**, 142, 171.
Hawkins, R. J., Tomlinson, W. J., Weiner, A. M., Heritage, J. P. and Thurston, R. N. (1988) *OSA Annual Meeting, 1988, Technical Digest Series, volume 11*, p. 66 (Optical Society of America: Washington, DC).
Hellwarth, R. W., Owyoung, A. and George, N. (1971) *Phys. Rev. A,* **4**, 2342.
Hodel, W. and Weber, H. P. (1987) *Opt. Lett.*, **12**, 924.
Knox, W. H., Downer, M. C., Fork, R. L. and Shank, C. V. (1984) *Opt. Lett.*, **9**, 552.
Krokel, D., Halas, N. J., Giuliani, G. and Grischkowsky, D. (1988) *Phys. Rev. Lett.*, **60**, 29.
Mitschke, F. M. and Mollenauer, L. F. (1986) *Opt. Lett.,* **11**, 659.
Mollenauer, L. F. and Smith, K. (1988) *Opt. Lett.*, **13**, 675.
Nakatsuka, H., Grischkowsky, D. and Balant, A. C. (1981) *Phys. Rev. Lett.*, **47**, 910.
Stolen, R. H., Gordon, J. P., Tomlinson, W. J. and Haus, H. A. (1989) *J. Opt. Soc. Am. B*, **6**, 1159.
Tai K., Hasegawa, A. and Bekki N. (1988) *Opt. Lett.,* **13**, 392.

Thurston, R. N. and Weiner, A. M. (1991) *J. Opt. Soc. Am. B,* **8**, 471.
Tomlinson, W. J., Hawkins, R. J., Weiner, A. M., Heritage, J. P. and Thurston, R. N. (1989) *J. Opt. Soc. Am. B*, **6**, 329.
Tomlinson, W. J., Stolen, R. H. and Shank, C. V. (1984) *J. Opt. Soc. Am. B*, **1**, 139.
Valdmanis J. A., R. L. Fork, R. L. and Gordon, J. P. (1985) *Opt. Lett.*, **10**, 131.
Weiner, A. M., Heritage, J. P., Hawkins, R. J., Thurston, R. N., Kirschner, E. M., Leaird, D. E. and Tomlinson, W. J. (1988a) *Phys. Rev. Lett.*, **61**, 2445.
Weiner, A. M., Heritage, J. P., Hawkins, R. J., Thurston, R. N., Kirschner, E. M., Leaird, D. E. and Tomlinson, W. J. (1988b) In T. Yajima, K. Yoshihara, C. B. Harris, and S. Shionoya, (eds) *Ultrafast Phenomena, volume, VI*, (Springer: Berlin), pp. 115–117.
Weiner, A. M., Heritage, J. P., Hawkins, R. J., Thurston, R. N., Kirschner, E. M., Leaird, D. E. and Tomlinson, W. J. (1988) unpublished.
Weiner, A. M., Heritage, J. P. and Salehi, J. A. (1988c) *Opt. Lett.*, **13**, 300.
Weiner, A. M., Heritage, J. P. and Kirschner, E. M. (1988d) *J. Opt. Soc. Am. B,* **5**, 1563.
Weiner, A. M., Thurston, R. N., Tomlinson, W. J., Heritage, J. P., Leaird, D. E., Kirschner, E. M. and Hawkins, R. J. (1989) *Opt. Lett.*, **14**, 868.
Zakharov, V. E. and Shabat, A. B. (1973) *Zh. Eksp. Teor. Fiz.*, **64**, 1627. (*Sov. Phys. – JETP*, **37** (1973) 823).
Zhao, W. and Bourkoff, E. (1989a) *Opt. Lett.* **14**, 703.
Zhao, W. and Bourhoff, E. (1989b) *Opt. Lett.*, **14**, 808.

11

Soliton-Raman effects

J. R. TAYLOR

11.1 Introduction

The earliest experimental schemes for generating optical solitons, see for example the chapter by Mollenauer in this book or the original paper by Mollenauer *et al.* (1980), relied on launching pulses with power and transform limited spectral characteristics which matched the soliton requirements for the particular optical fibre used. However, it was shown by Hasegawa and Kodama (1981), that a pulse with any reasonable shape could evolve into a soliton. In such a case, the energy not required to establish the soliton appears as a dispersive wave in the system.

An alternative mechanism for soliton generation was proposed by Vysloukh and Serkin (1983), based on stimulated Raman scattering in fibres, which was later verified by Dianov *et al.* (1985), through compression in multisoliton Raman generation from a pulsed laser source. Since then, there has been a considerable number of experimental reports of soliton generation through stimulated Raman scattering in various configurations and using several different pump sources, Islam *et al.* (1986), Zysset *et al.* (1986), Kafka and Baer, (1987), Gouveia-Neto *et al.* (1987), Vodopyanov *et al.* (1987), Nakazawa *et al.* (1988) and Islam *et al.* (1989). More generally, it has been shown theoretically by Blow *et al.* (1988a), that soliton formation is possible in the case where there is coupling between waves leading to energy transfer, specifically via a gain term in the non-linear Schrödinger (NLS) equation description of the system.

In this chapter, however, only the effect of Raman gain will be considered and the emphasis will be placed on the description of experimental schemes and processes used to generate simple wavelength tunable soliton systems. Some of the theoretical aspects of this work have been considered in the chapters by Hasegawa, Dianov *et al.* and Mamyshev and references therein.

11.2 Modulational instability

The process of modulational instability in optical fibres, as well as providing soliton-like pulse structures, can also act as the precursor to the soliton-Raman continuum, as will be described later. Although modulational instability had been modelled in various fields in physics such as fluids (Benjamin and Feir, 1967) and plasmas (Hasegawa, 1972) it was not until 1980 that the process was first analysed by Hasegawa and Brinkman for optical fibres. The conditions necessary for the observation of modulational instability are similar to those required for the generation of envelope solitons in fibres. As a result of the combined effects of anomalous dispersion and the non-linear optical Kerr effect, amplitude or phase modulations on a continuous wave exhibit an exponential growth rate. This is accompanied by sideband evolution at a frequency separation from the carrier which is proportional to the optical pump power.

It was shown by Hasegawa and Brinkman (1980), see also Chapter 1 by Hasegawa in this book, that for a sideband of frequency Ω which is less than a critical frequency Ω_c defined as

$$\Omega_c = \left[\frac{2\omega}{c}\frac{n_2}{K''}|E_0|^2\right]^{1/2}$$

where n_2 is the non-linear Kerr coefficient, ω the angular frequency and $K'' = \delta^2 K/\delta\omega^2$, then the perturbation would grow exponentially. For zero frequency shift the gain falls to zero. The maximum amplitude growth rate is proportional to the pump intensity and is given by $(\omega/2c)n_2|E_0|^2$. This maximum growth rate occurs for a frequency $\Omega = \Omega_c/2^{1/2}$. With increased pump power higher order sidebands evolve, which are separated from the carrier by integer multiples of the frequency of maximum growth rate, i.e. $n\Omega_c/2^{1/2}$.

If, for example, radiation at 1.32 μm with a power of 5 W is focused into a single mode fibre with an effective core area of 90 μm^2 and a group delay dispersion D of 5 ps nm^{-1} km^{-1} ($K'' = 4.62 \times 10^{-27}$ s^2 m^{-1}), then the growth with distance z will vary as $\exp(0.85 \times 10^{-3} Pz)$ where P is the pump power. Hence for a pump power of 5 W the gain length is 236 m and the maximum gain would occur at a frequency offset of 1.35×10^{12} Hz.

The first experimental observation of modulational instability in a single mode optical fibre was reported by Tai *et al*. (1986a). Using a mode-locked Nd:YAG laser at 1.32 μm, on the time scale of the expected modulation, given previously, the 100 ps pump pulses effectively appeared continuous. The use of mode-locked pulses also eliminated any potential problems due to loss caused by Brillouin scattering as would be possible from the purely c.w. laser. For powers of around 5 W launched into kilometre lengths of fibre, modulations on a picosecond time scale were observed.

Figure 11.1 shows a representative spectrum of modulational instability. This was obtained through launching an average power of 60 mW from a mode-locked c.w. pumped Nd:YAG laser at 1.32 μm into 500 m of single mode fibre which had a dispersion minimum wavelength around 1.30 μm. At a pulse repetition rate of 100 MHz, the peak power in the 100 ps pulses was approximately 6 W. At this power level the sideband evolution, separated by 67 cm^{-1} (2.01×10^{12} Hz) from the input carrier was clearly apparent. Also shown in the inset in Figure 11.1 is an intensity autocorrelation of the temporal output from the fibre. The 450 fs time separation of the soliton structures on the autocorrelation corresponded reasonably well to the measured frequency modulation. With the zero intensity level indicated, it can be seen that the soliton structures sat on top of the carrier which extended for 100 ps. As the signal power is increased, the frequency modulation increases and the time separation of the soliton structures decreases correspondingly. With an increased signal, higher order sidebands appear; however, with further increases other non-linearities begin to dominate. In particular, stimulated Raman generation occurs and the spectral modulation features are lost. As such, modulational instability can act as the precursor to the soliton Raman continuum as will be shown later and, although it is not a requirement, it does have the effect of lowering the threshold for the process.

Using picosecond pulse inputs and operating in a region of low dispersion, modulational instability pulse repetition rates of up to 5 THz have been observed, Sudo *et al.* (1989). As the dispersion is decreased towards zero, it may be expected that the modulational instability frequency would tend towards infinity. However, in this region, the high-order even dispersions, in particular fourth order, plays an important role in levelling off the frequency of modulation, Ito *et al.* (1989). The highest pulse repetition rate generated via modulational instability is 15 THz, but the medium was a colliding

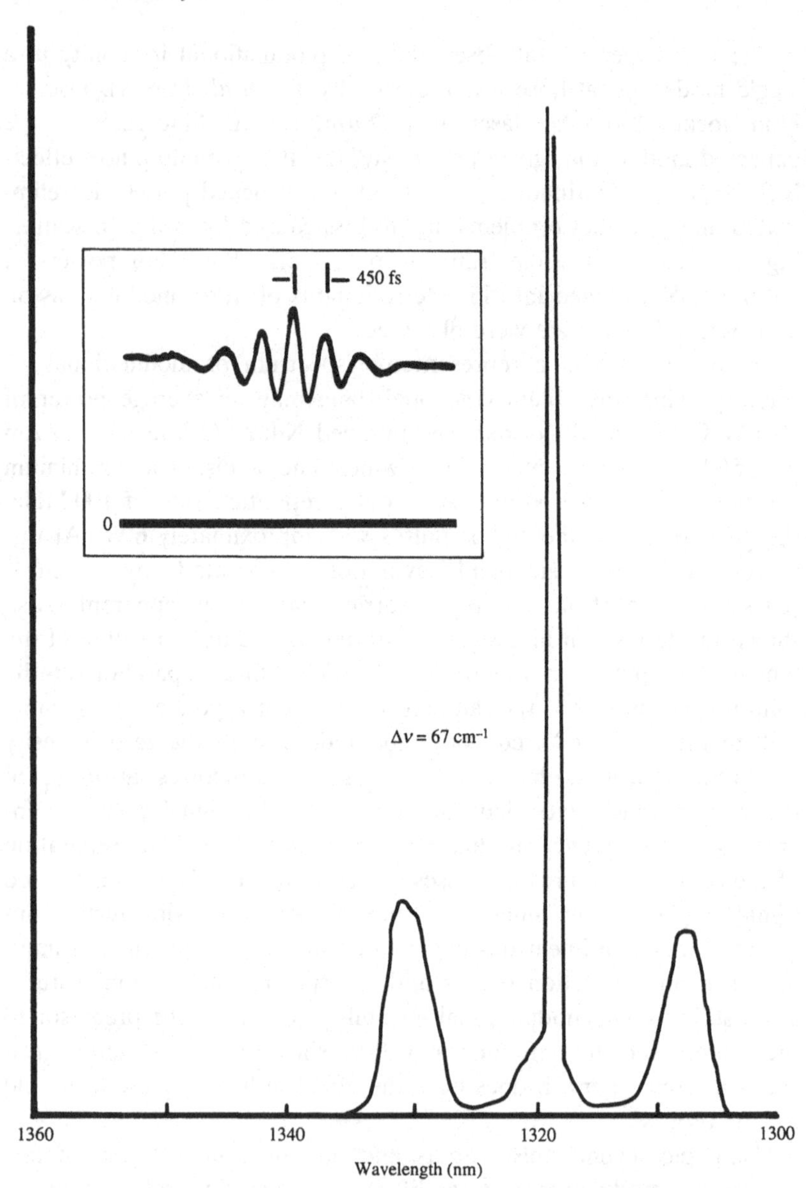

Figure 11.1. Spectral and (inset) temporal characteristics of modulational instability, obtained for a 6 W peak power, 1.32 μm pulse in 50 m of a single mode fibre.

pulse dye laser, not an optical fibre, where precise control of the dispersion and power permitted the generation of these high repetition rate soliton-like pulse sequences (Wang *et al*. 1989). Using a c.w. Nd:YAG laser at 1.319 μm, modulational instability around 100 GHz has been observed by Ito *et al.* (1989) in a 5 km length of low dispersion, small core fibre, in the first demonstration of the process using a purely c.w. pump.

11.2.1 *Induced modulational instability*

Modulational instability can be envisaged as a four-wave mixing process which is phase-matched through the mechanism of self-phase modulation. Exponential growth of the Stokes and anti-Stokes sidebands takes place at the expense of two photons from the beam converting to one at the upper and lower side band frequencies. Instead of allowing modulational instability to self-start from noise at the frequency separation at which the gain is a maximum, it is possible to seed the process through launching a probe signal together with the pump beam at a frequency separation which satisfies the gain condition. This process of induced modulational instability was first proposed and described by Hasegawa (1984). Experimentally the mechanism was verified by Tai *et al.* (1986b). In a similar experimental scheme pulse trains at repetition rates of up to 2 THz, have been observed by Greer *et al.* (1989). A pump and probe wave were launched simultaneously into a 500 m length of single mode, anomalously dispersive fibre. The probe signal was derived from a c.w. InGaAsP laser diode, which was 200 μm long, with plane uncoated facets and provided a multi-mode output around 1.307 μm and an average power of around 1 mW, of which 300 μm was launched into the fibre. This was substantially less than the 70 mW average pump power (7 W peak) derived from a c.w. pumped mode-locked Nd:YAG laser at 1.32 μm. The power from the YAG laser was adjusted to be just below modulational instability threshold, see Figure 11.2(*a*). As can be seen, the diode laser wavelength was in the region of the maximum modulational instability gain. When introduced on the anti-Stokes side of the pump, the diode laser probe signal led to enhanced modulation of the spectrum, see Figure 11.2(*b*), and the evolution of a periodic pulse break up in the time domain, which can be seen in the autocorrelation trace of Figure 11.3.

In the autocorrelation trace of Figure 11.3(*a*) trains of pulses separated by 4.6 ps, sat on top of an approximately 100 ps pulse corresponding to the YAG laser pump pulse. Each of the trains consisted of pulses of 130 fs separated by 490 fs see Figure 11.3(*b*). This 490 fs separation agrees reasonably well with the 475 fs separation

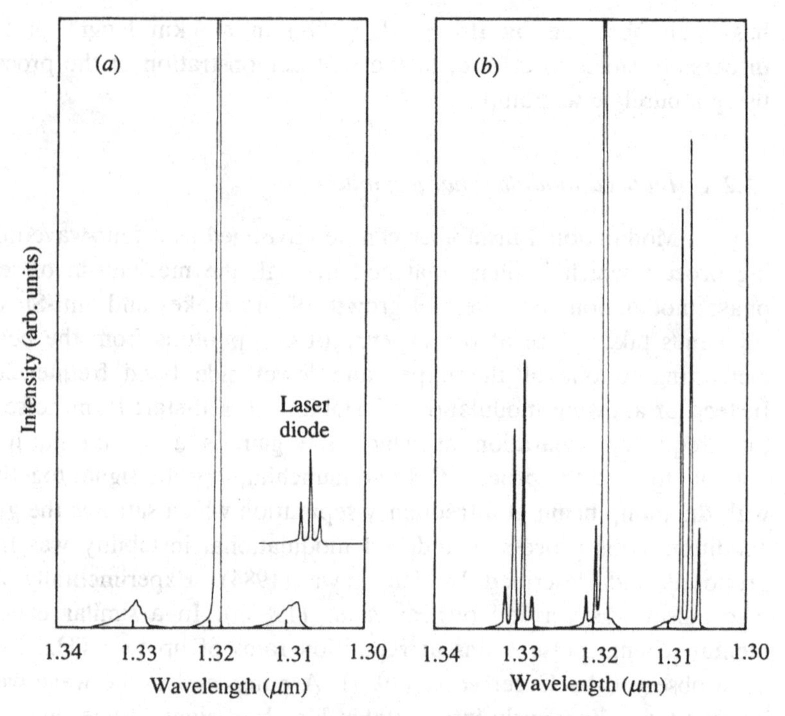

Figure 11.2. (*a*) Carrier exhibiting modulational instability plus applied laser diode signal, and (*b*) induced modulational instability 7 W peak power at 1.32 μW plus 300 μW average power at 1.307 μm from a c.w. diode laser.

expected from the 12 nm wavelength difference between the pump and probe wavelengths, while the 4.6 ps pulse train separation corresponds reasonably well with the 4.7 ps expected from the 1.2 nm between the modes of the diode laser. By tuning the laser diode either on the anti-Stokes or Stokes side of the pump, as demonstrated by Tai *et al.* (1986), it is possible to generate pulse trains at repetition rates determined by the frequency separation of pump and signal waves. The modulation depth of the autocorrelation at the peak was around 50% which indicated that the generated pulse trains were practically 100% modulated at the maximum.

Since a relatively small input signal can exhibit a rapid growth and develop deep modulation over moderate fibre lengths, induced modulational instability can be used as the mechanism for an ultrafast optical switch. Several schemes have been proposed and demonstrated, with small signal gains of more than 40 dB being observed (Islam *et al.* 1988; Soccolich and Islam, 1989).

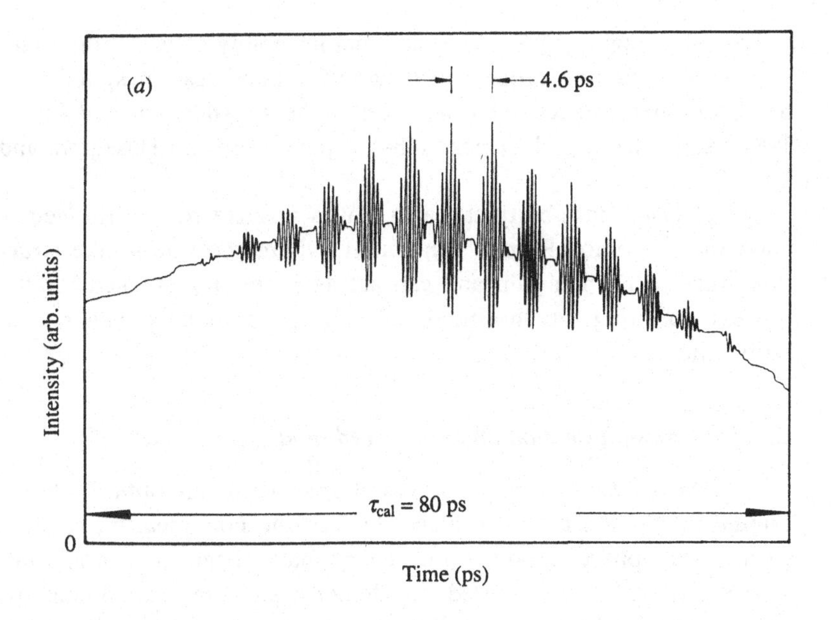

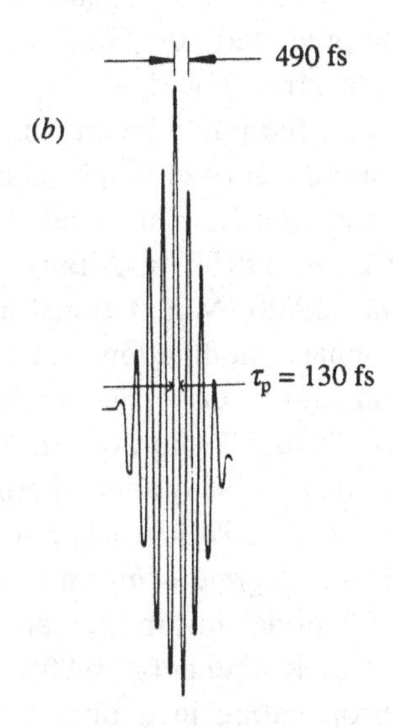

Figure 11.3. Background free autocorrelation trace of induced modulational instability (*a*) corresponding to the experimental conditions of Figure 11.2(*b*), and (*b*) pulse train on an expanded time scale.

The other regime where modulational instability plays an important role is as a limiting process in coherent communication systems, as has been characterised by several researchers (Anderson and Lisak, 1984; Hermansson and Yevick, 1984; Tajima, 1986 and Hasegawa and Tai, 1989).

Modulational instability can also play a positive role in the generation of the soliton-Raman continuum, where the soliton-like structures receive Raman gain and can act as a preferential seed for the process, lowering the threshold for pulse generation, as will be described later.

11.2.2 *Cross-phase modulation-induced modulational instability*

When two or more waves propagate in an optical fibre a mutual interaction can take place through the non-linearity, referred to as cross-phase modulation. There have been a considerable number of schemes devised to demonstrate this experimentally. Cross-phase modulation is inherent in the process of stimulated Raman generation in fibres. It was first predicted, by Gersten *et al.* (1980), that in the bulk, Raman spectra would be spectrally broadened by cross-phase modulation from the pump beam. In fibres, several researchers have investigated the effects of cross-phase modulation between the pump and a co-propagating Raman signal (Schadt and Jaskorzynska, 1987a, 1988; Islam *et al.* 1987a, b; Alfano *et al.*, 1987: Manassah, 1987 and Baldeck *et al.*, 1988). Modulational instability can also be induced through cross-phase modulation. As well as giving rise to solitary waves in the anomalously dispersive regime, as has been shown both theoretically (Schadt and Jaskorzynska, 1987b; Jaskorzynska and Schadt, 1988; Agrawal *et al.*, 1989) and experimentally (Gouveia-Neto *et al.*, 1988a, Greer *et al.*, 1990), it has also been proposed (Agrawal, 1987) that, even in the regime of normal dispersion, under certain circumstances modulational instability can occur and this has recently been demonstrated by Rothenberg (1990).

If one considers a weak signal co-propagating in a fibre with an intense pulsed pump beam, where the signal is in the anomalously dispersive regime, then the intensity-dependent refractive index gives rise, in the normal manner, to a time-dependent phase shift on the signal due to the pump field. The front edge of the signal is frequency down-shifted and the trailing edge is up-shifted. Together with anomalous dispersion, this gives rise to a temporal compression or 'soliton-like' shaping of the signal pulse. The efficiency of the process is clearly dependent on the degree of walk-off between pump and signal, and can be enhanced by arranging the pump to be in the normal dispersion regime, while the signal experiences anomalous dispersion, such that an effective matching of the group velocities takes place.

The system can be described by two coupled equations

$$\frac{\partial A_p}{\partial z} + \frac{1}{v_g^p}\frac{\partial A_p}{\partial t} + \frac{i}{2}\beta_p\frac{\partial^2 A_p}{\partial t^2} = i\frac{\omega_p}{c}n_2[|A_p|^2 + 2|A_s|^2]A_p$$

$$\frac{\partial A_s}{\partial z} + \frac{1}{v_g^s}\frac{\partial A_s}{\partial t} + \frac{i}{2}\beta_s\frac{\partial^2 A_s}{\partial t^2} = i\frac{\omega_s}{c}n_2[|A_s|^2 + 2|A_p|^2]A_s$$

where A is the envelope of the pulse, at frequency ω, and group velocity dispersion β, the subscripts s and p denoting signal and pump respectively. The phase shift experienced through cross-phase modulation is twice that experienced via self-phase modulation, since twice as many terms of $\chi^{(3)}$ contribute to the former. These equations have been solved numerically by several researchers previously mentioned, and the process has been well reviewed both by Agrawal (1989) and by Baldeck *et al.* (1989).

In a demonstration of modulational instability induced by cross-phase modulation, Gouveia-Neto *et al.* (1988a) used the 100 ps pulses from a mode-locked c.w. pumped Nd:YAG laser operating at 1.06 μm to provide the required cross-phase modulation on a weaker 1.32 μm signal, from a c.w. pumped mode-locked Nd:YAG laser. Pump and signal were launched into a length of single mode fibre, which had a dispersion minimum at 1.27 μm ensuring that the signal was in the anomalously dispersive regime. However, due to walk off, the effective length of the fibre was approximately 120 m.

Figure 11.4 shows the spectra of the signal pulses emerging from the fibre at an average power level of 20 mW. In the absence of the pump a self-phase-modulated spectral width of 0.66 μm was recorded, see Figure 11.4(*a*). On simultaneous launch of 30 mW of pump power at 1.06 μm, spectral broadening of the central feature took place plus the appearance of sidebands, the number and intensity of which were pump power dependent, see Figure 11.4(*b*). Associated with the spectral modulational instability, temporal modulation took place, as shown in Figure 11.5. The modulation depth of 60% at peak of the autocorrelation, indicated that at the centre of the pulse, complete modulation of the pulses took place. Pulses of 520 fs were generated separated by approximately 3.5 ps which corresponded well with the

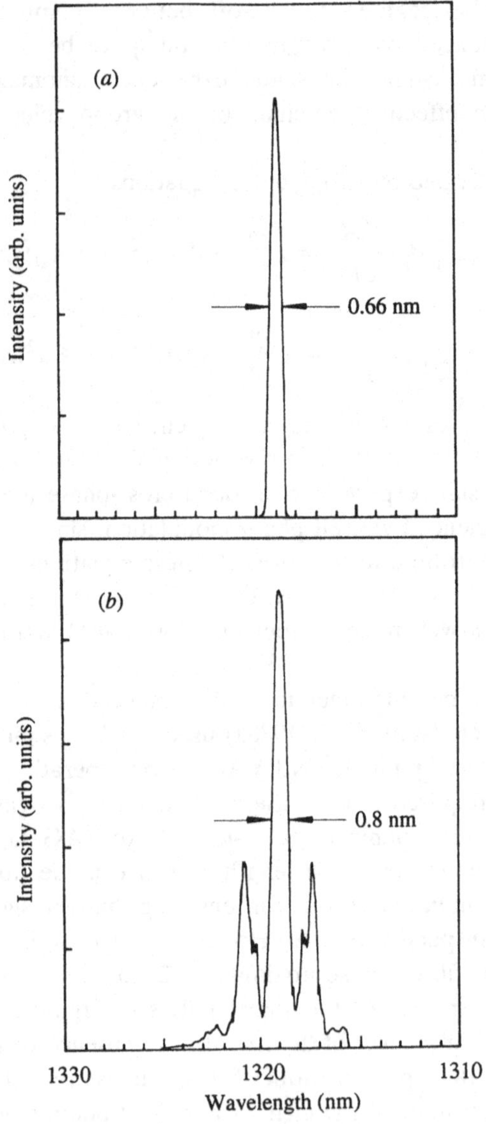

Figure 11.4. Spectra recorded for an average signal power of 20 mW at 1.319 μm and simultaneous average pump power of (*a*) 0, and (*b*) 30 mW at 1.06 μm showing the effect of cross-phase modulation-induced modulational instability.

290 GHz frequency separation of the induced modulational instability sidebands.

Greer *et al.* (1990) have also demonstrated this mechanism using a c.w. diode laser as the signal source. Again the pump signal was

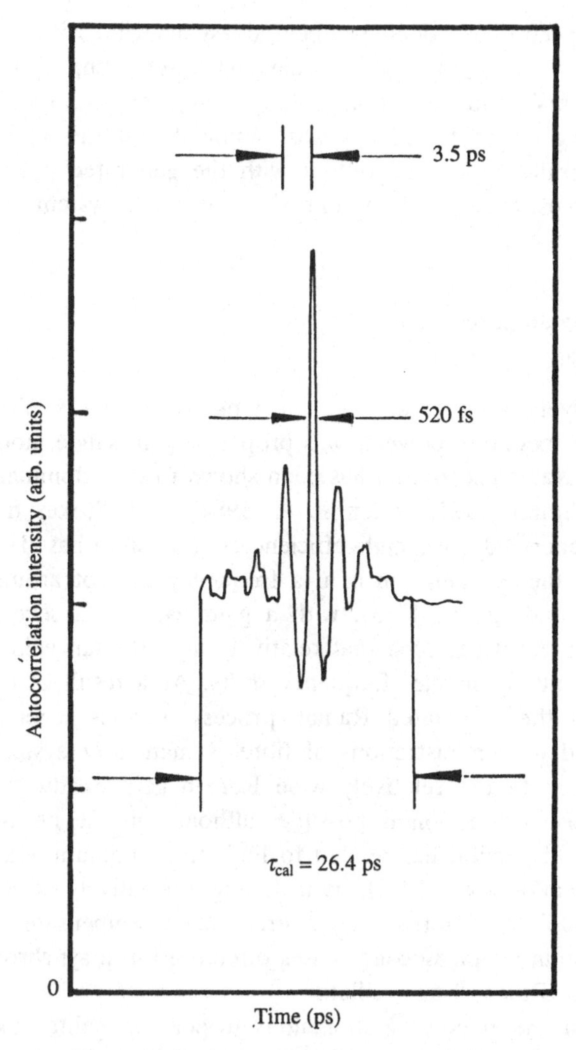

Figure 11.5. Autocorrelation trace of the output corresponding to the power levels indicated in Figure 11.4(b), exhibiting induced modulational instability.

derived from a 1.06 μm, mode-locked c.w. pumped, Nd:YAG laser, and both pump and signal were launched into a fibre with a dispersion minimum at 1.27 μm. By using a tunable c.w. laser diode, it was possible to investigate the effects of matching of the group velocities on the cross-phase modulation. The maximum spectral broadening was obtained for the optimum overlap of pump and signal disturbance in the fibre. On tuning the signal to either side of the match wavelength, the spectral broadening appeared asymmetric, with the asymmetry indicative of the walk off between signal and pump. For a signal level of 60 mW and an average pump power of 480 mW, a spectral broadening of up to 2.5 nm was obtained and the signal exhibited a temporally modulated output with the generated pulses being less than 100 ps, the resolution limit of the detection system.

11.3 Soliton-Raman generation

11.3.1 *Introduction*

For relatively long (the order of tens to hundreds picoseconds) pulses of moderate power levels propagating in single mode fibres, stimulated Raman scattering has been shown to be a dominant loss mechanism (Smith, 1972; Stolen *et al.*, 1984), with Stokes frequencies being generated with high efficiencies. For silica-based fibres, the peak of the Raman gain is at a frequency shift of around 440 cm^{-1} (Stolen and Ippen, 1973), with a gain bandwidth several hundred wavenumbers wide, such that relatively high Raman gain is possible for substantially smaller frequency shifts. As a result of the high efficiency of the stimulated Raman process, various schemes have been devised in demonstrations of fibre Raman laser systems (Stolen, 1980). Due to the relatively wide Raman gain bandwidth, subpicosecond pulse generation is possible, although in the normal dispersion regime dispersion has tended to limit the minimum available pulsewidth. Kafka *et al.* (1986), by including a negatively dispersive, grating pair delay line intra-cavity, were able to compensate for dispersion and obtain subpicosecond pulses directly from a synchronously pumped fibre Raman ring oscillator.

By operating in the region of anomalous dispersion, soliton-like shaping should take place for pulses of sufficient power, giving rise to ultrashort pulse generation. The first proposal for the use of the stimulated Raman process for the generation of solitons was made by Vysloukh and Serkin (1983, 1984). They showed that solitons with amplitudes greatly exceeding the amplitude of the pump could readily be formed at the Stokes frequency. Experimentally, this was first realised by Dianov *et al.* (1985) using a 30 ps pulse from a parametric oscillator to generate a stimulated Raman continuum of multi-solitons in a single-mode fibre. Pumping at 1.54 μm, for a peak power of 800 W incident on the 250 m fibre, a Stokes wing extending to beyond 1.6 μm and containing up to 50% of the total input energy was observed. In the time domain, spectral selection of the Raman-shifted components exhibited solitons of 200 fs duration without pedestals.

This technique has since been investigated by many authors primarily in single pass configurations (Gouveia-Neto *et al.*, 1987b; Beaud *et al.*, 1987, Vodopyanov *et al.* 1987, Islam *et al.*, 1989) although oscillator configurations have also been examined (Kafka and Baer, 1987; Islam *et al.*, 1987b; Gouveia-Neto *et al.*, 1987a). The method provides a simple means of generating a source of ultrashort pulse solitons tunable throughout a considerable spectral range.

11.3.2 *Single-pass soliton-Raman generation*

For a pump pulse in the region of the dispersion minimum, the interaction distance between the pump and first Stokes can be quite considerable, giving rise to high Raman conversion efficiencies. The exponential Stokes gain which is intensity dependent, is always in a region of strongly excited molecular vibrations of the host and consequently the overall gain is high. For pumping at 1.32 μm using 100 ps pulses, over 300 m of fibre the walk off can be much less than the pulsewidth. The clear depletion of the pump pulse which occurs can be seen in Figure 11.6, which shows synchroscan streak camera measurements of the power dependent transmission of a 1.32 μm pump pulse through 300 m of single-mode fibre. At an average (peak) power of 100 mW (10 W) below Raman threshold, the output pulsewidth is effectively that of the input, Figure 11.6(*a*). On increasing to 300 mW (30 W) and 700 mW (70 W), Figures 11.6(*b*) and (*c*) respectively, depletion can be seen. At the highest pump power, depletion to the base line of the pump pulse was apparent. In Figure 11.6(*c*) the sensitivity of the camera was increased to observe the remaining pump fragments and the time resolution was constant throughout. The duration of the frequency up-shifted pump trailing edge was less than that of the down-shifted leading edge indicating the relative motion of the Raman signal towards the rear of the pump.

In the minimum dispersion regime, in addition to modulational instability, cross-phase and induced modulation can contribute to the pulse formation and shaping mechanism. It was originally shown by Gouveia-Neto *et al.* (1988b, 1989a) that the soliton-Raman process could proceed through modulational instability when the pump wavelength was in the anomalously dispersive regime. This can be seen in Figure 11.7, showing the spectral development with input

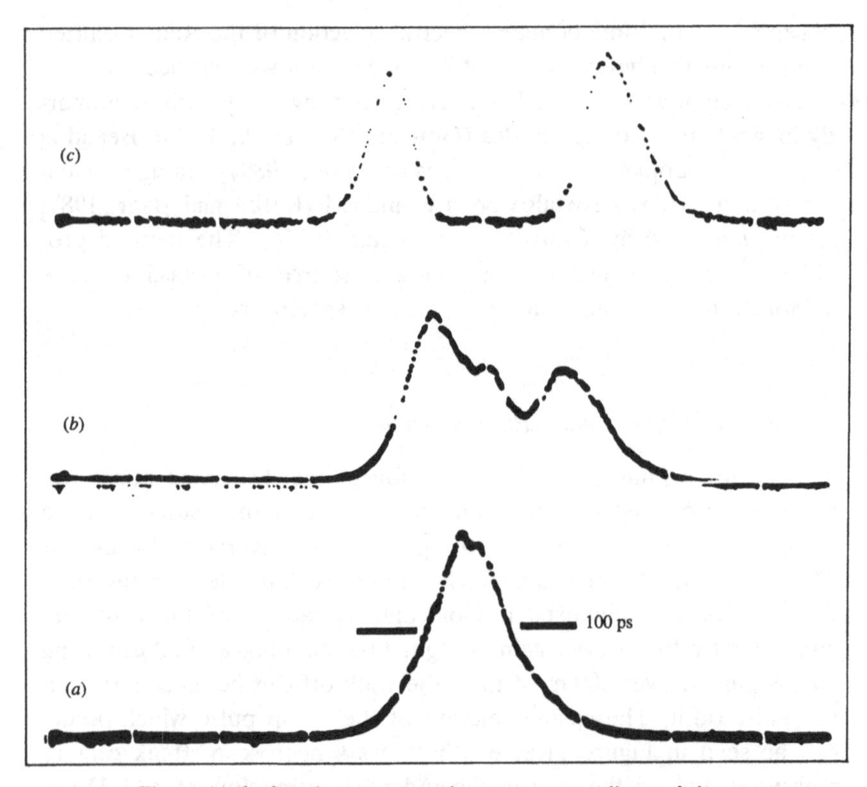

Figure 11.6. Synchroscan streak camera recordings of the power dependent temporal behaviour of a fundamental 100 ps pump pulse transmitted through 300 m of single mode fibre at average input levels of (*a*) 100 mW, (*b*) 300 mW and (*c*) 700 mW, showing pump depletion, time increases to the left.

power to a fibre from a mode-locked Nd:YAG laser operating at 1.32 μm. The fibre length was 500 m long and had a dispersion minimum around 1.3 μm.

At an average power of 60 mW, the distinctive sideband growth associated with modulational instability (Tai *et al.*, 1986a) is evident. The sidebands at a frequency separation of 67 cm^{-1} from the carrier corresponded to a modulation of approximately 2 THz (500 fs). This was in reasonable agreement with that measured using a standard background free autocorrelation technique as is shown in Figure 11.8(*a*). As the pump power was varied, the spectral separation of the sidebands varied, increasing with increased pump power and *vice*

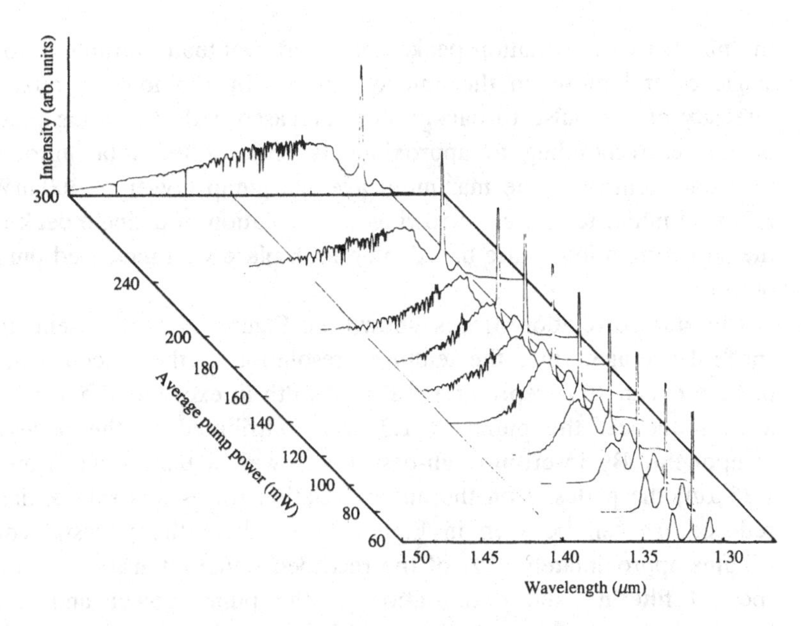

Figure 11.7. Variation of output spectra from a 500 m long fibre with input average power from a mode locked Nd:YAG laser showing the evolution from modulational instability to a soliton-Raman continuum.

versa. The peak of this modulation had a frequency separation distinct from the expected shift of the Raman maximum gain, around 440 cm^{-1}.

With increase in the pump power, the spectral separation between the pump and anti-Stokes components showed a decrease. This is consistent with a loss of fundamental power through the Raman gain process. The Stokes component exhibited a significant increase in spectral bandwidth and a dominant spectral shifting to longer wavelength. Up to the maximum average pump power indicated in Figure 11.7, this trend continued, such that for an average pump power of 300 mW, the spectral extent of the Stokes band was beyond 1.5 μm and had increased in a continuous manner.

In the time domain the evolution from modulational instability to Raman amplified solitary waves was apparent (see Figure 11.8). At an average power of 60 mW modulations on the 100 ps pump pulse occurred, which represented approximately 38% of the autocorrelation intensity. The individual pulses had durations of approximately 140 fs and were separated by 495 fs. With increasing power, the number of modulation peaks decreased eventually forming into a single central pulse on the autocorrelation. In addition the relative intensity of the pulse to background increased with increasing pump power, corresponding to approximately 63% of the total autocorrelation intensity at the maximum average pump power of 300 mW. After an initial temporal narrowing and evolution of a single peak on the autocorrelation, pulse broadening took place with increased pump power.

The autocorrelation traces shown in Figure 11.8 represent the spectral average, since the temporal resolution of the autocorrelator of better than 10 fs represented a bandwidth in excess of 200 nm. As a consequence, the pump at 1.32 μm contributed to the pedestal component. By inserting high-pass filters with a band edge around 1.35 μm, the pedestal on the autocorrelation traces was substantially reduced, as can be seen in Figure 11.9, where the pedestal contributes approximately 13% of the recorded signal intensity. Through spectral filtering and optimisation of the pump power and fibre length, the pedestal component could be reduced to as little as 1–2% of the overall autocorrelation signal (Gouveia-Neto *et al.,* 1987b, 1988b).

In terms of the overall energy content, the soliton component of the autocorrelation can contain as much as 30% of the total within the Raman band, with pulse powers of kilowatts being readily obtainable. From measurements of the peak power levels and from estimates of the fundamental soliton power (Mollenauer *et al.*, 1980) in the fibres used, infers that the output of the soliton Raman continuum consists of several solitons. This provides one contribution to the pedestal, since cross correlation between the randomly spaced soliton will enhance the signal level in the wings as an autocorrelation trace being recorded. The other contribution to the pedestal arises from straightforward stimulated Raman scattering. Not all the energy generated in the Raman band is used for soliton formation and amplification. That not required appears as a dispersive wave and being of reduced intensity and longer duration contributes with less weighting to the overall autocorrelation intensity as the pedestal.

The single peak, for example shown in Figure 11.9, of the autocorrelation trace is an integration of several single solitons. Using an electron optical streak camera, Grudinin *et al.* (1987) have shown that in the region of anomalous dispersion, the Stokes signal consists of several randomly spaced pulses, with a duration limited by the resolution of the streak camera. As the pump power was reduced, the

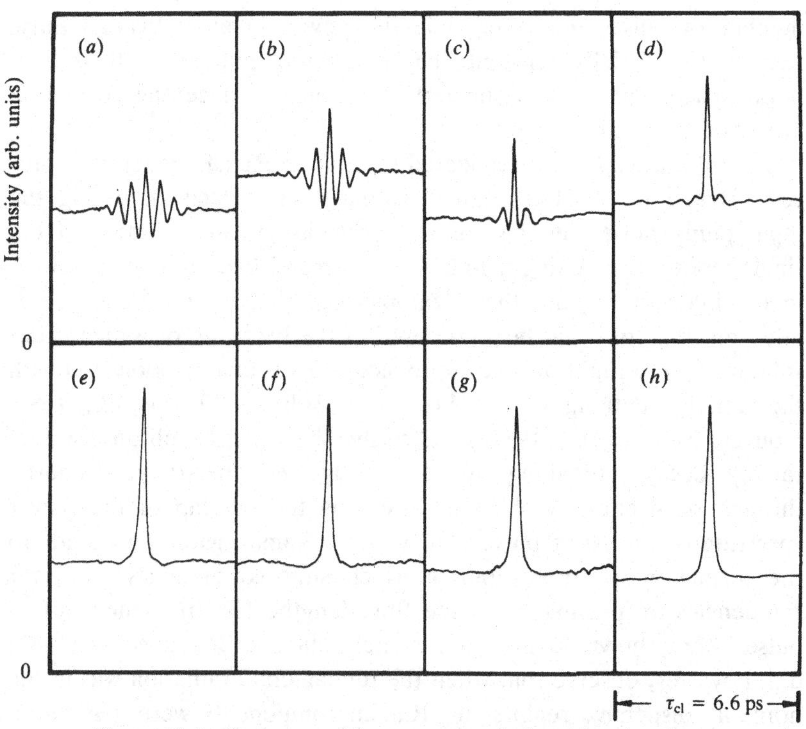

Figure 11.8. Background free autocorrelations showing the temporal development from modulational instability to soliton Raman continuum, associated with Figure 11.7 (*a*) 60 mW, (*b*) 80 mW, (*c*) 100 mW, (*d*) 120 mW, (*e*) 140 mW, (*f*) 160 mW, (*g*) 240 mW, and (*h*) 300 mW average power.

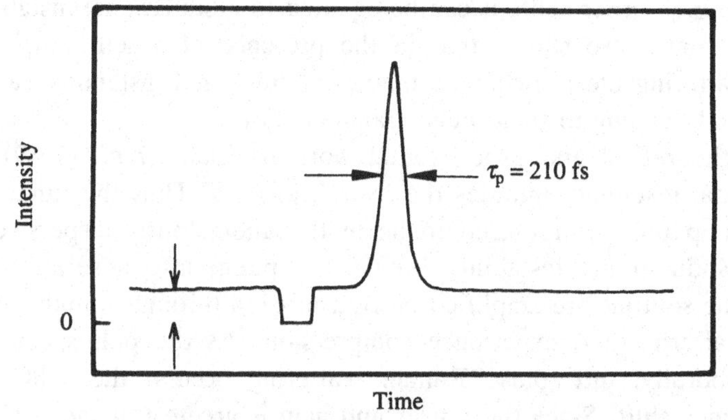

Figure 11.9. Background free autocorrelation trace of component spectrally filtered from a soliton-Raman continuum examining wavelengths greater than 1.35 μm at an average pump power of 240 mW.

number of pulses reduced, while the power in an individual pulse remained essentially constant. The number of pulses and their temporal spacing are not particularly controllable since the formation process evolves from noise.

Several groups have investigated the soliton-Raman generation process. Beaud *et al.* (1987) have examined the effects of tuning the input pump pulses, in the region of the dispersion minimum of the single-mode fibre. Using a tunable picosecond dye laser with powers in the kilowatt regime, they demonstrated that their Raman generation mechanism could be considered as the break up of a high order soliton. A very rapid pulse compression was observed, associated with the initial breathing of the high-order solitons (Tai *et al.*, 1986; Gouveia-Neto *et al.*, 1987e). The ultrashort pulse soliton formed, rapidly decays, breaking up temporally and spectrally, primarily through self-frequency shifting caused by the overlap of the power spectrum of the short pulse with its own Raman gain. This leads to the formation of Stokes pulses which shift continuously to lower frequencies on passing down the fibre length. The frequency shifted pulses were shown to be fundamental solitons. Beaud *et al.* (1987) did, however, observe that when the fundamental radiation was in the normally dispersive regime, no Raman components were generated. Islam *et al.* (1989) have also proposed that modulational instability acts as the precursor to the soliton-Raman self-generation process. The correlation between high order soliton decay and modulational instability has been examined theoretically by Nakazawa *et al.* (1989b). For relatively short fibre lengths, the amplitude ripple which appears on high order solitons can be equated to modulational instability. They have also shown that in the presence of a self-induced Raman scattering the growth rate of the modulational instability can be enhanced, leading to rapid pulse fragmentation.

In the theoretical and experimental work of Islam *et al.* (1989), modulational instability initiates the overall process. Thus the fundamental pump pulse is required to be in the anomalously dispersive regime. Modulational instability breaks the pump into a series of solitons, the solitons are amplified on transmission through stimulated Raman scattering and experience compression. As the pulses compress temporally, intra-pulse Raman scattering occurs, the soliton self-frequency shift. Since the power and gain is greatest at the peak of the pump pulse, it would be expected that the solitons would be first generated in this region. On experiencing the intra-pulse Raman frequency down-shifting, the associated change in the group velocity dispersion causes the pulse to move towards the rear of the pump pulse. As it does so, it collides with solitons formed later and to the rear of the pump. This generates high peak powers and shorter pulses, which gives rise to further spectral broadening. Short pulse solitons emerge from the collision, shifted in frequency and temporally separate from the remaining fragments. It can, therefore, be seen that above threshold, the soliton-Raman continuum consists of many short pulse solitons, the number and temporal separation of which are completely random. The spectral output from the fibre is the sum of many solitons plus non-soliton radiation and this explains the nature of the pedestal on the measured autocorrelations, see Figure 11.9. Consequently, by spectral filtering, it is possible to select fundamental solitons throughout the complete continuum. As the central wavelength of the bandpass filter is tuned to longer wavelengths, the peak-to-pedestal component of the autocorrelation increases (Dianov *et al.*, 1986; Gouveia-Neto *et al.*, 1988b, Islam *et al.*, 1989). At lower wavelengths, the randomly timed solitons to the front of the ensemble contribute to the pedestal, while at long wavelength the principal contribution is due to self-frequency-shifted, fundamental, well-separated solitons, spectrally distinct from the randomly spaced components. With increasing central wavelength, the duration of the soliton component also increased, which is related to soliton self-frequency shifting and the required increase in pulse length to counter increasing wavelength and dispersion to maintain fundamental soliton powers for the shifting pulse.

In the time domain, each pump pulse gives rise to an ensemble of solitons. However, since the soliton evolution primarily arises from noise, a timing jitter occurs between pulses. Keller *et al.* (1989) have examined the amplitude and phase noise of soliton-Raman sources, measuring a 5.2 ps r.m.s. timing jitter, using a conventional mode-locked Nd:YAG laser pump source. They also showed that the jitter increased with pump power and with fibre length.

Theoretically, Blow *et al.* (1987, 1988) have examined the most general cases of generation and stabilisation of solitons in a system described by the NLS equation incorporating an amplification term. An exponential compression takes place, saturating only when the bandwidth of the generated pulse approaches that of the amplifying medium. Blow and Wood (1989) have derived a single wave equation, which conserves photon number and generally describes transient Raman scattering in fibres. It predicts soliton Raman generation without the initial requirement of modulational instability. This is particularly relevant to the situation where the soliton-Raman continuum is produced using a pump pulse in the normal dispersion regime. The equation can also be used to correctly describe the growth of modulational and four-wave instabilities.

Experimentally, verification that the soliton-Raman continuum can evolve using pump pulses in the normal dispersion regime, has been undertaken by Gouveia-Neto and Taylor (1988). The experimental arrangement was identical to the conventional single pass scheme. A 1.32 μm c.w. pumped, mode-locked Nd:YAG was used as the pump source and a 500 m long single-mode fibre with a dispersion minimum at 1.38 μm was used as the non-linear medium. Figure 11.10 shows the spectral development of the soliton-Raman continuum with pump power. In comparison to the situation where the pump radiation was in the anomalous dispersion region, for equivalent fibre lengths and input pulse parameters, the situation where the pump was in the normal dispersion regime required substantially higher average pump powers to reach threshold. Compare, for example, the spectral extent of the Raman continuum in Figures 11.10(*a*), for an average pump power of 400 mW, and Figure 11.7 at a pump power of 300 mW, indicating that modulational instability clearly lowers the threshold condition for the soliton-Raman process. For pumping in the normally dispersive regime instabilities either in phase or amplitude on the essentially c.w. Raman signal, experience soliton-like shaping and Raman amplification in the anomalously dispersive regime. Building up from noise, the process proceeds then in a similar manner to that previously described, with the evolution of several solitons within the generated spectral contunuum. In the time domain, the autocorrelations show the generation of femtosecond solitons, with pedestal components accounting for 4–5% of the autocorrelation intensity. Again the contribution to the pedestal arises from the non-soliton dispersive signal plus that due to the cross correlation of low level solitons to the front of the pulse.

11.3.3 *Single pass cascade-Raman soliton generation*

The long wavelength limit of the soliton-Raman continuum is a function of the pump power, and is simply related to the degree of self-frequency shifting experienced by the generated solitons. With fibre lengths of around 500 m and pump powers of 50 W at 1.32 μm the soliton-Raman continuum extends to approximately 1.5 μm, with the maximum around 1.4 μm. In the normal dispersion regime,

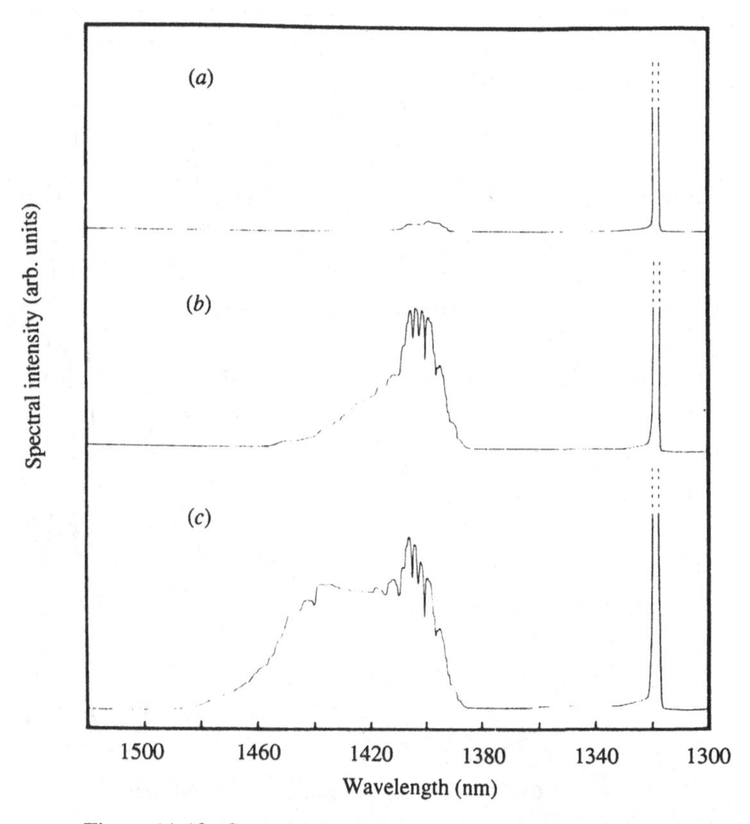

Figure 11.10. Spectral development of the soliton Raman continuum for a 1.32 μm pump pulse in the normally dispersive regime, $\lambda_0 = 1.38$ μm, at average pump powers of (*a*) 400 mW, (*b*) 550 mW and (*c*) 650 mW.

stimulated Raman scattering is a very efficient cascaded process, with high orders being readily generated for modest pump power levels (Stolen *et al.*, 1984). When soliton shaping occurs at the same time, the efficiency of the cascade process is considerably reduced. This is due to the effective reduction of the interaction length for the process as a result of the differences in the group velocity dispersions of the various Raman orders (Stolen and Johnson, 1986), and the relatively short fibre length required for total walk through of the 'pump' by the next highest order, when the 'pump' is a femtosecond soliton pulse.

To extend the available tuning range, for modest pump power levels, the cascade process can be utilised; however, the soliton-shaping mechanism should be restricted to operation on the highest

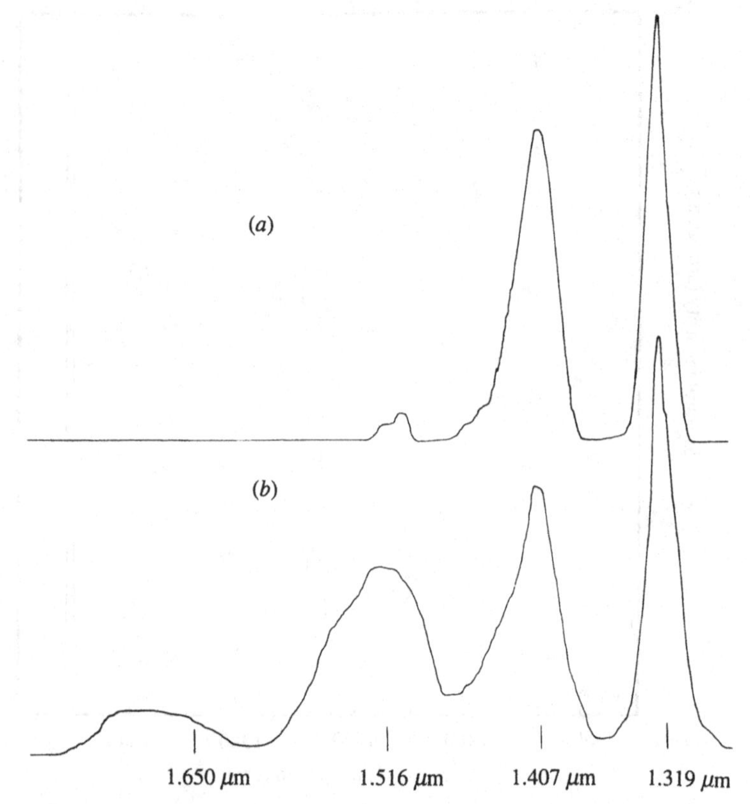

Figure 11.11. Single pass cascade Raman soliton generation spectra for a 600 m fibre with $\lambda_0 = 1.46\ \mu m$ and an average 1.32 μm pump power of (*a*) 450 mW and (*b*) 530 mW.

order Raman signal only. This can be achieved, through the use of dispersion shifted fibres (Ainslie and Day, 1986).

This technique was first demonstrated by Gouveia-Neto *et al.* (1987a), using a 1.32 μm, c.w. pumped mode-locked 100 ps Nd:YAG laser pump source, in conjunction with a 600 m length of single-mode fibre, with a dispersion minimum at 1.46 μm. At average pump powers below 400 mW, only the first Stokes Raman band was observed, generating pulses of around 100 ps centred at 1.4 μm. As the pump power was increased to 530 mW average power, radiation in the second Stokes band around 1.5 μm developed (see for example Figure 11.11) and autocorrelations revealed the evolution of pulses of 130 fs on top of pedestals, which accounted for 4–5% of the autocorrelation intensity, similar to those obtained in the first Stokes band, when the fibre dispersion minimum was in the region of 1.3 μm.

Using this technique, individual fundamental solitons, with peak powers in the kilowatt range can be readily generated, through spectral selection at the fibre output.

With much higher peak pump powers, cascading is more easily observed. For an input wavelength of 1.06 μm and pulse powers of about 5 kW, Grudinin *et al.* (1987) have observed soliton Raman generation in a fibre with a minimum dispersion around 1.32 μm and soliton generation at the fourth Stokes wavelength in a broad band which extended beyond 1.7 μm.

11.3.4 *Soliton-Raman oscillators*

In the near infrared and for normal dispersion, fibre Raman lasers, for many years have been used to provide simple sources of tunable radiation in various cavity configurations (Stolen, 1980). Pulsed outputs can be generated using various synchronously pumped schemes (Lin *et al.*, 1977; Lin and French, 1979; Stolen *et al.*, 1977), with wavelength tunability being obtained through dispersion tuning. In general, the typical pulsewidths generated are limited by dispersion. Through operating with the generated Raman signal in the anomalous dispersion regime, self-compression and soliton shaping can take place and sub 100 fs pulses can be generated.

One of the first soliton-Raman oscillator arrangements was reported by Zysset *et al.* (1986). Synchronously pumping, using the tunable pulses from a compressed dye laser 1.25–1.35 μm with peak powers of up to 1 kW in a 18 m fibre, 80–100fs soliton were generated in a simple linear cavity arrangement. It was found, however, that when the pump pulse was in the region of normal dispersion, no Raman signal was observed and the operation of this system is probably best described by multi-soliton break up and self-frequency shifting when the pump pulses were in the region of the dispersion minimum or were negatively dispersive.

In synchronously pumped fibre Raman laser systems as the pump and Stokes signals co-propagate, cross-phase modulation from the pump to the Stokes signal can dominate the interaction and impose a major shaping mechanism on the Stokes pulses. This behaviour has been extensively investigated by Islam *et al.* (1986, 1987). Using the so called FRASL (Fibre Raman Amplification Soliton Laser), ring laser systems incorporating single and two stage fibre arrangements were investigated. In the latter, the pulse amplification could be effectively separated from the pulse shaping. As would be expected, the pulse shaping was critically dependent on the pump and Stokes wavelengths. Where cross-phase modulation is present, complete walk through of pump by the signal should cancel the effect of cross-phase modulation. However, quite often depletion of the pump takes place, as the Stokes signal walks through, it accumulates a non-linear chirp which cannot be counteracted by simply propagating in a fibre. This consequently gives rise to a pedestal component in the ouput pulses from many fibre-Raman laser systems. Islam *et al.* have investigated the wavelength sensitivity of the system and have shown that for correct selection of fibre length and walk off, through dispersion tuning, the pedestal component can be virtually eliminated.

Probably the simplest laser configuration was that used by Kafka and Baer (1987) and Gouveia-Neto *et al.* (1987a) a schematic of which is shown in Figure 11.12. Both used a c.w. mode-locked Nd:YAG laser operating at 1.32 μm as the pump laser which was launched into a single mode, non-polarization-preserving fibre. Kafka and Baer used 1.1 km length, while Gouveia-Neto *et al.* initially used a 300 m length. The behaviour of the laser can depend on the fibre length, as will be described later. Light emitted from the output end of the fibre was redirected using plane mirrors back through the dichroic input beam splitter and the fibre laser optical path length was adjusted such that it was an exact multiple of the pump laser path length. Since the overall gain of the Raman laser is high and operating with a pump pulse in the region of minimum dispersion gives rise to long interaction lengths, high output coupling can be obtained from the system. After optimizing the component values in an arrangement similar to that of Figure 11.12, Kafka and Baer (1987) constructed a completely integrated fibre laser using fused fibre optic couplers, fusion spliced to the Raman active fibre in a loop geometry. This stable configuration produced similar output characteristics to the laser system incorporating bulk optics.

Figure 11.13 shows the spectral output from a 300 m fibre loop Raman laser: (*a*) in single pass and (*b*) with feedback into the loop, for an average pump power of 200 mW, approximately 20 W peak power. In single pass, the spectral output is characteristic of modulational instability, while with feedback, a Raman band appears centred around 1.38 μm ($\Delta\nu \simeq 345$ cm^{-1}) and the spectral separation of the modulational instability side bands decreased, characteristic of a loss of power from the pump, to the Raman signal.

In the time domain, autocorrelations of the output exhibited the pedestal component generally associated with the soliton-Raman pro-

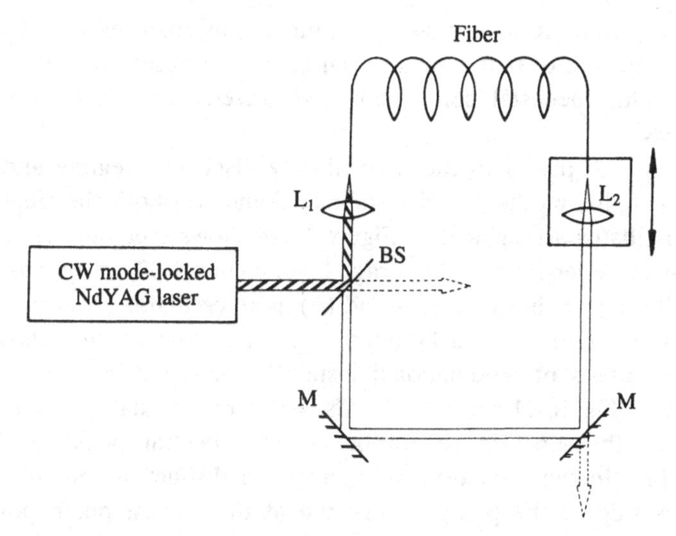

Figure 11.12. Schematic diagram of soliton-Raman fibre ring laser.

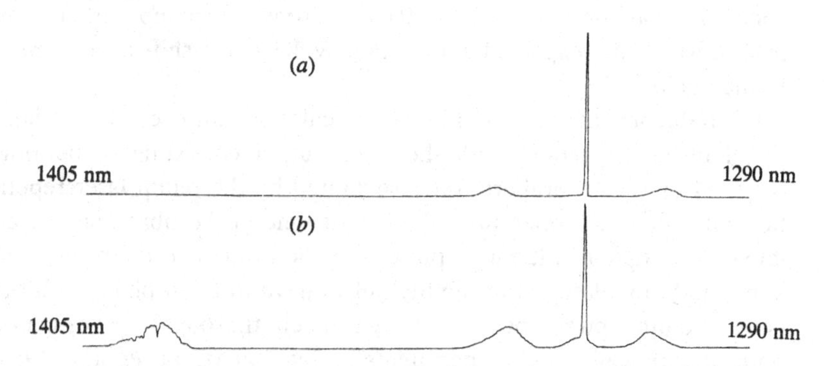

Figure 11.13. Spectra associated with a 300 m fibre long, soliton Raman ring laser at an average 1.32 μm pump power of 200 mW (*a*) single pass, and (*b*) with feedback.

cess. Depending on pump and cavity matching conditions, the pedestal could be as little as 1–2% of the peak intensity, although pedestals could be as much as 15–20%, indicating that at minimum, only 10–11% of the total energy was in the soliton component. Although generating femtosecond pulses, the synchronously pumped fibre Raman laser is not so critical to cavity length match to the pump laser as for example a femtosecond, synchronously pumped dye laser would be. In the latter case micrometre sensitivity is required, while with the fibre Raman system, centimetre mismatches can be tolerated, although over this range changes can be observed in the output pulsewidth, pedestal component and wavelength (Gouveia-Neto *et al.*, 1988c).

For pump pulses in the anomalously dispersive regime and short fibre lengths, modulational instability dominates both the single pass and oscillator arrangements. Figure 11.14 shows spectra obtained in a 100 m, fibre length for an average fundamental 1.32 μm, pump power of 750 mW; (*a*) in single pass and (*b*) in an oscillator arrangement. In the time domain, background free autocorrelations show the characteristics of modulational instability in single pass. In the example of Figure 11.14(*c*) the 220 fs period of modulation corresponds well to the 4.6 THz separation of the spectral peaks in Figure 11.14(*a*). In the oscillator arrangement, a distinct asymmetry to the Stokes side of the pump occurs and at the highest pump powers a single peak appears on the autocorrelation. This process can be viewed as the modulational instability preferentially seeding the Raman gain, with build up occurring on the modulational instability component at approximately 160 cm^{-1} from the pump rather than from noise at the expected approximately 400 cm^{-1} shift of maximum Raman gain.

On reduction of the pump power, oscillation can occur in modulational instability mode, with the laser output consisting of periodic trains of pulses at terahertz rates separated by the pump laser repetition rate. For such operation, the transit time in the fibre must be an integral multiple of the pump pulse repetition time and the re-injected pump and modulational instability pulses have to be in phase with the new incoming pump pulse. This has been thoroughly investigated both theoretically and experimentally by Nakazawa *et al.* (1988, 1989a, b). They define three regimes of operation:

(1) at low pump powers where the output characteristic is purely of modulational instability:
(2) medium pump powers, where a Stokes asymmetry is dominant in the modulational instability and the output frequency is defined by the sideband frequencies; and
(3) the regime of high pump power where the spectral asymmetry to the Stokes side is further enhanced and pump depletion takes place.

This latter regime corresponds to the appearance of a single peak on the autocorrelation traces and the temporal behaviour of this power-

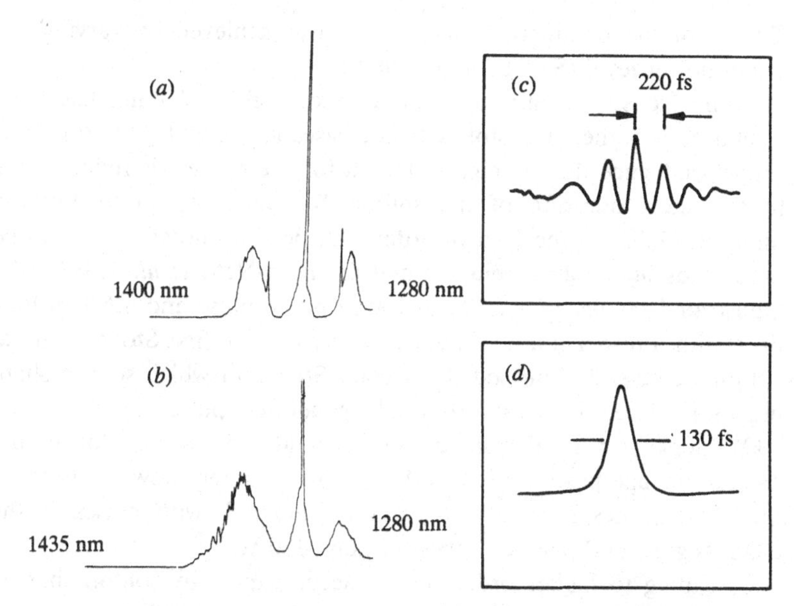

Figure 11.14. Operation of a 100 m long fibre soliton Raman ring laser on modulational instability bands at an average pump power of 750 mW (*a*) single pass, and (*b*) with feedback; (*c*) and (*d*) are the respective background free autocorrelation traces.

dependent trend is similar to that for the single pass condition shown in Figure 11.8. Nakazawa *et al.* have shown that in modulational instability oscillators, the Raman gain begins to dominate as the modulational frequency becomes large. This takes place for pump wavelengths close to the dispersion zero, and they have demonstrated this by using a tunable colour-centre laser operating in this region (Nakazawa *et al.*, 1988).

For pump pulses operating in the region of the dispersion zero and for relatively short fibre lengths, walk off between the pump, Stokes and anti-Stokes can be made extremely small and synchronisation between the three can be achieved. Soliton shaping can then take place for the Stokes component and the operation of this parametric soliton laser, which bears similarities to the modulational instability laser was first reported by Suzuki *et al.* (1989). As distinct from the fibre Raman soliton laser which can tune by changing the cavity length, the parametric soliton laser changes the soliton power with cavity length as would be expected in a synchronously pumped parametric four photon mixing process, where precise cavity matching is required to obtain a long interaction length between the waves.

Tuning of the parametric soliton laser was achieved by varying the pump power level (Suzuki *et al.*, 1989).

Similar to single pass generation, once soliton shaping has been established in the first Stokes band, cascading into higher orders is suppressed since the interaction length for the process is reduced due to the short duration of the soliton Raman pump pulse and the different GVDs of the Raman orders. Dispersion-shifted fibres can be used in oscillator arrangements and Gouveia-Neto *et al.* (1987c) demonstrated this using a 1.32 μm 100 ps pump pulse and a 600 m long fibre with a dispersion minimum at 1.46 μm. The first Stokes band at 1.4 μm, efficiently pumped the second Stokes in which soliton shaping took place. The laser routinely generated pulses of 200 fs and 2 kW peak power. Threshold operation at 1.5 μm was for fundamental average powers of 100 mW. At higher pump powers lasing on the third Stokes band at 1.6 μm was obtained, with pulses in the 200 fs regime and peak powers of around 120 W.

Cascading to higher order Stokes accompanied by soliton shaping has also been demonstrated by da Silva *et al.* (1988) using a c.w. pumped, mode-locked, 1.06 μm Nd:YAG laser, in association with a 1.5 km long fibre with a dispersion minimum at 1.27 μm. The experimental configuration was similar to that shown in Figure 11.12. For an average pump power of 1.8 W, laser action and soliton shaping was observed in a continuous band above 1.27 μm. Within this broad continuum the soliton pulsewidth increased and the peak-to-pedestal ratio increased with increasing wavelength, as would be associated with a soliton-Raman continuum.

11.4 Raman amplification

Over the lengths of fibre required for any practical soliton transmission system, propagation losses can be considerable, leading to temporal broadening. Amplification is therefore required to maintain the initial soliton properties and several authors have theoretically considered the soliton amplification process (Hasegawa and Kodama, 1982; Kodama and Hasegawa, 1982; Kodama and Hasegawa, 1983; Menyuk, 1985). The principal experimental techniques involve the use of either distributed or local amplification based primarily on Raman or erbium-doped fibre amplifiers. See, for example, Chapter 3 by Mollenauer and Chapter 6 by Nakazawa.

Numerical modelling of the stimulated Raman amplification process (Hasegawa, 1983; Dianov *et al.*, 1985b) has shown that large ampli-

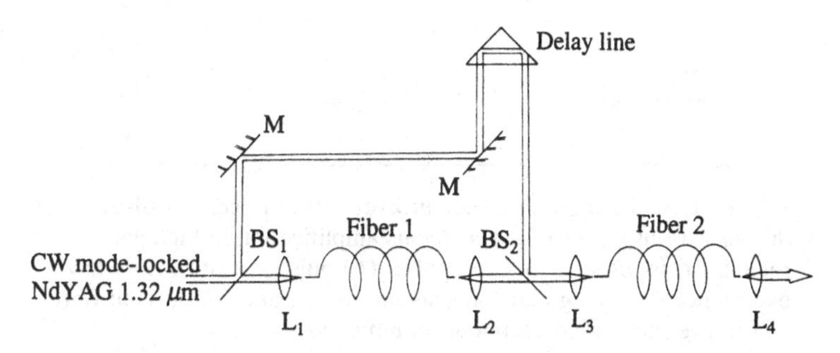

Figure 11.15. Schematic diagram of experimental arrangement used to investigate soliton reconstruction via synchronous Raman amplification.

fier separations can be used and that little change in the shape and duration of the propagating solitons can occur when Raman gain is used to compensate for propagation loss under optimum conditions. Using a counter-propagating pump geometry, Mollenauer *et al.* (1985) first demonstrated distortionless propagation of 10 ps fundamental solitons over 10 km of single-mode fibre using Raman gain to exactly balance the net energy loss on propagation. Based on an optimised co- and counter-propagating pump scheme (Mollenauer *et al.*, 1986), Raman gain has been used to demonstrate soliton transmission over more than 6000 km (Mollenauer and Smith, 1988). The Raman gain process has also been applied to solitons at rates of up to 3 GHz from semiconductor lasers over relatively short (~10 km) fibre lengths (Iwatsuki *et al.*, 1988). However, Raman gain schemes are now currently being superceded by erbium based amplification (see Chapter 6 by Nakazawa).

In the experimental demonstrations of Raman amplification previously described, operation was in the small gain regime where $\alpha_{\text{eff}} Z_0 < 0.05$, where Z_0 is the soliton period and α_{eff} the effective gain coefficient. This is since the Raman gain is merely required to compensate for the small propagation loss in the fibre. However, it is possible to regenerate solitons using Raman gain operating in the large gain regime where $\alpha_{\text{eff}} Z_0 \gg 0.05$ and large perturbations take place to the soliton energy and pulse width.

The experimental scheme used by Gouveia-Neto *et al.* (1988d) to investigate soliton reconstruction via Raman gain in the high-gain regime is shown in Figure 11.15. A c.w. mode-locked Nd:YAG at 1.32 μm was used both to generate the solitons and to provide the synchronous Raman gain. Single-pass soliton-Raman generation in

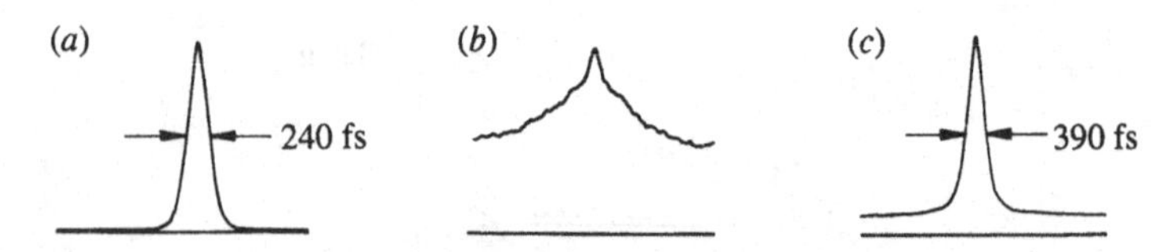

Figure 11.16. Background free autocorrelation traces of pulses during various stages of synchronous amplification in high gain regime (*a*) input signal soliton pulse, (*b*) pulse at exit of fibre having experienced coupling and transmission losses, and (*c*) pulse at fibre exit in presence of synchronous amplification.

fibre 1 provided the source of test solitons. Spectral selection before entering fibre 2 allowed a fundamental soliton to be launched. Due to energy loss, on relaunch into fibre 2 which had identical parameters to fibre 1, the solitons experience rapid temporal broadening (Gouveia-Neto *et al.*, 1987d). Using the temporal delay line, the synchronism of the soliton around 1.4 μm and the 100 ps pump pulse at 1.32 μm was adjusted to obtain maximum pulse overlap and hence maximum Raman gain in fibre 2. Figure 11.16 shows background free autocorrelation traces of the soliton throughout this Raman amplification process where $\alpha_{\text{eff}} Z_0$ was approximately 0.3. The input 1.4 μm soliton is shown in Figure 11.16(*a*) of 240 fs duration with negligible pedestal component. On propagation through 1.5 km of fibre the pulse broadened temporally beyond that which would be accurately measured by the scanning autocorrelator, see Figure 11.16(*b*). On introduction of 60 mW (~6 W peak) pump power, the emerging amplified soliton had recovered to 390 fs, Figure 11.16(*c*), with a pedestal containing the dispersive non-soliton component. This clearly demonstrated that solitons could be recovered operating in a highly perturbed regime of high gain.

Using a similar experimental arrangement, Gouveia-Neto *et al.* (1988e) have investigated the process of synchronous amplification of a modulational instability signal, which was generated in fibre 1. For a launched signal of 15 mW synchronous Raman amplification for pump powers of only 40 mW gave rise to the rapid evolution of a 140 fs soliton which exhibited a 2% pedestal component on the autocorrelation trace. The spectra of these pulses were similar to the early stages of the single pass generation mechanism (see Figure 11.7). However, the evolution to single pulse was much more rapid and the spectral extent was not so great.

The technique of synchronous Raman amplification has been applied to generate solitons from subfundamental soliton power pulses

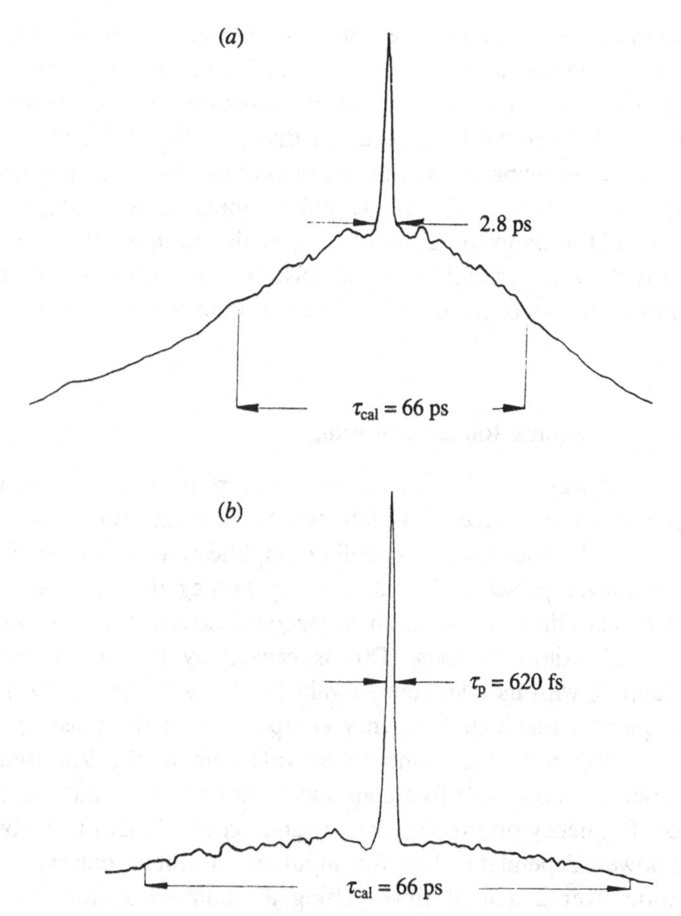

Figure 11.17. Background free autocorrelations associated with the synchronous Raman amplification of noise bursts (*a*) input pulse, and (*b*) output soliton plus dispersive pedestal component.

(Gouveia-Neto *et al.*, 1989b) and from noise bursts (Gouveia-Neto *et al.*, 1989c). In both experiments the fibre F_1 in Figure 11.15 was replaced by a synchronously pumped dispersion compensated fibre Raman ring laser operating around 1.4 μm (Gouveia-Neto *et al.*, 1989d). Through varying the synchronously pumped fibre laser cavity match, clean-pulses or bursts of noise could be generated. In the latter case, the duration of the coherence spike was simply related to the bandwidth of the intra-cavity filter, and the envelope of the noise was approximately that of the pump pulse.

Figure 11.17 shows autocorrelations of (*a*) the input, and (*b*) the output pulse in the presence of 80 mW average pump power at

1.32 μm providing gain at 1.4 μm. A soliton component evolved with increasing pump power, decreasing in duration and increasing in peak-to-pedestal intensity. However, estimates of the fundamental soliton power for the fibre revealed that for the case of noise inputs the solitons evolving via amplification consisted of several soliton-like components, whereas for single pulse inputs, only a single fundamental soliton evolved. This tends to limit the application of pulses generated via the amplification of noise bursts much in the same way as those generated via the single pass soliton-Raman evolution process.

11.5 Intra-pulse Raman scattering

As well as being advantageous by providing a simple means of generating a source of widely tunable fundamental solitons with femtosecond durations and enabling amplification of subfundamental soliton power pulses and counteracting propagation loss, stimulated Raman scattering can be disadvantageous, especially in the propagation of ultrashort solitons. This is caused by the power spectrum overlapping with its own Raman gain (Stolen and Ippen, 1973). As a consequence, the high frequency components of the pulse spectrum are able to act as the pump to provide gain to the low frequency components. This leads to a continuous shifting of the pulse spectrum to low frequency on propagation. Dianov *et al.* (1985a) first observed this power dependent shift for input multi-soliton pulses, 30 ps in duration over 250 m of fibre. Using a stabilised soliton laser as a source of pulses, Mitschke and Mollenauer (1986) carried out a controlled quantitative investigation of this effect which has been called the soliton self-frequency shift. The measured power-dependent spectral shift was in excellent agreement with the theoretical model developed by Gordon (1986), which predicted that the continuous down-shift in the pulse mean frequency was proportional for the inverse fourth power of the pulsewidth (or soliton power squared).

The self-compression of high order solitons provides a simple means of generating ultrashort pulses (Dianov *et al.*, 1986; Tai *et al.*, 1986; Gouveia-Neto *et al.*, 1988f). The multi-solitons are unstable (Golovchenko *et al.*, 1985) and with self-compression the soliton self-frequency shift gives rise to Stokes shifted components which evolve into fundamental solitons (Hodel and Weber, 1987). The spectral shift experienced by the fundamental soliton is a function of the input power, fibre length and dispersion and provides a means of

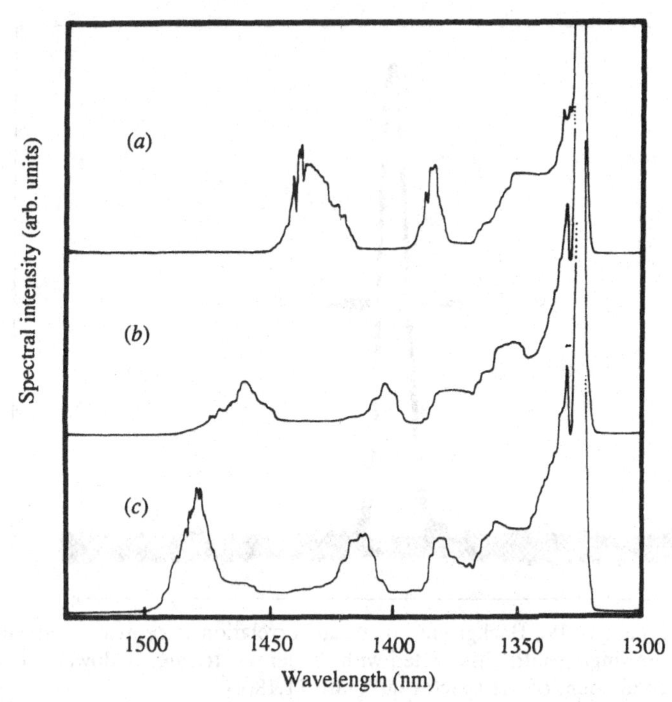

Figure 11.18. Spectra associated with the break up of a multi-soliton exhibiting the characteristics of soliton self-frequency with increasing average pump power (*a*) to (*c*).

generating a tunable, ultrashort pedestal free pulse (Gouveia-Neto *et al.*, 1988g).

The simple scheme demonstrated by Gouveia-Neto *et al.* (1988g) was based on a conventional optical fibre-grating pair compressor of the pulses from a 1.32 μm c.w. pump mode-locked Nd:YAG laser as the source of high order solitons. The 1.5 ps pulses obtained with peak powers of up to 2 kW were simply launched into various lengths of anomalously dispersive fibre at power levels corresponding to very high order solitons. Figure 11.18 shows the spectral variation with input power at the output of a 1.5 km long fibre with $\lambda_0 = 1.31$ μm. A clear, long wavelength shift with increasing power can be seen on the Stokes fragments which evolve from the input multi-soliton at 1.32 μm. In Figure 11.18 as the average power is increased from (*a*) 40 to (*b*) 50 mW and finally (*c*) 60 mW, the central wavelength of the longest soliton component shifted from 1430 nm, through 1460 nm to 1480 nm, respectively. Spectral selection of this long wavelength component, for example at 1480 nm in Figure 11.18(*c*) revealed it to be a

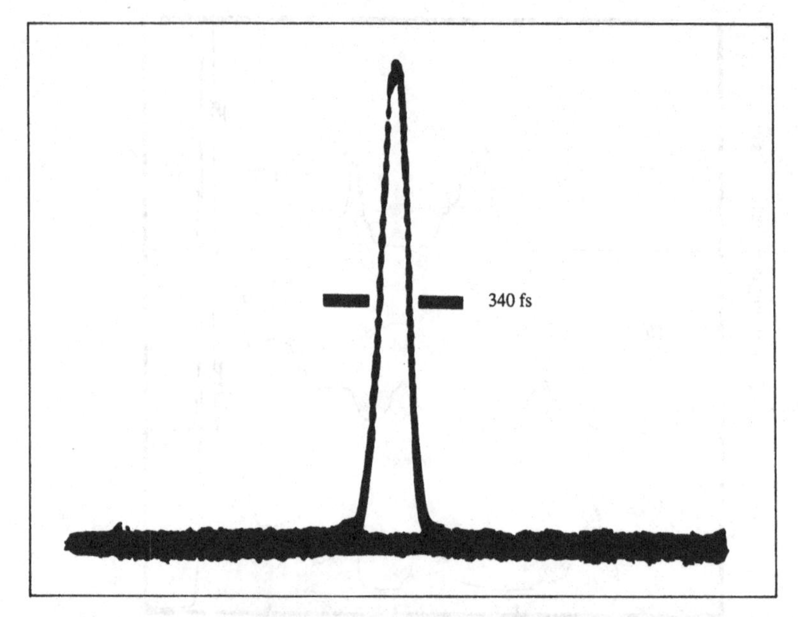

Figure 11.19. Background free autocorrelation trace associated with the single soliton associated with the largest frequency down-shifted component of the spectra in Figure 11.18(*c*).

single fundamental soliton, pedestal free of duration 340 fs, see Figure 11.19. This provides a simple means of generating fundamental solitons. By simply maintaining a fixed fibre length and varying the pump power, for example over more than 150 nm tunability was obtained, launching 1.5 ps pulses at 1.32 μm into a 3.3 km fibre length by increasing the average pump power up to 75 mW.

For pulse durations of the order of 100 fs propagating over fibre lengths of 100 m, the typical spectral shifts expected from the soliton self-frequency shift would be 30–100 THz. However, in the single-pass soliton-Raman generation process shifts of this magnitude are not generally observed. Blow *et al.* (1988b) have explained this suppression of the self-shifting in terms of bandwidth-limited amplification arising from the co-propagating pump pulse. Blow *et al*'s model included terms in the NLS equation associated with intra-pulse Raman scattering, and stimulated Raman scattering from the long pulse external pump. The former term gives rise to a continuous frequency down-shift while the latter gives a preferential amplification around the peak of the Stokes band, approximately to 440 cm^{-1} from the pump pulse. In the experimental configuration, the pump pulse is commonly a 100 ps pulse which should experience negligible self-fre-

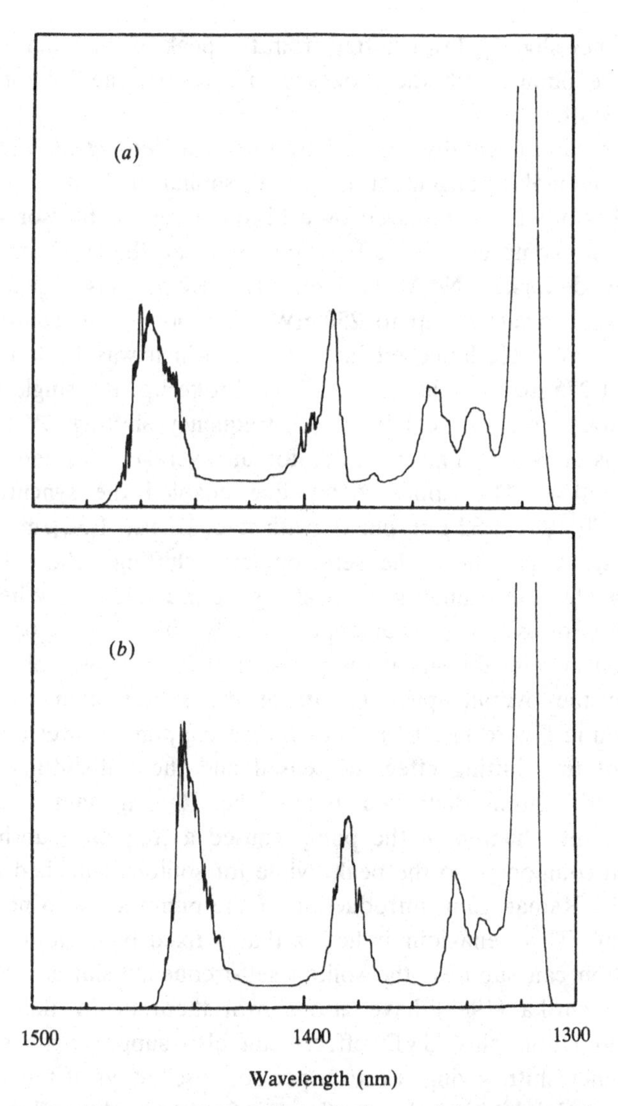

Figure 11.20. Spectra corresponding to (*a*) an average launched signal power of 70 mW (1.32 μm, 1.2 ps pulses) in the 1.5 km fibre and (*b*) 70 mW average power signal, plus 40 mW pump power showing suppression of the spectral shift of the solitons generated at 1.46 μm.

quency shifting. Therefore the Raman gain remains relatively fixed at a constant frequency shift from the fixed pump. Consequently the overall frequency shift will depend on the relative magnitudes of the two competing effects. In the presence of a relatively intense pump field Blow *et al.* show that the ultrashort Stokes pulses would be held at a frequency slightly down-shifted from the peak of the main Stokes band of the pump, with the Stokes gain arresting the soliton self-frequency shift.

This was experimentally verified by Gouveia-Neto *et al.* (1989e). The experimental arrangement used was similar to that of Figure 11.15, only fibre 1 was replaced by a fibre-grating compressor which was used to compress the 100 ps pulses from the 1.32 μm c.w. pumped mode-locked Nd:YAG laser. The 1.2 ps pulses generated with average powers of up to 250 mW corresponded to solitons of order up to $N = 17$, launched into fibre 2, which was 1.5 km long, with $\lambda_0 = 1.275$ μm. As the multi-solitons broke up, the single ultrashort solitons shed off, exhibited self-frequency shifting. A typical spectrum is shown in Figure 11.20, for an average launched signal power of 70 W. The optical delay line enabled the synchronous launch of 100 ps, 1.32 μm pulses with the 1.2 ps, 1.32 μm signal pulses which experienced the self-frequency shifting. These longer pulses provided the Raman gain, peaking around 1.42 μm. When the pump pulses of average power approximately 40 mW were launched in synchronism with the signal, suppression of the shift was observed, decreasing the overall spectral shift of the soliton component as can be seen in Figure 11.20(*b*). With decreasing pump power the suppression of the shifting effect decreased and the self-shifting dominated. For solitons launched below the Raman gain peak of the pump, introduction of the pump caused a frequency upshift of the soliton component to the peak, while for solitons launched at the peak of the Raman gain, introduction of the pump leads to no spectral shifting. This behaviour indicates that a fixed bandwidth-limited amplification can suppress the soliton self-frequency shifting. Schadt and Jaskorzynska (1988) have also shown theoretically that cross-phase modulation plus GVD effects can also suppress the soliton self-frequency shift, giving rise to a periodic oscillation of the soliton frequency and inhibiting the continuous frequency down-shift. The bandwidth-limited amplification of erbium in silica-based fibres centred around 1.53 μm should also allow suppression of the self-shifting effects.

11.6 Conclusion

In this chapter an experimental description has been given of soliton-Raman processes in optical fibres and related effects which give rise to the generation of ultrashort pulses. The single-pass soliton-Raman generation mechanism, provides one of the simplest means of generating wavelength tunable, femtosecond pulses, although the generation mechanism from noise, introduces problems of temporal jitter which severely limits the applicability of the source to pump-probe techniques where temporal jitter, pulse-to-pulse is not a problem. The multiple soliton nature of the output can be suppressed by wavelength seeding using diode laser signals to generate a single fundamental soliton, albeit over a rather restricted pump power regime and seeding using mode-locked diode laser signal should give rise to a reduction in the temporal jitter characteristics. Modulational instability plays a major role in the formation mechanism and although it is not a prerequisite, it does act to lower the overall threshold power for the soliton-Raman process. Cascade-Raman effects enable the tunability to be extended and operates in oscillator arrangements as well as single pass configurations. As well as being beneficial, for example, Raman amplification allows soliton generation from noise bursts, from modulational instability and from sub-fundamental soliton powered pulses, all of which have been described, intra-pulse it gives rise to frequency down-shifting which can be a distinct disadvantage in a soliton communication network, as has been described in other chapters of this book. However, it is possible to suppress, although not totally eliminate, the self-shifting using bandwidth-limited amplification, and this has been demonstrated with Raman amplification from long pump pulses, showing the versatility and breadth of Raman-based processes in single-mode fibre systems.

Acknowledgement

The author wishes to express his gratitude to his co-workers from the Femtosecond Optics Group, Imperial College, whose experimental work has been presented in this chapter, in particular to Dr A. S. Gouveia-Neto, Dr A. S. L. Gomes, Dr P. G. J. Wigley, Miss E. J. Greer and Mr D. M. Patrick. British Telecom Research Laboratories have financially supported this work and the author is grateful to Dr N. J. Doran, Dr K. J. Blow and Dr D. Wood of BTRL for many useful discussions.

References

Agrawal, G. P. (1987) *Phys. Rev. Lett.*, **59**, 880–3.
Agrawal, G. P. (1989) *Nonlinear Fiber Optics* (Academic Press: New York).

Agrawal, G. P., Baldeck, P. L. and Alfano, R. R. (1989) *Phys. Rev. A*, **39**, 3406–13.
Ainslie, B. J. and Day, C. R. (1986) *IEEE J. Lightwave. Technol.*, **4**, 967–79.
Alfano, R. R., Baldeck, P. L., Raccah, F. and Ho, P. (1987) *Appl. Opt.*, **26**, 3491–2.
Anderson, A. and Lisak, M. (1984) *Opt. Lett.*, **9**, 463–90.
Baldeck, P. L., Alfano, R. R. and Agrawal, G. P. (1988) *Appl. Phys. Lett.*, **52**, 1939–41.
Baldeck, P. L., Ho, P. P. and Alfano, R. R. (1989) In R. R. Alfano *The Supercontinuum Laser Source*, chap. 4, pp. 117–83. (Springer: Berlin).
Beaud, P., Hodel, W., Zysset, B. and Weber, H. P. (1987) *IEEE J. Quantum Electron.*, **23** 1938–46.
Benjamin, T. J. and Feir, J. E. (1967) *J. Fluid Mech.*, **27**, 417–30.
Blow, K. J., Doran, N. J. and Wood, D. (1987) *Opt. Lett.*, **12**, 1011–13.
Blow, K. J., Doran, N. J. and Wood, D. (1988a) *J. Opt. Soc. Am. B*, **5**, 381–90.
Blow, K. J., Doran, N. J. and Wood, D. (1988b), *J. Opt. Soc. Am. B*, **5**, 1301–04.
Blow, K. J. and Wood, D. (1989) *IEEE J. Quantum Electronic.*, **25**, 2665–73.
da Silva, V. L., Gomes, A. S. L. and Taylor, J. R. (1988) *Opt. Commun.*, **66**, 231–4.
Dianov, E. M., Karasik, A. Ya., Mamyshev, P. V., Prokhorov, A. M., Serkin, V. N., Stel'makh, M. F. and Fomichev, A. A. (1985a) *JETP Lett.*, **41**, 294–7.
Dianov, E. M., Nikonova, Z. S., Prokhorov, A. M. and Serkin, V. N. (1985b) *Sov. Phys. Dokl.*, **30**, 689–91.
Dianov, E. M., Nikonova, Z. S., Prokhorov, A. M. and Serkin, V. N. (1986) *Sov. Tech. Phys. Lett.*, **12**, 311–13.
Gersten, J., Alfano, R. R. and Belic, M. (1980) *Phys. Rev. A*, **21**, 1222–4.
Golovchenko, E. A., Dianov, E. M., Prokhorov, A. M. and Serkin, V. N. (1985) *JETP Lett.*, **42**, 87–91.
Gordon, J. P. (1986) *Opt. Lett.*, **11**, 662–4.
Gouveia-Neto, A. S. and Taylor, J. R. (1988) *Electron. Lett.*, **24**, 1544–6.
Gouveia-Neto, A. S., Gomes, A. S. L. and Taylor, J. R. (1987a) *Electron. Lett.*, **23**, 537–8.
Gouveia-Neto, A. S., Gomes, A. S. L. and Taylor, J. R. (1987b) *Opt. Lett.*, **12**, 1035–7.
Gouveia-Neto, A. S., Gomes, A. S. L., Taylor, J. R., Ainslie, B. J. and Craig, S. P. (1987c) *Opt. Lett.*, **12**, 927–9.
Gouveia-Neto, A. S., Gomes, A. S. L. and Taylor, J. R. (1987d) *Opt. Commun.*, **64**, 383–6.
Gouveia-Neto, A. S., Gomes, A. S. L. and Taylor, J. R. (1987e) *Opt. Lett.*, **12**, 395–7.
Gouveia-Neto, A. S., Faldon, M. E., Sombra, A. S. B., Wigley, P. G. J. and Taylor, J. R. (1988a) *Opt. Lett.*, **13**, 901–03.
Gouveia-Neto, A. S., Gomes, A. S. L. and Taylor, J. R. (1988b) *IEEE J. Quantum Electron.*, **24**, 332–40.
Gouveia-Neto, A. S., Gomes, A. S. L. and Taylor, J. R. (1988c) *Opt. Quantum Electron.*, **20**, 165–74.

Gouveia-Neto, A. S., Gomes, A. S. L., Taylor, J. R. and Blow, K. J. (1988d) *J. Opt. Soc. Am. B*, **5**, 799–803.
Gouveia-Neto, A. S., Faldon, M. E. and Taylor, J. R. (1988e) *Opt. Lett.*, **13**, 1029–31.
Gouveia-Neto, A. S., Gomes, A. S. L. and Taylor, J. R. (1988f) *J. Mod. Opt.*, **35**, 7–10.
Gouveia-Neto, A. S., Sombra, A. S. B. and Taylor, J. R. (1988g) *Opt. Commun.*, **68**, 139–42.
Gouveia-Neto, A. S., Faldon, M. E. and Taylor, J. R. (1989a) *Opt. Commun.*, **69**, 325–8.
Gouveia-Neto, A. S., Wigley, P. G. J. and Taylor, J. R. (1989b) *Opt. Commun.*, **72**, 119–22.
Gouveia-Neto, A. S., Wigley, P. G. J. and Taylor, J. R. (1989c) *Opt. Lett.*, **14**, 1122–4.
Gouveia-Neto, A. S., Wigley, P. G. J. and Taylor, J. R. (1989d) *Opt. Commun.*, **70**, 128–30.
Gouveia-Neto, A. S., Gomes, A. S. L. and Taylor, J. R. (1989e) *Opt. Lett.*, **14**, 514–16.
Greer, E. J., Patrick, D. M., Wigley, P. G. J. and Taylor, J. R. (1989) *Electron. Lett.*, **25**, 1246–7.
Greer, E. J., Patrick, D. M., Wigley, P. G. J. and Taylor, J. R. (1990) *Opt. Lett.*, pp, 1990.
Grudinin, A. B., Dianov, E. M., Korobkin, D. V., Prokhorov, A. M., Serkin, V. N. and Khaidarov, D. V. (1987) *JEPT Lett.*, **45**, 260–3.
Hasegawa, A. (1972) *Phys. Fluids*, **15**, 870–81.
Hasegawa, A. (1983) *Opt. Lett.*, **8**, 650–2.
Hasegawa, A. (1984a) *Opt. Lett.*, **9** 288–90.
Hasegawa, A. (1984b) *Appl. Opt.*, **23**, 3302–09.
Hasegawa, A. and Brinkman, W. F. (1980) *IEEE J. Quantum Electron.*, **16**, 694–7.
Hasegawa, A. and Kodama, Y. (1981) *Proc. IEEE*, **69**, 1145–50.
Hasegawa, A. and Kodama, Y. (1982) *Opt. Lett.*, **7**, 285–7.
Hasegawa, A. and Tai, K. (1989) *Opt. Lett.*, **14**, 512–13.
Haus, H. A. and Nakazawa, M. (1987) *J. Opt. Soc. Am. B*, **4**, 652–60.
Hermansson, B. and Yevick, D. (1984) *Opt. Commun.*, **52**, 99–102.
Hodel, W. and Weber, H. P. (1987) *Opt. Lett.*, **12**, 924–6.
Islam, M. N., Dijaili, S. P. and Gordon, J. P. (1988) *Opt. Lett.*, **13**, 518–20.
Islam, M. N., Mollenauer, L. F. and Stolen, R. H. (1986) In G. R. Fleming and A. E. Siegman (eds.) *Ultrafast Phenomena* volume V, pp. 46–50 (Springer: Berlin).
Islam, M. N., Mollenauer, L. F., Stolen, R. H., Simpson, J. R. and Shang, H. T. (1987a) *Opt. Lett.*, **12**, 625–7.
Islam, M. N., Mollenauer, L. F., Stolen, R. H., Simpson J. R. and Shang, H. T. (1987b) *Opt. Lett.*, **12**, 814–16.
Islam, M. N., Sucha, G., Bar Joseph, I., Wegener, M., Gordon, J. P. and Chemla, D. S. (1989) *J. Opt. Soc. Am. B*, **6**, 1149–58.
Ito, F., Kitayama, K. and Yoshinaga, H. (1989) *Appl. Phys. Lett.*, **54**, 2503–05.
Itoh, H., Davis, G. M. and Sudo, S. (1989) *Opt. Lett.*, **14**, 1368–70.
Iwatsuki, K., Takada, A. and Saruwatari, M. (1988) *Electron. Lett.*, **24**, 1572–4.

Jaskorzynska, B. and Schadt, D. (1988) *IEEE J. Quantum Electron.*, **24**, 2117–20.

Kafka, J. D. and Baer, T. (1987) **12**, 181–3.

Kafka, J. D., Head, D. F. and Baer, T. (1986) In G. R. Fleming and A. E. Siegman (eds.), *Ultrafast Phenomena*, volume V (Springer Series in Chemical Physics, **46**), pp. 51–53 (Springer: Berlin).

Keller, U., Li, K. D., Rodwell, M. and Bloom, D. M. (1989) *IEEE J. Quantum Electron.*, **25**, 280–8.

Kodama, Y. and Hasegawa, A. (1982) *Opt. Lett.*, **7**, 339–41.

Kodama, Y. and Hasegawa, A. (1983) *Opt. Lett.*, **8**, 342–4.

Lin, C. and French, W. G. (1979) *Appl. Phys. Lett.*, **34**, 666–8.

Lin, C., Stolen, R. H. and Cohen, L. G. (1977) *Appl. Phys. Lett.*, **31**, 97–9.

Manassah, J. T. (1987) *Appl. Opt.*, **26**, 3747–9.

Menyuk, C. R., Chen, H. H. and Lee, Y. C. (1985) *Opt. Lett.*, **10**, 451–3.

Mitschke, F. M. and Mollenauer, L. F. (1986) *Opt. Lett.*, **11**, 659–1.

Mollenauer, L. F., Gordon, J. P. and Islam, M. N. (1986) *IEEE J. Quantum Electron.*, **22**, 157–73.

Mollenauer, L. F. and Smith, K. (1988) *Opt. Lett.*, **13**, 675–7.

Mollenauer, L. F., Stolen, R. H. and Gordon, J. P. (1980) *Phys. Rev. Lett.*, **45**, 1095–8.

Mollenauer, L. F., Stolen, R. H. and Islam, M. N. (1985) *Opt. Lett.*, **10**, 229–31.

Nakazawa, M., Suzuki, K. and Haus, H. A. (1988) *Phys. Rev. A*, **38**, 5193–6.

Nakazawa, M., Suzuki, K. and Haus, H. A. (1989a) *IEEE J. Quantum Electron.*, **25**, 2036–44.

Nakazawa, M., Suzuki, K., Kubota, H. and Haus, H. A. (1989b) *Phys. Rev. A*, **39**, 5768–76.

Rothenberg, J. E. (1990) *Integrated Photonics Research, Optical Society of America Technical Digest Series, volume* 5, pp. 6–7 (OSA: Washington, DC).

Schadt, D. and Jaskorzynska, B. (1987a) *J. Opt. Soc. Am. B*, **4**, 856–62.

Schadt, D. and Jaskorzynska, B. (1987b) *Electron. Lett.*, **23**, 1090–1.

Schadt, D. and Jaskorzynska, B. (1988) *J. Opt. Soc. Am. B*, **5**, 2374–8.

Smith, R. G. (1972) *Appl. Opt.*, **11**, 2489–94.

Soccolich, C. E. and Islam, M. N. (1989) *Opt. Lett.*, **14**, 645–7.

Stolen, R. H. (1980) *Fiber and Integrated Optics*, **3**, 157–82.

Stolen, R. H. and Ippen, E. P. (1973) *Appl. Phys. Lett.*, **22**, 276–8.

Stolen, R. H. and Johnson, A. M. (1986) *IEEE J. Quantum Electron.*, **22**, 2154–60.

Stolen, R. H., Lee, C. and Jain, R. K. (1984) *J. Opt. Soc. Am. B*, **1**, 652–7.

Stolen, R. H., Lin, C. and Jain, R. K. (1977) *Appl. Phys. Lett.*, **30**, 340–2.

Sudo, S., Itoh, H., Okamoto, K. and Kubodera, K. (1989) *Appl. Phys. Lett.*, **54**, 993–4.

Suzuki, K., Nakazawa, M. and Haus, H. A. (1989) *Opt. Lett.*, **14**, 320–22.

Tai, K., Hasegawa, A. and Tomita, A. (1986a) *Phys. Rev. Lett.*, **56**, 135–8.

Tai, K., Tomita, A., Jewell, J. L. and Hasegawa, A. (1986b) **49**, 236–8.

Tai, K. and Tomita, A. (1986) *Appl. Phys. Lett.*, **48**, 1033–5.

Tajima, K. (1986) *IEEE J. Lightwave Technol.*, **4**, 900–04.

Vodopyanov, K. L., Grudinin, A. B., Dianov, E. M., Kulevskii, L. A.,

Prokhorov, A. M. and Khaidarov, D. V. (1987) *Sov. J. Quantum Electron.*, **17**, 1311–13.
Vysloukh, V. A. and Serkin, V. N. (1983) *JETP Lett.*, **38**, 199–202.
Vysloukh, V. A. and Serkin, V. N. (1984) *Bull. Acad. Sci. USSR, Phys. Ser.*, **48**, 125–9.
Wang, C-Y., Baldeck, P. L., Budansky, Y. and Alfano, R. R. (1989) *Opt. Lett.*, **14**, 497–9.
Zysset, B., Beaud, P., Hodel, W. and Weber, H. P. (1986) In G. R. Fleming and A. E. Siegman, (eds.), *Ultrafast Phenoma, volume V*, pp. 54–7 (Springer: Berlin).

Index